Biomass Transformation: Sustainable Development

Biomass Transformation: Sustainable Development

Topic Editors

Domenico Licursi
Juan J. Hernández

MDPI

Basel • Beijing • Wuhan • Barcelona • Belgrade • Novi Sad • Cluj • Manchester

Topic Editors
Domenico Licursi
Department of Chemistry and Industrial Chemistry
University of Pisa
Pisa
Italy

Juan J. Hernández
E.T.S. Ingenieria Industrial
University of Castilla-La Mancha
Ciudad Real
Spain

Editorial Office
MDPI AG
Grosspeteranlage 5
4052 Basel, Switzerland

This is a reprint of the Topic, published open access by the journals *Catalysts* (ISSN 2073-4344), *Energies* (ISSN 1996-1073), *Molecules* (ISSN 1420-3049), *Sustainability* (ISSN 2071-1050) and *Sustainable Chemistry* (ISSN 2673-4079), freely accessible at: https://www.mdpi.com/topics/7NH73W819U.

For citation purposes, cite each article independently as indicated on the article page online and as indicated below:

Lastname, A.A.; Lastname, B.B. Article Title. *Journal Name* **Year**, *Volume Number*, Page Range.

ISBN 978-3-7258-4271-1 (Hbk)
ISBN 978-3-7258-4272-8 (PDF)
https://doi.org/10.3390/books978-3-7258-4272-8

Contents

About the Editors

Domenico Licursi

Domenico Licursi obtained his master's degree in industrial chemistry in 2010 and PhD in Chemical Sciences from Galileo Galilei Graduate School in 2015. Currently, he is working as a Tenure Track Researcher in Industrial Chemistry at the Department of Chemistry and Industrial Chemistry, University of Pisa. For about 15 years, he is deepening the biomass conversion into added-value bioproducts (such as 5-hydroxymethylfurfural, furfural and levulinic acid), as well as the catalytic production of some biomass-derived oxygenated derivatives (such as alkyl levulinates, ketals, gamma-valerolactone, 2-methyltetrahydrofuran, and furan diols). His current research mainly concerns the hydrothermal carbonization and thermo-catalytic pyrolysis of waste biomasses to produce biochars and bio-oils, then developing the corresponding uses, thus always considering an industrial and applicative perspective. He is also working on the acid-catalysed synthesis of medium-chain esters from waste biomasses, as well as the hydrogenation/hydrogenolysis of (fatty) esters to the corresponding alcohols, working in the presence of mixed oxides as the heterogeneous catalysts. He is author of more than 50 publications indexed in the Web of Science (WoS), in high-impact journals for the scientific sector of industrial chemistry. He is an active member of the American Chemical Society (ACS), mainly within the Agrochemicals, Catalysis Science & Technology, and Cellulose & Renewable Materials divisions.

Juan J. Hernández

Juan J. Hernández got his PhD on Mechanical Engineering from the Polytechnic University of Valencia in 2000, after which he completed a post-doctoral stay at the University of Birmingham (UK, 2002–2003) funded by the MSCA Postdoctoral Fellowship program of the EU. He is professor in Thermal Engines at the University of Castilla-La Mancha since 2011. His research has been focused on the combustion process and the formation of pollutant emissions in internal combustion engines, with special interest in the use of advanced biofuels and e-fuels, and on the thermochemical conversion of biomass. His scientific career, with more than 100 contributions in the Web of Science (WoS), has allowed him to become one of the most representative researchers worldwide in his field, according to the ranking published by Stanford University. He has been the main researcher in several projects funded by competitive public calls, all of them focused on evaluating sustainable fuels and combustion modes for a cleaner transport sector as well as on biomass wastes transformation technologies leading to added-value biofuels and other co-products (the latter though the valorization of combustion and gasification residues). Prof. Juan J. Hernández has also been involved in several Erasmus+CBHE projects (as partner or coordinator) targeted to improve the teaching and research capacities on renewable energy technologies and energy efficiency of universities in non-EU countries. He has been a visiting professor at the University of Toronto (Canada, April-July 2016) and at the University of Michigan (USA, October 2021–July 2022). He is also a regular collaborator with various national and international agencies dedicated to evaluating research, such as the Spanish Research State Agency (AEI),

sustainability

Editorial

Biomass Transformation: Sustainable Development

Domenico Licursi [1,*] and Juan J. Hernández [2,*]

[1] Department of Chemistry and Industrial Chemistry, University of Pisa, Via Giuseppe Moruzzi 13, 56124 Pisa, Italy
[2] E.T.S. Ingenieria Industrial, University of Castilla-La Mancha, 13071 Ciudad Real, Spain
* Correspondence: domenico.licursi@unipi.it (D.L.); juanjose.hernandez@uclm.es (J.J.H.)

The development of new strategies for the thermochemical conversion of biomass into high value-added products, energy, and fuels represents an exciting and challenging topic, as demonstrated by the success of this Special Issue, in terms of the number of published manuscripts and visualizations. The authors' contributions fully exhibit that biomass transformation following sustainability criteria represents the best choice to be pursued and developed in the immediate future, also considering the profound and changing evolution of the global scenario, which must rely on easily scalable processing technologies. The authors who actively contributed to this Special Issue have amply shown relevant progress over the current state of the art, emphasizing that technological development in the field of biomass thermochemical conversion is possible.

The importance of valorizing waste biomasses according to the principles of sustainability and circular economy has been specifically highlighted in all publications of this Special Issue, which discuss interesting conversion possibilities. Some authors have advanced a broad and diversified discussion, bringing waste biomass to the center of the reader's attention. In this context, the conversion of biogenic residues from waste, agriculture, and wastewater sectors into energy is attracting increasing attention, as discussed by Pollack et al. [1], who have proposed a systematic approach to support the design of suitable energy systems for German rural areas. The authors underlined that these biogenic residues can be effectively considered a relevant component of the transition toward locally based, diversified, and climate-friendly energy systems. In another study, Zheng et al. [2] discussed the sustainable resource utilization of residual sewage sludge in China, focusing on building material utilization, energy utilization, and phosphorus recovery. Remarkably, each technology shows advantages and limitations, underscoring that the most suitable technology requires considering the characteristics of the starting feedstock and the application context to achieve the integration of pollution control and resource utilization. The exploitation of coconut waste is another noteworthy case study, as discussed by Viera et al. [3]. In their work, the authors evaluated different possibilities, including the use of bio-products such as ethanol, bio-oil, and organic fertilizers in the civil construction field, the environmental field as an adsorbent for pollutants, and in production. Remarkably, coconut waste represents a rich source of phenolic compounds, which can be exploited for high-value applications, such as their antioxidant and anti-inflammatory properties, and as active ingredients in the pharmaceutical industry.

Among the different exploitation possibilities, biofuel production is certainly attractive, as discussed in some noteworthy works in this Special Issue. In particular, Rojas-Flores et al. [4] proposed the use of Andean tuber (Olluco) waste as fuel for microbial fuel cells, demonstrating the achievement of promising performances, in terms of voltage, electrical current, and electrical potential, and generating enough electrical energy to light a LED light bulb. On this basis, Andean tuber would represent an attractive biomass to be grown in Peru, helping to solve the issue of the lack of electrical energy. In the wide field of biofuel production, bioethanol still represents a key topic, highlighting that current research is focused on the intensification of its production, mainly focusing on the work-up operations. In this context, Rola et al. [5] proposed a simulation of bioethanol production through lignocellulosic biomass fermentation, focusing on the distillation approach. Pressure swing distillation was effectively proposed to obtain fuel-grade ethanol (99.6 wt%), advantageously overcoming the issue of the formation of the ethanol–water azeotrope. In another work, Samuel et al. [6] studied biodiesel production from rubber seed oil by comparing the ethyl-based hydrodynamic cavitation reactor with the mechanical stirring one. The authors stressed the economic and environmental advantages of the ethyl-based hydrodynamic cavitation reactor, providing information for quickly scaling up the production of ethanolic biodiesel from non-edible and third-generation feedstocks. Remaining in the field of biodiesel production, Zhang et al. [7] discussed fatty acid methyl ester (FAME) synthesis from refined soybean oil and methanol, where this reaction was carried out in the presence of sodium silicates that modified calcium oxide as the solid base catalyst. If compared with the sole CaO catalyst, the two catalytic phases (Na_2CaSiO_4 and CaO) of the new synthesized catalyst synergistically act towards the improvement of both the activity and stability parameters. Within the biodiesel production chain from triglycerides, glycerol certainly represents the main by-product, requiring considerable attention to enhance its efficient exploitation for maximizing the sustainability of the entire process. In this context, the acid-catalyzed ketalization of glycerol to solketal is certainly an attractive approach, considering that these products are exploitable as green solvents and fuel additives, as well as in the pharmaceutical and food industries. In developing this approach, Rossa et al. [8] studied the ketalization reaction of glycerol with acetone, employing reduced graphene oxide as a new heterogeneous catalyst, whereas homogeneous *p*-toluenesulfonic acid was considered the reference. Remarkably, the catalytic activity of the synthesized graphene oxides was 30 times more efficient than the homogeneous *p*-toluenesulfonic acid, revealing their stability even after five consecutive cycles, maintaining the highest activity without any work-up treatment.

Another topic of great interest is related to biochar production and exploitation. Biochar is obtainable from thermochemical treatments of biomasses, often in high yields after the pyrolysis route, thus largely justifying the interest of the current research in improving its production and identifying new valorization strategies. In this Special Issue, Zhang et al. [9] investigated the pyrolysis of coffee husks into high-performance biochar-based fertilizers, which is a desirable approach for achieving more sustainable resource management and pollution control, at the same time fostering environmentally responsible agricultural practices. Remarkably, the appropriate optimization of the pyrolysis conditions allows for the optimal regulation of nutrient release within the pant-soil system. Further expanding upon this topic, in the review of Filho et al. [10], the synthesis and exploitation of the biochar derived from the pyrolysis of olive mill by-products (e.g., olive pomace and olive stone) were specifically discussed. For this purpose, the agronomic properties of biochars were considered, fully demonstrating their usefulness as plant nutrient reservoirs, good amendments to improve soil properties, and in long-term carbon sequestration. In another work, Meng et al. [11] proposed the co-pyrolysis of mushroom residue with pine sawdust or wheat straw, examining thermal characteristics, kinetic/thermodynamic aspects, and the structural evolution of chars during thermal treatment. The authors provided valuable insights for progressing the co-pyrolysis technology of edible fungi residue with wood/straw biomass. The application of biochar derived from *Agave angustifolia* bagasse as an adsorbent for ibuprofen removal from aqueous solutions was proposed and optimized by Ruiz-Velducea et al. [12], highlighting the significant potential of biochars in this application.

To develop the fractionation/conversion of the main biomass components into high added-value products, some authors have proposed particularly challenging approaches. Lozano-Pérez et al. [13] investigated liquid hot water (LHV) extraction and the hydrothermal carbonization (HTC) of coffee cherry for the production of valuable compounds, such as carbohydrates (e.g., glucose and xylose), furans (e.g., 5-hydroxymethylfurfural and furfural), and carboxylic acids (e.g., levulinic acid and formic acid). Both thermal approaches were optimized, by also considering the addition of appropriate acid or alkaline catalysts to improve their optimal formation. The authors demonstrated that the proposed hydrothermal treatments of coffee cherry may represent a significant opportunity for sustainable resource management and the development of a circular bioeconomy, also considering that the final solid (hydro)char can be immediately exploited as fuel and/or exploited for other applications, including those previously discussed in this editorial. Carboxylic acids are certainly high value-added biochemicals, as discussed by Aboudi et al. [14], who obtained short-chain fatty acids via an anaerobic fermentation path, employing sugar beet molasses as the starting feedstock and the semi-continuous feeding mode in completely stirred tank reactors. The hydraulic retention time was specifically considered by the authors, who pointed out that longer ones were more suitable for obtaining long carbon chain carboxylates, thus better exploiting the chain elongation of shorter carboxylic acids. Remaining in the context of the mild thermal processing of biomass, Kim et al. [15] claimed a simple two-stage extraction and recovery method for macromolecules from microalgae biomass (*Chlorella* sp. ABC-001), involving (1) acid pretreatment and (2) high-shear-assisted lipid extraction. After the optimization of the entire process, the highly efficient recovery of esterifiable lipids and glucose in the liquid phase was achieved, at the same time selectively keeping the proteins in the solid one. Remarkably, the proposed approach is attractive for the integrated production of biofuel, bioethanol, and animal feed. Another noteworthy example of biomass fractionation was provided by Mullen et al. [16], who developed the one-pot fractionation and depolymerization of the lignin fraction of biomass to obtain an oil rich in phenolic compounds and a cellulosic pulp. For this purpose, the lignin fraction of oak and switchgrass was fractionated and depolymerized in a solvent phase reaction, and the corresponding cellulosic pulps were tested for conversion to levoglucosan and BTEX via fast and catalytic pyrolysis over HZSM-5, respectively. The authors highlighted the advantages of using reductive conditions for improving levoglucosan yield via fast pyrolysis. Remarkably, within the catalytic pyrolysis over HZSM-5, the yield of BTEX decreased, in comparison with the raw biomass, demonstrating that the solvent phase fractionation of lignin is effective for the selective production of oxygenated compounds (e.g., levoglucosan). Interestingly, Du et al. [17] synthesized a sulfonyl-chloride-modified, lignin-based, porous carbon-supported metal phthalocyanine catalyst, which was proposed to replace the traditional Fenton's reagent for lignin degradation to small phenols. By employing the synthesized catalyst, the authors optimized H_2O_2 oxidation, thus maximizing the liquid product yield and selectivity to simple phenols. Remarkably, the corresponding char showed good energetic properties and a halved molecular weight compared to that of the pristine lignin. Lastly, the fractionation of high value-added components is also applicable to non-conventional biomasses, as discussed in the work of Shkuratov et al. [18], who isolated chitin nanowhiskers from crustacean biomass, carried out in the presence of sustainable and inexpensive ionic liquids, such as $[HN_{222}][HSO_4]$. The use of these extracting agents shows remarkable advantages over traditional chitin extraction processes, that is, reducing the overall production costs by simplifying the process, eliminating the need for initial purification steps, and reducing the consumption of chemicals and energy.

In summary, these eighteen papers detail innovative aspects in the field of biomass conversion strategies, with a noteworthy preference toward the conversion of waste, in agreement with the criteria of environmental, economic, and social sustainability. We would like to thank all the authors of our Special Issue for their valuable contributions, as well as all the reviewers and the entire Editorial Office of MDPI for having assisted and helped us during the peer-review process.

Funding: This research received no external funding.

Conflicts of Interest: The authors declare no conflicts of interest.

References

1. Pollack, M.; Lück, A.; Wolf, M.; Kraft, E.; Völker, C. Energy and business synergy: Leveraging biogenic resources from agriculture, waste, and wastewater in German rural areas. *Sustainability* **2023**, *15*, 16573. [CrossRef]
2. Zheng, W.; Shao, Y.; Qin, S.; Wang, Z. Future directions of sustainable resource utilization of residual sewage sludge: A review. *Sustainability* **2024**, *16*, 6710. [CrossRef]
3. Vieira, F.; Santana, H.E.P.; Jesus, M.; Santos, J.; Pires, P.; Vaz-Velho, M.; Silva, D.P.; Ruzene, D.S. Coconut waste: Discovering sustainable approaches to advance a circular economy. *Sustainability* **2024**, *16*, 3066. [CrossRef]
4. Rojas-Flores, S.; Cruz-Noriega, M.D.L.; Cabanillas-Chirinos, L.; Otiniano, N.M.; Soto-Deza, N.; Terrones-Rodriguez, N.; Cruz-Cerquin, M.D.L. Potential use of Andean tuber waste for the generation of environmentally sustainable bioelectricity. *Molecules* **2024**, *29*, 1978. [CrossRef] [PubMed]
5. Rola, K.; Gruber, S.; Goricanec, D.; Urbancl, D. Waste lignocellulosic biomass as a source for bioethanol production. *Sustain. Chem.* **2024**, *5*, 1–12. [CrossRef]
6. Samuel, O.D.; Aigba, P.A.; Tran, T.K.; Fayaz, H.; Pastore, C.; Der, O.; Erçetin, A.; Enweremadu, C.C.; Mustafa, A. Comparison of the techno-economic and environmental assessment of hydrodynamic cavitation and mechanical stirring reactors for the production of sustainable *Hevea brasiliensis* ethyl ester. *Sustainability* **2023**, *15*, 16287. [CrossRef]
7. Zhang, S.; Fu, J.; Xing, S.; Li, M.; Liu, X.; Yang, L.; Lv, P. Sodium silicates modified calcium oxide as a high-performance solid base catalyst for biodiesel production. *Catalysts* **2023**, *13*, 775. [CrossRef]
8. Rossa, V.; Vasconcelos, S.C.; Díaz, G.C.; Resende, J.A.; Reys Mattos, J.P.; Madriaga, V.G.C.; Massante, F.F.; Cruz, Y.R.; Nachez, J.L.; Xing, Y.; et al. Solketal production using eco-friendly reduced graphene oxide as the catalyst. *Catalysts* **2023**, *13*, 1427. [CrossRef]
9. Zhang, X.; Wang, Y.; Su, P.; Zeng, W.; Zhu, J.; Cai, Z. Sustainable biochar fertiliser production using melt adsorption and optimization. *Sustainability* **2025**, *17*, 1963. [CrossRef]
10. Filho, J.F.L.; da Silva, A.P.F.; Costa, S.T.; Gomes, H.T.; Figueiredo, T.D.; Hernández, Z. Biochars derived from olive mill byproducts: Typology, characterization, and eco-efficient application in agriculture—A systematic review. *Sustainability* **2024**, *16*, 5004. [CrossRef]
11. Meng, H.; Yang, H.; Wu, Z.; Li, D.; Wang, Z.; Wang, D.; Wang, H.; Li, H.; Li, J. Co-pyrolysis of mushroom residue blended with pine sawdust/wheat straw for sustainable utilization of biomass wastes: Thermal characteristics, kinetic/thermodynamic analysis, and structure evolution of co-pyrolytic char. *Sustainability* **2024**, *16*, 6677. [CrossRef]
12. Ruiz-Velducea, H.A.; Moreno-Vásquez, M.D.J.; Guzmán, H.; Esquer, J.; Rodríguez-Félix, F.; Graciano-Verdugo, A.Z.; Santos-Sauceda, I.; Quintero-Reyes, I.E.; Barreras-Urbina, C.G.; Vásquez-López, C.; et al. Valorization of *Agave angustifolia* bagasse biomass from the bacanora industry in Sonora, Mexico as a biochar material: Preparation, characterization, and potential application in ibuprofen removal. *Sustain. Chem.* **2024**, *5*, 196–214. [CrossRef]
13. Lozano-Pérez, A.S.; Guerrero-Fajardo, C.A. Liquid hot water (LHW) and hydrothermal carbonization (HTC) of coffee berry waste: Kinetics, catalysis, and optimization for the synthesis of platform chemicals. *Sustainability* **2024**, *16*, 2854. [CrossRef]
14. Aboudi, K.; Greses, S.; González-Fernández, C. Hydraulic retention time as an operational tool for the production of short-chain carboxylates via anaerobic fermentation of carbohydrate-rich waste. *Molecules* **2023**, *28*, 6635. [CrossRef] [PubMed]
15. Kim, D.; Kang, S.G.; Chang, Y.K.; Kwak, M. Two-step macromolecule separation process with acid pretreatment and high-shear-assisted extraction for microalgae-based biorefinery. *Sustainability* **2024**, *16*, 7589. [CrossRef]
16. Mullen, C.A.; Ellison, C.; Elkasabi, Y. Pyrolytic conversion of cellulosic pulps from "Lignin-First" biomass fractionation. *Energies* **2023**, *16*, 3236. [CrossRef]

17. Du, F.; Xian, X.; Tang, P.; Li, Y. Catalytic degradation of lignin over sulfonyl-chloride-modified lignin-based porous carbon-supported metal phthalocyanine: Effect of catalyst concentrations. *Molecules* **2024**, *29*, 347. [CrossRef] [PubMed]
18. Shkuratov, A.S.; Shibu, R.P.; Therasme, O.; Berton, P.; Shamshina, J.L. Sustainable production of chitin nanowhiskers from crustacean biomass using cost-effective ionic liquids: Strategies to avoid byproduct formation. *Sustain. Chem.* **2024**, *5*, 130–148. [CrossRef]

sustainability

MDPI

Article

Sustainable Biochar Fertiliser Production Using Melt Adsorption and Optimisation

Xia Zhang [1,2], Yiyang Wang [1,2], Panjie Su [1,2], Weida Zeng [1,2], Jingzhe Zhu [1,2] and Zongshou Cai [1,2,*]

1 Faculty of Mechanical and Electrical Engineering, Yunnan Agricultural University, Kunming 650201, China; zhxia8056@163.com (X.Z.); 18831159865@163.com (Y.W.); spj991106@163.com (P.S.); weidazeng@ynau.edu.cn (W.Z.); 15896370909@163.com (J.Z.)

2 Key Laboratory of Plateau Characteristic Modern Agriculture Energy Conservation and Carbon Reduction, Education Department of Yunnan Province, Kunming 650201, China

* Correspondence: 15887014091@163.com

Abstract: The global production of coffee generates substantial waste in the form of coffee husks, presenting an opportunity for sustainable resource utilisation. This study investigates the conversion of coffee husks into biochar-based fertilisers using an optimised pyrolysis process and the melt adsorption method, with the aim of improving slow-release nutrient characteristics. Coffee husks were pyrolysed under controlled conditions, and the resulting biochar was blended with urea in a 1:1 ratio to produce the fertiliser. The nutrient release rates of the fertiliser were evaluated through soil column leaching experiments. For slow-release fertilisers, less release in a period means better performance, and the optimal nutrient release rate is the minimum in the experiment. Single-factor experiments assessed the effects of critical process parameters, including raw material particle size (0.25–2.8 mm), pyrolysis atmosphere (CO_2 to N_2 ratios ranging from 0% to 100%), pyrolysis temperature (400–800 °C), heating rate (10–30 °C min^{-1}), and pyrolysis time (20–100 min). Orthogonal experiments were subsequently conducted to analyse the interactions between selected parameters of pyrolysis temperature, CO_2 to N_2 ratio, heating rate, and pyrolysis time using regression analysis. The optimal process conditions determined through MATLAB optimisation were a pyrolysis temperature of 591 °C, a heating rate of 15.4 °C·min^{-1}, and a pyrolysis time of 44.4 min, resulting in the minimum nutrient release rate of 40%. The sequence of process parameters influencing the slow-release characteristics was found to be heating rate > pyrolysis time > pyrolysis temperature > CO_2 to N_2 ratio. This study provides a framework for transforming agricultural residues into high-performance biochar-based fertilisers, aligning with sustainable resource management and pollution control strategies.

Keywords: coffee husks; biochar-based fertiliser; optimisation; pyrolysis; slow-release performance; melt adsorption method

1. Introduction

Fertiliser plays an indispensable role in agricultural production in a densely populated country like China. Meeting the demand for agricultural products for 1.4 billion people on limited arable land is a crucial challenge for China. The utilisation of fertiliser, as emphasised by the Food and Agriculture Organization of the United Nations (FAO), significantly enhances crop yields, accounting for 50–60% of global food production. Fertiliser usage has always been a vital assurance for China's grain production, leading it to have become the world's largest consumer since 1993, with consumption reaching 6.363×10^7 t in 2022 and application rates per unit area in China being one to three times higher than those of developed countries [1]. However, long-term extensive application of chemical fertilisers hampers their effective absorption and utilisation by crops, resulting in imbalanced soil nutrient structures, reduced fertility, and deterioration of physical properties and other negative consequences. Moreover, excessive chemical fertilisation can infiltrate groundwater and rivers through rainwater and farmland irrigation systems, causing pollution in certain areas. Therefore, reducing fertiliser loss and improving its efficiency is imperative.

Biochar-based fertiliser is a composite fertiliser that combines biochar and conventional fertilisers, with biochar serving as a carrier. It offers several advantages including the ability to delay nutrient release, enhance nutrient utilisation efficiency, and reduce the overall amount of fertiliser required. This innovative approach effectively addresses the issue of excessive chemical fertiliser application. In recent years, research and developments in the field of biochar-based fertilisers have gained significant attention worldwide. Studies have reported numerous benefits associated with these fertilisers, such as mitigating soil acidification and degradation, promoting soil microbial activity and biological processes [2–6], stimulating crop growth and development, while simultaneously reducing production costs and environmental pollution [7,8].

Currently employed methods for producing biochar-based fertilisers include direct mixing, mixed granulation or coating techniques, as well as adsorption methods [9,10]. The direct mixing method involves blending biochar with conventional fertilisers without undergoing any adsorption process, thereby limiting the typical slow-release properties associated with biochar-based fertilisers. In addition, the large volume and low-density characteristics of it make its transportation and usage inconvenient. Although mixed granulation or coating methods can improve the density of biochar-based fertilisers to some extent by using binders or coating materials, these two approaches pose environmental pollution concerns since these materials are difficult to degrade [11]. On the other hand, the adsorption method eliminates the need for any additives and offers numerous advantages in terms of environmental sustainability, increased density, compact size, and convenient application. The adsorption method can be classified into two categories: the solution adsorption method and the melt adsorption method. The process of solution adsorption involves the dissolution of urea in water, followed by the absorption of urea by biochar for a specific duration. Subsequently, the mixture is filtered and dried. Melt adsorption refers to melting urea at an elevated temperature so that biochar can fully absorb the liquid urea. In previous studies, the solution adsorption method was employed to prepare fertiliser based on coffee husk biochar while the optimal nutrient release rate within seven days was 45.4% [12]. However, the preparation of biochar-based fertiliser using the melt adsorption method has been sparsely documented.

High-quality biochar is a crucial material for the preparation of biochar-based fertilisers. Biochar is one of the pyrolysis products derived from biomass under anoxic or anaerobic conditions [13]. Biochar primarily contains five elements: hydrogen (H), oxygen (O), nitrogen (N), sulfur (S) and carbon (C), among which C is the main element and its content ranges from 30% to 90% (depending on the type of raw materials and pyrolysis temperature) [14]. The more carbon elements in biochar, the more pores there are in biochar. Due to its porous structure, biochar can effectively absorb nutrients from chemical fertilisers and gradually release them during agricultural production. The physical properties of biochar are influenced by various process parameters [14,15], including raw material size, pyrolysis temperature, heating rate, atmosphere ratio, and pyrolysis time. Rodriguez et al. (2020) reported that an increase in pyrolysis temperature led to a reduction in the yield of solid and volatile compounds [16]. Qin et al. (2020) found that elevating the temperature during pyrolysis could enhance the calorific value of biochar while increasing the pyrolysis pressure could boost its yield [17]. Chen et al. (2016) observed that at higher temperatures (400–600 °C), there was a decrease in carbon yield with an increase in heating rate [18]. Cao et al. (2018) demonstrated that within a temperature range of 200–700 °C, residence time had a negative effect on biochar yield for corn straw, rice husk, peanut shell, and tobacco straw [19]. Additionally, numerous studies have investigated the impacts of different pyrolysis atmospheres (such as pure N_2, pure O_2, pure CO_2 water vapor) on both the yield and physicochemical properties of biochar [20–23]. However, previous studies have predominantly focused on investigating the impacts of process parameters on biochar yield, calorific value, or other physical characteristics. Limited research has been conducted specifically on the slow-release properties of biochar-based fertilisers. At the same time, regarding biochar, research on many other influencing parameters has not been carried out. This research situation shows that it is very necessary to add influencing parameters to enrich models, which will help fill the gaps in the existing research on biochar and biochar-based fertilisers, so as to further explore the related phenomena and laws. Therefore, this paper aims to explore and optimise the influence of these pyrolysis process parameters on the slow-release properties of biochar-based fertilisers.

The careful selection of raw materials is essential in the preparation of biochar-based fertilisers. It has been reported that biomass with a higher lignin content is suitable for biochar production [24]. Coffee husks are abundant in organic matter, containing approximately 27.5% lignin [25]. According to the International Coffee Organization (ICO), global coffee production annually reaches 104.97 million tons. During coffee bean processing, approximately 15% to 20% consists of coffee husks, which are either discarded or landfilled, resulting in the emission of greenhouse gases such as methane. This not only leads to resource wastage but also contributes to environmental pollution. Therefore, this study selected coffee husks as the raw material for obtaining biochar.

The objective of this study was to investigate the impacts of process parameters for pyrolysing coffee husks on the slow-release characteristics of biochar-based fertiliser produced through the melt adsorption method. For the four-factor and three-level experiment in this study, in the most commonly used response surface optimisation method, the more factors there are, the more limited the approximation accuracy. Compared with the fitting approximation relationship of the response surface method, grid search searches for the optimal solution by traversing the given parameter combinations, and can systematically explore the parameter space to find the hyper-parameter combination with the best performance, thereby improving the accuracy and efficiency of the model. As the simplest and most direct hyper-parameter search method, although the computational cost is relatively high, especially when the number of hyper-parameters is large, its ability to find the optimal solution still gives it a wide range of application prospects in scenarios where high-precision models are required. The grid search in this study is implemented through MATLAB. This work directly supports the United Nations Sustainable Development Goals (SDGs), particularly Goal 12 on responsible consumption and production, and Goal 13 on climate action by mitigating environmental pollution and enhancing resource efficiency.

2. Materials and Methods

2.1. Materials and Equipment

The materials used were coffee husks (crushed, sieved, and dried) and urea (nitrogen element mass fraction greater than 46.20%). The urea was purchased from Tianjin Dengfeng Chemical Reagent Factory. The main equipment consisted of a pyrolysis furnace (OTF-100 1200X, Hefei Kejing Materials Technology Co., Ltd., Hefei, China), an electric constant temperature blast drying oven (101-0A, Shaoxing Yicheng Instrument Manufacturing Co., Ltd., Shaoxing, China), a vibrating sieve machine (TLS-200, Shaoxing Bowei Instrument Equipment Co., Ltd., Shaoxing, China), a portable pressure steam steriliser (YXQ-LS-18S

II (24SII), Shanghai Boxun Medical Biological Instrument Co., Ltd., Shanghai, China), an ultraviolet spectrophotometer (Jinghua752, Shanghai Jinghua Technology Instrument Co., Ltd., Shanghai, China), an ultrapure water machine (EU-10T, Nanjing Oukai Environmental Technology Co., Ltd., Nanjing, China), and an electronic analytical balance with 0.001 mg accuracy.

2.2. Preparation of Biochar

The coffee husks were initially crushed using a hammer mill, followed by passing through screens of varying sizes (ranging from 0.25 to 2.8 mm) to obtain material samples with different particle sizes. After being dried at 105 °C for 12 h in a drying oven, the material samples were placed into a tube furnace for pyrolysis under various process conditions. The temperature of the tube furnace was controlled by a temperature controller, and a mixed inert gas of CO_2 and N_2 was used as the pyrolysis atmosphere. The pyrolysis process was conducted within a temperature range of 400 °C to 800 °C, with heating rates controlled between 10 $°C{\cdot}min^{-1}$ and 30 $°C{\cdot}min^{-1}$, and varying pyrolysis times ranging from 20 to 100 min. The gas flows of CO_2 and N_2 were controlled by two flow controllers located at the upper outlet of the gas cylinders. The two gas streams were mixed and directed into the tube furnace through a three-way valve (Figure 1). The ratios of CO_2 to N_2 were adjusted at five different levels: 0%, 25%, 50%, 75%, and 100%. After the complete pyrolysis of the material samples under various process conditions, the tube furnace was cooled to ambient temperature, and the valves of the gas cylinders were securely closed. The biochar was subsequently extracted and stored in a hermetically sealed container.

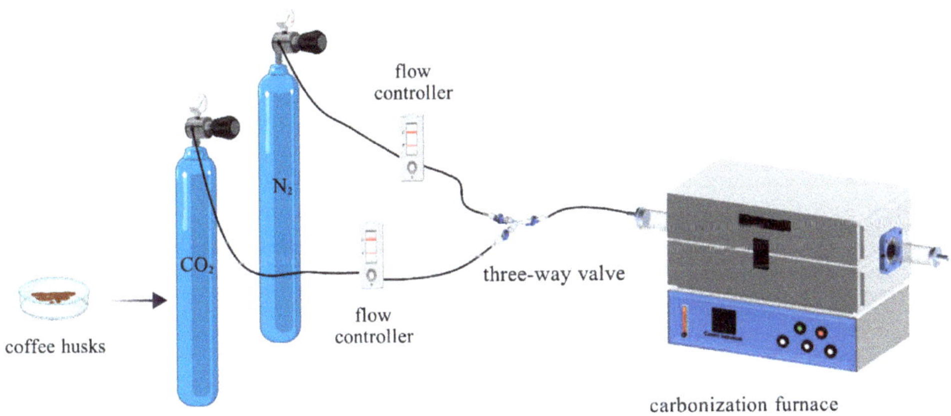

Figure 1. Preparation process of biochar.

2.3. Preparation of Biochar-Based Fertiliser

The biochar-based fertiliser was prepared using the melting adsorption method. Initially, coffee husk-derived biochar and urea were thoroughly mixed in a 1:1 ratio and placed inside a round-bottomed test tube. The test tube was then inserted into a beaker filled with quartz sand. Subsequently, the beaker was transferred to a drying oven and heated at 150 °C for one hour. After removing the beaker from the oven, it was allowed to cool down to room temperature. Throughout this process, the urea underwent heating and melting before being adsorbed by the biochar.

2.4. Experimental Design

In order to investigate the effects of process parameters for pyrolysing coffee husks on the slow-release characteristics of biochar-based fertiliser made by the melt adsorption method, single-factor tests were initially conducted. The pyrolysis parameters were the independent variables, while the nutrient slow-release rate of biochar-based fertiliser was the dependent variable measured by the method of a soil leaching test according to the national standard GB/T 23348-2009 of the People's Republic of China [26]. Soil leaching is an evaluation method for slow-release fertilisers [27]. Specific experiments were carried out in Zhao et al.'s research [28]. The testing conditions of this method were controlled. In combination with the actual situation, the experiment was carried out at 25 °C, and the influence of climate and humidity etc., on the experiment was not considered at this time.

According to previous studies, pyrolysis is the anoxic thermochemical decomposition of biomass at medium (300–800 °C) temperatures, with heating rates ranging from 10 °C/min to 30 °C/min, and a smaller particle size of less than 0.2 cm would be ideal [29]. The residence time is 5–240 min [30], and the ratio of CO_2 in N_2 is 0–100% [31]. Therefore, coffee husks were pyrolysed into biochar under various process conditions (Section 2.2) and then mixed with urea in a 1:1 ratio to produce biochar-based fertiliser using the melting adsorption method. The significances of each independent variable were assessed using a one-way ANOVA, and the optimal conditions for each independent variable were determined through the analysis of single-factor experiments. On the basis of single-factor experiments, the orthogonal experiments were conducted with four factors and three levels. These experimental factors included pyrolysis temperature (400, 500, 600 °C), ratio of CO_2/N_2 (0%, 25%, 50%), heating rate (15, 20, 25 °C min^{-1}), and pyrolysis time (40, 60, 80 min). Regression analysis and model equation were performed with the least squares method in MATLAB R2016b software, and the optimal pyrolysis parameters were obtained by the grid search.

2.5. Measurement of Nutrient Slow-Release Rate

The nutrient slow-release rate of biochar-based fertiliser was determined using the soil leaching test method specified in the national standard GB/T 23348-2009 of the People's Republic of China, as previously mentioned. The main experimental materials included PVC pipes, quartz sand, ammonia-free water, soil, 25 mL colourimetric tubes with glass stoppers, pipettes, a spectrophotometer, and 10 mm quartz cuvettes. The main experimental reagents consisted of a 40 g/L alkaline potassium persulfate solution and a 1 mol/L hydrochloric acid standard titration solution. Deionised ultrapure water was produced by the EU-10T laboratory ultrapure water machine. The test procedure comprised two distinct steps.

2.5.1. Preparation of Leaching Solution

The schematic diagram of the soil leaching test device is shown in Figure 2. The soil leaching column was constructed using a PVC tube with an inner diameter of 3.2 cm and a length of 15 cm. The bottom of the PVC tube was plugged with a filter cloth with a mesh size of 0.08 mm. The soil samples used in the soil leaching column were subjected to oven drying at a temperature of 105 °C for 12 h, followed by sieving through a mesh with a screen size of 0.6 mm. The soil leaching column was filled with 10 g of quartz sand, followed by the addition of 80 g of soil. Subsequently, 70 mL of ammonia-free water was added to saturate the soil sample and allowed to leach for 24 h. Following this, 200 mg biochar-based fertiliser and 5 g soil were sequentially introduced into the leach tube, followed by the addition of 10 g quartz sand to cover the surface layer of the soil in order to minimise the soil disturbance by water. Each leaching test lasted for 7 days, during which 30 mL of ammonia-free water was added to each column at specific intervals on the first, third, fifth, and seventh days. Subsequently, the leaching solution was collected from the bottom of the leaching column and placed into a beaker positioned beneath the column. The nitrogen content in the leaching solution was then determined using alkaline potassium persulfate oxidation ultraviolet spectrophotometry. Each set of experiments was conducted three times to ensure experimental accuracy and reliability.

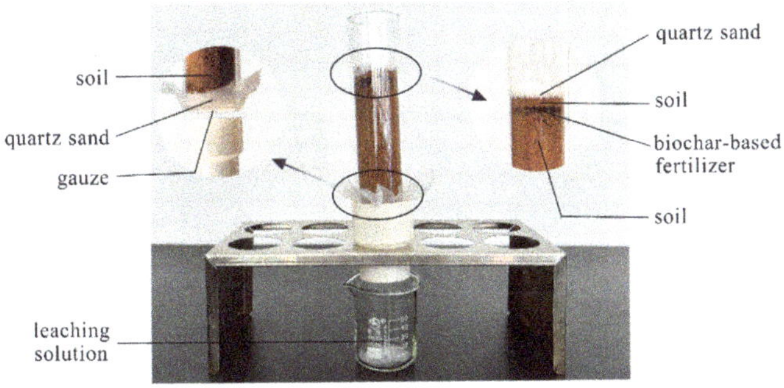

Figure 2. Soil column leaching device.

2.5.2. Determination of Nitrogen Content of Leaching Solution

The test procedure for determining the total nitrogen concentration in the leaching solution is illustrated in Figure 3. The volume of the original leaching solution was 20 to 30 mL, which was then diluted twice with ammonia-free water to reach a final volume of 1 L. A sample of the diluted solution (10 mL) was taken for testing. Initially, 5 mL of alkaline potassium persulfate solution was injected into a test tube containing the sample solution. Subsequently, the test tubes were sealed and placed in a steam steriliser at a constant pressure (1.1 to 1.4 kg/m^3) and temperature (120–124 °C) for 45 min. Afterwards, the tube was removed from the steam steriliser and allowed to cool down to room temperature. Then, 2 mL of hydrochloric acid (10%) was added to the test tube and further diluted with ammonia-free water up to a final volume of 25 mL. Finally, a portion (5 mL) of this test sample solution was transferred into a colourimetric dish where its absorbance at wavelengths 220 nm and 275 nm could be measured using an ultraviolet spectrophotometer.

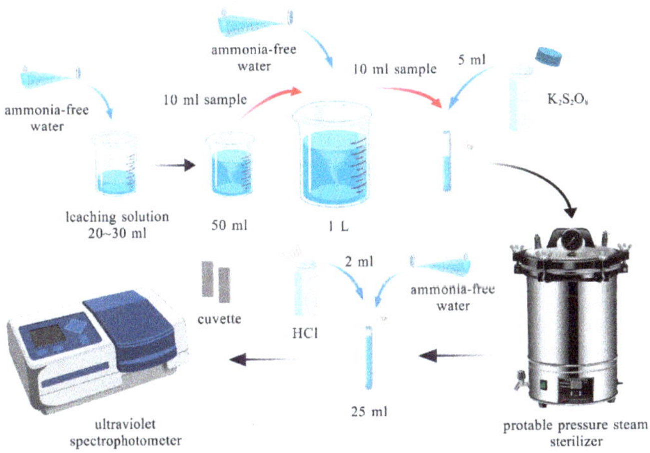

Figure 3. Measurement process of the absorbance of leaching solution.

The experiments were conducted to establish the relationship between the concentration of urea and the absorbance of the test sample solution by measuring the absorbance of five different concentrations of urea solution. Their relationship was represented by

Formula (1). The model demonstrated a significantly high coefficient of determination (R^2) at 0.9953.

$$\omega = 23.195x - 0.0873 \tag{1}$$

where ω represents the concentration of urea of solution (mg/L), and x indicates the absorbance of the sample solution.

The absorbance of the test sample solution was converted into the calculated concentration using Formula (1), and then Formula (2) was applied to determine the mass of nitrogen in the test sample solution.

$$m_t = n \times \omega \times V \tag{2}$$

where m_t indicates the mass of nitrogen in the solution (mg); n represents the dilution ratio of solution; ω denotes the calculated concentration of solution (mg/L); and V represents the volume of the solution (L).

The cumulative nutrient release rate of the leaching solution over a specified time was determined by calculating the percentage of nitrogen mass in the biochar-based fertiliser to the total nitrogen content of biochar-based fertiliser, as illustrated by Formula (3). The slower the rate of cumulative nutrient release, the more pronounced the slow-release characteristics of the biochar-based fertilisers in terms of their reduced speed of nutrient release.

$$c = \frac{\sum_0^i m_t}{m} \times 100\% \tag{3}$$

where c represents the cumulative nutrient release rate over days, (%); m_t denotes the mass of nitrogen in the solution, (mg); m represents the total mass of nutrients used in biochar-based fertiliser, (mg); and i indicates the number of days of the test.

3. Results and Analysis

3.1. *Single-Factor Experimental Results and Analysis*

The cumulative nutrient slow-release rates of the leaching solution for 7 days under different single independent variables are shown in Figure 4. All the nutrient slow-release rates exhibited relatively lower values on the 1st and 7th days compared to the other days while reaching their peaks on the 3rd day. The 3rd day was the middle day of the leaching test time, which proved that the choice of 7 days as the total time of the leaching test was reasonable. The following presented the results of the cumulative nutrient release rate of biochar-based fertilisers under different parameters.

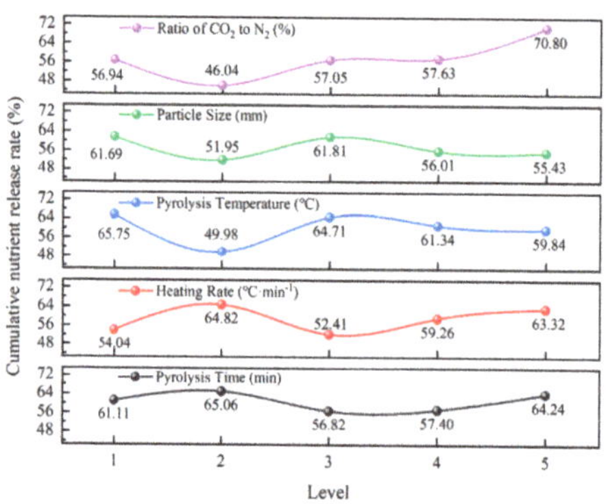

Figure 4. Cumulative nutrient release rate under different single independent variables.

When conducting a single-factor experiment, the control variable method is used to study the influence of pyrolysis parameters on the nutrient release rate. Other parameters are under controlled conditions, including a raw material particle size ranging from 0.5 to 1 mm, pyrolysis temperature set at 500 °C, heating rate maintained at 20 $°C·min^{-1}$, a pyrolysis time of 60 min, and a pure N_2 atmosphere. Each factor contains five levels, as shown in Table 1.

Table 1. Table of element levels.

Element	Level				
	1	2	3	4	5
Ratio of CO_2 to N_2/%	0	25	50	75	100
Particle size/mm	0.25−0.5	0.5−1.0	1.0−1.4	1.4−2.0	2.0−2.8
Pyrolysis temperature/°C	400	500	600	700	800
Heating rate/$°C·min^{-1}$	10	15	20	25	30
Pyrolysis time/min	20	40	60	80	100

3.1.1. Ratio of CO_2 to N_2

The influence trend of the ratio of CO_2 to N_2 on the nutrient release rate of the biochar-based fertilisers was as follows: with the increase of CO_2 concentration, the cumulative nutrient release rate first decreased and then increased, showing more significant changes in the range of 46.04% to 70.80% (Figure 4). Based on these experimental conditions, an optimal CO_2 to N_2 ratio observed was found to be around 25%. Among the five different ratios of CO_2 to N_2 tested, the nutrient slow-release rate was lowest at a ratio of 25%, measuring only 46.04%, while it reached its highest value of 70.80% at a ratio of 100%.

In biomass pyrolysis, CO_2 improves the production of pyrolysis gases (H_2, CO, and CH_4) by increasing the thermal cracking of tar, suppressing secondary coking, and enhancing coke gasification, thereby reducing the coke yield [23]. Moreover, the increase in CO_2 concentration (0–60%) will promote the progress of the reactions and simultaneously increase the yields of H_2 and CH_4, while the char yield of the biochar will decrease at 100% CO_2 [31]. The escape of gas will generate a large number of pores. This is consistent with the conclusion of this study, and also expands the research on the influence of CO_2 concentration on biochar performance in slow pyrolysis performance (400–700 °C).

3.1.2. Particle Size

As the particle size levels increase, the nutrient release rate of biochar-based fertilisers first decreased, then increased, then continued to decrease, and finally tended to be balanced (Figure 4). It could be observed that the cumulative nutrient release rate in the leaching solution remained relatively stable within a range of 61.81% to 51.95%. Among the five different particle sizes tested, the cumulative nutrient release rate was lowest for particle sizes ranging from 0.5 to 1.0 mm and highest for particle sizes ranging from 0.25 to 0.05 mm. The optimal particle size range was found to be between 0.5 to 1 mm under these fixed conditions. This is because the particle size mainly affects the biomass pyrolysis process in terms of heat transfer. Both smaller and larger particles will reduce the carbon content in biochar, which has also been confirmed in previous studies [32,33].

3.1.3. Pyrolysis Temperature

Under controlled conditions, the impact of varying pyrolysis temperatures on the nutrient slow-release rate of the nutrients was as follows: as the factor levels increase, the nutrient release rate of biochar-based fertilisers decreased, increased, continued to decrease, and finally tended to be balanced (Figure 4). It could be observed that the nutrient slow-release rate fluctuated between 49.98% and 65.75% as the pyrolysis temperature varied between 400 °C and 800 °C. Among the five different pyrolysis temperatures tested, the lowest nutrient slow-release rate was observed at a temperature of 500 °C, while the highest value of 65.75% was achieved at a temperature of 400 °C. Based on these experimental conditions, an optimal pyrolysis temperature of around 500 °C was determined. The most commonly used pyrolysis temperature is in the range of 300–900 °C. As the temperature rises, compounds at various stages, including moisture, hemicellulose, and lignin, decompose in turn and lignin starts to decompose above 500 °C [34–36]. The fact that the optimum in this study is at 500 °C also proves that biochar has strong performance at this stage.

3.1.4. Heating Rate

The influence trend of the heating rate was as follows: as the factor levels increase, the cumulative nutrient release rate first increased, then decreased, and then continuously increased, as shown in Figure 4. This is because, at a higher heating rate, the pyrolysis reaction of biomass is more intense, and more macromolecular substances are decomposed into small-molecular substances, resulting in a decrease in the solid yield. The results also found that the nutrient slow-release rate of nutrients ranged from 52.41% to 64.82% as the heating rate varied from 10 $°C \cdot min^{-1}$ to 30 $°C \cdot min^{-1}$ in increments of 5 $°C \cdot min^{-1}$. Among the five different heating rates tested, the lowest nutrient slow-release rate was observed at a heating rate of 20 $°C \cdot min^{-1}$. Meanwhile, the highest value of cumulative nutrient release rate (64.82%) was achieved at a heating rate of 15 $°C \cdot min^{-1}$. Under these experimental conditions, an optimal heating rate of around 20 $°C \cdot min^{-1}$ was indicated.

3.1.5. Pyrolysis Time

The influence trend of the pyrolysis time showed that as the factor levels increase, the cumulative nutrient release rate increased, decreased, and then increased (Figure 4). The results for the pyrolysis time experiment showed that the cumulative nutrient release rate in the leaching solution varied between 56.82% and 65.06% as the pyrolysis time gradually increased from 20 min to 100 min in increments of 20 min. Among the five different tested pyrolysis times, the lowest cumulative nutrient release rate was observed at a pyrolysis time of 60 min, while the highest was recorded at a pyrolysis time of 40 min. Therefore, based on these fixed conditions, an optimal pyrolysis time of 60 min was determined.

The extension of the pyrolysis time will gradually decompose the macromolecular substances inside the biomass, forming more pore structures [37]. Therefore, selecting appropriate condition levels for these factors can increase the specific surface area and porosity of biochar, thereby affecting its adsorption performance and chemical stability.

3.2. *Significant Analysis of Single Factor*

The significance of single independent variables was assessed using one-way ANOVA in Excel data analysis. The significant effects of single independent variables on cumulative nutrient release rate are summarised in Table 2. When the significance level was set at $\alpha = 0.05$, both the pyrolysis temperature and the ratio of CO_2 to N_2 exhibited extremely significant effects with p-values of less than 0.01. The p-values for heating rate, pyrolysis time, and particle size ranged between 0.01 and 0.05, indicating their considerable influence on the nutrient slow-release rate of biochar-based fertiliser. Notably, among all the independent variables considered, particle size showed the highest p-value suggesting its minimal impact on the nutrient slow-release rate of biochar-based fertiliser.

Table 2. Table of single factor variable ANOVA results.

Element	F	p-Value	F crit	Significant Ness
Ratio of CO_2 to N_2	17.828	0.0037	5.1922	**
Particle size	3.8261	0.0388	3.4781	*
Pyrolysis temperature	6.8801	0.0063	3.4781	**
Heating rate	5.0677	0.0171	3.4781	*
Pyrolysis time	5.9870	0.0100	3.4781	*

Note: * means significant; ** indicates extremely significant.

3.3. Orthogonal Experimental Design and Results

In order to further analyse the effects of pyrolysis process parameters and their interactions on the slow-release of biochar-based fertiliser, establish a mathematical regression model between the cumulative nutrient release rate and the main process parameters, and determine the sequence of each main process parameters and the optimal process parameter conditions, a four-factor and three-level orthogonal experiment design was also carried out, in which the pyrolysis temperature (X_1), ratio of CO_2 to N_2 ratio (X_2), heating rate (X_3), and pyrolysis time (X_4) were the independent variables, while the cumulative nutrient release rate (Y) was the dependent variable. The factors and levels of orthogonal tests were determined based on the results of single-factor experiments. These three levels, from the minimum to the maximum, were coded as −1, 0 and 1 in sequence, as presented in Table 3. Due to the relatively minor impact of raw material particle size on cumulative nutrient release rate compared to other process parameters, it was temporarily excluded from consideration in the orthogonal test conducted for this study. The experimental design and corresponding results are provided in Table 4.

Table 3. Factors and levels of orthogonal test.

Level	Independent Variable			
	Pyrolysis Temperature (X_1)/°C	Ratio of CO_2 to N_2 (X_2)/%	Heating Rate (X_3)/°C·min^{-1}	Pyrolysis Time (X_4)/min
−1	400	0	15	40
0	500	25	20	60
1	600	50	25	80

Table 4. Experimental design and results of orthogonal test.

Number	Pyrolysis Temperature (X_1)/°C	Ratio of CO_2 to N_2 (X_2)/%	Heating Rate (X_3)/°C·min^{-1}	Pyrolysis Time (X_4)/min	Cumulative Nutrient Release Rate (Y)/% *
1	−1	−1	−1	−1	52.93
2	−1	0	0	0	58.00
3	−1	1	1	1	48.59
4	0	−1	0	1	56.16
5	0	0	1	−1	53.11
6	0	1	−1	0	48.01
7	1	−1	1	0	51.46
8	1	0	−1	1	48.85
9	1	1	0	−1	48.70

Table 4. *Cont.*

Number	Pyrolysis Temperature (X_1)/°C	Ratio of CO_2 to N_2 (X_2)/%	Heating Rate (X_3)/°C·min^{-1}	Pyrolysis Time (X_4)/min	Cumulative Nutrient Release Rate (Y)/% *
10	−1	−1	0	0	58.70
11	−1	0	1	1	49.25
12	−1	1	−1	−1	53.06
13	0	−1	−1	1	56.87
14	0	0	0	−1	52.08
15	0	1	1	0	47.43
16	1	−1	1	−1	53.34
17	1	0	−1	0	43.72
18	1	1	0	1	48.47

Note: * was the average of 3 random tests.

3.4. Orthogonal Experimental Analysis

MATLAB R2016b software was used to conduct multiple linear regression analysis on the test data in Table 4. The cumulative nutrient release rate (Y) was taken as an objective function, and the pyrolysis temperature (X_1), ratio of CO_2 to N_2 (X_2), heating rate (X_3), and pyrolysis time (X_4) were taken as factor variables. When performing regression in MATLAB, consider all quadratic terms and interaction effects, import data, define the term model, and call the fitnlm function for non-linear fitting. The influence of four-factor variables, their interaction and quadratic terms on the value of the objective function were analysed. At the significant level of $\alpha = 0.05$, after analysis and elimination of non-significant items, the multiple linear regression equation was as follows:

$$Y = 54.486 - 0.272X_1 + 5.248X_3 + 0.703X_4 + 0.005X_1 \times X_3 - 0.039X_3 \times X_4 + 0.0001{X_2}^2 - 0.137{X_3}^2 \quad (4)$$

where Y is the cumulative nutrient release rate, (%); X_1 is the pyrolysis temperature, (°C); X_2 is the ratio of CO_2 to N_2, (%); X_3 is the heating rate, (°C·min^{-1}); X_4 is the pyrolysis time (min).

According to the significance test, at a significance level of $\alpha = 0.05$, the regression Equation (4) exhibited a significant relationship with a significance level of 0.004. Simultaneously, the determination coefficient R^2 for regression Equation (4) was found to be 0.996, indicating that this equation effectively built the relationship between the cumulative nutrient release rate and the ratio of CO_2 to N_2, the pyrolysis temperature, the heating rate, and the pyrolysis time. It could be seen from the regression Equation (4) that the absolute value of the coefficient of the first power of each factor variable was $X_3 > X_4 > X_1 > X_2$, indicating that the sequencing of the process parameters affecting the cumulative nutrient release rate of biochar-based fertiliser ranked as follows: heating rate > pyrolysis time > pyrolysis temperature > ratio of CO_2 to N_2. It should be noted that the significance of factor variable X_2 was greater than 0.05, suggesting that the impact of the ratio of CO_2 to N_2 on the cumulative nutrient release rate was not statistically significant in the presence of multiple factors. The interaction between pyrolysis temperature and heating rate affects the pore structure, elemental composition and functional groups of biochar, while the interaction between pyrolysis rate and time directly affects the pyrolysis efficiency, which in turn affects the carbon content. Therefore, the latter interaction has a stronger influence than the former.

According to the regression Equation (4) and the constraint conditions, the optimisation program of MATLAB was employed to calculate the optimal process parameter conditions for pyrolysis of coffee husks biochar-based fertiliser. The determined values were as follows: a pyrolysis temperature of 591.0 °C, a heating rate of 15.4 °C·min^{-1}, and a pyrolysis time of 44.4 min. Under these optimised conditions, the cumulative nutrient release rate of biochar-based fertiliser reached its minimum point at 40%, indicating an excellent slow-release characteristic. The optimal process parameters were validated through experimental verification, and the measured cumulative nutrient release rate was determined to be 41.25%, with an error of less than 5% compared to the calculated value. The reliability of the regression equation was confirmed.

4. Discussion

4.1. Comparison Assessment

In this study, the impacts of process parameters for pyrolysing coffee husks on the slow-release characteristics of biochar-based fertiliser produced through the melt adsorption method were investigated. The optimization of process parameters for preparing biochar-based fertiliser for coffee husks by the solution adsorption method had been explored in a previous study [12]. The main results of the two studies are summarized in Table 5.

Table 5. Comparison of the results of the study between the two preparation methods.

Comparative Element	Zhang et al. (2024) [12]	This Study
Raw material	coffee husks	coffee husks
Prepared method	Solution adsorption	Melt adsorption
Optimal pyrolysis temperature	809 °C	591 °C
Optimal heating rate	15.3 °C·min^{-1}	15.4 °C·min^{-1}
Optimal pyrolysis time	169.8 min	44.4 min
The minimum cumulative nutrient release rate	45.4%	40%

Comparing the two methods for preparing biochar-based fertiliser from Table 5, it can be seen that the minimum cumulative nutrient release rates of the two methods were 45.4% and 40%, respectively. This means that biochar-based fertiliser prepared by the melt adsorption method has better slow-release performance. From the perspective of the optimal process parameters of the two methods, it can be observed that the optimal pyrolysis temperatures of the solution adsorption and melt adsorption methods were 809 °C and 591 °C, respectively, the latter being only 73% of that of the former, which meant lower production energy consumption in the melt adsorption method. The optimal heating rates of the two methods was basically similar, which were 15.3 °C min^{-1} and 15.4 °C·min^{-1}, while the optimal pyrolysis time of the two methods were very different. The optimal pyrolysis time of solution adsorption and melt adsorption were 169.8 min and 44.4 min, respectively, meaning that the production efficiency of the melt adsorption method could be increased by 3.8 times of the solution adsorption method in the process of pyrolysis of biochar. In conclusion, biochar-based fertiliser prepared by melt adsorption has better slow-release property, lower energy consumption, and higher production efficiency. This could have more application value in industrial production of biochar-based fertiliser.

4.2. Limitations, Recommendations for Future Work, and Contribution to UNSDGs

This study demonstrates the optimisation of pyrolysis parameters for the conversion of coffee husks into biochar-based fertilisers, contributing to sustainable agriculture and resource management. The model obtained in this study is novel in using CO_2 as a pyrolysis medium. However, certain limitations must be addressed to enhance the robustness and applicability of the findings. First, the applicability of model (4) is limited as it depends on specific parameters and still lacks consideration of more pyrolysis parameters such as pyrolysis pressure. In addition, the range selected for the model is relatively small, requiring large-scale instruction control and not considering situations beyond 800 °C. Therefore, due to the lack of research, more pyrolysis parameters and higher pyrolysis temperatures have not been fully explored and understood. Secondly, the experiments were carried out under controlled laboratory conditions without considering a variety of complex natural conditions, and the influence of natural conditions in field trials still needs to be further explored. Finally, the interaction between urea and biochar groups at other temperatures has not been considered in terms of process. Moreover, high-temperature operation requires high energy and strict equipment requirements, and it is challenging to optimise and control process parameters.

Future research should focus on addressing these limitations through multi-site field trials conducted in diverse agro-climatic regions to validate the performance of biochar-based fertilisers under real-world conditions. A comprehensive lifecycle assessment encompassing environmental, economic, and social dimensions should be undertaken to evaluate the feasibility of large-scale implementation. Mechanistic studies investigating the interactions between biochar-based fertilisers, soil microbiota, and plant nutrient uptake over extended periods are recommended to provide deeper insights into their ecological implications. Expanding the scope to include other types of agricultural residues such as feedstocks would further demonstrate the generalisability of the findings. Furthermore, collaborations with policymakers and industry stakeholders are essential for developing practical guidelines, scaling production, and integrating biochar-based fertilisers into global agricultural practices.

The findings of this study align with and advance several United Nations Sustainable Development Goals (UNSDGs). By transforming coffee husk waste into biochar-based fertilisers, the research supports Goal 12: Responsible Consumption and Production, fostering sustainable waste management and promoting circular economy practices. The reduction in greenhouse gas emissions associated with decreased dependency on conventional chemical fertilisers contributes to Goal 13: Climate Action, while the potential of biochar to act as a carbon sink enhances its climate mitigation capacity. Additionally, by improving soil health, reducing nutrient leaching, and promoting sustainable land use, this study addresses Goal 15: Life on Land, aiding in the restoration of degraded soils and the preservation of terrestrial ecosystems. These contributions highlight the relevance of the study to global sustainability efforts and highlight its potential to inform environmental management policies and practices.

5. Conclusions

In this study, the coffee husks were pyrolysed into biochar under various process conditions, followed by blending with urea in a 1:1 ratio to produce biochar-based fertiliser using the melting adsorption method. The nutrient release rates of the bio-charcoal-based fertiliser were determined through soil column leaching experiments. By utilising orthogonal experimental design and regression analysis, a regression model was established to investigate the correlation between the cumulative nutrient release rate and the parameters of the ratio of CO_2 to N_2, pyrolysis temperature, heating rate, and pyrolysis time. Furthermore, the sequencing of the process parameters affecting the cumulative nutrient release rate of biochar-based fertiliser was analysed. The optimal pyrolysis process parameters for coffee husk with minimal nutrient slow-release rate were finally obtained by utilising the MATLAB optimisation program. The specific conclusions are as follows:

In the single-factor experimental studies, all process parameters including particle size, atmosphere ratio, pyrolysis temperature, heating rate, and pyrolysis time exerted a significant influence on the cumulative nutrient release rate of biochar-based fertiliser.

Among them, pyrolysis temperature and CO_2 to N_2 ratio exhibited a considerable effect while particle size had a minimal impact. When the CO_2 to N_2 ratio was 25%, the pyrolysis temperature was set at 500 °C, the heating rate was maintained at 20 $°C \cdot min^{-1}$, and the pyrolysis duration lasted for 60 min. It was observed that biochar-based fertiliser exhibited superior slow-release performance.

Through multi-factor orthogonal test analysis and regression analysis, the multiple linear regression equation between the cumulative nutrient release rate (Y) and the factor variables of the ratio of CO_2 to N_2 (X_1), pyrolysis temperature (X_2), heating rate (X_3) and pyrolysis time (X_4), exhibited a significant relationship with a significance level of 0.004. The determination coefficient R^2 for the regression equation was 0.996. The order of the process parameters affecting the nutrient slow-release rate of biochar-based fertiliser was as follows: heating rate > pyrolysis time > pyrolysis temperature > ratio of CO_2 to N_2. The impact of the ratio of CO_2 to N_2 on the cumulative nutrient release rate was not statistically significant in the presence of multiple factors. The best cumulative nutrient release rate of biochar-based fertiliser was 40% when the process parameters were as follows: a pyrolysis temperature of 591.0 °C, a heating rate of 15.4 $°C \cdot min^{-1}$, and a pyrolysis time of 44.4 min. The discrepancy between the theoretical and experimental values was found to be less than 5%. The regression equation demonstrates a high level of agreement with the experimental data.

Overall, this research advances sustainable agricultural practices by providing practical solutions for converting agricultural residues into high-value biochar-based fertilisers, aligning with global efforts to enhance resource efficiency, reduce environmental pollution, and improve soil health. By optimising slow-release nutrient properties, the study supports sustainable waste management and the development of scalable strategies to improve agricultural productivity. The findings establish a scientific basis for integrating biochar-based fertilisers into environmentally sustainable agricultural systems, addressing critical challenges in resource management and pollution control, and highlighting the transformative potential of biochar technologies in supporting global sustainability initiatives and fostering environmentally responsible agricultural practices.

Author Contributions: Conceptualization, X.Z. and W.Z.; methodology, Y.W. and X.Z.; software, Y.W.; validation, Y.W.; formal analysis, X.Z., Y.W. and P.S.; investigation, Y.W.; resources, X.Z., Y.W. and J.Z.; data curation, Y.W.; writing—original draft preparation, Y.W.; writing—review and editing, X.Z. and Y.W.; visualization, Y.W.; supervision, P.S., J.Z., Z.C. and W.Z.; project administration, Y.W. and P.S.; funding acquisition, Z.C. All authors have read and agreed to the published version of the manuscript.

Funding: This work is partially supported by Central agricultural production development advantageous characteristic industry cluster construction (30000221100000920058), National Natural Science Foundation of China (No. 52166010, 51706195) and the Yunnan Provincial Project Fund (No. 202101AT070202).

Institutional Review Board Statement: Not applicable.

Informed Consent Statement: Not applicable.

Data Availability Statement: The original contributions presented in this study are included in the article. Further inquiries can be directed to the corresponding author.

Acknowledgments: The authors would like to thank the Huang Danian teacher team of agricultural biological environment and energy engineering.

Conflicts of Interest: The authors declare that they have no known competing financial interests or personal relationships that might influence the work reported in this paper.

References

1. Zhang, Y.; Fan, X.; Mao, Y.; Wei, Y.; Xu, J.; Wu, L. The Coupling Relationship and Driving Factors of Fertilizer Consumption, Economic Development and Crop Yield in China. *Sustainability* **2023**, *15*, 7851. [CrossRef]
2. Jin, X.; Cai, J.; Yang, S.; Li, S.; Shao, X.; Fu, C.; Li, C.; Deng, Y.; Huang, J.; Ruan, Y. Partial substitution of chemical fertilizer with organic fertilizer and slow-release fertilizer benefits soil microbial diversity and pineapple fruit yield in the tropics. *Appl. Soil Ecol.* **2023**, *189*, 104974. [CrossRef]
3. Li, X.; Qiao, L.; Huang, Y.; Li, D.; Xu, M.; Ge, T.; Meersmans, J.; Zhang, W. Manuring improves soil health by sustaining multifunction at relatively high levels in subtropical area. *Agric. Ecosyst. Environ.* **2023**, *353*, 108539. [CrossRef]
4. Llovet, A.; Vidal-Durà, A.; Alcañiz, J.M.; Ribas, A.; Domene, X. Biochar addition to organo-mineral fertilisers delays nutrient leaching and enhances barley nutrient content. *Arch. Agron. Soil Sci.* **2023**, *69*, 2537–2551. [CrossRef]
5. Lu, Z.; Zhou, Y.; Li, Y.; Li, C.; Lu, M.; Sun, X.; Luo, Z.; Zhao, J.; Fan, M. Effects of partial substitution of chemical fertilizer with organic manure on the activity of enzyme and soil bacterial communities in the mountain red soil. *Front. Microbiol.* **2023**, *14*, 1234904. [CrossRef]
6. Yu-Xin, W.; Xie-Yang, L.; Hui-Hui, D.; Xiao-Song, H.; Yue, S.; Song, X.; Zhan-Bin, H. The mechanism of microbial community succession and microbial co-occurrence network in soil with compost application. *Sci. Total Environ.* **2024**, *906*, 167409.
7. Hei, Z.; Peng, Y.; Hao, S.; Li, Y.; Yang, X.; Zhu, T.; Müller, C.; Zhang, H.; Hu, H.; Chen, Y. Full substitution of chemical fertilizer by organic manure decreases soil N2O emissions driven by ammonia oxidizers and gross nitrogen transformations. *Glob. Change Biol.* **2023**, *29*, 7117–7130. [CrossRef]
8. Thakur, B.K.; Sharma, S.; Sharma, A.; Singh, K.K.; Pal, P.K. Integration of biochar with nitrogen in acidic soil: A strategy to sequester carbon and improve the yield of stevia via altering soil properties and nutrient recycling. *J. Environ. Manag.* **2023**, *345*, 118872. [CrossRef]
9. Cen, Z.; Wei, L.; Muthukumarappan, K.; Sobhan, A.; McDaniel, R. Assessment of a biochar-based controlled release nitrogen fertilizer coated with polylactic acid. *J. Soil Sci. Plant Nutr.* **2021**, *21*, 2007–2019. [CrossRef]
10. Chandra, S.; Medha, I.; Bhattacharya, J. Potassium-iron rice straw biochar composite for sorption of nitrate, phosphate, and ammonium ions in soil for timely and controlled release. *Sci. Total Environ.* **2020**, *712*, 136337. [CrossRef] [PubMed]
11. Xiang, A.; Qi, R.; Wang, M.; Zhang, K.; Jiang, E.; Ren, Y.; Hu, Z. Study on the infiltration mechanism of molten urea and biochar for a novel fertilizer preparation. *Ind. Crops Prod.* **2020**, *153*, 112558. [CrossRef]
12. Zhang, X.; Su, P.; Wang, W.; Yang, W.; Ge, Y.; Jiang, K.; Huang, J. Optimized carbonization of coffee shell via response surface methodology: A circular economy approach for environmental remediation. *Environ. Pollut.* **2024**, *346*, 123018. [CrossRef]
13. Lehmann, J.; Rillig, M.C.; Thies, J.; Masiello, C.A.; Hockaday, W.C.; Crowley, D. Biochar effects on soil biota–a review. *Soil Biol. Biochem.* **2011**, *43*, 1812–1836. [CrossRef]
14. Cha, J.S.; Park, S.H.; Jung, S.-C.; Ryu, C.; Jeon, J.-K.; Shin, M.-C.; Park, Y.-K. Production and utilization of biochar: A review. *J. Ind. Eng. Chem.* **2016**, *40*, 1–15. [CrossRef]
15. Chandra, S.; Bhattacharya, J. Influence of temperature and duration of pyrolysis on the property heterogeneity of rice straw biochar and optimization of pyrolysis conditions for its application in soils. *J. Clean. Prod.* **2019**, *215*, 1123–1139. [CrossRef]
16. Rodriguez, J.A.; Lustosa Filho, J.F.; Melo, L.C.A.; de Assis, I.R.; de Oliveira, T.S. Influence of pyrolysis temperature and feedstock on the properties of biochars produced from agricultural and industrial wastes. *J. Anal. Appl. Pyrolysis* **2020**, *149*, 104839. [CrossRef]
17. Qin, L.; Wu, Y.; Hou, Z.; Jiang, E. Influence of biomass components, temperature and pressure on the pyrolysis behavior and biochar properties of pine nut shells. *Bioresour. Technol.* **2020**, *313*, 123682. [CrossRef]
18. Chen, D.; Li, Y.; Cen, K.; Luo, M.; Li, H.; Lu, B. Pyrolysis polygeneration of poplar wood: Effect of heating rate and pyrolysis temperature. *Bioresour. Technol.* **2016**, *218*, 780–788. [CrossRef]
19. Cao, T.; Chen, F.w.; Meng, J. Influence of pyrolysis temperature and residence time on available nutrients for biochars derived from various biomass. *Energy Sources Part A Recovery Util. Environ. Eff.* **2018**, *40*, 413–419. [CrossRef]
20. Zhang, H.; Ni, J.; Qian, W.; Yu, S.; Xiang, Y.; Yang, L.; Chen, W. Pyrolysis Atmospheres and Temperatures Co-Mediated Spectral Variations of Biochar-Derived Dissolved Organic Carbon: Quantitative Prediction and Self-Organizing Maps Analysis. *Molecules* **2023**, *28*, 2247. [CrossRef]
21. Yi, Z.; Li, C.; Zhang, L.; Zhang, S.; Gao, W.; Wang, S.; Li, B.; Hu, X. Impacts of CO_2 atmosphere on property of the biochar from pyrolysis of lignin. *J. Anal. Appl. Pyrolysis* **2022**, *167*, 105689. [CrossRef]
22. Yu, S.; Wu, L.; Ni, J.; Zhang, H.; Wei, R.; Chen, W. The chemical compositions and carbon structures of pine sawdust-and wheat straw-derived biochars produced in air-limitation, carbon dioxide, and nitrogen atmospheres, and their variation with charring temperature. *Fuel* **2022**, *315*, 122852. [CrossRef]
23. Premchand, P.; Demichelis, F.; Chiaramonti, D.; Bensaid, S.; Fino, D. Biochar production from slow pyrolysis of biomass under CO_2 atmosphere: A review on the effect of CO_2 medium on biochar production, characterisation, and environmental applications. *J. Environ. Chem. Eng.* **2023**, *11*, 110009. [CrossRef]

24. Tian, H.; Wei, Y.; Cheng, S.; Huang, Z.; Qing, M.; Chen, Y.; Yang, H.; Yang, Y. Optimizing the gasification reactivity of biochar: The composition, structure and kinetics of biochar derived from biomass lignocellulosic components and their interactions during gasification process. *Fuel* **2022**, *324*, 124709. [CrossRef]
25. Van, N.D.; Thu, D.C.T.; Minh, V.C.N.; Minh, N.H.; Thi, P.T.; Tuyet-Mai, T.-T.; Quang, N.L. Data on chemical composition of coffee husks and lignin microparticles as their extracted product. *Data Brief* **2023**, *51*, 109781.
26. *GB/T 23348-2009*; Slow release fertilizer. General Administration of Quality Supervision, Inspection and Quarantine of the People's Republic of China. Standardization Administration of the People's Pepublic of China, SAC: Beijing, China, 2009.
27. Duan, Q.; Jiang, S.; Chen, F.; Li, Z.; Ma, L.; Song, Y.; Yu, X.; Chen, Y.; Liu, H.; Yu, L. Fabrication, evaluation methodologies and models of slow-release fertilizers: A review. *Ind. Crops Prod.* **2023**, *192*, 116075. [CrossRef]
28. Zhao, X.; Yan, X.; Wang, S.; Xing, G.; Zhou, Y. Effects of the addition of rice-straw-based biochar on leaching and retention of fertilizer N in highly fertilized cropland soils. *Soil Sci. Plant Nutr.* **2013**, *59*, 771–782. [CrossRef]
29. Pahnila, M.; Koskela, A.; Sulasalmi, P.; Fabritius, T. A Review of Pyrolysis Technologies and the Effect of Process Parameters on Biocarbon Properties. *Energies* **2023**, *16*, 6936. [CrossRef]
30. Yaashikaa, P.R.; Kumar, P.S.; Varjani, S.; Saravanan, A.J.B. A critical review on the biochar production techniques, characterization, stability and applications for circular bioeconomy. *Biotechnol. Rep.* **2020**, *28*, e00570. [CrossRef] [PubMed]
31. Liu, Z.; Zhang, F.; Liu, H.; Ba, F.; Yan, S.; Hu, J. Pyrolysis/gasification of pine sawdust biomass briquettes under carbon dioxide atmosphere: Study on carbon dioxide reduction (utilization) and biochar briquettes physicochemical properties. *Bioresour. Technol.* **2018**, *249*, 983–991. [CrossRef]
32. Akhtar, J.; Amin, N.S. A review on operating parameters for optimum liquid oil yield in biomass pyrolysis. *Renew. Sustain. Energy Rev.* **2012**, *16*, 5101–5109. [CrossRef]
33. Haykiri-Acma, H. The role of particle size in the non-isothermal pyrolysis of hazelnut shell. *J. Anal. Appl. Pyrolysis* **2006**, *75*, 211–216. [CrossRef]
34. Tomczyk, A.; Sokołowska, Z.; Boguta, P. Biochar physicochemical properties: Pyrolysis temperature and feedstock kind effects. *Rev. Environ. Sci. Bio/Technol.* **2020**, *19*, 191–215. [CrossRef]
35. Lee, X.J.; Lee, L.Y.; Gan, S.; Thangalazhy-Gopakumar, S.; Ng, H.K. Biochar potential evaluation of palm oil wastes through slow pyrolysis: Thermochemical characterization and pyrolytic kinetic studies. *Bioresour. Technol.* **2017**, *236*, 155–163. [CrossRef] [PubMed]
36. Eliana, C.; Gabriel, G.; Jorge, P.F.; Ana, M. The effect of biochar and compost from urban organic waste on plant biomass and properties of an artificially copper polluted soil. *Int. Biodeterior. Biodegrad.* **2017**, *124*, 223–232.
37. Maulinda, L.; Husin, H.; Arahman, N.; Rosnelly, C.M.; Syukri, M.; Nurhazanah; Nasution, F.; Ahmadi. The Influence of Pyrolysis Time and Temperature on the Composition and Properties of Bio-Oil Prepared from Tanjong Leaves (*Mimusops elengi*). *Sustainability* **2023**, *15*, 13851. [CrossRef]

sustainability

MDPI

Article

Two-Step Macromolecule Separation Process with Acid Pretreatment and High-Shear-Assisted Extraction for Microalgae-Based Biorefinery

Donghyun Kim [1], Seul-Gi Kang [2], Yong Keun Chang [2] and Minsoo Kwak [2,*]

[1] Department of Chemical Engineering, College of Engineering, Qatar University, Doha P.O. Box 2713, Qatar; dkim@qu.edu.qa

[2] Department of Chemical and Biomolecular Engineering, KAIST, 291 Daehak-ro, Yuseong-gu, Daejeon 34141, Republic of Korea; ychang@kaist.ac.kr (Y.K.C.)

* Correspondence: minsoo.kwak@kaist.ac.kr

Abstract: A simple two-stage extraction and recovery method for macromolecules from microalgae biomass, termed CASS (concentrating the microalgae solution, acid pretreatment, high-shear-assisted lipid extraction, and separation), was developed. This method effectively processed the wet biomass of *Chlorella* sp. ABC-001 at a moderately low biomass concentration (50 g/L). The optimal conditions were acid pretreatment with 5 wt.% H_2SO_4 at 100 °C for 1 h, followed by high-shear extraction using hexane at 3000 rpm for 30 min. The acid pretreatment hydrolyzed carbohydrates and phospholipids, disrupting the cell wall and membrane, while high-shear mixing enhanced mass transfer rates between solvents and lipids, overcoming the hydraulic barrier at the cell surface. Within 10 min after completing the process, the extraction mixture achieved natural phase separation into water, solvent, and biomass residue layers, each enriched with carbohydrates, lipids, and proteins, respectively. The CASS process demonstrated high esterifiable lipid yields (91%), along with substantial recovery of glucose (90%) and proteins (100%). The stable phase separation prevented emulsion formation, simplifying downstream processing. This study presents the results on cell disruption, optimal acid treatment concentration, and high-shear mixing to achieve macromolecule separation, expanding the lipid-centric microalgal process to a comprehensive biorefinery concept.

Keywords: microalgae; biorefinery; macromolecule; acid pretreatment; high-shear mixer

1. Introduction

Recent trends in Renewable Fuel Standards (RFS) indicate a growing emphasis on increasing the use of renewable fuels in the transportation sector [1]. The RFS determines a minimum volume of renewable fuel to be blended into U.S. transportation fuel each year, with specific classifications for different types of biofuels like conventional biofuel, advanced biofuel, cellulosic biofuel, and biomass-based diesel [2]. In addition, ASTM D7566 recently approved blends of up to 10% converted triglyceride oil derived from microalgae for Sustainable Aviation Fuel (SAF) [3]. Microalgae have been studied as a third-generation biofuel feedstock, offering advantages such as higher growth rates, biomass yield, and oil productivity on an areal basis compared to other counterparts [4–6]. Nonetheless, the higher production costs than conventional fossil fuels have hindered its commercial viability. To establish a more sustainable production model, it is crucial to utilize non-lipid macromolecules such as carbohydrates and proteins [7]. This approach expands the scope of microalgae-based processes from a sole focus on lipid-centric biofuel production to encompass comprehensive biorefinery concepts [8].

A complete utilization of biomass macromolecules faces numerous challenges due to the complex biochemical composition of microalgae [9]. Carbohydrates and proteins, potential carbon and nitrogen sources for the subsequent cultivation of microbes, are accumulated during the late exponential phase of microalgae growth [10]. Due to their hydrophilicity, carbohydrates and proteins may remain in the biomass residue and water phase, leading to potential issues during lipid extraction [11]. Those macromolecules, especially proteins, can cause the emulsification of organic and water phases with the biomass residue, resulting in a significant loss in lipid yield and additional downstream processing costs [12]. On the other hand, phospholipids, another type of well-known bio-surfactant, not only cause emulsification but also act as poison for catalysts in the lipid conversion process, e.g., hydrotreating or hydrocracking reactions [13]. Other inorganic compounds such as Ca^{2+}, Mg^{2+}, and K^{+}, which are remaining nutrients, also exist in extracted lipids and induce catalytic poisoning [14]. To increase the lipid extraction efficiency and economic feasibility of microalgal biorefinery, it is necessary to efficiently separate macromolecules and inorganic compounds from extracted crude lipids.

Thermochemical routes for the cell disruption process have been widely studied to increase lipid extraction yield from wet biomass [15]. Acid pretreatment under high temperatures induces an acidic hydrolysis reaction of various cellular molecules, including the cellulosic cell wall, carbohydrates, proteins, and polar lipids, e.g., phospholipids and glucolipids [16–18]. This method has been employed for a relatively low concentration biomass (<60 g/L) [19–26] because highly concentrated biomass requires much higher acidic conditions, which may reduce process operability by corrosion [27]. However, after acid pretreatment at low biomass concentrations, the lipid recovery process with non-polar solvents like hexane may take a long time. This is because the water contents in the biomass form a hydrophilic barrier, preventing hexane from reaching the lipid droplets and requiring extended process times to capture the lipids [4]. In the previous reports, our group developed an energy-efficient lipid extraction process using a high-shear mixer from wet algal biomass [12]. The high-shear mixer (HSM), which is a unique mixing equipment that consists of a rotor–stator assembly, is capable of producing high rotor tip speed (ranging from 10 to 50 m/s), high shear rates (ranging from 20,000 to 100,000 s^{-1}), and highly localized energy dissipation rates near the mixing head [28]. HSM has been proven to minimize the water barriers and dramatically increase mass transfer rate through strong turbulent mixing. However, the authors focused on lipid extraction from the concentrated biomass (>250 g/L) without utilizing other biomass components, which requires more studies about major nutrients (carbohydrates and protein) recovery and impurity removals to develop final concepts of bio-algal refineries [12]. Furthermore, highly concentrated biomass is difficult to achieve due to the energy-intensive harvesting technology, which increases the cost of the overall process.

Herein, we have developed a process that combines thermochemical treatment and high-shear mixing for the fractionation of macromolecules and impurities, aiming for a complete utilization of biomass. *Chlorella* sp. ABC-001, which has a unique characteristic of accumulating both carbohydrates and lipids during the stationary growth phase under nitrogen-depleted conditions, was used as a model biomass [29]. The CASS process proposed in this paper comprises four stages: (1) concentrating the microalgae solution, (2) acid pretreatment, (3) high-shear-assisted lipid extraction, and (4) separation (Figure 1). The application of high shear during lipid extraction makes the process operable at a moderate biomass concentration (50 g/L), which can be achieved by energy-efficient membrane harvesting technology [30,31]. This approach hypothesizes as follows: (1) Hydrolysis breaks down the carbohydrate contents of the microalgae, e.g., cellulosic cell wall and starch, into smaller sugars, e.g., monosaccharides, in the aqueous phase. (2) The acids detach the hydrophilic heads from the phospholipids, leaving behind nonpolar hydrocarbon tails, which are then separated into the aqueous phase. (3) Both high lipid extraction and glucose recovery yield are achieved by high-shear mixing, which enhances the mass transfer rate between the water/nonpolar solvent and biomass, overcoming the hydraulic barrier at the surface of the cells when the solvents contact the concentrated wet biomass. This study evaluates the cell disruption, separation of macromolecules, and their compositions observed in each stage, suggesting a sustainable biorefinery concept based on the CASS process.

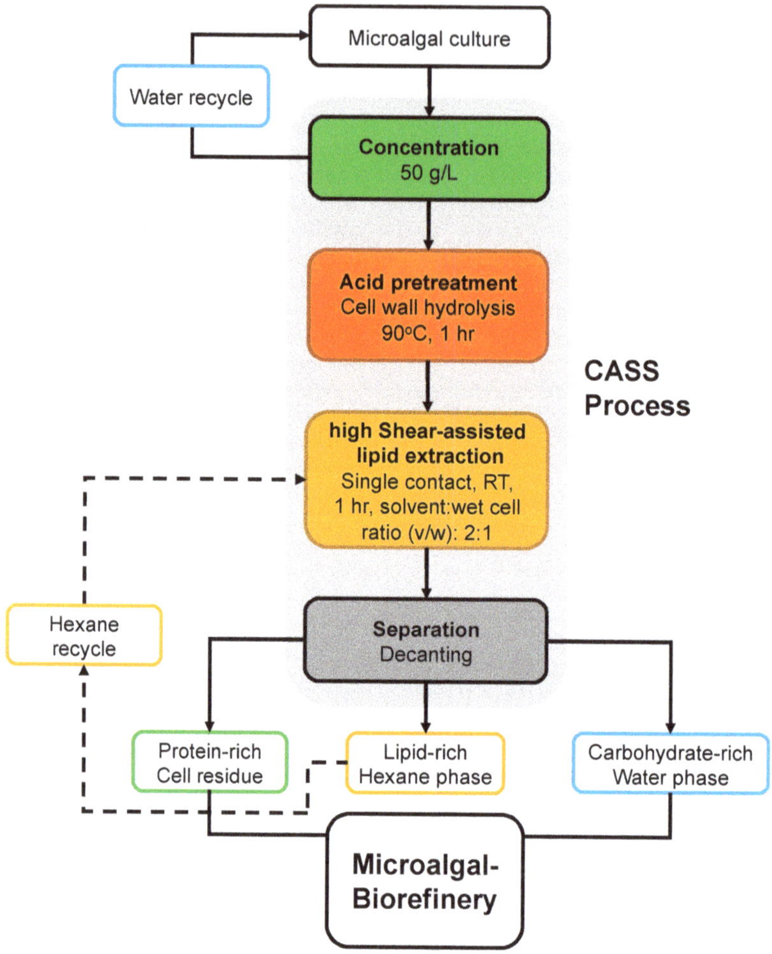

Figure 1. Microalgal-based biorefinery based on the CASS process.

2. Materials and Methods

2.1. Materials

Chlorella sp. ABC-001 was selected as the model strain due to its high content of lipids and total carbohydrates compared to other microalgae strains [29]. The microalgae was cultivated in an indoor photobioreactor (PBR) system by KIER (Korea Institute of Energy Research). The biomass was harvested and dewatered by centrifuge and freeze-dryer, respectively, and stored in a refrigerator (4 °C) until use. Bulk N-hexane (>95%) for lipid extraction was purchased from Samchun Chemical (Pyeongtaek, Republic of Korea). Chloroform, methanol (HPLC grade), sulfuric acid (99.8%), FAME mix, and heptadecanoic acid (≥99.5%) were purchased from Merck (Darmstadt, Germany). Nitrogen (N_2) for hexane evaporation and pure compressed air, helium (He), and hydrogen (H_2) gases for gas chromatography (GC) were purchased from Samo Gas (Daejoen, Republic of Korea).

2.2. Biochemical Composition Analysis

The composition of *Chlorella* sp. ABC-001 was analyzed prior to the extraction experiment. The total amount of lipids was quantified by the modified Folch method in gravimetric yield [32]. A total of 50 mg of fully dried cells and 10 mL of chloroform/methanol

(2:1 v/v) solution were treated in a sonication bath for 1 h at maximum amplitude. After pretreatment, the solution was mixed with 5 mL of water for phase separation and separated by centrifuge. Then, 3 mL of chloroform was decanted into the pre-weighed aluminum dish and evaporated under a fume hood. Residues were weighed to calculate the total lipid contents by the following equation.

$$\text{Total lipid content (wt.\%)} = (W_l - W_d) \times \frac{V_t}{V_d \times W_b} \quad (1)$$

W_l, W_d, and W_b represent the weight of the aluminum dish with lipids, aluminum dish, and biomass, respectively. V_t and V_d represent the total volume and decanted volume of chloroform, respectively.

The total esterifiable lipids were analyzed by an in situ transesterification method [33]. A total of 20 mg of dry biomass was mixed with chloroform/methanol (2:1 v/v) and vortexed for 5 min with an additional 1 mL of methanol, internal standard (C17:0, 0.5 mg), and 0.3 mL of sulfuric acid. The mixture was heated at 100 °C for 20 min to convert lipids into fatty acid methyl ester (FAME). After phase separation by adding distilled water, the organic phase containing FAME was filtrated with a 0.22 µm polytetrafluoroethylene (PTFE) filter. The filtrate was analyzed to characterize FAME content by gas chromatography 6890 (Agilent Technologies, Santa Clara, CA, USA) with a flame-ionized detector (GC-FID) and an INNOWAX capillary column (30 m × 0.32 mm × 0.5 µm, Agilent, USA).

Total carbohydrate content was analyzed by colorimetric method using phenol–sulfuric acid, using glucose as a standard [34]. A total of 1 mL of algal suspension solution (0.5 g/L) was mixed with 1 mL of phenol (5% w/v) and sulfuric acid (95%, 5 mL) and cooled in a dark room for 30 min. The total carbohydrate content was determined by UV absorbance at 470 nm. The calibration curve was obtained using a glucose standard. The monosaccharide content was analyzed using high-performance liquid chromatography [35]. A total of 0.6 g of dry biomass was mixed with 60 mL of 2 N nitric acid. The mixture was then heated at 90 °C for 3 h and 1 mL was sampled every 30 min. Each sample was filtered and soluble carbohydrates (glucose, fucose, xylose, galactose, mannose, rhamnose) were analyzed by high-performance liquid chromatography with refractive index detection (HPLC-RID) with HPX-87H column, while 0.05 mM sulfuric acid was used as the mobile phase. The monosaccharide recovery was estimated as follows:

$$\text{Monosaccharide recovery (\%)} = \frac{\text{Hydrolzed soluble carbohydrate content}}{\text{Total soluble carbohydrate content}} \times 100 \quad (2)$$

where total soluble carbohydrate content represents the content of monosaccharide in dry biomass hydrolyzed by the method described.

Ash content was measured gravimetrically after calcination. Before calcination, an aluminum dish was preheated at 550 °C for 4 h. A total of 0.5 g of dried biomass was heated with the pretreated aluminum dish in a furnace at 575 °C for 4 h, and the residue was measured by a gravimetrical-like total lipid analysis method [36]. The protein content was calculated by subtracting all the other components (total lipids, carbohydrates, ash) from 100%. All experiments for the biochemical composition analysis were carried out in triplicate.

2.3. Acid Pretreatment of Wet Biomass

Acid pretreatment of concentrated wet biomass (50 g/L) was carried out to produce hydrolyzed wet biomass. Firstly, 0.5 g of dried biomass was dispersed in 10 mL of distilled water in a 50 mL Duran bottle to prepare the model concentrated wet biomass. Sulfuric acid (99.8%) was then added to the wet biomass at specific concentrations, and mixing was performed with a magnetic stirrer by using a conventional heating plate. After mixing, the sample was cooled to room temperature before adding 15 mL n-hexane. The mixture was stirred again at 1000 rpm at room temperature for 12 h to fully extract lipids from the microalgal cells. The organic solvent layer (n-hexane containing lipids) and aqueous phase (containing cell debris) were separated by centrifugation at 4000 rpm for 10 min. A known volume of lipid-containing solvent was transferred to a pre-weighed aluminum dish and evaporated in a fume hood. The extracted lipid content was analyzed by a gravimetric method, and lipid recovery was calculated using the following equation.

$$\text{Lipid recovery yield } (\%) = \frac{\text{Extracted lipid content}}{\text{Total lipid content}} \times 100 \quad (3)$$

The effects of acid concentration, acid treatment time, and temperature on lipid recovery were investigated. Acid concentrations were varied from 0.5 to 5 vol.% (0.5, 1, 3, 5), equivalent to 0.18, 0.36, 1.08 and 1.8 N, respectively, at 100 °C for 1 h. The acid treatment time was varied from 5, 10, 30, and 60 min under a static condition of 5 vol.% of sulfuric acid at 100 °C. The temperature of acid treatment was varied from 70, 80, 90, 100, and 120 °C under a static condition of 5 vol.% of sulfuric acid and 60 min of treatment time. To verify the effects of treatment temperature on cell morphology, microscopic images of disrupted cells were obtained with a DM2500 microscope equipped with a DFC425C camera (Leica Microsystems, Wetzlar, Germany). All experiments were carried out in duplicates.

2.4. *High Shear-Assisted Lipid Extraction*

Following the acid pretreatment of wet biomass, a lipid extraction process using a high-shear mixer (HSM) was developed to enhance the lipid extraction yield and reduce the extraction time. A conventional lab-scale HSM, L5M-A (Silverson, Chesham, UK), was used in this research. The diameter of the rotor and the rotor–stator gap distance were 15.7 and 0.075 mm, respectively. A commercial vertical screen equipped with the L5M-A, which had 12 slits, each measuring 2 × 10 mm, was used. The surface of the HSM was coated with ethylene tetrafluoroethylene (ETFE) EC-6515 (Dalkin, Osaka, Japan) to reduce the attachment of biomass.

The extraction efficiency of the conventional stirrer and HSM was compared with varying rotational speed and treatment time. The conventional stirrer was operated at 500, 750, and 1000 rpm for 0, 1.5, 3, 6, 9, and 12 h. The HSM was operated at 1000, 2000, and 3000 rpm for 0, 5, 10, 30, 60, and 90 min. At each time point, a known volume of lipid-containing hexane was sampled and decanted into a 15 mL vertical glass tube. The solvent was then evaporated using a nitrogen purge, and the lipids remaining in the tube were converted into FAME by the in situ transesterification method and analyzed by GC (using the same methods as previously mentioned in Section 2.2). Esterifiable lipid yield was calculated using the following equation.

$$\text{Esterifiable lipid yield } (\%) = \frac{\text{Extracted esterifiable lipids}}{\text{Total esterifiable lipids}} \times 100 \quad (4)$$

2.5. *Characterization of Oil, Water, and Solid Phase Products*

Detailed characterization of biomass, oil phase, and water phase was performed to find a distribution of macromolecules. For the oil phase and water phase, lipid and carbohydrate characterization were performed, respectively. The solid phase, lipid-extracted algae (LEA), was separated and analyzed for lipid content and esterifiable content by the method described in Section 2.2. The ash content of oil, water, and solid phase was analyzed by the method described in Section 2.2. The carbohydrate and protein content of the solid phase was calculated by subtracting extracted macromolecule contents in the oil and water phase, respectively. Elemental analysis (C, H, O, N, S) was performed on raw biomass, extracted oil, and LEA using an elemental analyzer, FLASH 2000 series (Thermo Scientific, Waltham, MA, USA).

Inorganic element analysis (P, Ca, Na, K, Cu, Zn, Mg) of oil, water, and solid phase was performed using inductively coupled plasma–optical emission spectroscopy (ICP-OES). For oil and water phase analysis, the solvent and water were fully evaporated using a vacuum evaporator. Pre-weighed samples were pretreated with nitric acid to remove any organic content. A 40 mg oil sample was diluted with V-SOLVTM (VHG Labs, Manchester, UK) solution and CONOSTAN Multi-Element Standard (150-021-015) was used as the standard. ICP-OES analysis conditions for quantifying metal and phosphorus contents are described as follows: plasma power 1400 W, pump speed 15 rpm, coolant flow 12 L/min, auxiliary flow 1.5 L/min, nebulizer flow 0.6 L/min, oxygen flow 0.2 L/min. The initial content was determined by analyzing the crude oil extracted by the modified Folch method described in Section 2.2.

3. Results and Discussion

3.1. Biochemical Composition of *Chlorella* sp. ABC-001

Table 1 shows the biochemical composition of *Chlorella* sp. ABC-001 used for the CASS process. The lipid content of the biomass was 35.7 wt.%, which contained 29.8 wt.% of esterifiable lipids and 4.9 wt.% of non-esterifiable lipids (e.g., chlorophyll, hydrocarbon). The carbohydrate content of the biomass, which is another major nutrient of ABC-001, was 40.1 wt.%, which contained 19.1 wt.% of glucose that could be converted into valuable biochemicals, e.g., bioethanol and biopolymer, through fermentation [37]. The rest of the biomass consisted of 18.6 wt.% of proteins and 5.5 wt.% of ashes. These high esterifiable lipids and monosaccharide carbohydrate contents are attributed to the characteristics of the *Chlorella* sp. ABC-001 under nitrogen-depleted conditions [29], which showed great potential for microalgal biorefinery feedstock.

Table 1. Biochemical composition of *Chlorella* sp. ABC-001. All the values are presented as the average of triplicated samples with standard deviations (n = 3).

Composition		Biomass (wt.%)	Crude Lipids [2] (wt.%)	CASS Process [3] Oil Phase (wt.%)	Water Phase (wt.%)	Solid Phase [4] (wt.%)
Macromolecule composition [1]	Lipid	35.7 ± 0.6	35.7 ± 0.6	30.1 ± 0.2	N/A	2.2 ± 0.01
	Esterifiable Lipid (FAME)	29.8 ± 0.2	29.8 ± 0.2	27.0 ± 0.5		0.5 ± 0.2
	Carbohydrate	40.1 ± 1.3	N/A	N/A	29.1 ± 0.8	2.9 ± 0.1
	Glucose	19.1 ± 0.1			17.2 ± 0.3	0.2 ± 0.1
	Protein	18.7 ± 2.3			N/A	19.0 ± 0.4
	Ash	5.5 ± 0.1	0.1 ± 0.1	<0.1	4.9 ± 0.0	0.5 ± 0.0
Elemental composition	C	51.8 ± 0.1	74.1 ± 0.6	75.4 ± 0.4	N/A	44.5 ± 0.1
	H	8.2 ± 0.1	11.8 ± 0.1	12.2 ± 0.1		6.5 ± 0.1
	O	29.7 ± 0.1	11.8 ± 0.1	11.4 ± 0.1		27.4 ± 0.0
	N	2.6 ± 0.0	0.4 ± 0.1	0.2 ± 0.0		8.2 ± 0.1
	S	0.2 ± 0.1	N/D	N/D		2.5 ± 0.1

[1] Macromolecule composition is based on the original dry biomass. [2] The composition of crude lipids extracted by the Folch Method will be discussed in Section 3.4. [3] The composition of the extraction mixture by the CASS process will be discussed in Section 3.4. [4] LEA (Lipid-extracted algae).

3.2. Acid Pretreatment of Concentrated Wet Biomass

The effect of acid concentration, pretreatment time, and temperature was investigated using concentrated wet biomass (50 g/L) in terms of lipid extraction yield (Figure 2). To clarify the effects of acid treatment on the wet biomass, lipid recovery was performed using a conventional stirrer under overnight conditions to achieve mass transfer equilibrium. The lipid yield increased as the acid concentration, pretreatment time, and temperature increased and reached a saturation of around 90%. For acid concentration, only 0.5 wt.% sulfuric acid achieved 76% lipid yield and showed saturated behavior over 3 wt.% sulfuric acid, which is equivalent to 1.08 N concentration (Figure 2a). Lipid yield increased rapidly throughout the treatment and reached saturation point after 30 min (Figure 2b). Lipid yield was saturated at a temperature higher than 100 °C at 5 wt.% sulfuric acid condition (Figure 2c). The maximum lipid yield of 90.6% was achieved under the condition of 5 wt.%

sulfuric acid at 120 °C for 1 h. However, the difference compared to the second-highest temperature condition (100 °C) was insignificant, with a yield of 89.3%. Zhang et al. (2018) reported that they extracted 414.1 mg/g cell lipids with 0.5 M sulfuric acid at 120 °C for 1 h treatment from *Chlorella vulgaris* [38]. Martins et al. (2020) used 20 g/L of *Chlorella vulgaris* and achieved over 90% of lipid extraction yield with 5 vol.% sulfuric acids at 120 °C for 10 min treatment [39]. The results showed consistency with previous studies that concluded acid pretreatment requires more than 100 °C at diluted concentration.

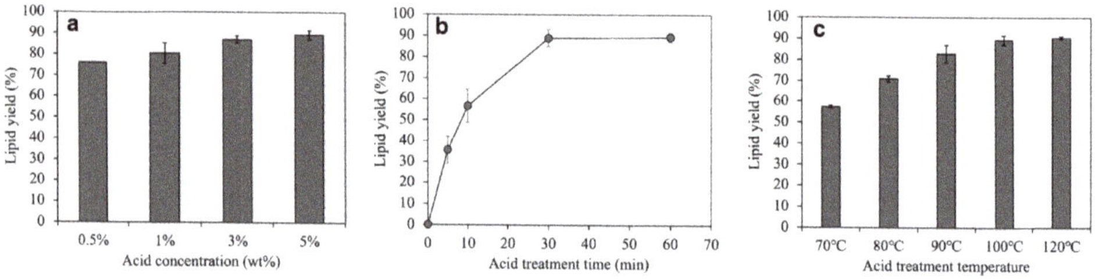

Figure 2. (**a**) Effect of acid concentration, (**b**) acid treatment time, and (**c**) acid treatment temperature on lipid extraction recovered by using a conventional stirrer at 1000 rpm for 12 h.

Providing enough thermal energy is key to inducing effective hydrolysis reactions. Cell morphology was observed through microscopic images after acid treatment at various temperature conditions (Figure 3). Before the acid treatment, the cell envelope exhibited a uniform circular shape with evenly distributed sizes of 3–5 μm (Figure 3 (a)). As the temperature increased from 70 to 80 °C, cells began to agglomerate together without significant damage observed. Partial or complete destruction of cell walls was first found at 90 °C (Figure 3 (b)) and completely disrupted over 100 °C (Figure 3 (c)). At 120 °C, most cells were ruptured, resulting in the drainage of intracellular fluids from the cell envelope (Figure 3 (d)). The lipid droplets that remain in the intracellular fluids can easily be uptaken by organic solvents, even with low biomass concentration in a water medium. These results correspond with lipid yield results, showing that the destruction of cell walls is required to achieve a high lipid extraction yield of over 90%. Based on the above results, the acid treatment condition was fixed at 100 °C, 5 wt.% for 1 h on further research.

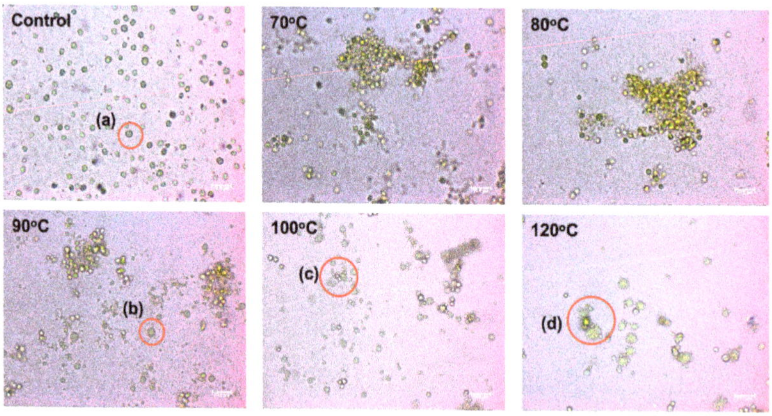

Figure 3. Microscopic image of *Chlorella* sp. ABC-001 before (control) and after acid treatment at various temperature conditions (5 wt.%, 1 h treatment). The scale bar is 10 μm. Morphological changes are highlighted in red circles. (a) A normal cell; (b) the overall size of the cell is enlarged, and its shape starts to deviate from a sphere; (c) disrupted cells are agglomerated with cell debris; and (d) a complete cell rupture is observed.

3.3. Effects of Mixing Efficiency on the Lipid Recovery Process

The governing factor of the lipid recovery process is the mass transfer phenomenon, which can be defined as the mass transfer efficiency of lipid droplets into organic solvents through a water barrier [40]. The mass transfer rate is governed by diffusion, which is affected by surface area, concentration gradient between bulk solvent and biomass, and intrinsic diffusivity of lipids through the water layer. Even after the cell wall was ruptured by acid pretreatment, the polar water barrier hindered the extraction of lipids into solvents, reducing the recovery rate, which resulted in low process efficiency. Since our primary goal was to recover each of the biological components, especially the lipid, as much as possible economically, decreasing the time and increasing the yield were both important. Thus, a mechanical method of generating high-shear mixing was introduced to effectively deliver enough mixing energy to a water–organic solvent mixture.

Figure 4 shows the effects of the mixing efficiency of various rotational speeds of the HSM, and the conventional stirrer was compared in terms of lipid extraction yield and extraction time. The results demonstrated that increasing mixing speed affected lipid extraction yield and rate. In the case of the conventional stirrer, 500 and 750 rpm could not achieve a mass transfer saturation point due to their low mixing efficiency (Figure 4a). For 1000 rpm conditions, the saturation limit was achieved at about 85% after 6 h of treatment. On the other hand, the HSM exhibited higher lipid extraction efficiency than the stirrer at the same 1000 rpm, achieving the same level of extraction yield at 90 min, which matched the yield of the conventional stirrer at 180 min (Figure 4b). Even though the rotor diameter of the HSM and the stirrer was the same, the HSM's unique rotor–stator geometry generated additional shear force and turbulent flow, dramatically increasing mass transfer efficiency. As the rotational speed of the HSM increased, the saturation curve was achieved within 30 min at 3000 rpm and reached the highest yield of 90.7%.

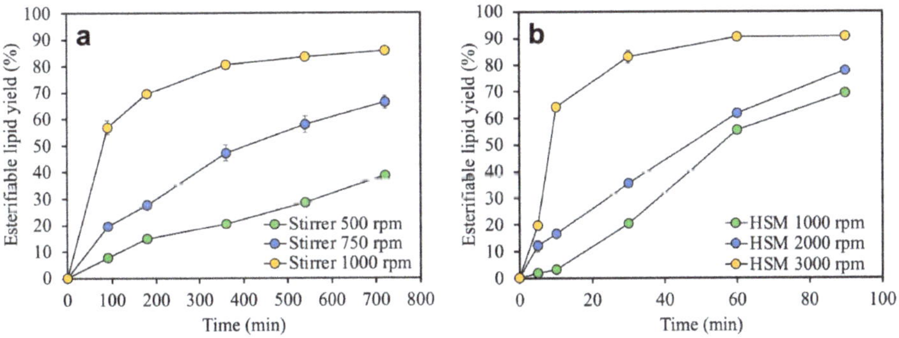

Figure 4. Effect of rotational speed and mixing time on the esterifiable lipid recovery yield using (**a**) a conventional stirrer and (**b**) a high-shear mixer after acid treatment (5 wt.%, 100 °C, 1 h).

The HSM was designed to be operated at high rpm conditions that could not be reached when using the conventional stirrer, and therefore, employing the HSM in the mixing process is beneficial in homogenizing immiscible fluids. The water and hexane need to contact the biomass as much as possible to recover responsible components for each phase. A number of papers provided proof-of-concept and optimization studies for high-shear-assisted solvent-based extraction in highly concentrated biomass conditions (>200 g dry cell/L) without water medium [12,41]. The extraction process with highly concentrated biomass could achieve a high yield with low solvent usage and energy consumption. However, the microalgae concentrating process requires extremely high energy and is time-consuming, necessitating proper selection and optimization of harvesting and extraction technology. Throughout the investigation presented in this study, the high-shear-assisted extraction process has expanded its utility for recovering biological components from wet biomass at low concentrations.

3.4. Separation of Macromolecules in Post-Extracted Mixture: Biorefinery Concept

The previous section revealed that the CASS process can potentially extract lipids from low-concentration biomass with acid pretreatment and high-shear-assisted recovery. Since ABC-001 still contains significant sources of carbohydrates and proteins, it is necessary to clarify the extraction behavior of these macromolecules within the system to validate the potential of an algae-based biorefinery. Therefore, a comprehensive analysis of the mixtures in three phases (liquid–liquid–solid), consisting of water, an organic solvent containing lipids, and treated biomass, was performed from the perspective of macromolecules and other impurities, namely inorganic compounds. From the biomass, over 90% of the lipids and fermentable glucose were transferred into the organic solvent and water phases, respectively. All the protein contents were preserved (100%) in the lipid-extracted algae (LEA) layer (Figure 5). Notably, this phase separation was achieved naturally within 10 min after the CASS process. Details on each phase will be discussed in the following sub-chapters.

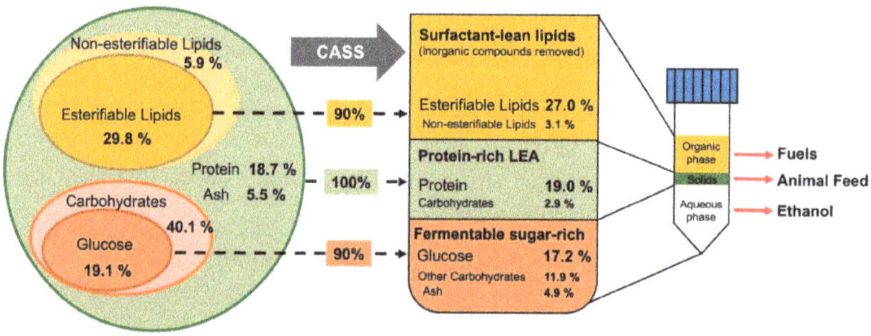

Figure 5. The composition and proposed use of three layers resulting from the natural separation of the extraction mixture obtained from the complete CASS process.

3.4.1. Lipid Phase: Esterifiable Lipids Recovery

After the acid treatment and lipid recovery, the extracted mixture naturally separated into water, biomass residue, and hexane layers from bottom to top. The macromolecule analysis results are presented in Table 1, based on the weight percentage of the original biomass. The results indicated that lipids, carbohydrates, and proteins are distributed into oil, water, and solid phases, respectively. About 35.7 wt.% of lipids in the original biomass were distributed, with 30.1 wt.% in the oil phase and 2.2 wt.% in the lipid-extracted algae (LEA), respectively. In the case of fatty acid content, 27.0 wt.% was recovered by the oil phase, and only 0.5 wt.% remained in the solid phase, which is about 90% of the recovery yield from the original biomass.

Notably, the purity of the lipids extracted by the CASS process (89.7%) was higher than that of the conventional Folch method (83.5%). The lipids extracted by hexane usually have higher FAME contents than lipids extracted by the Folch method due to the higher polarity of chloroform/methanol than hexane, which could extract unwanted carbohydrates and protein-associated lipids [42]. Furthermore, acid treatment hydrolyzes galactolipids (MGDG, DGDG) and phospholipids (PE, PC), which are major components of chloroplasts and cell membranes, respectively, into polar functional groups and fatty acids [41].

Compared to the crude lipid extraction methods, the acid hydrolysis of polar lipids could cause a minor loss of total lipid contents in the CASS process. Total carbohydrate content increased slightly more than the original biomass, which indicates that the hydrolyzed galactose group was transferred into the water phase. The elemental analysis results further support the hydrolysis of unwanted polar lipids. In the oil phase, carbon and hydrogen contents increased, and the oxygen and nitrogen contents decreased compared to the crude lipids. This observation is further supported by the hydrogen/carbon atomic ratio (H/C), an important descriptor that characterizes the quality of petroleum, which also can be applied in biomass-derived fuels in a slightly modified form, 'effective hydrogen-to-carbon ratio (H/C_{eff})', as defined as follows [43]:

$$H/C_{eff} = (n(H) - 2 \times n(O) - 3 \times n(N) - 2 \times n(S))/n(C) \quad (5)$$

where n(H), n(O), n(N), n(S), and n(C) represent the molar concentrations of hydrogen, oxygen, nitrogen, sulfur, and carbon, respectively. The H/C ratio of the oil phase from the CASS process was 1.71, whereas the crude lipids from the Folch method had a ratio of 1.66, implying that the quality of extracted lipids has been improved. When compared to the other type of extraction process, e.g., hydrothermal liquefaction (HTL), the CASS process achieved a higher value (1.71) than a previously reported study for the same species (1.56–1.63) [44], indicating that the lipid quality has been enhanced to be more suitable for fuels.

3.4.2. Water Phase: Fermentable Glucose Recovery

The water phase contained 29.1 wt.% of total carbohydrates and 4.9 wt.% of inorganic compounds. The fermentable glucose was 17.2 wt.%, found in the water phase. This amount corresponds to 90% of the original biomass, and only 0.2 wt.% remained in the LEA. These results indicate that acid hydrolysis converted carbohydrates into fermentable glucose that is easily soluble in the water phase. The other carbohydrates remained in the algal cells, about 2.9 wt.% of the original biomass, corresponding to 7.2% of the original carbohydrate content. The final concentration of fermentable glucose in the water phase was 8.6 g/L. Meanwhile, the typical glucose concentration for yeast fermentation is more than 20 g/L [45]. This difference can be adjusted by concentrating the water phase or increasing biomass concentration. However, careful considerations should be taken when deciding on the adjustment since the capital costs may increase.

The other study using algal biomass first conducted lipid extraction, then carbohydrate hydrolysis, using lipid-extracted algae by applying post-acid treatments [46]. This concept has the advantage of uncoupling lipid extraction and carbohydrate hydrolysis, providing the flexibility to optimize each process independently, resulting in high yields of lipids and carbohydrates. On the other hand, the CASS process has the advantage of simultaneous lipid extraction and carbohydrate hydrolysis, achieving process simplification and requiring only two continuous mixing tanks and a three-phase separator, compared to the previous study that required more than two CSTRs and two separators for the continuous process [46]. Also, the earlier study applied a hexane/ethanol mixture (additional polar solvents) to achieve a high lipid yield, resulting in increased operating and raw material costs [39]. In the CASS process, high-shear mixing provided enough lipid extraction capabilities only with a non-polar solvent (hexane), without the assistance of polar solvents (e.g., methanol and ethanol).

3.4.3. Solid Phase: Maintaining Protein Integrity

The solid phase is composed mainly of proteins, 56 wt.% of the LEA, corresponding to 18.8 wt.% of the original biomass. This indicates that most proteins remain in the biomass without being extracted. These results align with the elemental analysis results, given that LEA's nitrogen and sulfur content was 3.15 and 12.5 folds higher than the original biomass, respectively. Hydrolysis of protein into amino acids requires much harsher conditions than the CASS process, where the classical hydrolysis method used 6 N hydrochloric acid at 110 °C for 24 h treatment [47]. The original protein content appears preserved in LEA, making it suitable for nutritious animal and fish feeds [48].

3.4.4. Impurities Removal Effects on Extracted Lipids

The inorganic compounds, namely ashes, were mainly dissolved in water (90%), and the remaining amount was left in the LEA (10%). The detailed analysis of inorganic compounds (P, Ca, Na, K, Cu, Zn, Mg), performed using ICP analysis, is presented in Table 2. The results showed that most of the inorganic compounds were slipped into the water phase, the rest remained in the LEA, and only a small portion of phosphorus and copper were found in the oil phase. The phosphorus, mainly from phospholipids comprising cell membranes, was separated into water by acid hydrolysis. Phospholipids are the main impurities from bio-based feedstock that cause catalytic poisoning in the hydrodeoxygenation [49] and isomerization processes for biofuel production [50]. In the conventional process, phosphorus and other impurities are removed by a degumming process, which adds water and acid into crude lipids to separate impurities as a form of gum [51]. In the CASS process, acid treatment resulted in the pre-removal of impurities into the water phase, integrating lipid extraction and the degumming process.

Table 2. Inorganic compositions of *Chlorella* sp. ABC-001, crude lipids extracted by the Folch method, and products separated by the CASS process. All the values are presented as the average of triplicated samples with standard deviations (n = 3).

Content	Biomass (wt. ppm)	Crude Lipid [1] (wt. ppm)	CASS Process		
			Oil Phase (wt. ppm)	Water Phase (wt. ppm)	Solid Phase [2] (wt. ppm)
P	27,021 ± 49	1130 ± 56	206.5 ± 1.9	24,318 ± 155	3346 ± 155
Ca	3168 ± 69	74.5 ± 5.1	N/D	2738 ± 21	461.2 ± 5.3
Na	623 ± 5	N/D	N/D	792.8 ± 6.7	137.4 ± 3.3
K	19,844 ± 509	65.3 ± 1.8	N/D	17,344 ± 242	312.9 ± 2.0
Cu	889 ± 37	271.8 ± 8.2	24.9 ± 1.3	700.7 ± 0.4	352.1 ± 5.1
Zn	232 ± 53	17.4 ± 1.8	N/D	298.1 ± 0.7	N/D
Mg	3325 ± 44	N/D	N/D	2891 ± 5	574.0 ± 10.2

[1] The composition of crude lipids extracted by the Folch method. [2] LEA.

The CASS process induced a stable phase separation without forming an emulsion. Emulsion formation, a critical issue in the two-step cell disruption and lipid extraction process, causes a loss of lipids and carbohydrates by trapping them within the emulsion layer and significantly increases the costs of additional separation [52]. High biomass concentrations and the severity of cell disruption increase the chance of emulsion formation due to the secretion of bio-surfactants, which mainly consist of phospholipids, small carbohydrates, and proteins [53]. In fact, ABC-001 also formed an emulsion when concentrated biomass was directly mixed with hexane using a high-shear mixer without acid pretreatment, indicating the presence of bio-surfactants. As demonstrated in this study, hydrolysis of phospholipids and carbohydrates prevented emulsion formation, obtaining a clear phase separation after vigorous high-shear mixing. Applying acid pretreatment at the low biomass concentration was beneficial not only for the proper distribution of macromolecules, but also for preventing emulsion formation during lipid recovery.

3.5. Comparison with Other Biorefinery Processes

Many researchers have reported several methods to fractionate macromolecules (Table 3). Most of them used a combination of acid hydrolysis (pretreatment) and solvent extraction, similar to our approach, focusing on the simultaneous production of biofuel and bioethanol. Nonetheless, there is a significant difference between CASS and the other studies. This study presents all the components' conversion ratios in different phases, discussing the quality of products to take practical issues such as emulsification and catalytic poisoning into account.

Table 3. Comparison of the microalgal biorefinery processes.

Process	Strains	Biomass Conc. (g/L)	Operating Conditions	Macromolecule Yield (%)	References
High-pressure homogenizer	*Nannochloropsis* sp.	110–230	(1) Cell weakening (35 °C, 7–24 h) (2) High-pressure homogenizer (800–1000 bar, 1 pass) (3) Lipid recovery (2 h, biomass (paste):solvent 5:2 *w*/*w*)	Lipid: 25 Carbohydrate in water: 41 Protein in biomass: 51	[54]
Acid hydrolysis	*Tribonema* sp.	50	(1) Acid hydrolysis (3 wt%, sulfuric acid, 121 °C, 45 min) (2) Lipid recovery (ethanol-hexane (1:3, *v*/*v*) on shaker incubation, 2.5 h, 50 °C, biomass (paste):solvent = 1:6 *w*/*v*)	Biodiesel: 98.5 Monosaccharides: 81.5	[55]
Acid hydrolysis	*Chlorella* sp.	250	(1) Acid hydrolysis (2 wt%, sulfuric acid, 155 °C, 10 min) (2) Lipid recovery (hexane, 2 h, biomass (paste):solvent 1:1 *v*/*v*)	Lipid: 22.2 Glucose: 81.2	[56]
	Scenedesmus sp.			Lipid: 92.5 Glucose: 73.1	
Acid hydrolysis	*Chlorella* sp. ABC-001	50	(1) Acid hydrolysis (0.1 N, sulfuric acid, 170 °C, 4 min) (2) Phase separation using centrifugation (1000 rpm, 5 min) (3) Lipid recovery (hexane with vortexing, 20 min)	Lipid: 100 Monosaccharides: 89	[57]
Hydrothermal liquefaction	*Chlorella* sp. KR1	50	(1) Hydrothermal liquefaction (180 °C, 1 h) (2) Lipid recovery (hexane with sonication, 2 h)	Lipid: 87 Glucose: 70	[49]
CASS	*Chlorella* sp. ABC-001	50	(1) Acid hydrolysis (5 wt.% sulfuric acid, 100 °C, 1 h) (2) Lipid recovery (hexane with high-shear mixing, 3000 rpm, 30 min, biomass (wet):solvent = 1:1 *v*/*v*)	Lipid: 90.6 Glucose: 90.0 Protein: 100	This study

Halim et al. (2016) employed a high-pressure homogenizer to disrupt the strong cell wall of *Nanochloropsis* sp. and monitored the yield of the macromolecules' recovery, which included 25% of lipids, 41% of carbohydrates, and 51% of proteins [54]. Although a comprehensive biorefinery concept was proposed and analyzed, most macromolecule components, especially lipids, were trapped in the emulsion layer. Since protein plays a significant role in emulsion formation, species enriched in carbohydrates and lipids while lacking in protein have been favored for the complete fractionation and utilization of macromolecules. Wang et al. (2014) used *Tribonema* sp., which has very low protein contents, for biodiesel and monosaccharide production [55]. Laurence et al. (2015) used *Chlorella* sp., which is also low in protein, but obtained low lipid recovery since they employed the conventional stirrer for the extraction [56].

In the same context, *Chlorella* sp. ABC-001, which accumulates high levels of lipids and carbohydrates, has been studied for biorefinery applications. Seon et al. (2023) adopted acid hydrolysis under lower acidic concentration (0.1 N) at higher temperatures (170 °C) than this study, followed by centrifugation of hydrolysate for volume reduction, and then solvent extraction [57]. The yields were high (lipids of 100% and monosaccharides of 89%), but this process required additional steps for phase separation and holds a potential concern for corrosion issues due to high temperatures. Jin et al. (2017) used hydrothermal liquefaction (HTL) instead of acid hydrolysis and then recovered lipids using solvents from *Chlorella* sp. KR1 [49], which had a similar compositional character to ABC-001. The lipid yield was comparable to this study (87%), but conversion of carbohydrates (70%) was low for bioethanol fermentation.

The novelties of the CASS process and this study are (1) the application of a high-shear mixer to disrupt the water barrier in mildly concentrated wet biomass, (2) a comprehensive analysis of macromolecule conversion, and (3) the demonstration of acid hydrolysis's effects on both carbohydrate saccharification and phospholipid breakdown. The use of high-shear mixing enabled the efficient processing of low-concentration wet biomass at a relatively low biomass-to-solvent ratio (1:1 *v*/*v*), resulting in a high lipid yield exceeding 90%. The effects of acid pretreatment were thoroughly investigated, with the expanded scope of analysis encompassing overall macromolecule yield and quality. Carbohydrate hydrolysis produced a glucose yield of 90%, surpassing other studies. Additionally, phospholipid hydrolysis transferred hydrophilic polar heads to the aqueous phase, thereby preventing emulsion formation in the solvent phase and minimizing product loss. As discussed in the previous paragraph, emulsion formation is a significant operational challenge that has been extensively recognized. This issue was effectively mitigated by applying hydrolysis prior to high-shear mixing. Notably, protein remains undegraded under mild acidic conditions, allowing it to be preserved in the solid phase (LEA) for potential use in nutritional feeds. Moreover, the removal of inorganic compounds from the lipids reduces the risk of catalytic poisoning in the following processes. The CASS process is well suited for providing high yields while maintaining product integrity as a bridging extraction process between harvesting/dewatering and catalytic conversion in the whole biorefinery concept. Furthermore, the CASS process is scalable, as high-shear mixers are commercially available for large-scale production and have proven effective [28]. As the diameter of the rotor increases, so does the Reynolds number, enhancing turbulent mixing efficiency relative to energy consumption [12].

4. Conclusions

The CASS process was developed by combining acid pretreatment and high-shear-assisted extraction to efficiently separate macromolecules from microalgae biomass. Acid treatment effectively hydrolyzed phospholipids (biosurfactants) and removed inorganic compounds (potential catalytic poisons), enabling the recovery of more than 90% of esterifiable lipids with improved quality in the following high-shear-assisted extraction using hexane. Simultaneously, over 90% of the glucose was hydrolyzed into the water phase at a concentration of 8.7 g/L, making it suitable for bioethanol fermentation. Due to the mild acid concentration, proteins were preserved in the solid lipid-extracted algae phase without any loss (100%). The extraction mixture did not suffer from emulsification but formed stable layers of solvent, water, and solids, each rich in lipids, glucose, and protein, without requiring additional separation processes. The CASS process shows potential for scalable biorefineries to fractionate nutrient-replete ABC-001 biomass for the co-production of biofuel, bioethanol, and nutritious animal feed. Future research could investigate the applicability of this process to other microalgal species, conduct techno-economic analysis for different target products under various cultivation scenarios, and include sensitivity analysis to assess the impact of biomass composition on economic feasibility.

Author Contributions: D.K.—Conceptualization, data curation, writing—original draft, writing—review and editing, funding acquisition; S.-G.K.—conceptualization, formal analysis, methodology, writing—original draft; Y.K.C.—project administration, resources, supervision; M.K.—conceptualization, data curation, writing—original draft, writing—review and editing, visualization, supervision. All authors have read and agreed to the published version of the manuscript.

Funding: This work was supported by Qatar University QUT2RP-CENG-24/25-501. The findings achieved herein are solely the responsibility of the authors.

Institutional Review Board Statement: Not applicable.

Informed Consent Statement: Not applicable.

Data Availability Statement: Data are contained within the article.

Conflicts of Interest: The authors declare no conflicts of interest.

References

1. Spandagos, C. Achieving decarbonization goals through biofuels: Policy challenges and opportunities in the European Union and the United States. In *Advances in Biofuels Production, Optimization and Applications*; Elsevier: Amsterdam, The Netherlands, 2024; pp. 269–283.
2. Bracmort, K. *The Renewable Fuel Standard (RFS): An Overview*; Congressional Research Service: Washington, DC, USA, 2023.
3. Bhatt, A.H.; Zhang, Y.; Milbrandt, A.; Newes, E.; Moriarty, K.; Klein, B.; Tao, L. Evaluation of performance variables to accelerate the deployment of sustainable aviation fuels at a regional scale. *Energy Convers. Manag.* **2023**, *275*, 116441. [CrossRef]

4. Halim, R.; Danquah, M.K.; Webley, P.A. Extraction of oil from microalgae for biodiesel production: A review. *Biotechnol. Adv.* **2012**, *30*, 709–732. [CrossRef] [PubMed]
5. Chisti, Y. Biodiesel from microalgae. *Biotechnol. Adv.* **2007**, *25*, 294–306. [CrossRef] [PubMed]
6. Wijffels, R.H.; Barbosa, M.J. An Outlook on Microalgal Biofuels. *Science* **2010**, *329*, 796–799. [CrossRef]
7. Klein, B.C.; Chagas, M.F.; Davis, R.E.; Watanabe, M.D.; Wiatrowski, M.R.; Morais, E.R.; Laurens, L.M. A systematic multicriteria-based approach to support product portfolio selection in microalgae biorefineries. *Chem. Eng. J.* **2024**, *481*, 148462. [CrossRef]
8. Kruger, J.S.; Wiatrowski, M.; Davis, R.E.; Dong, T.; Knoshaug, E.P.; Nagle, N.J.; Laurens, L.M.; Pienkos, P.T. Enabling production of algal biofuels by techno-economic optimization of co-product suites. *Front. Chem. Eng.* **2022**, *3*, 803513. [CrossRef]
9. Olguín, E.J.; Sánchez-Galván, G.; Arias-Olguín, I.I.; Melo, F.J.; González-Portela, R.E.; Cruz, L.; De Philippis, R.; Adessi, A. Microalgae-based biorefineries: Challenges and future trends to produce carbohydrate enriched biomass, high-added value products and bioactive compounds. *Biology* **2022**, *11*, 1146. [CrossRef]
10. Andreeva, A.; Budenkova, E.; Babich, O.; Sukhikh, S.; Dolganyuk, V.; Michaud, P.; Ivanova, S. Influence of carbohydrate additives on the growth rate of microalgae biomass with an increased carbohydrate content. *Mar. Drugs* **2021**, *19*, 381. [CrossRef]
11. Sed, G.; Cicci, A.; Jessop, P.G.; Bravi, M. A novel switchable-hydrophilicity, natural deep eutectic solvent (NaDES)-based system for bio-safe biorefinery. *RSC Adv.* **2018**, *8*, 37092–37097. [CrossRef]
12. Kwak, M.; Roh, S.; Yang, A.; Lee, H.; Chang, Y.K. High shear-assisted solvent extraction of lipid from wet biomass of *Aurantiochytrium* sp. KRS101. *Sep. Purif. Technol.* **2019**, *227*, 115666. [CrossRef]
13. Ramasubramania, I. The issue of reducing or removing phospholipids from total lipids of a microalgae and an oleaginous fungus for preparing biodiesel. *Biofuels* **2016**, *7*, 55–72.
14. Arora, P. Deactivation of Catalysts and Reaction Kinetics for Upgrading of Renewable Oils. Ph.D. Thesis, Chalmers Tekniska Hogskola, Gothenburg, Sweden, 2019.
15. Nitsos, C.; Filali, R.; Taidi, B.; Lemaire, J. Current and novel approaches to downstream processing of microalgae: A review. *Biotechnol. Adv.* **2020**, *45*, 107650. [CrossRef]
16. Aveldano, M.I.; Horrocks, L.A. Quantitative release of fatty acids from lipids by a simple hydrolysis procedure. *J. Lipid Res.* **1983**, *24*, 1101–1105. [CrossRef] [PubMed]
17. Rinaldi, R.; Schüth, F. Acid hydrolysis of cellulose as the entry point into biorefinery schemes. *ChemSusChem Chem. Sustain. Energy Mater.* **2009**, *2*, 1096–1107. [CrossRef]
18. Wijaya, Y.P.; Putra, R.D.D.; Widyaya, V.T.; Ha, J.-M.; Suh, D.J.; Kim, C.S. Comparative study on two-step concentrated acid hydrolysis for the extraction of sugars from lignocellulosic biomass. *Bioresour. Technol.* **2014**, *164*, 221–231. [CrossRef] [PubMed]
19. Sukias, J.; Craggs, R. Enhanced methane yields from microalgal digestion with various pre-treatments. In Proceedings of the 7th IWA Specialist Group Conference on Waste Stabilization Ponds, Bangkok, Thailand, 25–27 September 2006; pp. 25–27.
20. Sposob, M.; Kim, D.-H.; Yun, G.-S.; Yun, Y.-M. Assessment of the relationship between solubilization and biogas production on anaerobic digestion of pretreated lipid-extracted microalgae waste. *Biomass Bioenergy* **2020**, *141*, 105702. [CrossRef]
21. Santos, N.O.; Oliveira, S.M.; Alves, L.C.; Cammarota, M.C. Methane production from marine microalgae *Isochrysis galbana*. *Bioresour. Technol.* **2014**, *157*, 60–67. [CrossRef]
22. Samson, R.; Leduy, A. Influence of mechanical and thermochemical pretreatments on anaerobic digestion of *Spirulina maxima* algal biomass. *Biotechnol. Lett.* **1983**, *5*, 671–676. [CrossRef]
23. Rincón-Pérez, J.; Razo-Flores, E.; Morales, M.; Alatriste-Mondragón, F.; Celis, L.B. Improving the biodegradability of Scenedesmus obtusiusculus by thermochemical pretreatment to produce hydrogen and methane. *BioEnergy Res.* **2020**, *13*, 477–486. [CrossRef]
24. Marques, A.d.L.; Pinto, F.P.; Araújo, O.Q.d.F.; Cammarota, M.C. Assessment of methods to pretreat microalgal biomass for enhanced biogas production. *J. Sustain. Dev. Energy Water Environ. Syst.* **2018**, *6*, 394–404. [CrossRef]
25. Juárez, J.M.; Pastor, E.R.; Sevilla, J.M.F.; Torre, R.M.; García-Encina, P.A.; Rodríguez, S.B. Effect of pretreatments on biogas production from microalgae biomass grown in pig manure treatment plants. *Bioresour. Technol.* **2018**, *257*, 30–38. [CrossRef] [PubMed]
26. Cheng, Q.; Deng, F.; Li, H.; Qin, Z.; Wang, M.; Li, J. Nutrients removal from the secondary effluents of municipal domestic wastewater by Oscillatoria tenuis and subsequent co-digestion with pig manure. *Environ. Technol.* **2018**, *39*, 3127–3134. [CrossRef] [PubMed]
27. Phwan, C.K.; Chew, K.W.; Sebayang, A.H.; Ong, H.C.; Ling, T.C.; Malek, M.A.; Ho, Y.-C.; Show, P.L. Effects of acids pre-treatment on the microbial fermentation process for bioethanol production from microalgae. *Biotechnol. Biofuels* **2019**, *12*, 191. [CrossRef]
28. Zhang, J.; Xu, S.; Li, W. High shear mixers: A review of typical applications and studies on power draw, flow pattern, energy dissipation and transfer properties. *Chem. Eng. Process. Process Intensif.* **2012**, *57–58*, 25–41. [CrossRef]
29. Cho, J.M.; Oh, Y.K.; Park, W.K.; Chang, Y.K. Effects of Nitrogen Supplementation Status on CO(2) Biofixation and Biofuel Production of the Promising Microalga *Chlorella* sp. ABC-001. *J. Microbiol. Biotechnol.* **2020**, *30*, 1235–1243. [CrossRef]
30. Bilad, M.; Arafat, H.A.; Vankelecom, I.F. Membrane technology in microalgae cultivation and harvesting: A review. *Biotechnol. Adv.* **2014**, *32*, 1283–1300. [CrossRef]
31. Kim, D.; Kwak, M.; Kim, K.; Chang, Y.K. Turbulent jet-assisted microfiltration for energy efficient harvesting of microalgae. *J. Membr. Sci.* **2019**, *575*, 170–178. [CrossRef]
32. Folch, J.; Lees, M.; Sloane Stanley, G.H. A simple method for the isolation and purification of total lipides from animal tissues. *J. Biol. Chem.* **1957**, *226*, 497–509. [CrossRef]

33. Kwak, M.; Kang, S.G.; Hong, W.-K.; Han, J.-I.; Chang, Y.K. Simultaneous cell disruption and lipid extraction of wet *Aurantiochytrium* sp. KRS101 using a high shear mixer. *Bioprocess Biosyst. Eng.* **2018**, *41*, 671–678. [CrossRef]
34. DuBois, M.; Gilles, K.A.; Hamilton, J.K.; Rebers, P.A.; Smith, F. Colorimetric Method for Determination of Sugars and Related Substances. *Anal. Chem.* **1956**, *28*, 350–356. [CrossRef]
35. Seon, G.; Joo, H.W.; Kim, Y.J.; Park, J.; Chang, Y.K. Hydrolysis of lipid-extracted *Chlorella vulgaris* by simultaneous use of solid and liquid acids. *Biotechnol. Prog.* **2019**, *35*, e2729. [CrossRef]
36. Laurens, L.M.L.; Dempster, T.A.; Jones, H.D.T.; Wolfrum, E.J.; Van Wychen, S.; McAllister, J.S.P.; Rencenberger, M.; Parcher, K.J.; Gloe, L.M. Algal Biomass Constituent Analysis: Method Uncertainties and Investigation of the Underlying Measuring Chemistries. *Anal. Chem.* **2012**, *84*, 1879–1887. [CrossRef]
37. Cho, E.J.; Trinh, L.T.P.; Song, Y.; Lee, Y.G.; Bae, H.-J. Bioconversion of biomass waste into high value chemicals. *Bioresour. Technol.* **2020**, *298*, 122386. [CrossRef]
38. Zhang, C.; Tang, X.; Yang, X. Overcoming the cell wall recalcitrance of *heterotrophic Chlorella* to promote the efficiency of lipid extraction. *J. Clean. Prod.* **2018**, *198*, 1224–1231. [CrossRef]
39. Martins, L.B.; Soares, J.; da Silveira, W.B.; Sousa, R.d.C.S.; Martins, M.A. Dilute sulfuric acid hydrolysis of *Chlorella vulgaris* biomass improves the multistage liquid-liquid extraction of lipids. *Biomass Convers. Biorefinery* **2021**, *11*, 2485–2497. [CrossRef]
40. Ranjan, A.; Patil, C.; Moholkar, V.S. Mechanistic assessment of microalgal lipid extraction. *Ind. Eng. Chem. Res.* **2010**, *49*, 2979–2985. [CrossRef]
41. Kwak, M.; Kim, D.; Kim, S.; Lee, H.; Chang, Y.K. Solvent screening and process optimization for high shear-assisted lipid extraction from wet cake of *Nannochloropsis* sp. *Renew. Energy* **2020**, *149*, 1395–1405. [CrossRef]
42. Gong, M.; Hu, Y.; Yedahalli, S.; Bassi, A. Oil extraction processes in microalgae. *Recent Adv. Renew. Energy* **2017**, *1*, 377–411.
43. Huber, G.W.; Corma, A. Synergies between bio-and oil refineries for the production of fuels from biomass. *Angew. Chem. Int. Ed.* **2007**, *46*, 7184–7201. [CrossRef] [PubMed]
44. Jin, M.; Oh, Y.-K.; Chang, Y.K.; Choi, M. Optimum utilization of biochemical components in *Chlorella* sp. KR1 via subcritical hydrothermal liquefaction. *ACS Sustain. Chem. Eng.* **2017**, *5*, 7240–7248. [CrossRef]
45. Zhang, Q.; Wu, D.; Lin, Y.; Wang, X.; Kong, H.; Tanaka, S. Substrate and Product Inhibition on Yeast Performance in Ethanol Fermentation. *Energy Fuel* **2015**, *29*, 1019–1027. [CrossRef]
46. Yun, J.-H.; Nam, J.-W.; Yang, J.H.; Lee, Y.J.; Cho, D.-H.; Choi, H.I.; Hong, J.S.; Ahn, K.H.; Kim, H.-S. Toward a zero-waste microalgal biorefinery: Complete utilization of defatted *Chlorella* biomass as a sole heterotrophic substrate for *Chlorella* sp. HS2 and an improved composite filler. *Chem. Eng. J.* **2024**, *480*, 147998. [CrossRef]
47. Gehrke, C.W.; Wall Sr, L.L.; Absheer, J.S.; Kaiser, F.E.; Zumwalt, R.W. Sample preparation for chromatography of amino acids: Acid hydrolysis of proteins. *J. Assoc. Off. Anal. Chem.* **1985**, *68*, 811–821. [CrossRef]
48. Singh, A.K.; Prajapati, K.S.; Shuaib, M.; Kushwaha, P.P.; Kumar, S. Microbial proteins: A potential source of protein. In *Functional Foods and Nutraceuticals: Bioactive Components, Formulations and Innovations*; Springer: Cham, Switzerland, 2020; pp. 139–147.
49. Abdus Salam, M.; Creaser, D.; Arora, P.; Tamm, S.; Lind Grennfelt, E.; Olsson, L. Influence of bio-oil phospholipid on the hydrodeoxygenation activity of $NiMoS/Al_2O_3$ catalyst. *Catalysts* **2018**, *8*, 418. [CrossRef]
50. van der Bij, H.E.; Weckhuysen, B.M. Phosphorus promotion and poisoning in zeolite-based materials: Synthesis, characterisation and catalysis. *Chem. Soc. Rev.* **2015**, *44*, 7406–7428. [CrossRef]
51. Paisan, S.; Chetpattananondh, P.; Chongkhong, S. Assessment of water degumming and acid degumming of mixed algal oil. *J. Environ. Chem. Eng.* **2017**, *5*, 5115–5123. [CrossRef]
52. Law, S.Q.; Chen, B.; Scales, P.J.; Martin, G.J. Centrifugal recovery of solvent after biphasic wet extraction of lipids from a concentrated slurry of *Nannochloropsis* sp. biomass. *Algal Res.* **2017**, *24*, 299–308. [CrossRef]
53. Law, S.Q.; Mettu, S.; Ashokkumar, M.; Scales, P.J.; Martin, G.J. Emulsifying properties of ruptured microalgae cells: Barriers to lipid extraction or promising biosurfactants? *Colloids Surf. B Biointerfaces* **2018**, *170*, 438–446. [CrossRef] [PubMed]
54. Halim, R.; Webley, P.A.; Martin, G.J. The CIDES process: Fractionation of concentrated microalgal paste for co-production of biofuel, nutraceuticals, and high-grade protein feed. *Algal Res.* **2016**, *19*, 299–306. [CrossRef]
55. Wang, H.; Ji, C.; Bi, S.; Zhou, P.; Chen, L.; Liu, T. Joint production of biodiesel and bioethanol from filamentous oleaginous microalgae *Tribonema* sp. *Bioresour. Technol.* **2014**, *172*, 169–173. [CrossRef]
56. Laurens, L.; Nagle, N.; Davis, R.; Sweeney, N.; Van Wychen, S.; Lowell, A.; Pienkos, P. Acid-catalyzed algal biomass pretreatment for integrated lipid and carbohydrate-based biofuels production. *Green Chem.* **2015**, *17*, 1145–1158. [CrossRef]
57. Seon, G.; Kim, M.; Lee, Y.W.; Cho, J.M.; Kim, H.; Park, W.-K.; Chang, Y.K. Development of an integrated biomass refinery process for whole cell biomass utilization of *Chlorella* sp. ABC-001. *Chem. Eng. J.* **2023**, *451*, 138543. [CrossRef]

sustainability

MDPI

Review

Future Directions of Sustainable Resource Utilization of Residual Sewage Sludge: A Review

Weicheng Zheng [1], Yuchao Shao [2,*], Shulin Qin [1] and Zhongquan Wang [1]

1 Hangzhou Research Institute of China Coal Technology & Engineering Group, Hangzhou 311201, China; weicheng.zheng@foxmail.com (W.Z.); hzqsl722@163.com (S.Q.); wzhquan@126.com (Z.W.)
2 State Key Laboratory of Environmental Criteria and Risk Assessment, Chinese Research Academy of Environmental Sciences, Beijing 100012, China
* Correspondence: shaoyc20@mails.tsinghua.edu.cn; Tel.: +86-18800101194

Abstract: With the simultaneous increase in wastewater generation and wastewater treatment rates in China, the annual production of residual sewage sludge (RSS) has been steadily rising, exceeding 70 million tons with an 80% moisture content. The sustainable resource utilization of RSS will be the predominant disposal method instead of sanitary landfilling in China. This review aimed to systematically analyze the major sustainable resource utilization technologies for RSS. Firstly, the basic characteristics of RSS in China were analyzed. A comparative analysis was conducted to assess the advantages, disadvantages, and applicability of three primary sustainable resource utilization technologies for RSS: building materials, energy utilization, and phosphorus recovery, aiming to provide clear insights for the development of future strategies. The research findings revealed that no single method can economically and environmentally address all challenges in the utilization of RSS resources. It is necessary to adopt appropriate resource utilization technologies according to the characteristics of RSS from wastewater treatment, achieving integration of pollution control and resource utilization. This review can provide scientific guidance for future sustainable utilization of RSS resources.

Keywords: residual sewage sludge; sustainable resource utilization; building materials; energy utilization; carbonization; phosphorus recovery

1. Introduction

As the global issue of energy resource shortages becomes increasingly severe, the resource utilization of waste has become an urgent imperative [1–4]. Residual sewage sludge (RSS) is generated during the biological treatment of wastewater. It consists of a complex and heterogeneous substance primarily composed of water, organic (e.g., pathogenic bacteria, parasites) constituents, and inorganic (e.g., inorganic particles, heavy metals, salts) constituents, with extracellular polymeric substance as its framework [5–7]. The main elements in RSS are C, N, and P (nutrients), but these can act as pollutants if released into the environment [8]. This is problematic as there are already other pollutants in the RSS (i.e., pathogenic bacteria, heavy metals). Therefore, RSS is both a pollutant and a potential resource. However, the resources within RSS cannot be directly utilized and require specific technological means to achieve resource utilization. As a result, sludge resource utilization technologies have garnered the attention of many researchers and continue to evolve [9].

Currently, the activated sludge process and its variations are the most widely used technologies in the field of wastewater treatment [10,11]. The activated sludge process transfers pollutants from wastewater into RSS, inevitably leading to the generation of substantial amounts of RSS. In 2020, it was estimated that the wastewater treatment capacity of China's sewage treatment plants reached 190 million cubic meters per day; the production of wet sludge with an 80% moisture content in cities was approximately 66.64 million tons (data were collected from the Ministry of Housing and Urban-Rural Development,

China). It is projected that the wet RSS production will reach 90 million tons by 2025. In recent years, with increased environmental inspections in China, the construction of water pollution prevention and control facilities has entered a rapid development phase. Currently, the treatment rate for municipal sewage and industrial wastewater has reached as high as 97.89% reported by the Ministry of Housing and Urban-Rural Development, China. However, the disposal of RSS has not received adequate attention in various regions. This situation has resulted in a lag in the construction of sludge disposal facilities compared to wastewater treatment facilities. The most widely used sludge disposal method in China is sanitary landfilling [12–14], which not only consumes a significant amount of public resources but also wastes the valuable resources contained within the sludge. The resource utilization of RSS in China has significant potential for development.

This article begins by describing the basic characteristics of RSS from wastewater treatment plants in China and systematically elaborates on the research progress of major resource utilization technologies for RSS (i.e., building materials, energy utilization, and phosphorus recovery). It analyzes the advantages and disadvantages of each technology, aiming to provide a reference for further research on sludge sustainable resource utilization and disposal.

2. Chemical Properties of Residual Sewage Sludge

Residual sewage sludge (RSS) originates from urban sewage treatment plants and industrial wastewater treatment plants. The chemical components of RSS from both sources are essentially the same, but the contents of each compound are different. For instance, industrial RSS usually contains more recalcitrant organic compounds, heavy metals, oils, and other specific pollutants present in industrial wastewater. A research group conducted investigations on the chemical compounds of RSS produced by 90 sewage treatment plants from 25 cities in China, and it was discovered that there were significant differences in the composition of different types of RSS [15]. The representative indicators of RSS, including nutritional content (nitrogen, phosphorus, and potassium) and heavy metal concentrations, are shown in Table 1 (data are sourced from reference [15]). The maximum and minimum values of these indicators were significantly different. For example, the concentration of Cr in some RSS could reach as high as 2960 mg/kg, while the minimum value was only 9.6 mg/kg; the organic matter in RSS was in the range of 13.4% to 51.4%. The high heavy metal contents in RSS are caused by the discharge of heavy metals from specific industrial activities such as metallurgy, chemical manufacturing, and electronics production. Additionally, the organic matter in industrially-derived RSS is typically lower than in domestically-derived RSS.

Table 1. Analysis of nutrient and heavy metal contents in residual sewage sludge.

Indicator	Average Value	Maximum Value	Minimum Value
Organic matter (%)	51.4	77.0	13.4
Total nitrogen (%)	3.6	7.2	0.3
Total phosphorus (%)	2.3	14.7	0.0
Total potassium (%)	1.4	7.4	0.1
Cd (mg/kg)	2.8	49.6	/
Hg (mg/kg)	5.7	13.6	0.0
Pb (mg/kg)	93.0	729.0	2.7
Cr (mg/kg)	266.7	2960.0	9.6

Table 1. Cont.

Indicator	Average Value	Maximum Value	Minimum Value
As (mg/kg)	29.0	301.0	0.0
Ni (mg/kg)	120.2	615.4	2.7
Zn (mg/kg)	1794.7	14,285.0	15.9
Cu (mg/kg)	493.3	10,531.1	16.9
B (mg/kg)	41.7	98.6	/

The chemical composition of RSS is different from sludge obtained in different operating stages of municipal wastewater plants. It can be observed that the total solid content in RSS was low, while the volatile solid and protein content was relatively high, compared to primary sludge and digested sludge (Table 2; data are sourced from reference [15,16]). By consolidating the statistical findings presented in Table 1, it becomes evident that the organic matter content in RSS generated across various regions or from diverse wastewater sources consistently exceeds 50%, with nutrient elements such as nitrogen and phosphorus consistently surpassing 2%. This underscores a notable degree of resource potential within RSS. However, RSS also contains significant heavy metal components that cannot be ignored. The concentration of heavy metals in RSS is closely related to the composition characteristics of the treated wastewater. Therefore, different resource utilization methods should be adopted for RSS having different characteristics.

Table 2. Typical composition of sludge solids in municipal wastewater plants.

Composition (wt%)	Primary Sludge		Digested Sludge		Residual Sludge
	Range	Typical Value	Range	Typical Value	Range
Total solids	5.0~9.0	6.0	2~5	4.0	0.8~1.2
Volatile solids	60~80	65	30~60	40	59~88
Oils and fats	6.0~30	/	5~20	18	/
Protein	20~30	25	15~20	18	32~41
Cellulose	8.0~15	10	8.0~15	10	/
Nitrogen	1.5~4.0	2.5	1.6~3.0	3.0	2.4~5.0
Phosphorus	0.8~2.8	1.6	1.5~4.0	2.5	2.8~11
Potassium	0~1.0	0.4	0~3.0	1.0	0.5~0.7

3. Resource Utilization Technologies for Residual Sewage Sludge Disposal

The composition of RSS can be divided into inorganic matter (e.g., calcium, silicon, aluminum, and iron) and organic matter (e.g., carbon, nitrogen, and phosphorus). Different components in RSS can be targeted for resource utilization using various technologies, including sludge building material utilization for recovering inorganic matter, sludge energy utilization for recovering organic matter, sludge phosphorus recovery for reclaiming phosphorus resources, and sludge carbonization for recovering carbon elements (Figure 1).

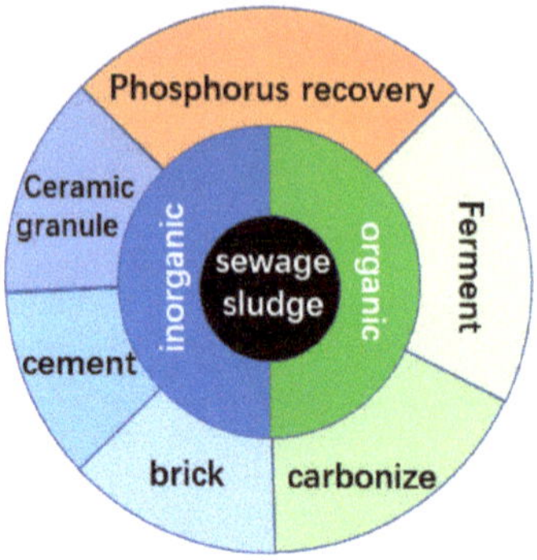

Figure 1. The primary pathways for resource utilization of residual sewage sludge.

3.1. *Residual Sewage Sludge for Building Material Utilization*

RSS contains elements such as calcium, silicon, aluminum, and iron, which are similar to the components of building materials. Therefore, RSS has the potential for utilization in building materials. The Chinese government has been promoting policies and regulations to encourage the sustainable utilization of resources and the development of circular economy practices, providing a supportive regulatory environment for the application of RSS in construction. Currently, there are three main directions for the utilization of RSS in building materials: brick production [17], cement production [18], and ceramic granule production [19] (Figure 2).

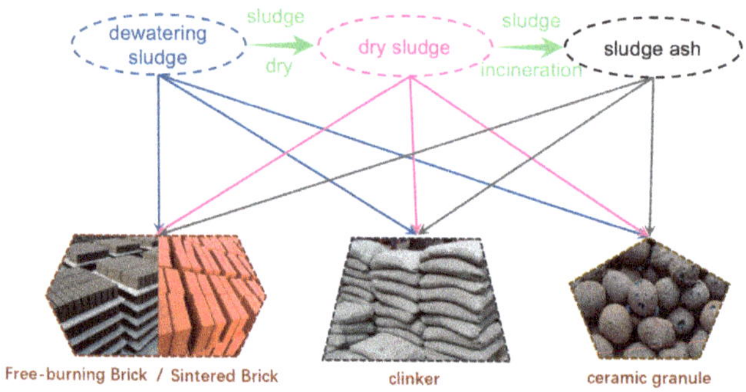

Figure 2. Utilization pathways of residual sewage sludge in construction materials.

3.1.1. Sludge Brick-Making

Brick factories typically use clay as the primary raw material for brick production. The major components of clay are SiO_2, Al_2O_3, and small amounts of calcium, magnesium, iron, carbonate minerals, etc. [20]. These substances, at sintering temperatures of around 1000 °C, react with aluminum, calcium, magnesium, iron, sodium, and silicon to form new silicate materials. Easily fusible materials create a liquid phase that binds the remaining solid solution (eutexia), solid solution, and new minerals during cooling, forming stable crystals. The inorganic composition of RSS is similar to that of clay. The SiO_2, Al_2O_3, and CaO contents in sewage sludge and sludge incineration ash are shown in Figure 3 [21]. In 1987, scholars used dried urban RSS mixed with clay to produce qualified sintered bricks [22]. These researchers were pioneers in RSS brick-making technology, and their early work primarily focused on exploring the feasibility of the technology. In 2003, researchers conducted a systematic study of RSS brick-making [23]. The results indicated that key factors affecting brick quality were the sludge-to-clay ratio and firing temperature. Increasing RSS content reduced the brick's shrinkage rate, water absorption rate, and compressive strength. They also conducted tests on the leaching of toxins from the RSS, revealing low levels of metal leaching.

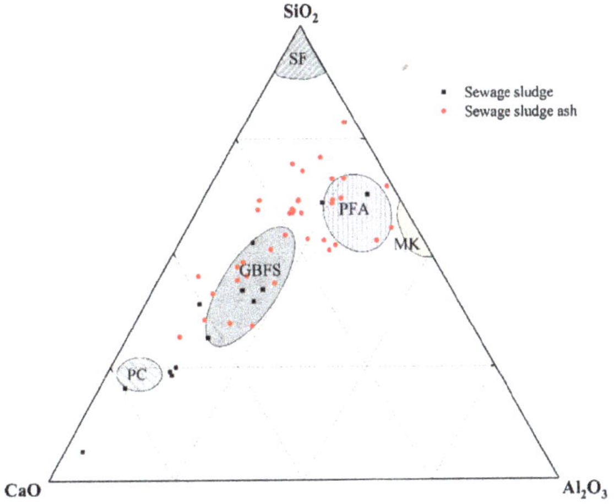

Figure 3. Ternary plot of SiO_2, Al_2O_3, and CaO contents for sewage sludge and incinerated sludge ash. PC = portland cement, GBFS = granulated blast furnace slag, PFA = pulverized fuel ash, MK = metakaolin, SF = silica fume. Reproduced from the work by [21] with permission from Elsevier, copyright 2020.

Another method of making bricks from RSS is by using RSS ash produced after incineration. In developed countries such as Europe, the United States, and Japan, incineration is a common method for RSS disposal, and the main components of RSS ash are SiO_2 and Al_2O_3, which are similar to the components of clay [21]. To utilize the abundant ash resources generated from RSS incineration, RSS ash is used in brick-making. A research group found that adding RSS ash to clay can lead to a decrease in the compressive strength of bricks, but it increases the water absorption rate [22]. When 10% of RSS ash was added, the compressive strength of bricks only decreased by 1.72%, and the water absorption rate increased from 0.03% to 0.07%. When the addition amount was increased to 50%, the compressive strength decreased by 20.4%, and the water absorption rate increased to 1.7% [22]. Another study attempted to make bricks entirely from RSS ash [24]. This required controlling the particle size of the RSS ash to below 20 μm, pressing it using a high pressure of 98 Mpa, and firing it at temperatures above 1000 °C. However, due to its good water absorption and high calcium content, the bricks tended to develop moss on the surface and turn white due to leaching of calcium carbonate [24]. These issues could only be addressed by increasing the firing temperature and applying chemical coatings, which resulted in no economic advantage compared to regular bricks. Some researchers have also studied the production of unfired bricks using RSS. For example, a group used construction waste, RSS, and fly ash as raw materials, with quality proportions of 60%, 2%, and 15%, respectively, along with 5% desulfurized gypsum as an activator [25]. After the shaping, curing, and aging processes, unfired bricks were produced having a compressive strength of 13.75 Mpa. Although this method has the limitation of the RSS disposal capacity, if the bricks were successful, then scaling production would still assist in overall RSS utilization. One method is not going to solve RSS utilization, but a combination of methods that are proven to work will.

3.1.2. Cement Produced from Sludge

Cement is one of the most important materials in urban construction. The global cement production in 2022 was approximately 4.31 billion tons, with China's capacity reaching 2.36 billion tons, accounting for a staggering 55% share (data were collect from the National Bureau of Statistics, Beijing, China). Cement kilns used in cement production are characterized by ultra-high temperatures (>1200 °C), large capacity, and prolonged residence times for flue gases [26]. In the early 1990s, developed countries in Europe and North America proposed a technical route for the co-disposal of waste in cement kilns. This involves the environmentally friendly and resourceful treatment of waste materials that meet the combustion conditions in cement kilns. Lime production primarily requires raw materials rich in calcium and aluminum, such as limestone and clay [27]. RSS can provide SiO_2, Al_2O_3, and CaO materials as well as fuel for cement production. Therefore, using RSS to produce cement not only achieves complete resource utilization of the RSS but also addresses the increasing volume of RSS by incorporating it into the vast cement industry market.

It is critical to control the sludge content in the sludge-infused cement preparation process. A research group studied the impact of substituting different levels of RSS ash for cement on the heat of hydration and the 7-day compressive strength of cement mortar samples [28]. The results showed that RSS ash in cement affected the hydration process, reducing the rate of heat generation during hydration, particularly in the early setting period. The maximum reduction in the 7-day compressive strength of cement mortar samples was 20% when the RSS ash content was increased. In the process of using RSS to produce cement, the migration of heavy metals in RSS has also received considerable attention. Heavy metals in RSS can be divided into two categories: easily volatile or semi-volatile heavy metals like mercury, lead, and cadmium, which can be trapped by the waste gas treatment system or released into the atmosphere along with flue gases, and non-volatile heavy metals like chromium and nickel, which can enter the solid phase. Researchers studied the physicochemical interactions between textile wastewater sludge and portland cement and analyzed changes in the leachability of metals [29]. It was found that the incorporation of textile wastewater sludge into cement mortar resulted in the formation of more weak interfaces and increased porosity. However, leaching tests showed that all toxic metals were stabilized, and the leaching rate of metals from portland cement-textile wastewater sludge mortar was very low, posing little to no environmental risk [29]. Another group finely ground powder from heavy metal-containing sludge incineration and added it to portland cement [23]. The concentrations of leachable heavy metals in cementitious materials with different addition levels were lower than the limits specified in Chinese standard GB 30760—2014 (the technical specifications for the co-disposal of solid waste in cement kilns) [30]. They also investigated the leaching of heavy metals in cement specimens with different water-to-cement ratios and ages. The leaching rate of chromium in specimens with a water-to-cement ratio of 0.5 and 40% addition of sludge ash powder reached 34.2%, and the leaching concentration of all heavy metals was 0.11 mg/L, indicating a potential risk of exceeding the limits [31]. In conclusion, using RSS for cement production is a technically feasible, environmentally friendly, and economically viable direction for RSS recycling. However, careful consideration is needed when selecting this technology to treat RSS from specific industries with higher heavy metal content.

3.1.3. Sludge Ceramsite Production

Ceramsite is an artificial lightweight aggregate made from various inorganic minerals such as clay, shale, slate, coal gangue, and fly ash. It is produced through molding and calcination, possessing advantages like porosity, lightweightness, high water absorption, fire resistance, freezing resistance, and corrosion resistance [32]. Ceramsite finds wide applications in construction materials, cultivation substrates, water treatment filters, land reclamation, and many other fields [32]. The traditional raw materials for ceramsite production, such as fly ash and clay, mainly consist of SiO_2 and Al_2O_3, which are also the major inorganic components of sludge [21]. Therefore, utilizing sludge for ceramsite production is feasible in terms of material composition. The preparation process of sludge ceramsite is shown in Figure 4. The sludge is dewatered, dried, molded, and calcined to produce sludge granules [33]. Studies have shown that ceramsite exhibits better compressive strength and more densified surfaces when the mass fraction of SiO_2 is between 30% and 45% [34]. However, the mass fraction of SiO_2 in urban sludge in China is significantly lower than this range [35]. This leads to lower strength and increased fragility of ceramsite produced solely from sludge. Therefore, when using sludge for ceramsite production, it is often necessary to add other raw materials to optimize the material composition and obtain high-quality qualified ceramsite [36]. Clay and fly ash are the most commonly used auxiliary materials when using sludge for ceramsite production. Their addition can promote ceramsite formation, increase strength, and improve overall ceramsite quality. This is because the calcium in these additives can encourage the formation of more calcium feldspar during the firing process. Additionally, the higher content of Al_2O_3 can facilitate ceramsite bonding, reduce surface cracks in ceramsite, and lower its apparent density. Silicon, aluminum, and iron in these materials can serve as the framework for ceramsite [34,37,38]. In the preparation process of ceramsite, the most significant factors affecting ceramsite production are the calcination (preheating and sintering) processes. If a relatively long preheating at a low temperature is carried out, it allows for the sufficient decomposition of organic matter in the sludge, which is beneficial for the bloating process of ceramsite, thereby reducing the apparent density of the ceramsite. The sintering temperature is crucial in ceramsite production. When sintered at a low temperature (below 1000 °C), the resulting ceramsite has more surface cracks and higher apparent density. However, when sintered at a high temperature (above 1200 °C), a larger amount of liquid phase is generated inside the ceramsite, with lower viscosity and higher fluidity. The enclosed gas is prone to escape, and the internal pores of the ceramsite are filled by the liquid phase, leading to an increase in apparent density [39–41].

Figure 4. Preparation process of sludge ceramsite. Reproduced from the work by [33] with permission from Elsevier (Amsterdam, The Netherlands), copyright 2021.

In conclusion, utilizing RSS along with certain other inorganic materials to produce ceramsite is promising for RSS utilization. It not only reduces sludge volume but also substitutes for some of the raw materials and fuel used in ceramsite production, achieving the resource utilization of sludge. Moreover, the leaching of heavy metals such as copper, lead, and cadmium in ceramsite is much lower than national standards [42]. In recent years, research on sludge ceramsite has focused more on ceramsite modification. The main objectives of modification are to increase specific surface area, reaction sites, and adsorption capacity. This is often achieved through methods such as metal compound modification, magnetic modification, charge modification, and alkali modification, to obtain ceramsite with enhanced application value [43–45].

3.2. Residual Sewage Sludge Utilization for Energy

The solid phase of RSS contains a very high proportion of organic components, which has the potential for energy utilization. However, RSS contains a large amount of moisture, making it difficult to be directly utilized for energy. In order to develop more green energy, China continues to explore new technologies and methods to provide more options and possibilities for RSS resource utilization and has already made significant progress in RSS energy utilization research. Various RSS energy utilization technologies have emerged, including common methods such as anaerobic fermentation [46,47] and co-incineration of RSS [48,49].

3.2.1. Sludge Anaerobic Fermentation

Anaerobic fermentation of RSS is the process of utilizing anaerobic microorganisms in sludge to hydrolyze and convert organic substances such as proteins and polysaccharides into small water-soluble organic molecules [50]. These molecules are then further converted into methane, carbon dioxide, and other products by specific anaerobic microorganisms. This process not only reduces the volume of sludge but also recovers methane, achieving RSS reduction and resource utilization. The biochemical reactions involved in the fermentation process are illustrated in Figure 5 [51]. The anaerobic fermentation of RSS is typically divided into three stages: hydrolysis, acidogenesis, and methanogenesis. Among these stages, hydrolysis is considered the limiting factor for the overall fermentation rate because the rate of hydrolysis of extracellular polymeric substances and cell membranes is relatively slow. Therefore, pretreatment of RSS before anaerobic fermentation is often conducted to enhance the release and degradation of organic matter in sludge, thereby expediting the fermentation process. Common pretreatment techniques include hydrothermal treatment and acid/alkali treatment [52–56]. Hydrothermal treatment initially aimed to improve the dewaterability of sludge, but it was later found that this process increased both the rate of anaerobic fermentation and methanogenesis. A previous study has indicated the optimal temperature for sludge hydrothermal treatment falls within the range of 140 to 160 °C [57]. Temperatures higher than this range can significantly increase the production of recalcitrant compounds, leading to the inhibition of the fermentation process. Another study investigated the influence of low-temperature hydrothermal treatment on anaerobic digestion of RSS [58]. The researchers performed hydrothermal treatment of RSS at 180 °C for 30 min, resulting in methane production increasing from 142.7 mL CH_4/g-VS to 343.1 mL CH_4/g-VS compared to the control group. Because of its remarkable enhancement in biogas production rates, the "hydrothermal pretreatment + anaerobic fermentation" process has gained widespread application [59].

The main influencing factors in the anaerobic fermentation process of RSS include the C/N ratio, pH, temperature, and sludge retention time. Depending on the fermentation temperature, fermentation is typically categorized as low-temperature fermentation (4~20 °C), mesophilic fermentation (20~45 °C), or thermophilic fermentation (45~60 °C) [60]. Temperature affects the activity of different microorganisms, resulting in variations in fermentation products. Researchers demonstrated that when the anaerobic fermentation temperature was increased from 35 °C to 55 °C, the proportion of acetic acid in volatile fatty acids in the fermentation broth decreased from 85.6% to 65.8%, while the proportion of propionic acid increased from approximately 0% to 15.5% [61]. Similar to the mechanism of temperature influence, changes in pH also affect the dominant microbial community during fermentation [62]. The optimal pH for anaerobic digestion is typically in the range of 6.8 to 7.4, but organic acids are generated during fermentation. Therefore, anaerobic sludge digestion systems are usually operated to maintain a pH around 8. The initial C/N ratio of sludge is generally around 12, while the optimal C/N ratio for anaerobic fermentation of sludge falls between 20 and 30 [63]. A C/N ratio that is too low can result in excessive ammonia production, inhibiting fermentation, while a too-high C/N ratio can lead to a decrease in the nitrogen content of the sludge system, affecting microbial growth [63]. Therefore, in practical applications, organic matter such as kitchen waste [64] or other organic materials [65] is often added for co-fermentation to adjust to the appropriate C/N ratio and enhance biogas production.

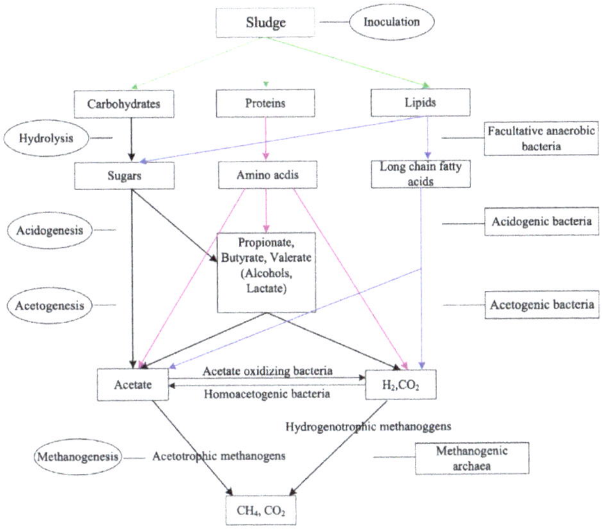

Figure 5. Biochemical reactions during the sludge fermentation process. Reproduced from the work by [51] with permission from Elsevier, copyright 2018.

In summary, anaerobic fermentation of RSS for biogas production is a promising method for RSS utilization. The key challenge in the application of this technology lies in controlling fermentation parameters to achieve efficient biogas production. Additionally, attention should be paid to substances in the sludge such as ammonia, sulfides, metal ions, hydrogen, and fatty acids [66], as they can have inhibitory effects on the anaerobic fermentation process.

3.2.2. Sludge Co-Incineration

Burning RSS and coal together, at temperatures exceeding a thousand degrees Celsius, can completely oxidize and decompose the organic matter present in the RSS [67]. Additionally, burning the organic matter in the RSS can generate heat, making it an efficient and convenient method for the harmless and resourceful disposal of RSS [67].

When co-burning RSS and coal, there are specific requirements regarding the moisture content of the RSS. The moisture content significantly impacts the heating value of the RSS and can have adverse effects on boiler combustion efficiency and stability if it is too high. High moisture content can also lead to reduced system temperatures and blockages in the feeding system. Even after mechanical dewatering, the moisture content of RSS can still be around 80%. Further reducing the moisture content through techniques like sludge drying consumes a considerable amount of energy and can affect boiler flue gas emissions. Therefore, many researchers have investigated the impact of different moisture contents of RSS and coal blends on boilers [68–70]. Studies have examined the influence of RSS blending ratios and RSS moisture content on 100 MW coal-fired power plant boilers [68]. They found that as the blending ratio increased, the heat flux on the furnace wall in the primary combustion zone significantly decreased. Beyond a 10% blending rate, it had a noticeable impact on the combustion stability within the furnace. When 20% RSS was co-burned, the average heat flux decreased by approximately 17%, as shown in Figure 6. A group researched the combustion of coal powder and RSS in a tangentially fired coal-fired boiler [69]. They found that RSS with a moisture content of 56%, when co-burned at a blending rate exceeding 20%, significantly affected fuel characteristics, reducing ignition performance and increasing nitrogen oxide emissions. Co-burning RSS also has a significant impact on excessive flue gas emissions. Some researchers discovered that when the RSS mixing ratio was lower than 20%, the emission of gaseous pollutants such as SO_2, NOx, and dioxin increased linearly with the sludge mixing ratio [70]. However, it could also meet the national standard for the pollution control of municipal solid waste incineration in China. Another group studied the combustion characteristics of brown coal mixed with RSS in a circulating fluidized bed and found that when brown coal was co-burned with RSS, CO and hydrocarbon emissions were lower compared to burning brown coal alone; SO_2 emissions decreased with an increase in the RSS percentage in the mixture, while there was a slight increase in NOx emissions [71]. Researchers also conducted co-combustion of coal and RSS in a circulating fluidized bed boiler [72]. The results indicated that during the co-burning process, the CO volume fraction in the flue gas increased with an increase in the sludge co-burning ratio. In contrast, the NO volume fraction significantly decreased, and there was a slight increase in the SO_2 volume fraction. When the RSS co-burning ratio increased from 0% to 100%, the carbon content in the fly ash increased from 8.09% to 28.26%, and the combustion efficiency decreased from 99.23% to 87.76%. Compared to coal combustion alone, co-burning RSS reduced the fly ash melting temperature [72]. The different research studies mentioned above show that co-burning RSS has varying effects on the emissions of nitrogen oxides and sulfur compounds in exhaust gases, depending on the characteristics and proportion of the sludge used.

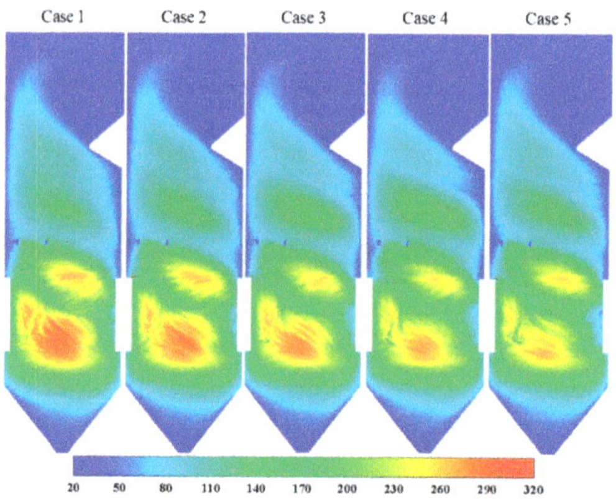

Figure 6. Heat flux distribution on the right water-wall (unit: kW/m^2). Reproduced from the work by [68] with permission from Elsevier, copyright 2017.

Different types of boilers have varying fuel requirements. In China, coal powder boilers, which have high fuel quality requirements, are the most widely used in power plants. Therefore, the practical application of co-burning RSS is limited, and RSS needs to be dried before being mixed and co-burned with coal powder. In contrast, circulating fluidized bed boilers have a significant advantage in fuel adaptability. They are commonly used for co-burning RSS because they can handle sludge with relatively high moisture content. Additionally, their furnace temperature is lower, allowing for better control of the formation of nitrogen oxides and dioxins and other pollutants [73]. Therefore, circulating fluidized bed boilers are more suitable for co-burning sludge. In summary, co-burning RSS effectively addresses the challenge of degrading toxic and harmful substances in RSS while fully utilizing the organic matter's heat value. It achieves both the harmless disposal and resource utilization of RSS, making it a relatively straightforward and feasible technology for RSS utilization. However, co-burning RSS requires the selection of circulating fluidized bed boilers and imposes higher demands on flue gas treatment.

3.2.3. Residual Sewage Sludge Carbonization

RSS carbonization mainly involves two pathways: pyrolytic carbonization to form biochar and hydrothermal carbonization to produce hydrochar [74,75]. Pyrolytic carbonization refers to the process in which RSS is heated in an oxygen-free environment, causing organic materials to undergo decomposition, resulting in residues (biochar), bio-oil, and a mixture of gases. On the other hand, RSS hydrothermal carbonization refers to a series of reactions that occur when RSS is subjected to water, a temperature of approximately 150 °C to 300 °C, and saturated vapor pressure. These reactions involve hydrolysis, decarboxylation, condensation, and aromatization processes.

RSS pyrolytic carbonization originated in developed countries such as Japan, the United States, and Europe in the 1980s [76]. After several decades of technological iterations, it has become mature. China introduced this technology from Japan in 2005. However, at that time, there was a lack of awareness regarding RSS treatment and disposal, and the imported equipment was expensive, making it challenging to promote its application. During the sludge pyrolytic carbonization process, the decomposition of organic materials generates combustible gases, and the heat produced during combustion can be used to dry the RSS. Thus, the entire carbonization process requires only a small amount of additional energy to maintain the reaction. Carbonization is typically classified into low-, medium-, and high-temperature carbonization, and their technical differences and characteristics are outlined in Table 3. The advantages of RSS pyrolytic carbonization technology include lower energy consumption compared to traditional drying and incineration methods, pollution-free combustion of pyrolysis gas, solidification of heavy metals with no leaching risk, and high utilization and value of the carbonized products [76,77].

Table 3. Characteristics of sludge pyrolysis carbonization.

Characteristics	Low-Temperature Carbonization	Medium-Temperature Carbonization	High-Temperature Carbonization
Carbonization temperature (°F)	≤600	800~1000	1200~1800
Additional pressure	/	/	6~8 Mpa
Pretreatment requirement	No drying	Drying to reduce the sludge moisture content (<90%)	Drying to reduce the sludge moisture content (<30%)
Operating odor	Micro-odor	Odorless	Odorless
Biochar calorific value (Kcal/kg)	3600~4900	2600~4300	2000~3000
Application	Wide	Less	Seldom

RSS hydrothermal carbonization is a novel approach for RSS resource utilization. It carbonizes sludge in water, eliminating the need for dewatering pretreatment, which significantly reduces the cost of RSS preprocessing. Additionally, it substantially decreases the generation of greenhouse gases during the pyrolysis process, making it a highly promising RSS utilization technology [77]. Elemental analysis revealed that 80% of the carbon elements and 40% of the nitrogen and sulfur remained in the hydrochar, while the dehydration and decarboxylation reactions of organic compounds during the hydrothermal reaction resulted in H/C and O/C ratios decreasing to 1.53 and 0.39, respectively [78]. With increasing carbonization time, the oxygen-containing functional groups in RSS-derived hydrochar decreased, and its aromatic structures increased. The organic composition of RSS and the hydrothermal conversion process are highly complex. A summary of the migration and transformation pathways of major organic compounds in sludge during hydrothermal treatment is shown in Figure 7 [79–82]. The key factors influencing the calorific value of hydrochar are reaction temperature, time, and product moisture content [83]. As the reaction temperature and time increase, the calorific value increases due to the higher carbon content in the products, with the moisture content of the final product having a more significant impact on its calorific value.

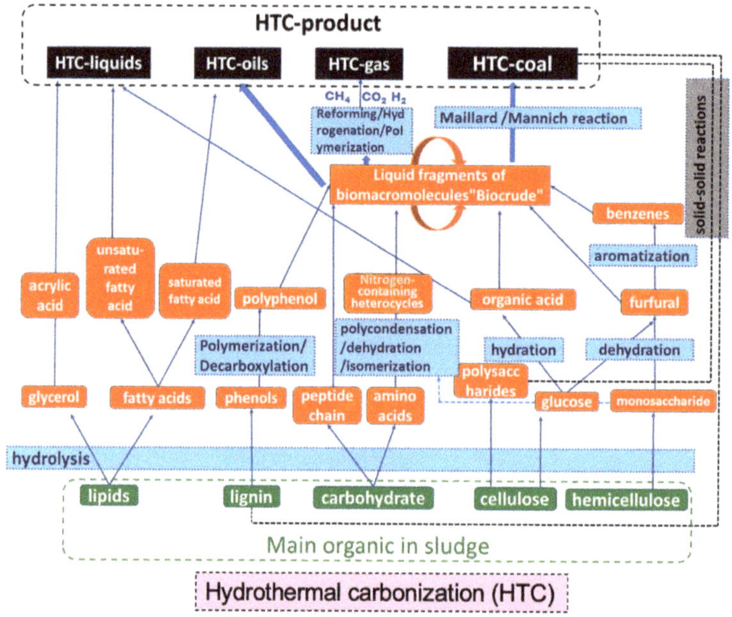

Figure 7. Organic matter conversion pathway during hydrothermal carbonization of sludge. Reproduced from the work by [81] with permission from Elsevier, copyright 2021.

Due to its higher calorific value, the products of RSS carbonization can be used as fuel, and RSS-derived hydrochar also possesses a relatively high surface area, making it suitable as an adsorbent material [84–86]. Currently, many researchers have conducted modifications on RSS-derived hydrochar, leading to the development of a variety of applications. These include adsorption of specific pollutants in wastewater, heavy metal adsorption, catalysis, soil improvement, and enhancement of fermentation processes [87–90]. With ongoing research into carbonization reactions and the continuous expansion of its application scenarios, RSS carbonization technology holds prospects for wide-ranging applications.

3.3. Phosphorus Recovery from Residual Sewage Sludge

Phosphorus is an essential element required for the metabolic processes of both animals and plants, making it crucial for modern agricultural production. Phosphorus primarily exists in nature in the form of phosphate minerals, with approximately 90% of all globally mined phosphorus ores being used for phosphorus fertilizer production. Therefore, phosphorus is a strategic resource, and researchers have focused their attention on phosphorus resources in wastewater [91]. Through the action of phosphorus-accumulating bacteria, about 90% of the phosphorus in wastewater ends up in sludge, making sludge an important secondary source of phosphorus. Figure 8 presents common processes for recovering phosphorus from sludge and its thermal treatment products [92].

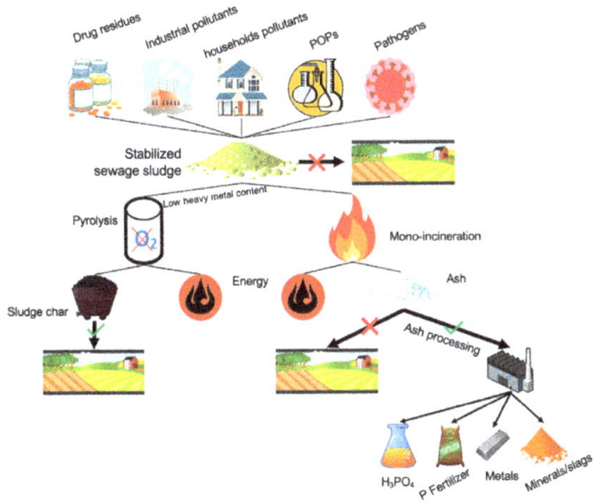

Figure 8. Summary of common process and methods for recovering and recycling P from sewage sludge. Reproduced from the work by [92] with permission from Elsevier, copyright 2022.

RSS has a high moisture content, and phosphorus is mainly present in the form of organic phosphorus. The phosphorus content per unit mass of RSS is very low, making direct extraction of phosphorus from raw RSS highly challenging. However, after RSS incineration, the mass and volume of the RSS are greatly reduced, and the concentration of phosphorus in the RSS ash is significantly increased. Studies have shown that RSS ash can contain around 80% of the phosphorus from the original sludge [93]. Therefore, phosphorus is typically recovered from the ash fraction of RSS incineration. During the incineration process, the incineration temperature has a significant impact on the content and form of phosphorus in RSS ash [94]. When the incineration temperature is above 850 °C, the phosphorus exists in the form of volatile oxides and is carried away by the flue gas as it cools, leading to a substantial reduction in phosphorus elements remaining in the bottom ash [94]. Additionally, numerous studies have indicated that the proportion of bioavailable phosphorus in incineration ash produced at temperatures above 700 °C is relatively low [95,96]. Therefore, strict control of the incineration temperature and the amount of fly ash produced during the incineration process is necessary.

The recovery of phosphorus from RSS ash mainly involves wet chemical and thermal chemical methods. The wet chemical method involves adding acid to the RSS ash to dissolve the phosphorus. A research team achieved a phosphorus leaching rate of 94.6% from RSS ash and 91.1% from dried RSS using 0.4 mol/L hydrochloric acid (75 mL/g) and 0.6 mol/L hydrochloric acid (25 mL/g) at room temperature [97]. However, heavy metals may also leach out during this process [98]. Subsequently, separation of the heavy metals and phosphorus from the phosphorus-rich leachate is required. Continuous precipitation methods, known for their ease of operation and cost advantages, are widely used. In continuous precipitation, the pH of the phosphorus-rich leachate is adjusted by adding an alkaline substance, causing the phosphorus to precipitate in the form of aluminum-phosphorus, while the heavy metals remain in the solution [99]. Thermochemical methods are another important means of phosphorus recovery. Chlorides (e.g., $CaCl_2/MgCl_2$) are added to RSS ash, and at high temperatures, chloride ions form volatile compounds with heavy metals such as cadmium, copper, lead, tin, zinc, which are separated from the RSS ash along with the flue gas [100]. Most of the remaining phosphorus in the ash fraction is in forms like Ca-P and Mg-P, which can be directly used in phosphate fertilizer production and are readily absorbed by plants [101]. Additionally, chloride incorporation can enhance the bioavailability of phosphorus. For instance, $CaCl_2$ doping can convert $AlPO_4$ into Ca_2PO_4Cl, which has higher biological availability [100]. However, thermochemical methods require higher energy consumption and larger investments, limiting their practical application.

Recovering phosphorus resources from RSS ash is an economically viable RSS resource utilization technology. However, this method is most effective when the RSS is incinerated separately, and the resulting ash fraction has a higher phosphorus content. It also requires specific incineration temperature conditions. Therefore, in practice, only a portion of RSS incineration ashes are suitable for phosphorus recovery.

4. The Comparison of Resource Utilization Technologies for Residual Sewage Sludge

RSS utilization technologies each have their own advantages and limitations (Table 4). Brick and cement production can comprehensively utilize RSS, but they are constrained by blending ratios and metal leaching issues, requiring pretreatment and potentially affecting product quality. Lightweight aggregate production is widely applied but also faces limitations in strength and blending ratios. In terms of energy utilization, anaerobic digestion and co-combustion technologies can convert RSS into energy, but they come with time costs, space requirements, and secondary pollution issues. Pyrolysis carbonization and hydrothermal carbonization technologies offer economic potential and the advantage of not requiring pretreatment, but the former requires high investment and solutions for exhaust gas issues, while the latter faces challenges of equipment corrosion and moderate investment costs. Phosphorus recovery technologies, through wet chemical and thermochemical methods, can achieve resource recovery but are affected by incineration temperatures, raw material limitations, and heavy metal volatilization issues. Overall, when choosing RSS utilization technologies, it is essential to comprehensively consider their characteristics, environmental standards, economic benefits, and operational feasibility. Future research will focus on improving efficiency, reducing costs, and minimizing environmental impacts.

Table 4. The comparison of resource utilization technologies for residual sewage sludge (RSS) disposal.

RSS Utilization Methods		Advantages	Restrictions
Building material utilization	Brick-making	• Full utilization • Low levels of metal leaching	➢ Limited amount of doped ➢ Pretreatment required ➢ Brick quality
	Cement production	• Inexpensive • Full utilization	➢ Limited amount of doped ➢ Low strength of cement mortar ➢ High levels of metal leaching
	Ceramsite production	• Ceramic particles have a wide range of applications • Low levels of metal leaching	➢ Low strength of ceramic granules ➢ Limited amount of doped
Utilization for energy	Anaerobic fermentation	• Low carbon • Energy production	➢ Long required time ➢ Large footprint ➢ Secondary pollution
	Co-incineration	• Fully decomposed • Energy utilization • Heavy metal stabilization (except mercury)	➢ Tail gas ➢ Limited amount of doped (or pretreatment required) ➢ Equipment requirements
	Pyrolytic carbonization	• Economic output • Full utilization • Heavy metal stabilization (except mercury)	➢ Pretreatment required ➢ High investment ➢ Difficulties in the operating process ➢ Tail gas
	Hydrothermal carbonization	• Process moderate • No preprocessing required • No exhaust • Economic benefits	➢ Equipment corrosion ➢ Medium investment

Table 4. *Cont.*

RSS Utilization Methods		Advantages	Restrictions
Phosphorus recovery (from ash fraction of RSS incineration)	Wet chemical methods	• Process is simple • Resource recovery	➢ Incineration temperature requirements ➢ Metal leaching ➢ Raw material restrictions
	Thermal chemical methods	• Resource recovery • Full utilization	➢ Tail gas ➢ Raw material restrictions ➢ Heavy metal volatilization ➢ High incineration temperatures

5. Conclusions and Perspectives

This review systematically analyzed the fundamental characteristics of RSS and the sustainable resource utilization technologies available for it. The review found that although RSS contains a significant amount of organic matter and nutrients, its high moisture content and heavy metal concentrations limit its direct use. Three main sustainable resource utilization technologies—building material utilization, energy utilization, and phosphorus recovery—each have their own advantages and limitations. Selecting the appropriate technology requires considering the characteristics of the RSS and the application context to achieve an integration of pollution control and resource utilization.

Future research should focus on technological innovation and policy support to improve the efficiency and economic viability of RSS utilization. Other widely studied RSS treatment technologies, such as fermentation for carbon source production, can also be considered. Interdisciplinary collaboration and life-cycle assessment will be crucial, along with enhancing public awareness and international cooperation to share best practices. Additionally, risk management and environmental safety must be integral throughout the development and application of RSS utilization technologies to ensure their sustainability.

Author Contributions: Conceptualization, Y.S.; methodology, Y.S.; validation, W.Z.; investigation, W.Z.; data curation, W.Z.; writing—original draft preparation, W.Z.; writing—review and editing, W.Z., Y.S., S.Q. and Z.W. All authors have read and agreed to the published version of the manuscript.

Funding: This research received no external funding.

Institutional Review Board Statement: Not applicable.

Informed Consent Statement: Not applicable.

Data Availability Statement: The original contributions presented in this study are included in the article. Further inquiries can be directed to the corresponding author.

Conflicts of Interest: Authors Weicheng Zheng, Shulin Qin and Zhongquan Wang were employed by the company Hangzhou Research Institute of China Coal Technology & Engineering Group. The remaining author declares that the research was conducted in the absence of any commercial or financial relationships that could be construed as a potential conflict of interest.

References

1. Abidli, A.; Huang, Y.; Ben Rejeb, Z.; Zaoui, A.; Park, C.B. Sustainable and Efficient Technologies for Removal and Recovery of Toxic and Valuable Metals from Wastewater: Recent Progress, Challenges, and Future Perspectives. *Chemosphere* **2022**, *292*, 133102. [CrossRef] [PubMed]
2. Capodaglio, A.G.; Callegari, A. Energy and Resources Recovery from Excess Sewage Sludge: A Holistic Analysis of Opportunities and Strategies. *Resour. Conserv. Recycl. Adv.* **2023**, *19*, 200184. [CrossRef]
3. Shao, Y.C.; Li, Z.H.; Long, Y.Y.; Zhao, J.; Huo, W.Z.; Luo, Z.R.; Lu, W.J. Direct Humification of Biowaste with Hydrothermal Technology: A Review. *Sci. Total Environ.* **2024**, *908*, 168232. [CrossRef] [PubMed]
4. Cai, W.; Wang, X.; Zhu, Z.; Kumar, R.; Nana Amaniampong, P.; Zhao, J.; Hu, Z.-T. Synergetic Effects in the Co-pyrolysis of Lignocellulosic Biomass and Plastic Waste for Renewable Fuels and Chemicals. *Fuel* **2023**, *353*, 129210. [CrossRef]

5. Xu, Q.; Luo, L.; Li, D.; Johnravindar, D.; Varjani, S.; Wong, J.W.C.; Zhao, J. Hydrochar Prepared from Digestate Improves Anaerobic Co-digestion of Food Waste and Sewage Sludge: Performance, Mechanisms, and Implication. *Bioresour. Technol.* **2022**, *362*, 127765. [CrossRef] [PubMed]
6. Ahmad, T.; Ahmad, K.; Alam, M. Sustainable Management of Water Treatment Sludge through 3'R' Concept. *J. Clean. Prod.* **2016**, *124*, 1–13. [CrossRef]
7. Zaker, A.; Chen, Z.; Wang, X.L.; Zhang, Q. Microwave-Assisted Pyrolysis of Sewage Sludge: A Review. *Fuel Process. Technol.* **2019**, *187*, 84–104. [CrossRef]
8. Heimersson, S.; Svanström, M.; Laera, G.; Peters, G. Life Cycle Inventory Practices for Major Nitrogen, Phosphorus and Carbon Flows in Wastewater and Sludge Management Systems. *Int. J. Life Cycle Assess.* **2016**, *21*, 1197–1212. [CrossRef]
9. Raheem, A.; He, J. Opportunities and Challenges in Sustainable Treatment and Resource Reuse of Sewage Sludge: A Review. *Chem. Eng. J.* **2018**, *337*, 616–641. [CrossRef]
10. He, H.J.; Chen, Y.J.; Li, X.; Cheng, Y.; Yang, C.P.; Zeng, G.M. Influence of Salinity on Microorganisms in Activated Sludge Processes: A Review. *Int. Biodeterior. Biodegrad.* **2017**, *119*, 520–527. [CrossRef]
11. Ouyang, J.; Li, C.Y.; Wei, L.; Wei, D.; Zhao, M.; Zhao, Z.; Zhang, J.; Chang, C.C. Activated Sludge and Other Aerobic Suspended Culture Processes. *Water Environ. Res.* **2020**, *92*, 1717–1725. [CrossRef]
12. Ding, Y.; Zhao, J.; Liu, J.-W.; Zhou, J.; Cheng, L.; Zhao, J.; Shao, Z.; Iris, Ç.; Pan, B.; Li, X.; et al. A Review of China's Municipal Solid Waste (MSW) and Comparison with International Regions: Management and Technologies in Treatment and Resource Utilization. *J. Clean. Prod.* **2021**, *293*, 126144. [CrossRef]
13. Remmas, N.; Manfe, N.; Zerva, I.; Melidis, P.; Raga, R.; Ntougias, S. A Critical Review on the Microbial Ecology of Landfill Leachate Treatment Systems. *Sustainability* **2023**, *15*, 949. [CrossRef]
14. Tan, X.; Wang, L. Investigation and Analysis on the Treatment and Disposal Methods of Typical Sewage Treatment Plant Sludge in China's Key River Basins. *China Water Wastewater* **2022**, *38*, 1–8.
15. Wang, T. Analysis on Urban Sludge Nutrition Element and Heavy Metal Content in China. *China Environ. Prot. Ind.* **2015**, *04*, 42–45.
16. Hong, E.; Yeneneh, A.M.; Sen, T.K.; Ang, H.M.; Kayaalp, A. A comprehensive review on rheological studies of sludge from various sections of municipal wastewater treatment plants for enhancement of process performance. *Adv. Colloid Interface Sci.* **2018**, *257*, 19–30. [CrossRef]
17. Thukkaram, S.; Kumar, A.A. Characteristics of Sewage Sludge and Its Potential Applications in the Construction Industry: A Review. *Int. J. Environ. Waste Manag.* **2021**, *28*, 17–40. [CrossRef]
18. Lynn, C.J.; Dhir, R.K.; Ghataora, G.S.; West, R.P. Sewage Sludge Ash Characteristics and Potential for Use in Concrete. *Constr. Build. Mater.* **2015**, *98*, 767–779. [CrossRef]
19. Bieszczad, A.; Popardowski, E.; Lubinska, W.; Gliniak, M.; Nawalany, G.; Sokolowski, P. Possibility of Using Waste Materials as Substitutes for Gravel or Water in Concrete Mix. *Materials* **2023**, *16*, 1810. [CrossRef]
20. Wang, S.; Gainey, L.; Mackinnon, I.D.R.; Allen, C.; Gu, Y.; Xi, Y. Thermal Behaviors of Clay Minerals as Key Components and Additives for Fired Brick Properties: A Review. *J. Build. Eng.* **2023**, *66*, 105802. [CrossRef]
21. Chang, Z.Y.; Long, G.C.; Zhou, J.L.; Ma, C. Valorization of Sewage Sludge in the Fabrication of Construction and Building Materials: A Review. *Resour. Conserv. Recycl.* **2020**, *154*, 14. [CrossRef]
22. Tay, J.H. Bricks Manufactured from Sludge. *J. Environ. Eng.-ASCE* **1987**, *113*, 278–284. [CrossRef]
23. Weng, C.H.; Lin, D.F.; Chiang, P.C. Utilization of Sludge as Brick Materials. *Adv. Environ. Res.* **2003**, *7*, 679–685. [CrossRef]
24. Okuno, N.; Takahashi, S. Full Scale Application of Manufacturing Bricks from Sewage. *Water Sci. Technol.* **1997**, *36*, 243–250. [CrossRef]
25. Xu, Z.; Zhang, M.; Li, J. Experimental Study on No-roasting Bricks Made from Sludge and Building Rubbish. *Non-Met. Mines* **2011**, *34*, 11–15.
26. Ramasamy, V.; Kannan, R.; Muralidharan, G.; Sidharthan, R.K.; Veerasamy, G.; Venkatesh, S.; Amirtharajan, R. A Comprehensive Review on Advanced Process Control of Cement Kiln Process with the Focus on MPC Tuning Strategies. *J. Process Control* **2023**, *121*, 85–102. [CrossRef]
27. Tang, J.; Ma, S.H.; Li, W.F.; Yang, H.; Shen, X.D. Research Progress on the Hydration of Portland Cement with Calcined Clay and Limestone. *Mater. Sci. Forum.* **2021**, *1036*, 240–246. [CrossRef]
28. Haustein, E.; Kurylowicz-Cudowska, A.; Luczkiewicz, A.; Fudala-Ksiazek, S.; Cieslik, B.M. Influence of Cement Replacement with Sewage Sludge Ash (SSA) on the Heat of Hydration of Cement Mortar. *Materials* **2022**, *15*, 1547. [CrossRef]
29. Jian, Z.B.; Li, J.S.; Xing, X.D.; Sun, P.C. Recycling Hazardous Textile Effluent Sludge in Cement-Based Construction Materials: Physicochemical Interactions between Sludge and Cement. *J. Hazard. Mater.* **2020**, *381*, 9. [CrossRef]
30. *GB/T 30760-2014*; Technical Specification for Coprocessing of Solid Waste in Cement Kiln. National Standard of the People's Republic of China: Beijing, China, 2014.
31. Gao, Q.; Zhang, Y.; Ba, M.; Zhang, D.; Cheng, W.; Zhou, S. Effect of Heavy Metal Sludge Ground Powder on Properties of Portland Cement-Based Materials. *Bull. Chin. Ceram. Soc.* **2022**, *41*, 450–460.
32. He, B.; Wang, G. Is Ceramsite the Last Straw for Sewage Sludge Disposal: A Review of Sewage Sludge Disposal by Producing Ceramsite in China. *Water Sci. Technol.* **2019**, *80*, 1–10. [CrossRef]

33. Wang, H.; Xu, J.; Liu, Y.; Sheng, L. Preparation of Ceramsite from Municipal Sludge and Its Application in Water Treatment: A Review. *J. Environ. Manag.* **2021**, *287*, 112374. [CrossRef]
34. Xu, G.; Zou, J.; Li, G. Ceramsite Made with Water and Wastewater Sludge and Its Characteristics Affected by SiO_2 and Al_2O_3. *Environ. Sci. Technol.* **2008**, *42*, 7417–7423. [CrossRef]
35. Zhang, X.; Li, J.; Feng, L.; Wang, J. Research Progress on Preparation Technology and Application of Municipal Sludge Ceramsite. *Inorg. Chem. Ind.* **2022**, *54*, 28–38. [CrossRef]
36. Jia, G.; Wang, Y.; Yang, F.; Ma, Z. Preparation of CFB Fly Ash/Sewage Sludge Ceramsite and the Morphological Transformation and Release Properties of Sulfur. *Constr. Build. Mater.* **2023**, *373*, 130864. [CrossRef]
37. Zou, J.L.; Xu, G.R.; Li, G.B. Ceramsite Obtained from Water and Wastewater Sludge and Its Characteristics Affected by Fe_2O_3, CaO, and MgO. *J. Hazard. Mater.* **2009**, *165*, 995–1001. [CrossRef]
38. Xu, G.R.; Zou, J.L.; Li, G.B. Ceramsite Obtained from Water and Wastewater Sludge and Its Characteristics Affected by $(Fe_2O_3+CaO+MgO)/(SiO_2+Al_2O_3)$. *Water Res.* **2009**, *43*, 2885–2893. [CrossRef]
39. Nie, J.; Wang, Q.; Gao, S.; Poon, C.S.; Zhou, Y.; Li, J. Novel Recycling of Incinerated Sewage Sludge Ash (ISSA) and Waste Bentonite as Ceramsite for Pb-Containing Wastewater Treatment: Performance and Mechanism. *J. Environ. Manag.* **2021**, *288*, 112382. [CrossRef]
40. Xu, G.R.; Zou, J.L.; Li, G.B. Effect of Sintering Temperature on the Characteristics of Sludge Ceramsite. *J. Hazard. Mater.* **2008**, *150*, 394–400. [CrossRef]
41. Liang, C.; Lin, S.; Liu, G. Study on the Preparation of Biochar Ceramsite Based on Sewage Sludge and the Characterization of Its Properties. *Appl. Sci.* **2021**, *11*, 5522. [CrossRef]
42. Wang, C.; Duan, D.; Huang, D.; Chen, Q.; Tu, M.; Wu, K.; Wang, D. Lightweight Ceramsite Made of Recycled Waste Coal Gangue & Municipal Sludge: Particular Heavy Metals, Physical Performance and Human Health. *J. Clean. Prod.* **2022**, *376*, 134309. [CrossRef]
43. Liu, F.; Liang, X.; He, S.; Li, F.; Jin, Y.; Zhao, Z.; Zhu, L. Performance of a New Low-Cost Zn/Fe-Layered Double Hydroxide-Modified Ceramsite for the Removal of P from Agricultural Runoff. *Ecol. Eng.* **2021**, *159*, 106117. [CrossRef]
44. Qian, W.; Huang, H.; Diao, Z.; Li, H.; Liu, H.; Ye, M.; Deng, Y.; Xu, Z. Advanced Treatment of Dye Wastewater Using a Novel Integrative Fenton-like/MnO_2-Filled Upward Flow Biological Filter Bed System Equipped with Modified Ceramsite. *Environ. Res.* **2021**, *194*, 110641. [CrossRef] [PubMed]
45. Shao, Q.; Zhang, Y.; Liu, Z.; Long, L.; Liu, Z.; Chen, Y.; Hu, X.-M.; Lu, M.; Huang, L.-Z. Phosphorus and Nitrogen Recovery from Wastewater by Ceramsite: Adsorption Mechanism, Plant Cultivation and Sustainability Analysis. *Sci. Total Environ.* **2022**, *805*, 150288. [CrossRef] [PubMed]
46. Chow, W.L.; Chong, S.; Lim, J.-W.; Chan, Y.; Chong, M.; Tiong, J.; Chin, J.; Pan, G.-T. Anaerobic Co-Digestion of Wastewater Sludge: A Review of Potential Co-Substrates and Operating Factors for Improved Methane Yield. *Processes* **2020**, *8*, 39. [CrossRef]
47. Wang, Z.Y.; Li, X.; Liu, H.; Zhou, T.; Qin, Z.H.; Mou, J.H.; Sun, J.; Huang, S.Y.; Chaves, A.V.; Gao, L.; et al. Bioproduction and Applications of Short-Chain Fatty Acids from Secondary Sludge Anaerobic Fermentation: A Critical Review. *Renew. Sust. Energ. Rev.* **2023**, *183*, 16. [CrossRef]
48. Wang, C.; Chen, M.; Zhao, P.; Zhou, L.; Hou, Y.; Zhang, J.; Lyu, Q.; Che, D. Investigation on Co-Combustion Characteristics and NOx Emissions of Coal and Municipal Sludge in a Tangentially Fired Boiler. *Fuel* **2023**, *340*, 127608. [CrossRef]
49. Roy, M.M.; Dutta, A.; Corscadden, K.; Havard, P.; Dickie, L. Review of Biosolids Management Options and Co-Incineration of a Biosolid-Derived Fuel. *Waste Manag.* **2011**, *31*, 2228–2235. [CrossRef]
50. Battista, F.; Strazzera, G.; Valentino, F.; Gottardo, M.; Villano, M.; Matos, M.; Silva, F.; Reis, M.A.M.; Mata-Alvarez, J.; Astals, S.; et al. New Insights in Food Waste, Sewage Sludge and Green Waste Anaerobic Fermentation for Short-Chain Volatile Fatty Acids Production: A Review. *J. Environ. Chem. Eng.* **2022**, *10*, 13. [CrossRef]
51. Wang, P.; Wang, H.; Qiu, Y.; Ren, L.; Jiang, B. Microbial Characteristics in Anaerobic Digestion Process of Food Waste for Methane Production–A Review. *Bioresour. Technol.* **2018**, *248*, 29–36. [CrossRef]
52. Khanh Nguyen, V.; Kumar Chaudhary, D.; Hari Dahal, R.; Hoang Trinh, N.; Kim, J.; Chang, S.W.; Hong, Y.; Duc La, D.; Nguyen, X.C.; Hao Ngo, H.; et al. Review on Pretreatment Techniques to Improve Anaerobic Digestion of Sewage Sludge. *Fuel* **2021**, *285*, 119105. [CrossRef]
53. Shao, Y.; Bao, M.; Huo, W.; Ye, R.; Ajmal, M.; Lu, W. From Biomass to Humic Acid: Is There an Accelerated Way to Go? *Chem. Eng. J.* **2023**, *452*, 139172. [CrossRef]
54. Shao, Y.; Bao, M.; Huo, W.; Ye, R.; Liu, Y.; Lu, W. Production of Artificial Humic Acid from Biomass Residues by a Non-Catalytic Hydrothermal Process. *J. Clean. Prod.* **2022**, *335*, 130302. [CrossRef]
55. Shao, Y.; Geng, Y.; Li, Z.; Long, Y.; Ajmal, M.; Lu, W.; Zhao, J. Unlocking the Potential of Humic Acid Production through Oxygen-Assisted Hydrothermal Humification of Hydrochar. *Chem. Eng. J.* **2023**, *472*, 145098. [CrossRef]
56. Shao, Y.; Huo, W.; Ye, R.; Liu, Y.; Ajmal, M.; Lu, W. Hydrothermal Humification of Lignocellulosic Components: Who Is Doing What? *Chem. Eng. J.* **2023**, *457*, 141180. [CrossRef]
57. Devos, P.; Haddad, M.; Carrère, H. Thermal Hydrolysis of Municipal Sludge: Finding the Temperature Sweet Spot: A Review. *Waste Biomass Valorization* **2021**, *12*, 2187–2205. [CrossRef]

58. Mohammad Mirsoleimani Azizi, S.; Dastyar, W.; Meshref, M.N.A.; Maal-Bared, R.; Ranjan Dhar, B. Low-Temperature Thermal Hydrolysis for Anaerobic Digestion Facility in Wastewater Treatment Plant with Primary Sludge Fermentation. *Chem. Eng. J.* **2021**, *426*, 130485. [CrossRef]
59. Kim, D.; Lee, K.; Park, K.Y. Enhancement of Biogas Production from Anaerobic Digestion of Waste Activated Sludge by Hydrothermal Pre-Treatment. *Int. Biodeterior. Biodegrad.* **2015**, *101*, 42–46. [CrossRef]
60. Chen, H.; Chang, S. Impact of Temperatures on Microbial Community Structures of Sewage Sludge Biological Hydrolysis. *Bioresour. Technol.* **2017**, *245*, 502–510. [CrossRef]
61. Cho, H.U.; Kim, Y.M.; Choi, Y.-N.; Kim, H.G.; Park, J.M. Influence of Temperature on Volatile Fatty Acid Production and Microbial Community Structure during Anaerobic Fermentation of Microalgae. *Bioresour. Technol.* **2015**, *191*, 475–480. [CrossRef]
62. Zheng, X.; Su, Y.; Li, X.; Xiao, N.; Wang, D.; Chen, Y. Pyrosequencing Reveals the Key Microorganisms Involved in Sludge Alkaline Fermentation for Efficient Short-Chain Fatty Acids Production. *Environ. Sci. Technol.* **2013**, *47*, 4262–4268. [CrossRef] [PubMed]
63. Zhang, C.J.; Pan, J.G.; Zhang, A.A.; Jiang, H.M.; Ji, X.; Cai, L. Optimize the Effects of C/N Ratio and pH on Sludge Anaerobic Fermentation Gas Production Utilizing CCD. *Hubei Agric. Sci.* **2016**, *55*, 1422–1424. [CrossRef]
64. Zhang, Y.; Bai, J.; Zuo, J. Production of Medium Chain Fatty Acids through Co-Fermentation of Food Waste and Sewage Sludge without External Electron Donors. *J. Environ. Chem. Eng.* **2022**, *10*, 108688. [CrossRef]
65. Yin, Y.; Hu, Y.; Wang, J. Co-Fermentation of Sewage Sludge and Lignocellulosic Biomass for Production of Medium-Chain Fatty Acids. *Bioresour. Technol.* **2022**, *361*, 127665. [CrossRef] [PubMed]
66. Yuan, H.; Zhu, N. Progress in Inhibition Mechanisms and Process Control of Intermediates and By-Products in Sewage Sludge Anaerobic Digestion. *Renew. Sustain. Energy Rev.* **2016**, *58*, 429–438. [CrossRef]
67. Hong, J.M.; Xu, C.Q.; Hong, J.L.; Tan, X.F.; Chen, W. Life Cycle Assessment of Sewage Sludge Co-Incineration in a Coal-Based Power Station. *Waste Manag.* **2013**, *33*, 1843–1852. [CrossRef] [PubMed]
68. Tan, P.; Ma, L.; Xia, J.; Fang, Q.; Zhang, C.; Chen, G. Co-Firing Sludge in a Pulverized Coal-Fired Utility Boiler: Combustion Characteristics and Economic Impacts. *Energy* **2017**, *119*, 392–399. [CrossRef]
69. Zhang, C.; Zhu, T.Y.; Yin, L.B.; Fang, Q.Y.; Zhan, Z.G.; Xu, Q.S.; Cheng, G. Field Test and Numerical Simulation for Co-combustion of Sludge in a 100 MW Coal Fired Boiler. *J. Combust. Sci. Technol.* **2015**, *21*, 114–123.
70. Qu, Z.; Wei, X.; Chen, W.; Wang, F.; Wang, Y.; Long, J. Co-Combustion Characteristics of Municipal Sewage Sludge and Coal in a Lab-Scale Fluidized Bed Furnace. *Energies* **2023**, *16*, 2374. [CrossRef]
71. Toraman, O.Y.; Topal, H.; Bayat, O.; Atimtay, A.T. Emission Characteristics of Co-Combustion of Sewage Sludge with Olive Cake and Lignite Coal in a Circulating Fluidized Bed. *J. Environ. Sci. Health Part A Tox. Hazard Subst. Environ. Eng.* **2004**, *39*, 973–986. [CrossRef]
72. Jiang, M.; Zhang, Z.; Sun, G.; Duan, Y.; Duan, L. Co-combustion characteristic of sewage sludge and coal in 0.3MWth circulating fluidized bed. *Clean Coal Technol.* **2022**, *28*, 130–138. [CrossRef]
73. Mathekga, H.I.; Oboirien, B.O.; North, B.C. A review of oxy-fuel combustion in fluidized bed reactors. *Int. J. Energy Res.* **2016**, *40*, 878–902. [CrossRef]
74. Mian, M.M.; Alam, N.; Ahommed, M.S.; He, Z.B.; Ni, Y.H. Emerging Applications of Sludge Biochar-Based Catalysts for Environmental Remediation and Energy Storage: A Review. *J. Clean. Prod.* **2022**, *360*, 22. [CrossRef]
75. Wang, L.P.; Chang, Y.Z.; Li, A.M. Hydrothermal Carbonization for Energy-Efficient Processing of Sewage Sludge: A Review. *Renew. Sust. Energ. Rev.* **2019**, *108*, 423–440. [CrossRef]
76. Urban, D.L.; Antal, M.J. Study of the Kinetics of Sewage Sludge Pyrolysis Using DSC and TGA. *Fuel* **1982**, *61*, 799–806. [CrossRef]
77. Beesley, L.; Moreno-Jiménez, E.; Gomez-Eyles, J.L.; Harris, E.; Robinson, B.; Sizmur, T. A Review of Biochars' Potential Role in. the Remediation, Revegetation and Restoration of Contaminated Soils. *Environ. Pollut.* **2011**, *159*, 3269–3282. [CrossRef] [PubMed]
78. Thengane, S.K.; Kung, K.S.; Gomez-Barea, A.; Ghoniem, A.F. Advances in Biomass Torrefaction: Parameters, Models, Reactors, Applications, Deployment, and Market. *Prog. Energy Combust. Sci.* **2022**, *93*, 101040. [CrossRef]
79. Peng, C.; Zhai, Y.; Zhu, Y.; Xu, B.; Wang, T.; Li, C.; Zeng, G. Production of Char from Sewage Sludge Employing Hydrothermal Carbonization: Char Properties, Combustion Behavior and Thermal Characteristics. *Fuel* **2016**, *176*, 110–118. [CrossRef]
80. He, C.; Giannis, A.; Wang, J.-Y. Conversion of Sewage Sludge to Clean Solid Fuel Using Hydrothermal Carbonization: Hydrochar Fuel Characteristics and Combustion Behavior. *Appl. Energy* **2013**, *111*, 257–266. [CrossRef]
81. Khan, N.; Mohan, S.; Dinesha, P. Regimes of Hydrochar Yield from Hydrothermal Degradation of Various Lignocellulosic Biomass: A Review. *J. Clean. Prod.* **2021**, *288*, 125629. [CrossRef]
82. Li, Y.; Liu, H.; Xiao, K.; Jin, M.; Xiao, H.; Yao, H. Combustion and Pyrolysis Characteristics of Hydrochar Prepared by Hydrothermal Carbonization of Typical Food Waste: Influence of Carbohydrates, Proteins, and Lipids. *Energy Fuels* **2020**, *34*, 430–439. [CrossRef]
83. Li, Y.; Liu, H.; Xiao, K.; Liu, X.; Hu, H.; Li, X.; Yao, H. Correlations between the Physicochemical Properties of Hydrochar and Specific Components of Waste Lettuce: Influence of Moisture, Carbohydrates, Proteins and Lipids. *Bioresour. Technol.* **2019**, *272*, 482–488. [CrossRef] [PubMed]
84. Qin, J.; Rui, Z.; Yang, R.; Fang, J.; Zhang, Y.; Li, X.; Gao, J. Sludge Char-to-Fuel Approaches Based on the Hydrothermal Fueling IV: Fermentation. *Water Sci. Technol.* **2021**, *84*, 880–891. [CrossRef]

85. Zhao, P.; Shen, Y.; Ge, S.; Yoshikawa, K. Energy Recycling from Sewage Sludge by Producing Solid Biofuel with Hydrothermal Carbonization. *Energy Convers. Manag.* **2014**, *78*, 815–821. [CrossRef]
86. Zhao, L.; Sun, Z.F.; Pan, X.W.; Tan, J.Y.; Yang, S.S.; Wu, J.T.; Chen, C.; Yuan, Y.; Ren, N.Q. Sewage Sludge Derived Biochar for Environmental Improvement: Advances, Challenges, and Solutions. *Water Res. X* **2023**, *18*, 12. [CrossRef] [PubMed]
87. Nguyen, L.H.; Van, H.T.; Chu, T.H.H.; Nguyen, T.H.V.; Nguyen, T.D.; Hoang, L.P.; Hoang, V.H. Paper Waste Sludge-Derived Hydrochar Modified by Iron (III) Chloride for Enhancement of Ammonium Adsorption: An Adsorption Mechanism Study. *Environ. Technol. Innov.* **2021**, *21*, 101223. [CrossRef]
88. Chu, Q.; Xue, L.; Singh, B.P.; Yu, S.; Müller, K.; Wang, H.; Feng, Y.; Pan, G.; Zheng, X.; Yang, L. Sewage Sludge-Derived Hydrochar That Inhibits Ammonia Volatilization, Improves Soil Nitrogen Retention and Rice Nitrogen Utilization. *Chemosphere* **2020**, *245*, 125558. [CrossRef] [PubMed]
89. Wang, F.; Zhang, M.; Liu, X.; Li, Z.; Zhu, H.; Lian, F.; Liu, X.; Li, L.; Wu, X.; Sun, H. Unraveling the Critical Role of Iron-Enriched Sludge Hydrochar in Mediating the Fenton-like Oxidation of Triclosan. *Environ. Pollut.* **2023**, *321*, 121205. [CrossRef] [PubMed]
90. Xu, S.; Wang, C.; Duan, Y.; Wong, J.W.-C. Impact of Pyrochar and Hydrochar Derived from Digestate on the Co-Digestion of Sewage Sludge and Swine Manure. *Bioresour. Technol.* **2020**, *314*, 123730. [CrossRef]
91. Tangredi, A.; Barca, C.; Ferrasse, J.-H.; Boutin, O. Effect of Process Parameters on Phosphorus Conversion Pathways During Hydrothermal Treatment of Sewage Sludge: A Review. *Chem. Eng. J.* **2023**, *463*, 142342. [CrossRef]
92. Hušek, M.; Moško, J.; Pohořelý, M. Sewage SludgeTreatment Methods and P-recovery Possibilities: Current State-of-the-art. *J. Environ. Manag.* **2022**, *315*, 115090. [CrossRef] [PubMed]
93. Belevi, H.; Moench, H. Factors Determining the Element Behavior in Municipal Solid Waste Incinerators. 1. Field Studies. *Environ. Sci. Technol.* **2000**, *34*, 2501–2506. [CrossRef]
94. Li, R.; Teng, W.; Li, Y.; Wang, W.; Cui, R.; Yang, T. Potential Recovery of Phosphorus during the Fluidized Bed Incineration of Sewage Sludge. *J. Clean. Prod.* **2017**, *140*, 964–970. [CrossRef]
95. Herzel, H.; Krüger, O.; Hermann, L.; Adam, C. Sewage Sludge Ash—A Promising Secondary Phosphorus Source for Fertilizer Production. *Sci. Total Environ.* **2016**, *542*, 1136–1143. [CrossRef] [PubMed]
96. Luyckx, L.; Van Caneghem, J. Recovery of Phosphorus from Sewage Sludge Ash: Influence of Incineration Temperature on Ash Mineralogy and Related Phosphorus and Heavy Metal Extraction. *J. Environ. Chem. Eng.* **2021**, *9*, 106471. [CrossRef]
97. Semerci, N.; Ahadi, S.; Coşgun, S. Comparison of Dried Sludge and Sludge Ash for Phosphorus Recovery with Acidic and Alkaline Leaching. *Water Environ. J.* **2021**, *35*, 359–370. [CrossRef]
98. Li, Y.; Cui, R.; Yang, T.; Zhai, Z.; Li, R. Distribution Characteristics of Heavy Metals in Different Size Fly Ash from a Sewage Sludge Circulating Fluidized Bed Incinerator. *Energy Fuels* **2017**, *31*, 2044–2051. [CrossRef]
99. Takahashi, M.; Kato, S.; Shima, H.; Sarai, E.; Ichioka, T.; Hatyakawa, S.; Miyajiri, H. Technology for Recovering Phosphorus from Incinerated Wastewater Treatment Sludge. *Chemosphere* **2001**, *44*, 23–29. [CrossRef] [PubMed]
100. Xu, Y.; Chen, J.; Yang, F.; Fang, Y.; Qian, G. Transformation of Phosphorus by $MgCl_2$ and $CaCl_2$ during Sewage Sludge Incineration. *Environ. Sci. Pollut. Res.* **2021**, *28*, 60268–60275. [CrossRef]
101. Peplinski, B.; Adam, C.; Michaelis, M.; Kley, G.; Emmerling, F.; Simon, F. Reaction Sequences in the Thermo-Chemical Treatment of Sewage Sludge Ashes Revealed by X-ray Powder Diffraction—A Contribution to the European Project SUSAN. *Z. Krist.* **2009**, *30*, 459–464. [CrossRef]

sustainability

MDPI

Article

Co-Pyrolysis of Mushroom Residue Blended with Pine Sawdust/Wheat Straw for Sustainable Utilization of Biomass Wastes: Thermal Characteristics, Kinetic/Thermodynamic Analysis, and Structure Evolution of Co-Pyrolytic Char

Haiyu Meng [1,*], Heng Yang [1], Zhiqiang Wu [2], Danting Li [1], Zhe Wang [1], Dongqi Wang [1], Hui Wang [1], Huaien Li [1] and Jiake Li [1,*]

[1] Department of Municipal and Environmental Engineering, School of Water Resources and Hydro-Electric Engineering, Xi'an University of Technology, Xi'an 710048, China; 17629019398@163.com (H.Y.); 19891178388@163.com (D.L.); wangzhewater@xaut.edu.cn (Z.W.); wangdq@xaut.edu.cn (D.W.); wanghui306@xaut.edu.cn (H.W.); lhuaien@mail.xaut.edu.cn (H.L.)

[2] Shaanxi Key Laboratory of Energy Chemical Process Intensification, School of Chemical Engineering and Technology, Xi'an Jiaotong University, Xi'an 710049, China; zhiqiang-wu@mail.xjtu.edu.cn

* Correspondence: hymeng321@163.com (H.M.); xaut_ljk@163.com (J.L.); Tel./Fax: +86-29-83239907 (H.M.)

Abstract: Co-pyrolysis technology is considered to be one of the most promising methods for the sustainable utilization of biomass wastes, as it can realize waste reduction and convert wastes into high-value-added products with little impact on the environment. The evaluation of thermal characteristics and product properties is necessary for understanding this technique. In this paper, thermal characteristics and kinetic and thermodynamic analysis during the co-pyrolysis of mushroom residue (MR) with pine sawdust (PS) or wheat straw (WS) were investigated in a TGA. The carbon structure and surface textures of co-pyrolytic char were explored using Raman spectroscopy and a scanning electron microscope. As the PS or WS mass ratio increased, the devolatilization index increased obviously, indicating that volatile release was promoted and concentrated. Weak interactions were observed between 250 and 400 °C during the co-pyrolysis process, which primarily affected the mass transfer, resulting in a change in the thermal decomposition temperatures and rates. The interactions had no prominent influence on the volatiles' yields. The non-additive performance of average activation energies for the blends was observed due to the interactions, and the lowest average activation energy was obtained when the PS or WS mass ratio was 50%. The lower average pre-exponential factor of the blends indicated the reduced complicacy of the pyrolysis reaction. The relatively small deviation between the activation energy and enthalpy change (4.94–5.18 kJ·mol^{-1}) signified the energy sensitivity of product formation. PS promoted the formation of small aromatic rings (<6 fused rings) in co-pyrolytic chars, whereas WS favored the production of larger rings ($\geq$6 fused rings). The surface textures of the co-pyrolytic chars became porous, and the greater fractal dimensions of the surface morphology for the co-pyrolytic chars indicated that the char surface became irregular and rough.

Keywords: sustainable utilization of biomass waste; co-pyrolysis; thermal characteristics; kinetic/thermodynamic analysis; char structure

1. Introduction

Sustainable energy development, environmental contamination, and greenhouse gas reduction render it necessary to develop renewable energy for the partial substitution of fossil energy [1,2]. In this situation, biomass, the only renewable carbon resource with "zero carbon" characteristics, has been extensively researched for producing clean fuels, high-value chemicals, carbon-based functional materials, and so on [3–6]. Edible fungi residue (EFR), the primary by-product of edible fungi cultivation, is the spent cultivation substrate after several mature edible fungi harvests. The production amount of wet EFR was approximately five times the quantity of edible fungi produced [7,8]. The cultivation substrates of edible fungi are predominantly composed of lignocellulose-based feedstocks such as straw, sawdust, bagasse, corncob, wood chips, etc. [9,10]. Thus, EFRs are regarded as secondary biomass resources because they have a certain amount of cellulose, hemicellulose, and lignin [11]. Nowadays, the Chinese edible fungi cultivation industry is the world's largest, and the total annual amount of EFR exceeds 13 million tons in China [12]. Presently, EFRs can be utilized as alternative fuels, animal feeds, fertilizers, cultivation substrates for new edible fungi, and bioremediation agents for the adsorption and degradation of contaminants [10,13]. However, due to various reasons, most of the EFRs are not appropriately treated (directly discarding, landfilling, open burning, or compositing), which causes the waste of resources and secondary environmental problems [14,15]. Therefore, the need to develop reasonable and efficient technology for the resource utilization of EFRs is urgent, which is meaningful for environmental protection and fossil fuel substitution.

Pyrolysis technology has been considered to be one of the most promising pathways to generate high-value-added products including combustible gas, bio-oil, and porous char from biomass [4,16]. Many efforts have been devoted to exploring the pyrolysis characteristics of different kinds of EFRs [17,18]. Researchers found that the volatile product yield of spent mushroom substrate (SMS) pyrolysis was larger than 60%, and the produced bio-oil had a promising potential of energy resource [19–21]. Pyrolytic char derived from waste mushroom substrate contained abundant nutrients that could be used to sustain and improve the soil nutritional status, but the surface area of pyrolytic char produced at 400–700 °C was very small (1.05–4.04 $m^2 \cdot g^{-1}$) [22]. The catalytic pyrolysis or microwave catalytic pyrolysis of EFR were also investigated, and the produced bio-oil had a high hydrocarbon content [11,23,24].

Recently, co-pyrolysis technology has received much attention because it can achieve the simultaneous disposal of multi-component feedstocks. Furthermore, the pyrolysis performance and product quality will probably be improved due to the interactions between different feedstocks [25,26]. Currently, co-pyrolysis technology is being extensively investigated with biomass and other carbon-based feedstocks such as coal, biomass, sludge, waste plastics, etc. [1,27]. Some investigations into the co-pyrolysis of EFRs with other feedstocks have been conducted. Li et al. found the existence of good interactions between mushroom residue and three kinds of waste plastics, and the activation energy of mushroom residue pyrolysis declined by 31.23–46.44% [14]. Wu et al. observed the occurrence of interactions promoting the generation of volatile products (by 9% at 50% EFR) for the co-pyrolysis of *Flammulina velutipes* residue with coal and obtained high-reactivity char containing more small aromatic ring structures (3–5 rings) [28]. Huang et al. found that there existed obvious interactions between SMS and textile dyeing sludge during the co-pyrolysis process, which promoted the volatile yield (obtaining the most significant effect at 30% SMS) and improved the aromatics content by 4.13% in the bio-oil [29]. Sewu et al. obtained a superior biochar with good adsorption performance by the co-pyrolysis of SMS blended with marine macroalgae (seaweed kelp), of which the adsorptive capacity was 2.2 times higher than that of SMS char [30]. Jiang et al. found that the co-pyrolysis of oil shale semi-coke and SMS can lead to an increase in bio-oil yield (obtaining the maximum yield of 8.59% at 50% SMS) and produced remarkable effects on the compositions of bio-oil [31]. The existing research indicates that the co-pyrolysis of EFR with some feedstocks (coal, waste plastics, sludge, macroalgae, and oil shale) indeed changes the pyrolysis performance and product quality, which results from the differences in the chemical components and structure of different feedstocks. As the main agricultural and forestry biomass, the co-pyrolysis of EFR with straw/wood is a convenient and widely adaptable method for the collaborative disposal of biomass waste, but no relevant research has been reported. There still exists a tremendous knowledge disparity in existing investigations about the thermal characteristics and interactions during the co-pyrolysis of EFR and wood or straw.

Furthermore, to gain a deeper understanding of this technique, the carbon structure and surface textures of co-pyrolytic char, two important factors affecting its application performance in the subsequent application scenarios, have also gained attention. It was reported that the surface textures of chars from EFR and coal mixtures became homogeneous [28], while the surface textures grew coarser and uneven during the co-pyrolysis process with marine macroalgae or textile dyeing sludge [29,30]. However, research on the carbon structure and surface textures of co-pyrolytic char from EFR and wood/straw blends has not been conducted. Compared to EFR char, the structure of co-pyrolytic char may be changed because of the different thermal characteristics of each component. Raman spectroscopy has been proved to be an appropriate method to investigate the carbon structure of pyrolytic chars from carbonaceous solid feedstocks (coal, biomass, etc.) [32,33]. Scanning electron microscopy (SEM) has been popularly applied to analyze the surface textures of pyrolytic chars. Further quantitative knowledge about surface textures can be obtained by the fractal analysis of micrographs, using the fractal dimension [34]. Consequently, obtaining knowledge about the carbon structure and surface textures of co-pyrolytic chars from EFR and wood/straw blends is necessary for the sake of deeply understanding the impact of possible interactions and evaluating the application performance of co-pyrolytic chars.

In this context, this research aims to explore the thermal characteristics, kinetic/thermodynamic analysis, and structure evolution of chars during the co-pyrolysis of EFR and wood/straw biomass. The influences of biomass type and blending ratio on the thermal characteristics were explored using a thermogravimetric analyzer, and the interactions between feedstocks were analyzed. The activation energies were computed using the iso-conversional method, and the thermodynamic parameters were calculated. The carbon microcrystalline structure of pyrolytic chars was measured applying Raman spectroscopy. The surface texture of residual chars was analyzed employing scanning electron microscopy and was quantificationally described by the fractal dimension. The results will provide useful knowledge about thermal characteristics and interactions during the co-pyrolysis of edible fungi residue and wood/straw biomass.

2. Experimental Section

2.1. Materials

The yield of mushrooms is the highest among edible fungi in China; thus, mushroom residue (MR) was chosen as the typical representative of EFR in this study. The MR was from an edible fungi (*Lentinus edodes*) cultivation base in Southern Shaanxi, China. As important forestry and agricultural biomass waste in China, pine sawdust (PS) and wheat straw (WS) were used to conduct the co-pyrolysis experiments. Air-dried samples were crushed by a crusher and then screened by a standard vibrating screen machine. Samples of less than 125 μm were selected to carry out the co-pyrolysis experiments. The blending ratios of MR were 30%, 50%, and 70%, which were marked as MRPS3-7, MRPS1-1, and MRPS7-3, respectively, for the MRPS blends. The MRWS blends were marked in the same way as the MRPS blends. The analysis results of the basic characteristics of the MR, PS, and WS are listed in Table 1. The proximate analysis of raw materials was carried out according to the GB/T 28731-2012 criterion. The MR showed a relatively high content of volatile matter, which was consistent with the spent cultivation substrates of other edible fungi [11,28]. Compared with previous research [26], the WS used in our study had a relatively high ash content, which probably affected the activation energies of the pyrolysis process.

Table 1. Proximate and ultimate analysis of MR and wood/straw biomass.

Material	MR	PS	WS
	Proximate analysis/wt %, ad		
Moisture	8.24	7.86	7.15
Ash	7.79	1.40	17.12
Volatile matter	72.16	77.24	61.81
Fixed carbon	11.81	13.50	13.92
	Ultimate analysis/wt %, daf		
Carbon	47.92	50.03	47.10
Hydrogen	5.05	5.37	5.20
Nitrogen	0.85	0.20	1.56
Sulfur	0.20	0.03	0.57
Oxygen [c]	45.98	44.37	45.57
H/C atomic ratio	1.26	1.29	1.32
$Q_{gr,d}/MJ\cdot kg^{-1}$	17.10	20.47	16.64

ad: air-dried; daf: dry ash-free; c: calculated by difference.

2.2. Pyrolysis Experiments

A thermogravimetric analyzer (STA2500, NETZSCH, Selb, Germany) was employed to explore the thermal characteristics of individual and blended samples. A sample of about 7 mg was used for each pyrolysis experiment. The purge gas and protective gas were high-purity nitrogen (99.999%), of which the flow rates were, respectively, 60 and 20 mL·min^{-1}. The purge gas mainly discharged volatile products from the instrument, and the protective gas played a role in protecting the balance. Thermogravimetric tests were executed from room temperature to 950 °C at heating rates of 10, 20, and 40 °C·min^{-1}. Each test was performed at least two times to assure the repeatability of the thermogravimetric analysis and the accuracy of the data. The pyrolytic chars produced at 10 °C·min^{-1} were collected and further analyzed.

To quantitatively describe the volatile releasing characteristics of the pyrolysis process, the devolatilization index (D_i) was calculated as follows [34,35]:

$$D_i = R_{max}/(T_{in}T_{max}\Delta T_{1/2}) \quad (1)$$

where R_{max} is the maximum weight loss rate, T_{in} refers to the initial devolatilization temperature, T_{max} represents the temperature of the maximum weight loss rate, and $\Delta T_{1/2}$ is the temperature interval when the instantaneous weight loss rate (R_d) is half of R_{max}. The T_{in} values were obtained from the thermogravimetric (TG) and differential thermogravimetric (DTG) profiles based on the TG-DTG graphing method [36]. The greater the D_i value is, the more easily and violently the volatile is discharged.

The interactions between the MR and PS/WS during the co-pyrolysis process were evaluated using the deviation (ΔW) between the experimental and calculated results. The computing formula of ΔW is as follows [37]:

$$\Delta W = W_{\text{Experimental}} - W_{\text{Calculated}} \quad (2)$$

where $W_{\text{Experimental}}$ is the experimental weight loss or weight loss rate of the blended sample. $W_{\text{Calculated}}$ is the calculated result based on the weighted-average sum of experimental values from independent pyrolysis. The $W_{\text{Calculated}}$ is calculated as follows:

$$W_{\text{Calculated}} = x_M W_M + x_B W_B \quad (3)$$

where x_M and x_B are the mass ratios of each component in the mixtures, and W_M and W_B are the weight losses or weight loss rates of individual feedstocks obtained under the same experimental conditions as those of MRPS or MRWS blends.

Moreover, the root mean square (RMS) of ΔW was used to evaluate the intensity of interactions. The larger the RMS is, the stronger the interactions are. The value of RMS

should be zero if there were no interactions. The *RMS* is calculated based on the following formula [38]:

$$RMS = \sqrt{\frac{\sum_{i=1}^{n}\left(W_{\text{Experimental}} - W_{\text{Calculated}}\right)^2}{n}} \quad (4)$$

where n is the number of ΔW data points.

2.3. Kinetic Method

The activation energy can be conveniently computed using the iso-conversional method since this method does not need to presume the reaction mechanism [39]. The activation energies at different conversions are acquired on the basis of thermogravimetric experimental data for at least three heating rates. The Kissinger-Akahira-Sunose (KAS) method was applied for the computation of activation energy in this paper. The formula is as follows [27,40]:

$$\ln\frac{\beta}{T_\alpha^2} = -\frac{E_\alpha}{RT_\alpha} + \ln\frac{AR}{E_\alpha G(\alpha)} \quad (5)$$

where β is the heating rate ($°C\cdot min^{-1}$), α is the conversion of tested samples, T_α is the temperature corresponding to the conversion of α (K), R is the universal gas constant (8.314 $J\cdot mol^{-1}\cdot K^{-1}$), A is the pre-exponential factor (s^{-1}), E_α is the activation energy at each conversion ($kJ\cdot mol^{-1}$), and G(α) is the integration function of the reaction mechanism which is only determined by the conversion. The activation energy can be acquired from the slope of fitting a straight line between $\ln(\beta/T_\alpha{}^2)$ and $1/T_\alpha$.

The pre-exponential factor (A_α) at each conversion is computed based on the activation energies using the formula below [39]:

$$A_\alpha = \frac{\beta E_\alpha \exp(E_\alpha/RT_{\text{max}})}{RT_{\text{max}}^2} \quad (6)$$

where T_{max} refers to the temperature of the maximum weight loss rate (K).

2.4. Thermodynamic Parameter Calculation

The assessment of thermodynamic parameters can help to understand the pyrolysis process from this perspective of energy conversion. The thermodynamic parameters, enthalpy change (ΔH), Gibbs free energy (ΔG), and entropy change (ΔS), are computed following the formula below [41]:

$$\Delta H = E_\alpha - RT \quad (7)$$

$$\Delta G = E_\alpha + RT_{\text{max}} \ln(K_B T_{\text{max}}/hA) \quad (8)$$

$$\Delta S = (\Delta H - \Delta G)/T_{\text{max}} \quad (9)$$

where E_α is calculated by the KAS method, and K_B and h are the Boltzmann constant (1.381×10^{-23} $J\cdot K^{-1}$) and Plank's constant (6.626×10^{-34} $J\cdot S^{-1}$), respectively.

2.5. Char Characterization

The Raman spectrum of pyrolytic chars was acquired employing a Raman spectrometer (InVia Qontor, Renishaw, Gloucestershire, UK). The spectrum in the first-order region (800–1800 cm^{-1}) was curve-fitted with 10 Gaussian bands, according to the reported method [35,42]. A curve-fitting example of a Raman spectrum for co-pyrolytic char is presented in Figure 1. In this study, major bands (G_R, V_L, V_R, and D bands) were analyzed to understand the carbon microcrystalline structures of pyrolytic chars. The surface texture was measured using scanning electron microscopy (JEOL JSM-6700F) and quantitatively explored by the fractal analysis.

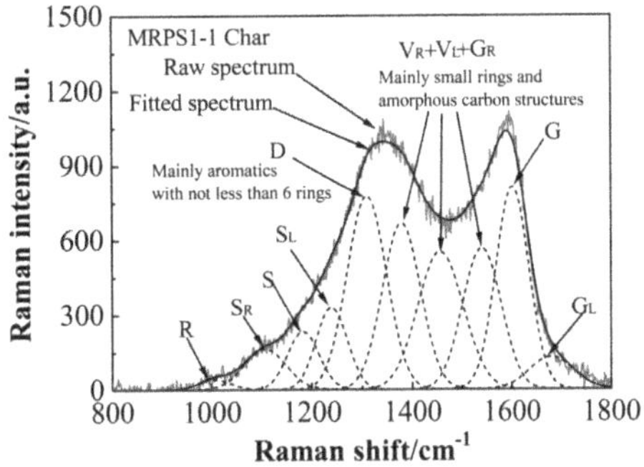

Figure 1. Curve fitting of the Raman spectrum for MRPS1-1 char.

3. Results and Discussion

3.1. Thermal Characteristics

3.1.1. Individual Feedstocks

The thermogravimetric (TG) and differential thermogravimetric (DTG) curves of the MR, PS, and WS at three heating rates are illustrated in Figure 2. The thermal decomposition of the MR mainly consisted of three steps. The first stage was from 50 °C to 210 °C due to the removal of free water and bound water from the MR [14]. The moisture removal process of the MR was held for a longer time duration, which was attributed to the existence of abundant capillary channels in MR particles. The water vapor generated surface tension in these channels when migrating from the inside of particles to the surface of particles, hindering the evaporation of moisture and extending the dehydration stage of the MR [19]. The second step from 210 °C to 540 °C was the main decomposition stage. The weight loss rate increased rapidly with the increase in temperature, and the maximum decomposition rate occurred in this stage. The most remarkable mass loss was primarily owing to the intense decomposition of cellulose, hemicellulose, and part of the lignin; meanwhile, a large amount of gaseous products was released, including CO_2, CO, and other hydrocarbon gas [21,28]. The DTG profile of the MR showed a shoulder in the weight loss peak of this stage. This phenomenon indicated that the thermal decomposition of MR components did not take place independently, and the decomposition of hemicellulose and cellulose partially overlapped. Previous research has also reported the same conclusions [19,20]. The MR exhibited a broad and blunt mass loss peak due to the longer duration of the lignin decomposition process. The third step was the continued decomposition of lignin and the carbonization stage from 540 °C to 950 °C accompanied by a slight mass reduction [40]. As the volatile substances were released continuously, the char was formed through polycondensation reactions. The DTG pattern of the MR illustrated two additional peaks in the temperature interval of 470–525 °C and 650–725 °C, which were probably caused by the dehydration of calcium hydroxide and the decomposition of calcium carbonate [18,20]. The observed thermal characteristics of the MR in this study were basically consistent with previous investigations on the pyrolysis characteristics of other spent mushroom substrates [14,19,21].

The thermal behavior of the PS and WS was different from that of the MR due to their different components. The pyrolysis process of the PS was composed of two stages, except the dewatering phase below 150 °C. In the first stage from 220 °C to 470 °C, the volatiles in the PS were primarily discharged because of the decomposition of hemicellulose, cellulose, and part of the lignin. The lignin decomposition and carbonization process mainly took place in the second stage above 500 °C. Hemicellulose and cellulose can basically complete the pyrolysis process in the temperature range of about 230–400 °C due to the high reactivity, while the decomposition of lignin started at a relatively high temperature and continued to the end of the pyrolysis process because of its cross-link structures [43]. As seen in Figure 2e–f, the evolution of the thermal behavior curves for the WS was different from that of the PS. Although the decomposition process of the WS also consisted of two weight loss stages, the initial devolatilization temperature of the WS was lower than that of the PS, and the release of volatiles was more concentrated. The thermal characteristics of the PS and WS were consistent with previous reports [44,45]. With the increase in the heating rate, the mass loss rate increased significantly for these three samples, and the thermal behavior profiles shifted to a higher temperature region due to the increased delay in the heat and mass transfer at a higher heating rate [14]. Furthermore, it can be obviously found that the major pyrolysis stage of the MR partially overlapped with the main volatile releasing stage of the PS and WS, which was due to their similar lignocellulose structures.

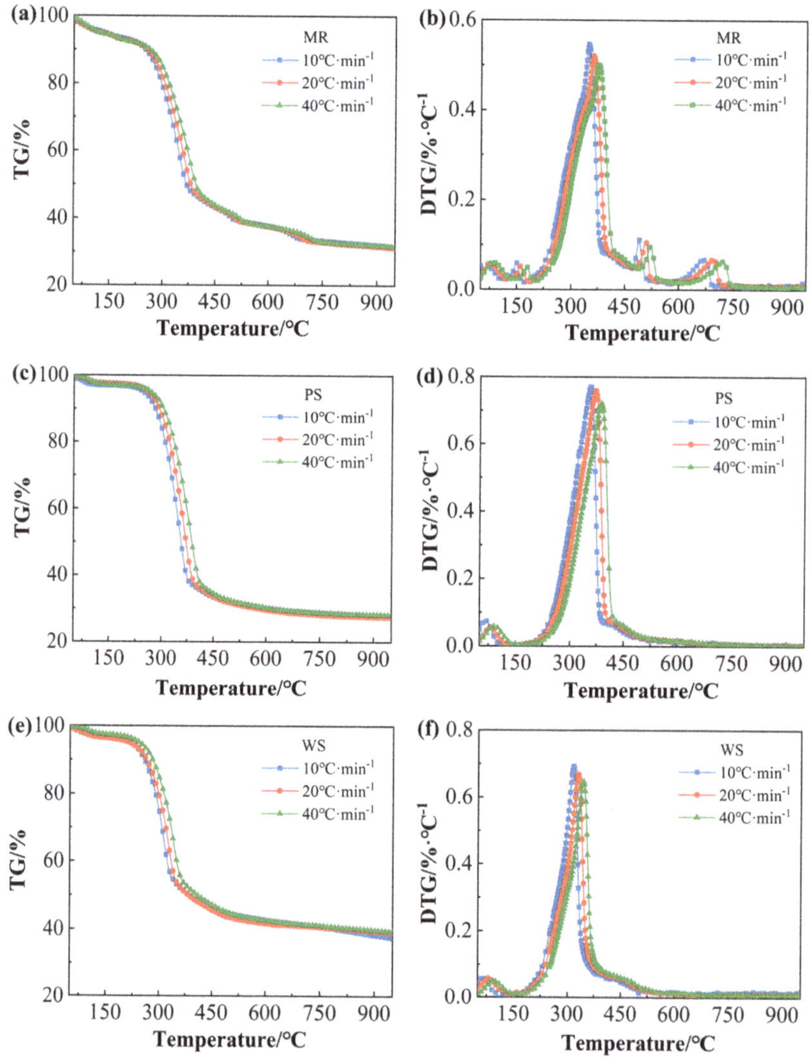

Figure 2. TG and DTG curves of MR, PS, and WS at three heating rates: (**a**) MR TG, (**b**) MR DTG, (**c**) PS TG, (**d**) PS DTG, (**e**) WS TG, and (**f**) WS DTG.

Several parameters for thermal characteristics are presented in Table 2. The initial devolatilization temperature of the MR was greater than that of the WS and lower than that of the PS. Compared to the PS and WS, the MR showed a lower maximum mass loss rate and larger temperature intervals of volatile release, which suggested that the MR had more temperature-resistant substances. The D_i values of the MR, PS, and WS were 10.85, 17.65, and 26.77, respectively. The characteristic parameters demonstrated that the thermal decomposition of the MR was relatively difficult. The char yields of the MR, PS, and WS were, respectively, 31.20%, 27.37%, and 39.04%. The higher char yield of the WS compared to the other two samples was attributed to its higher ash content.

Table 2. Thermal behavior parameters of the samples at β = 20 °C·min^{-1}.

Parameters	MR	30% PS	50% PS	70% PS	PS	30% WS	50% WS	70% WS	WS
T_{in}/°C	292	299	304	307	307	281	281	282	281
T_1/°C	361	363	368	369	370	338	334	332	329
R_1/%·min^{-1}	10.41	11.93	12.82	13.69	15.24	10.19	10.69	12.21	13.61
T_2/°C	508	507	505	504		502	500	475	
R_2/%·min^{-1}	2.19	1.53	1.21	1.01		1.60	1.35	1.17	
T_3/°C	689	683	668	662		678	672	657	
R_3/%·min^{-1}	1.35	0.88	0.83	0.59		0.89	0.84	0.59	
T_{max}/°C	361	363	368	369	370	338	334	332	329
R_{max}/%·min^{-1}	10.41	11.93	12.82	13.69	15.24	10.19	10.69	12.21	13.61
$\Delta T_{1/2}$/°C	91	84	81	79	76	95	88	62	55
D_i/10^{-7}%·min^{-1}·°C^{-3}	10.85	13.09	14.15	15.30	17.65	11.29	12.94	21.03	26.77
Char yield/%	31.20	29.41	28.81	28.02	27.37	33.12	34.31	36.60	39.04

T_{in}: initial devolatilization temperature; T_1: temperature of the first DTG peak; R_1: weight loss rate of the first DTG peak; T_2: temperature of the second DTG peak; R_2: weight loss rate of the second DTG peak; T_3: temperature of the third DTG peak; R_3: weight loss rate of the third DTG peak; T_{max}: temperature of the maximum weight loss rate; R_{max}: maximum weight loss rate; $\Delta T_{1/2}$: temperature interval when R_d/R_{max} = 1/2; D_i: devolatilization index.

3.1.2. Blended Feedstocks

The thermal characteristic patterns of the MRPS mixtures and MRWS mixtures at β of 10, 20, and 40 °C·min^{-1} are presented in Figure 3. It can be found that the pyrolysis characteristics of the mixtures presented the thermal behaviors of individual samples. The co-pyrolysis process of all the mixtures primarily consisted of three mass loss phases, which were the dehydration process, volatile releasing stage, and carbonization stage. These three stages were the result of the simultaneous thermal decomposition of the MR and PS/WS. Most of the volatiles were released in the second steps, and the weight loss rate was influenced by the blending ratio. As the MR blending ratio declined, the weight loss rate of the second phase increased, attributed to the higher pyrolysis rate of the PS/WS. However, the DTG peaks between 650 and 725 °C caused by the decomposition of calcium carbonate in the MR were reduced significantly and almost disappeared with a 30% MR mass ratio. Moreover, the weight loss rate of the MRPS blends and MRWS blends increased with the increment in heating rates.

As shown in Table 2, the char yields of the MRPS blends declined with the PS addition because the PS had a higher volatile content and a lower ash content. As the PS blending ratio increased from 30% to 70%, the initial devolatilization temperatures of the MRPS mixtures increased by 7–15 °C. Meanwhile, the largest weight loss rate caused by volatile release increased from 11.93%·min^{-1} to 13.69%·min^{-1}, and the peak temperature increased from 363 °C to 369 °C. However, the dehydration rate of calcium hydroxide and the decomposition rate of calcium carbonate mainly caused by the MR were reduced with the increase in the PS mixing ratio, and the corresponding temperature declined synchronously. Co-pyrolysis with PS increased the initial devolatilization temperature and the peak temperature of volatile release but reduced the peak temperature of calcium hydroxide dehydration and carbonate decomposition. The D_i values of the MRPS blends increased gradually as the PS mass ratio increased, and the duration of volatile release was reduced. The D_i values were, respectively, 13.09, 14.15, and 15.30 for MRPS7-3, MRPS1-1, and MRPS3-7 and basically linear with the PS mass ratio. Furthermore, the D_i values of the MRPS mixtures were greater than that of the MR, demonstrating that the releasing performance of volatile substances may be promoted by co-pyrolysis with PS.

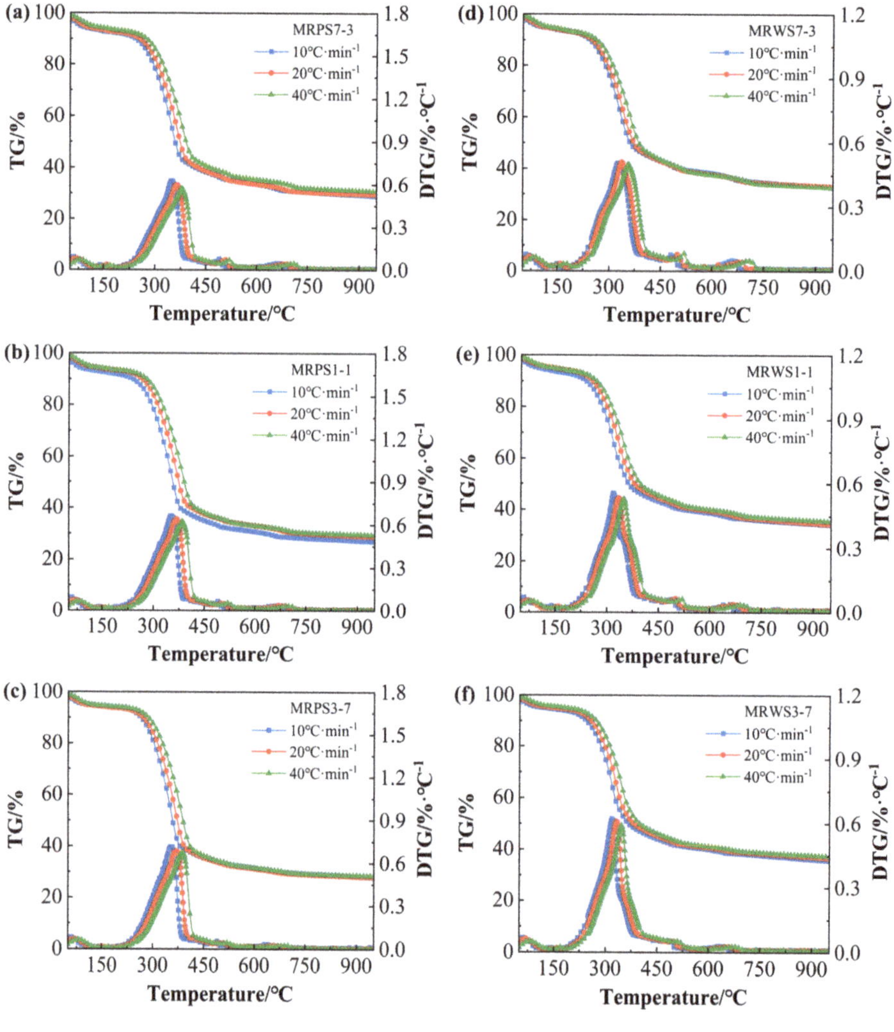

Figure 3. TG and DTG curves of MRPS and MRWS blends at three heating rates: (**a**) MRPS7-3, (**b**) MRPS1-1, (**c**) MRPS3-7, (**d**) MRWS7-3, (**e**) MRWS1-1, and (**f**) MRWS3-7.

The parameters for the thermal behavior of the MRWS blends showed a different variation trend from the MRPS blends. The char yields of the MRWS blends increased from 33.12% to 36.60% as the WS mass ratio was increased from 30% to 70%, which was because of the lower volatile content and higher ash content of the WS. With the increment in the WS blending ratio, the largest weight loss rate of volatile release increased slightly, whereas the initial devolatilization temperature, the peak temperature of the volatile releasing step, and the peak temperature of calcium hydroxide dehydration and carbonate decomposition declined obviously. The D_i of the MRWS blends also demonstrated higher values than that of the MR, increasing gradually from 11.29 to 21.03 as the WS mass ratio was increased from 30% to 70%. The addition of WS influenced the D_i values nonlinearly, and the D_i value of MRWS3-7 increased significantly due to the remarkable increment in the maximum weight loss rate and the obvious reduction in the volatile releasing duration.

In summary, co-pyrolysis with PS or WS significantly affected the thermal characteristics of the MR and boosted the volatile releasing performance of the MR. The observed changes in the thermal behavior parameters for the blends indicated that there probably existed some interactions during the co-pyrolytic process of MRPS blends and MRWS blends.

3.2. Interactions between MR and PS/WS

Figure 4 presents the comparative analysis of experimental and calculated results for the MRPS blends. It can be observed that the experimental weight losses at the final pyrolysis temperature showed no obvious difference from the calculated results, indicating that co-pyrolysis did not affect the pyrolysis reaction degree of the MRPS blends. To deeply analyze the influence of co-pyrolysis on the thermal characteristics, the deviations of the experimental and calculated results are presented in Figure 4d,e. The analysis results suggested that weak interactions occurred between the MR and PS during the co-pyrolysis process. The ΔW of weight loss below 210 °C was larger than zero, suggesting that the dehydration process was accelerated. The PS blending reduced the number of capillary channels in the samples; thus, the diffusion resistance of the water vapor was reduced, and the dehydration rate increased. The positive values of ΔW for DTG curves below 210 °C also proved this influence. As the temperature increased to the range of 270–400 °C, the ΔW of weight loss presented a variation trend of first declining and then improving, which was related to the change in the volatile releasing rate. In the meantime, the ΔW of DTG first showed smaller negative values and then became larger positive values, indicating that the experimental rate of volatile release first declined slightly and then increased obviously compared with the calculated rate. As the pyrolysis temperature continuing to increase, the ΔW values of weight loss kept a slight decreasing tendency until the pyrolysis reaction was completed, and the ΔW values of DTG were close to zero except the step of calcium hydroxide dehydration and calcium carbonate decomposition. In these two steps, the ΔW values of weight loss first increased slightly and then decreased, indicating that these two steps were advanced, which was consistent with the variation tendency of the thermal characteristic parameters. At the final pyrolysis temperature, the ΔW of weight loss for the MRPS blends was less than 1%. The *RMS* for the ΔW values of weight loss was, respectively, 1.493, 1.040, and 1.371 for MRPS7-3, MRPS1-1, and MRPS3-7. These values were further proof of the occurrence of weak interactions between the MR and PS during co-pyrolysis. The interactions mainly affected the performance of volatile release in the main weight loss step and the decomposition characteristics of inorganic minerals in the subsequent phase, resulting in the variation in characteristic temperatures and peak values of DTG curves, as described above. However, the interactions had no significant influence on the total volatile yields for the co-pyrolysis process of MRPS blends.

Figure 5 shows the comparative analysis of experimental and calculated results for the MRWS blends. Weak interactions also took place between the MR and WS during co-pyrolysis, while the impact of interactions on the volatile release was different from MRPS blends. When the temperature was in the range of 250–400 °C, the deviation curves of both the weight loss and DTG values showed a complicated variation tendency. Compared with the MR, the initial devolatilization temperature of the MRWS blends as well as the peak temperature of volatile release decreased rapidly, respectively, from 292 °C to 281–282 °C and from 361 °C to 332–338 °C. These temperatures were very close to those of the WS, suggesting that the volatile release of the MR was prominently increased with the addition of WS. Moreover, the interactions produced a more obvious influence on weight loss and DTG values for MRWS7-3. Similar to the MRPS blends, the calcium hydroxide dehydration and calcium carbonate decomposition steps were also advanced. When the pyrolysis reaction was completed, the ΔW values of weight loss for all the MRWS blends were also less than 1%, indicating that the volatile yields of the co-pyrolytic process were not significantly promoted. The *RMS* for the ΔW values of weight loss was, respectively, 1.066, 0.485, and 0.471 for MRWS7-3, MRWS1-1, and MRWS3-7, demonstrating the relatively more remarkable interactions at a 30% WS blending ratio.

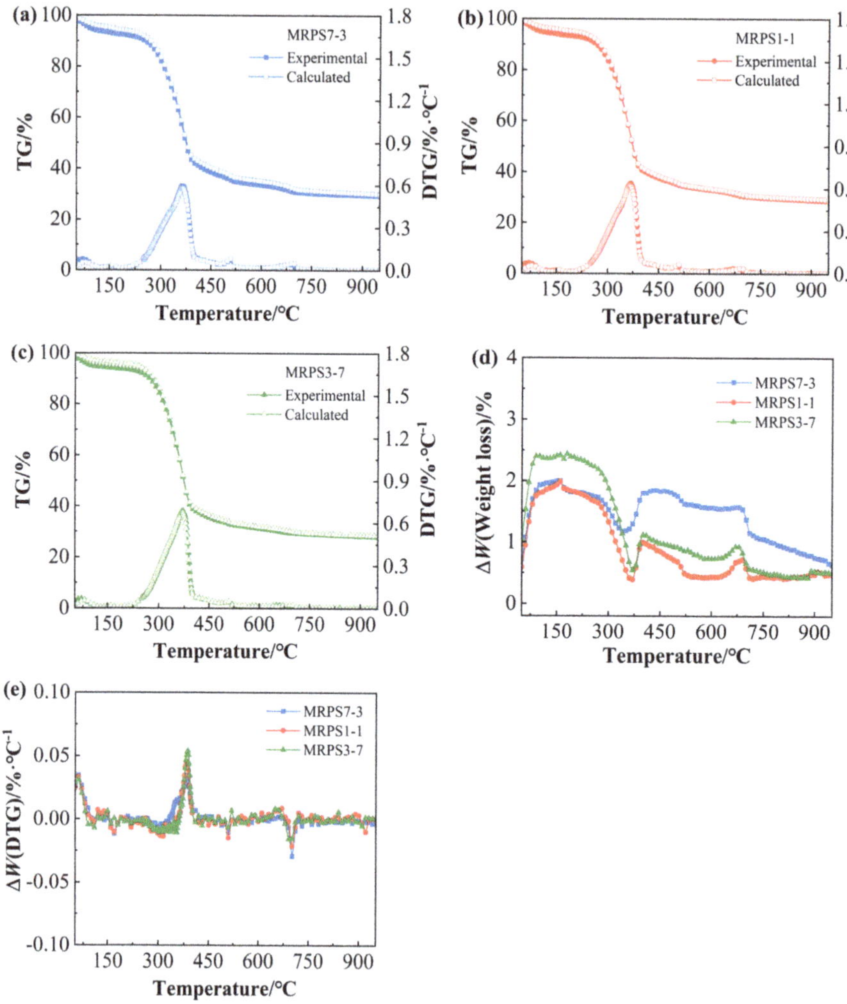

Figure 4. Comparative analysis of the experimental and calculated results for MRPS blends at $\beta = 20\ ^{\circ}\mathrm{C{\cdot}min^{-1}}$: (**a**) MRPS7-3, (**b**) MRPS1-1, (**c**) MRPS3-7, (**d**) ΔW of weight loss, and (**e**) ΔW of DTG.

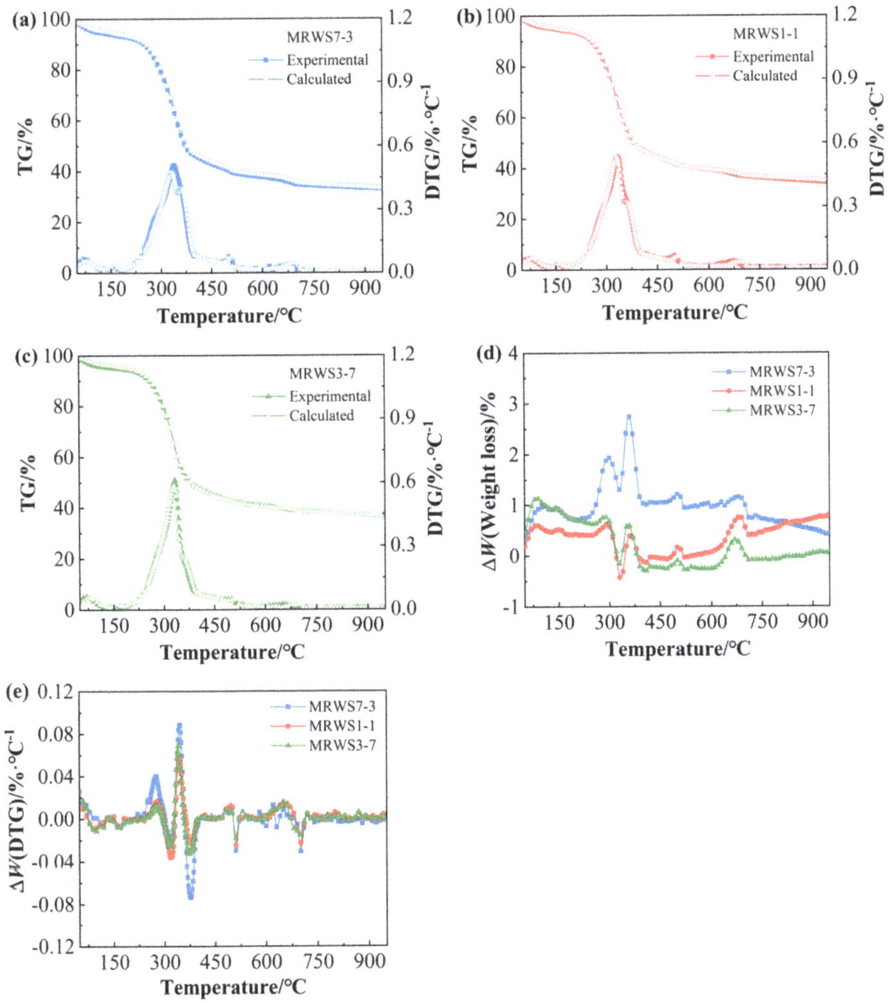

Figure 5. Comparative analysis of the experimental and calculated results for MRWS blends at $\beta = 20\ ^{\circ}C\cdot min^{-1}$: (**a**) MRWS7-3, (**b**) MRWS1-1, (**c**) MRWS3-7, (**d**) ΔW of weight loss, and (**e**) ΔW of DTG.

The different influence of interactions on the thermal characteristics of the MRPS blends and MRWS blends could be due to the differences in pyrolysis characteristics between PS and WS. For these three samples, the order of the initial devolatilization temperature for individual pyrolysis is PS > MR > WS. For the co-pyrolysis of MRWS blends, the WS pyrolysis started first at the lower temperature. Compared to MR pyrolysis, intense volatile release caused by WS decomposition would create a large number of pores in the samples, which can promote the diffusion of volatiles produced by MR pyrolysis. Thus, the volatile release of the MRWS blends was advanced, resulting in the obvious diminution of the initial devolatilization temperatures and the peak temperatures of volatile release. In contrast, the MR pyrolysis began first for the co-pyrolysis of MRPS blends. Compared with PS pyrolysis, the volatile releasing rate of the MR was lower, and the diffusion of volatile matter to the outside of particles was relatively slow. Consequently, the macromolecular volatile substances produced by MR pyrolysis would adhere to the surface of PS particles which restrained the further release and weakened the diffusion of volatiles produced by PS pyrolysis [29]. Moreover, as the pyrolysis temperature increased, the pyrolysis rate of the PS increased significantly, which accelerated the volatile release. At the same time, copious pores were formed in the samples. For both MRPS blends and MRWS blends, the decomposition and release of volatiles in the main weight loss stage produced a porous char structure, which advanced the diffusion of products generated by the calcium hydroxide dehydration and calcium carbonate decomposition. Many studies have found that the co-pyrolysis of coal with hydrogen-rich feedstocks (biomass, waste plastics, etc.) can accelerate the volatile yields attributed to the hydrogen transfer effect [35,46]. In the present study, due to the similar H/C atomic ratio of these three samples (shown in Table 1), the interactions promoting the volatile yields caused by the hydrogen transfer effect were not observed. Due to the lower initial devolatilization temperature and high volatile releasing rate, the WS pyrolysis produced many pore channels in the samples, promoting the release of volatiles at lower temperatures. Thus, compared with the MR, the initial devolatilization temperatures decreased by about 10 °C, and the temperatures of the maximum weight loss rate were reduced by 23–29 °C for the MRWS blends. On the contrary, in the initial stage of MRPS pyrolysis, the adhesion of MR pyrolysis substances on the PS particles would block part of the pore channels in the samples, delaying the volatile release of the blends. Furthermore, the volatile release was accelerated due to the PS pyrolysis at higher temperatures. Therefore, compared with the MR, the initial devolatilization temperatures increased by 7–15 °C, and the peak temperatures of volatile release increased by 2–8 °C for MRWS blends. In summary, the weak interactions between the MR and PS or WS were mainly caused by the formation or blocking of pore channels, primarily affecting the mass transfer of the co-pyrolysis process. Under the influence of these interactions, the thermal decomposition temperatures and rates of the blends changed, but the yields of volatile products showed no significant change.

3.3. Kinetic Analysis

The activation energies of the MR, PS, WS, and blends were computed using the KAS method, and the calculation results are listed in Tables 3 and 4. Most of the correlation coefficients (R^2) showed values higher than 0.9600, indicating that the fitting results were satisfactory and the calculation results of activation energies were reliable. The experimental data in the conversion scope of below 0.2 and above 0.8 were neglected because of the poor linearity for all the tested samples [35]. Figure 6 also presents the fitting straight lines between $\ln(\beta/{T_\alpha}^2)$ and $1/T_\alpha$.

Table 3. Kinetic parameters of MR, PS, and their blends calculated by KAS method.

Samples	α	E_α/kJ·mol^{-1}	A_α/s^{-1}			R^2
			10 °C·min^{-1}	20 °C·min^{-1}	40 °C·min^{-1}	
MR	0.2	190.72	1.02×10^{14}	9.75×10^{13}	8.07×10^{13}	0.9999
	0.3	186.76	4.64×10^{13}	4.51×10^{13}	3.79×10^{13}	0.9999
	0.4	180.38	1.30×10^{13}	1.30×10^{13}	1.12×10^{13}	0.9999
	0.5	174.20	3.81×10^{12}	3.88×10^{12}	3.45×10^{12}	0.9999
	0.6	174.96	4.44×10^{12}	4.50×10^{12}	3.99×10^{12}	0.9998
	0.7	169.19	1.41×10^{12}	1.46×10^{12}	1.33×10^{12}	0.9999
	0.8	379.73	1.50×10^{30}	7.20×10^{29}	2.60×10^{29}	0.9049
	Avg.	**207.99**	$\mathbf{2.14 \times 10^{29}}$	$\mathbf{1.03 \times 10^{29}}$	$\mathbf{3.72 \times 10^{28}}$	
MRPS7-3	0.2	164.54	5.27×10^{11}	5.29×10^{11}	4.69×10^{11}	0.9942
	0.3	162.99	3.87×10^{11}	3.91×10^{11}	3.49×10^{11}	0.9953
	0.4	159.41	1.90×10^{11}	1.94×10^{11}	1.76×10^{11}	0.9998
	0.5	160.91	2.56×10^{11}	2.60×10^{11}	2.35×10^{11}	0.9988
	0.6	161.22	2.72×10^{11}	2.77×10^{11}	2.49×10^{11}	0.9999
	0.7	155.41	8.55×10^{10}	8.89×10^{10}	8.22×10^{10}	0.9974
	0.8	196.15	2.81×10^{14}	2.48×10^{14}	1.90×10^{14}	0.9638
	Avg.	**165.80**	$\mathbf{4.03 \times 10^{13}}$	$\mathbf{3.57 \times 10^{13}}$	$\mathbf{2.74 \times 10^{13}}$	

Table 3. *Cont.*

Samples	α	E_α/kJ·mol^{-1}	A_α/s^{-1}			R^2
			10 °C·min^{-1}	20 °C·min^{-1}	40 °C·min^{-1}	
MRPS1-1	0.2	144.87	9.93×10^{9}	8.98×10^{9}	8.44×10^{9}	0.9960
	0.3	151.37	3.63×10^{10}	3.18×10^{10}	2.89×10^{10}	0.9999
	0.4	147.68	1.74×10^{10}	1.55×10^{10}	1.44×10^{10}	0.9987
	0.5	155.50	8.27×10^{10}	7.08×10^{10}	6.32×10^{10}	0.9999
	0.6	149.69	2.60×10^{10}	2.29×10^{10}	2.11×10^{10}	0.9988
	0.7	155.77	8.72×10^{10}	7.46×10^{10}	6.66×10^{10}	0.9999
	0.8	157.25	1.17×10^{11}	9.94×10^{10}	8.81×10^{10}	0.9998
	Avg.	**151.73**	$\mathbf{5.38 \times 10^{10}}$	$\mathbf{4.63 \times 10^{10}}$	$\mathbf{4.15 \times 10^{10}}$	
MRPS3-7	0.2	178.80	5.99×10^{12}	6.09×10^{12}	4.63×10^{12}	0.9877
	0.3	167.00	5.88×10^{11}	6.24×10^{11}	5.03×10^{11}	0.9953
	0.4	161.58	2.02×10^{11}	2.19×10^{11}	1.81×10^{11}	0.9962
	0.5	156.09	6.85×10^{10}	7.56×10^{10}	6.45×10^{10}	0.9969
	0.6	156.22	7.03×10^{10}	7.75×10^{10}	6.61×10^{10}	0.9990
	0.7	155.77	6.43×10^{10}	7.11×10^{10}	6.07×10^{10}	0.9999
	0.8	161.93	2.17×10^{11}	2.34×10^{11}	1.94×10^{11}	0.9974
	Avg.	**162.49**	$\mathbf{1.03 \times 10^{12}}$	$\mathbf{1.06 \times 10^{12}}$	$\mathbf{8.14 \times 10^{11}}$	
PS	0.2	152.90	3.47×10^{10}	3.88×10^{10}	3.38×10^{10}	0.9998
	0.3	149.51	1.78×10^{10}	2.02×10^{10}	1.78×10^{10}	0.9998
	0.4	150.21	2.04×10^{10}	2.31×10^{10}	2.03×10^{10}	0.9987
	0.5	150.80	2.30×10^{10}	2.59×10^{10}	2.27×10^{10}	0.9998
	0.6	150.34	2.10×10^{10}	2.37×10^{10}	2.08×10^{10}	0.9999
	0.7	149.48	1.77×10^{10}	2.00×10^{10}	1.77×10^{10}	0.9998
	0.8	144.43	6.53×10^{9}	7.53×10^{9}	6.83×10^{9}	0.9998
	Avg.	**149.67**	$\mathbf{2.02 \times 10^{10}}$	$\mathbf{2.27 \times 10^{10}}$	$\mathbf{2.00 \times 10^{10}}$	

Table 4. Kinetic parameters of MR, WS, and their blends calculated by KAS method.

Samples	α	E_α/kJ·mol^{-1}	A_α/s^{-1}			R^2
			10 °C·min^{-1}	20 °C·min^{-1}	40 °C·min^{-1}	
MR	0.2	190.72	1.02×10^{14}	9.75×10^{13}	8.07×10^{13}	0.9999
	0.3	186.76	4.64×10^{13}	4.51×10^{13}	3.79×10^{13}	0.9999
	0.4	180.38	1.30×10^{13}	1.30×10^{13}	1.12×10^{13}	0.9999
	0.5	174.20	3.81×10^{12}	3.88×10^{12}	3.45×10^{12}	0.9999
	0.6	174.96	4.44×10^{12}	4.50×10^{12}	3.99×10^{12}	0.9998
	0.7	169.19	1.41×10^{12}	1.46×10^{12}	1.33×10^{12}	0.9999
	0.8	379.73	1.50×10^{30}	7.20×10^{29}	2.60×10^{29}	0.9049
	Avg.	**207.99**	$\mathbf{2.14 \times 10^{29}}$	$\mathbf{1.03 \times 10^{29}}$	$\mathbf{3.72 \times 10^{28}}$	
MRWS7-3	0.2	183.13	9.46×10^{13}	8.83×10^{13}	5.61×10^{13}	0.9845
	0.3	176.55	2.43×10^{13}	2.33×10^{13}	1.54×10^{13}	0.9709
	0.4	171.56	8.68×10^{12}	8.49×10^{12}	5.77×10^{12}	0.9761
	0.5	165.35	2.41×10^{12}	2.41×10^{12}	1.70×10^{12}	0.9802
	0.6	159.63	7.37×10^{11}	7.55×10^{11}	5.51×10^{11}	0.9834
	0.7	168.18	4.32×10^{12}	4.28×10^{12}	2.97×10^{12}	0.9833
	0.8	234.74	3.83×10^{18}	2.92×10^{18}	1.36×10^{18}	0.9195
	Avg.	**179.88**	$\mathbf{5.47 \times 10^{17}}$	$\mathbf{4.17 \times 10^{17}}$	$\mathbf{1.95 \times 10^{17}}$	
MRWS1-1	0.2	149.19	1.24×10^{11}	1.11×10^{11}	1.04×10^{11}	0.9955
	0.3	155.00	4.18×10^{11}	3.65×10^{11}	3.32×10^{11}	0.9985
	0.4	151.20	1.88×10^{11}	1.68×10^{11}	1.55×10^{11}	0.9986
	0.5	152.10	2.27×10^{11}	2.02×10^{11}	1.86×10^{11}	0.9998
	0.6	152.75	2.61×10^{11}	2.30×10^{11}	2.12×10^{11}	0.9987
	0.7	161.29	1.56×10^{12}	1.32×10^{12}	1.16×10^{12}	0.9935
	0.8	176.48	3.73×10^{13}	2.93×10^{13}	2.40×10^{13}	0.9213
	Avg.	**156.86**	$\mathbf{5.73 \times 10^{12}}$	$\mathbf{4.53 \times 10^{12}}$	$\mathbf{3.74 \times 10^{12}}$	

Table 4. *Cont.*

Samples	α	E_α/kJ·mol^{-1}	A_α/s^{-1}			R^2
			10 °C·min^{-1}	20 °C·min^{-1}	40 °C·min^{-1}	
MRWS3-7	0.2	157.07	6.44×10^{11}	6.22×10^{11}	5.87×10^{11}	0.9997
	0.3	161.54	1.64×10^{12}	1.56×10^{12}	1.44×10^{12}	0.9980
	0.4	155.97	5.12×10^{11}	4.96×10^{11}	4.70×10^{11}	0.9985
	0.5	156.17	5.34×10^{11}	5.17×10^{11}	4.90×10^{11}	0.9999
	0.6	162.13	1.86×10^{12}	1.75×10^{12}	1.62×10^{12}	0.9998
	0.7	186.60	3.08×10^{14}	2.62×10^{14}	2.16×10^{14}	0.9953
	0.8	287.37	3.68×10^{23}	2.01×10^{23}	1.06×10^{23}	0.9904
	Avg.	**180.98**	$\mathbf{5.26 \times 10^{22}}$	$\mathbf{2.87 \times 10^{22}}$	$\mathbf{1.51 \times 10^{22}}$	
WS	0.2	152.31	2.81×10^{11}	2.75×10^{11}	2.63×10^{11}	0.9638
	0.3	159.17	1.19×10^{12}	1.13×10^{12}	1.05×10^{12}	0.9901
	0.4	159.20	1.20×10^{12}	1.14×10^{12}	1.06×10^{12}	0.9853
	0.5	151.77	2.51×10^{11}	2.46×10^{11}	2.36×10^{11}	0.9878
	0.6	157.79	8.91×10^{11}	8.51×10^{11}	7.94×10^{11}	0.9962
	0.7	161.00	1.75×10^{12}	1.65×10^{12}	1.52×10^{12}	0.9748
	0.8	202.17	9.83×10^{15}	7.72×10^{15}	5.89×10^{15}	0.9920
	Avg.	**163.35**	$\mathbf{1.41 \times 10^{15}}$	$\mathbf{1.10 \times 10^{15}}$	$\mathbf{8.42 \times 10^{14}}$	

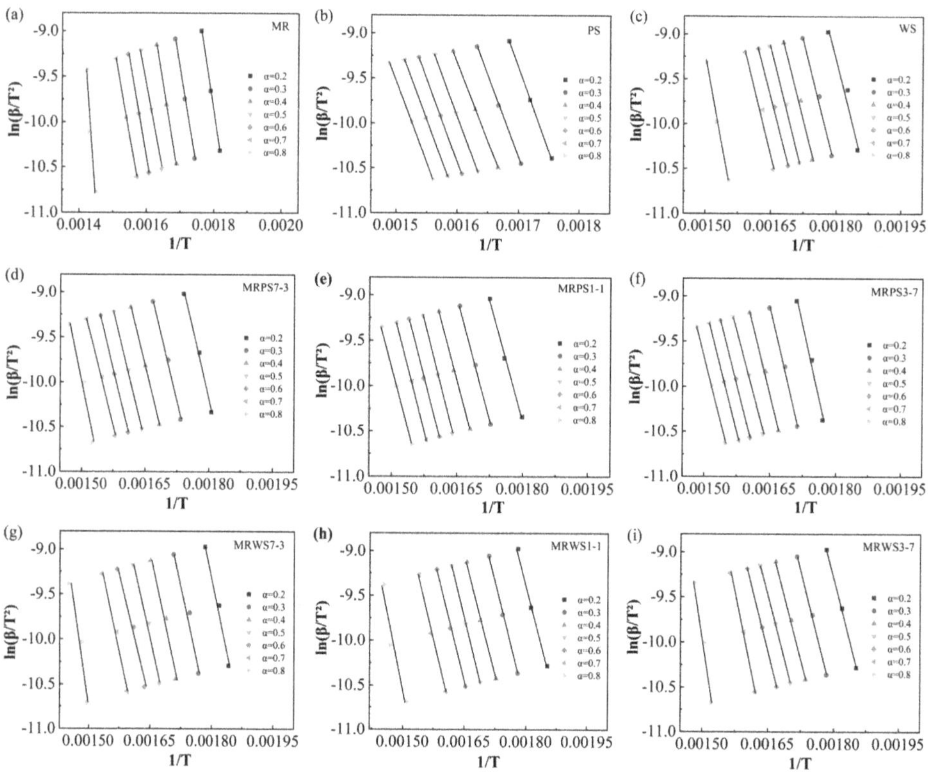

Figure 6. Linear relationship in the plot of $\ln(\beta/T_\alpha{}^2)$ versus $1/T_\alpha$: (**a**) MR, (**b**) PS, (**c**) WS, (**d**) MRPS7-3, (**e**) MRPS1-1, (**f**) MRPS3-7, (**g**) MRWS7-3, (**h**) MRWS1-1, and (**i**) MRWS3-7.

The activation energies versus the conversion for the MRPS blends and MRWS blends are illustrated in Figure 7. The activation energies changed with the conversion, indicating that the pyrolysis of individual and blended samples underwent multi-step kinetics in the various reaction stages [27]. With the increment in conversion, the activation energies of the MR first declined and then increased. However, the variation patterns of the PS and WS were different. The activation energies of the PS were reduced slightly as the conversion was increased, while the activation energies of the WS showed a continuous and slight increasing trend. The increasing rates of activation energies were most conspicuous at $\alpha = 0.7$ for the MR and WS. The obvious changes in activation energies probably corresponded to the char formation phase in the presence of a high content of minerals, which had a negative impact on the diffusion of heat and the volatile release [47]. The average activation energy of the MR, PS and WS was 207.99, 149.67, and 163.35 kJ·mol^{-1}, respectively. The similar change trend of activation energies with the conversion for MR has been reported, and the obtained average activation energy was almost the same as in previous research [17]. Jerzak et al. investigated the activation energies of oat straw pyrolysis using three iso-conversional methods (Friedman, Flynn-Wall-Ozawa, and Kissinger-Akahira-Sunose) [48]. It was found that the activation energies of oat straw increased slightly with the conversion and showed an obvious upward trend at a high conversion ($\alpha = 0.9$). The change trend of activation energies for WS pyrolysis in our study was basically consistent with oat straw pyrolysis.

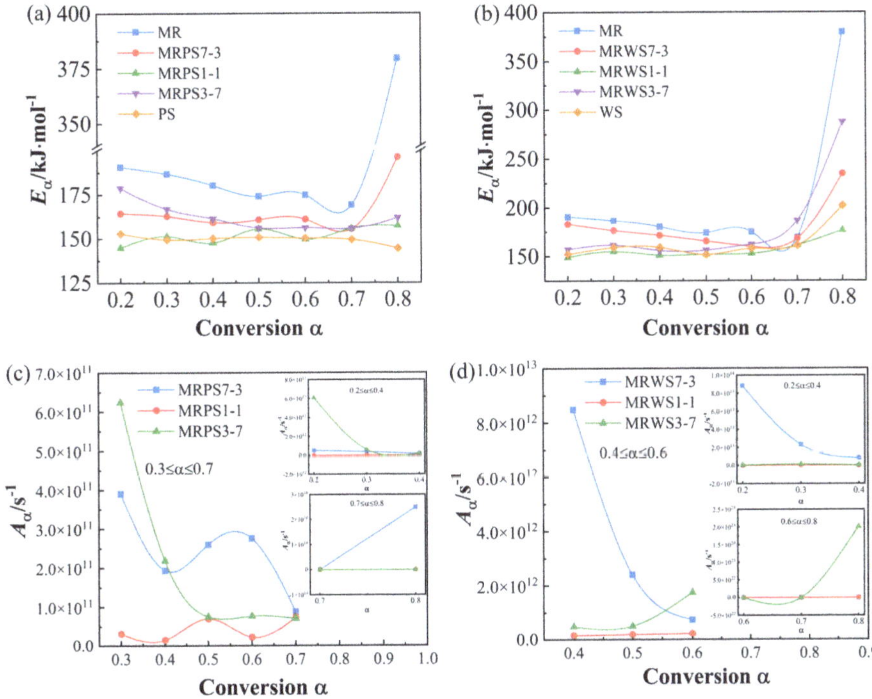

Figure 7. Variation in the kinetic parameters versus conversion for co-pyrolysis process: (**a**,**c**) E_α and A_α of MRPS blends and (**b**,**d**) E_α and A_α of MRWS blends.

The activation energies of the MRPS blends were less than those of the MR at different conversions and presented a change trend of first reducing and then rising with the increase in conversion. The average activation energy of MRPS7-3, MRPS1-1, and MRPS3-7 was, respectively, 165.80, 151.73, and 162.49 kJ·mol^{-1}, indicating that PS blending lowered the average activation energy. The activation energies of MRWS7-3 first declined and then increased with the increment in conversion. However, the activation energies of MRWS1-1 and MRWS3-7 increased continuously with the increase in the conversion. The average activation energy of MRWS blends was also reduced compared to MR, respectively, 179.88, 156.86, and 180.98 kJ·mol^{-1} for MRWS7-3, MRWS1-1, and MRWS3-7. The average activation energies for both MRPS blends and MRWS blends first declined and then increased with the increasing PS or WS blending ratio. The lowest average activation energy was found for MRPS1-1 and MRWS1-1. Furthermore, the predicted average activation energy of the blends was calculated by the weighted average based on the average activation energy of individual feedstocks and the blending ratio. It can be found that the average activation energies of MRPS blends and MRWS blends were lower than the predicted average activation energies. The non-additive performance on the average activation energies of MRPS blends and MRWS blends confirmed the existence of interactions between individual feedstocks during the co-pyrolysis process. A similar conclusion was obtained by previous research, which reported that the interactions between waste plastics and mushroom residue reduced the activation energies of the co-pyrolysis process [14].

The pre-exponential factor (A_α) was calculated based on Kissinger's method (Equation (6)), and the results are also presented in Tables 3 and 4. For the pyrolysis process, the pre-exponential factor was an indicator reflecting the complicacy of a sample's surface structure or pyrolysis reaction [44,49]. Consequently, a larger value of the pre-exponential factor meant that the complexity of the reaction increased. It can be observed that the A_α values varied from 1.33×10^{12} to 1.50×10^{30} s^{-1} for the MR, 6.53×10^{9} to 3.88×10^{10} s^{-1} for the PS, and 2.36×10^{11} to 9.83×10^{15} s^{-1} for the WS. The widely distributed values especially for the MR indicated the complicated composition of the mushroom residue and the complex reactions during the pyrolysis process. As seen from Figure 7c,d, the A_α values ranged from 8.22×10^{10} to 2.81×10^{14} s^{-1} for MRPS7-3, 8.44×10^{9} to 1.17×10^{11} s^{-1} for MRPS1-1, and 6.07×10^{10} to 6.09×10^{12} s^{-1} for MRPS3-7. The values of A_α varied from 5.51×10^{11} to 3.83×10^{18} s^{-1} for MRWS7-3, $1.04 \times 10^{11} s^{-1}$ to 3.73×10^{13} s^{-1} for MRWS1-1, and 4.70×10^{11} to 3.68×10^{23} s^{-1} for MRWS3-7. These results indicated that both individual and blended samples experienced sophisticated multiphasic degradation reactions at different conversions, and a range of parallel reactions constituted the overall pyrolysis process. For most samples, the highest value of A_α was observed at $\alpha = 0.8$, indicating that the reactions were most complicated in the final stage of the pyrolysis process. The average A_α value ranged from 3.72×10^{28} to 2.14×10^{29} s^{-1} for the MR, 2.00×10^{10} to 2.27×10^{10} s^{-1} for the PS, and 8.42×10^{14} to 1.41×10^{15} s^{-1} for the WS throughout the conversion scope of 0.2–0.8. The larger A_α values for the MR suggested the higher complexity of the pyrolysis reaction. In comparison with the MR, the average A_α values of both the MRPS blends and MRWS blends decreased significantly, signifying that the complicacy of the pyrolysis reaction was reduced due to the addition of PS or WS.

3.4. Thermodynamic Analysis

The thermodynamic parameters for the pyrolysis of the MR, PS, WS, and blends were computed according to Equations (7)–(9), respectively. The calculation results for the MRPS blends and MRWS blends are, respectively, demonstrated in Tables 5 and 6. The enthalpy change (ΔH) at different heating rates was very close, which suggested an ignorable influence of the heating rate on enthalpy change. However, the Gibbs free energy (ΔG) illustrated a slight increase with the increment in the heating rate.

Table 5. Thermodynamic parameters of MR, PS, and their blends at three heating rates.

Samples	α	10 °C·min^{-1}			20 °C·min^{-1}			40 °C·min^{-1}		
		ΔH/ kJ·mol^{-1}	ΔG/ kJ·mol^{-1}	ΔS/ J·mol^{-1}·K^{-1}	ΔH/ kJ·mol^{-1}	ΔG/ kJ·mol^{-1}	ΔS/ J·mol^{-1}·K^{-1}	ΔH/ kJ·mol^{-1}	ΔG/ kJ·mol^{-1}	ΔS/ J·mol^{-1}·K^{-1}
MR	0.2	186.15	180.06	9.78	186.07	180.18	9.29	186.00	181.08	7.57
	0.3	181.99	180.17	2.92	181.90	180.29	2.54	181.82	181.19	0.97
	0.4	175.46	180.35	−7.86	175.37	180.48	−8.06	175.28	181.38	−9.41
	0.5	169.15	180.53	−18.30	169.05	180.66	−18.32	168.95	181.57	−19.45
	0.6	169.79	180.51	−17.23	169.69	180.64	−17.26	169.58	181.55	−18.43
	0.7	163.90	180.68	−26.97	163.79	180.81	−26.84	163.68	181.73	−27.81
	0.8	373.99	176.50	317.44	373.91	176.55	311.22	373.89	177.36	302.73
	Avg.	**202.92**	**179.83**	-	**202.83**	**179.94**	-	**202.74**	**180.84**	-

Table 5. *Cont.*

Samples	α	10 °C·min^{-1} ΔH/ kJ·mol^{-1}	ΔG/ kJ·mol^{-1}	ΔS/ J·mol^{-1}·K^{-1}	20 °C·min^{-1} ΔH/ kJ·mol^{-1}	ΔG/ kJ·mol^{-1}	ΔS/ J·mol^{-1}·K^{-1}	40 °C·min^{-1} ΔH/ kJ·mol^{-1}	ΔG/ kJ·mol^{-1}	ΔS/ J·mol^{-1}·K^{-1}
MRPS 7-3	0.2	159.93	181.14	−34.03	159.86	181.58	−34.15	159.76	182.79	−35.32
	0.3	158.19	181.19	−36.90	158.11	181.63	−36.98	158.00	182.85	−38.10
	0.4	154.47	181.30	−43.06	154.37	181.75	−43.04	154.26	182.97	−44.01
	0.5	155.85	181.26	−40.78	155.75	181.70	−40.80	155.63	182.92	−41.84
	0.6	156.06	181.25	−40.42	155.95	181.69	−40.46	155.83	182.90	−41.51
	0.7	150.15	181.44	−50.21	150.04	181.88	−50.06	149.91	183.10	−50.90
	0.8	190.70	180.23	16.79	190.63	180.65	15.68	190.50	181.84	13.27
	Avg.	**160.76**	**181.12**	-	**160.67**	**181.55**	-	**160.56**	**182.77**	-
MRPS 1-1	0.2	140.25	182.12	−67.08	140.14	183.81	−68.12	140.05	185.32	−68.79
	0.3	146.56	181.89	−56.61	146.46	183.58	−57.90	146.35	185.08	−58.85
	0.4	142.73	182.02	−62.94	142.62	183.71	−64.09	142.51	185.22	−64.89
	0.5	150.43	181.75	−50.18	150.32	183.43	−51.65	150.20	184.94	−52.77
	0.6	144.53	181.95	−59.95	144.40	183.64	−61.20	144.29	185.14	−62.08
	0.7	150.51	181.74	−50.04	150.39	183.43	−51.52	150.27	184.93	−52.66
	0.8	151.87	181.69	−47.78	151.75	183.38	−49.33	151.62	184.87	−50.52
	Avg.	**146.70**	**181.88**	-	**146.58**	**183.57**	-	**146.47**	**185.07**	-
MRPS 3-7	0.2	174.10	182.92	−13.99	174.04	183.00	−13.96	173.94	184.78	−16.43
	0.3	162.14	183.27	−33.53	162.06	183.37	−33.18	161.95	185.16	−35.15
	0.4	156.60	183.45	−42.60	156.51	183.54	−42.10	156.39	185.34	−43.85
	0.5	151.00	183.63	−51.77	150.90	183.73	−51.12	150.78	185.53	−52.64
	0.6	151.04	183.62	−51.71	150.93	183.72	−51.06	150.81	185.52	−52.59
	0.7	150.51	183.64	−52.58	150.39	183.74	−51.93	150.27	185.54	−53.43
	0.8	156.56	183.43	−42.65	156.45	183.53	−42.17	156.32	185.32	−43.94
	Avg.	**157.42**	**183.42**	-	**157.33**	**183.52**	-	**157.21**	**185.31**	-
PS	0.2	148.16	184.05	−56.87	148.06	184.15	−56.11	147.96	185.95	−57.45
	0.3	144.63	184.17	−62.65	144.52	184.27	−61.80	144.41	186.07	−63.01
	0.4	145.22	184.15	−61.67	145.11	184.24	−60.85	145.00	186.04	−62.08
	0.5	145.71	184.12	−60.86	145.60	184.22	−60.06	145.48	186.02	−61.32
	0.6	145.17	184.14	−61.75	145.05	184.24	−60.93	144.93	186.04	−62.18
	0.7	144.23	184.17	−63.28	144.11	184.27	−62.44	143.98	186.07	−63.66
	0.8	139.10	184.35	−71.70	138.97	184.45	−70.72	138.83	186.26	−71.73
	Avg.	**144.60**	**184.16**	-	**144.49**	**184.26**	-	**144.37**	**186.06**	-

Table 6. Thermodynamic parameters of MR, WS, and their blends at three heating rates.

Samples	α	10 °C·min^{-1} ΔH/ kJ·mol^{-1}	ΔG/ kJ·mol^{-1}	ΔS/ J·mol^{-1}·K^{-1}	20 °C·min^{-1} ΔH/ kJ·mol^{-1}	ΔG/ kJ·mol^{-1}	ΔS/ J·mol^{-1}·K^{-1}	40 °C·min^{-1} ΔH/ kJ·mol^{-1}	ΔG/ kJ·mol^{-1}	ΔS/ J·mol^{-1}·K^{-1}
MR	0.2	186.15	180.06	9.78	186.07	180.18	9.29	186.00	181.08	7.57
	0.3	181.99	180.17	2.92	181.90	180.29	2.54	181.82	181.19	0.97
	0.4	175.46	180.35	−7.86	175.37	180.48	−8.06	175.28	181.38	−9.41
	0.5	169.15	180.53	−18.30	169.05	180.66	−18.32	168.95	181.57	−19.45
	0.6	169.79	180.51	−17.23	169.69	180.64	−17.26	169.58	181.55	−18.43
	0.7	163.90	180.68	−26.97	163.79	180.81	−26.84	163.68	181.73	−27.81
	0.8	373.99	176.50	317.44	373.91	176.55	311.22	373.89	177.36	302.73
	Avg.	**202.92**	**179.83**	-	**202.83**	**179.94**	-	**202.74**	**180.84**	-
MRWS 7-3	0.2	178.62	173.04	9.30	178.56	173.29	8.62	178.47	175.53	4.66
	0.3	171.84	173.23	−2.31	171.79	173.48	−2.77	171.68	175.72	−6.41
	0.4	166.70	173.37	−11.12	166.64	173.62	−11.43	166.52	175.87	−14.83
	0.5	160.38	173.55	21.99	160.30	173.81	−22.10	160.18	176.06	−25.21
	0.6	154.55	173.73	−32.01	154.47	173.99	−31.94	154.33	176.25	−34.77
	0.7	162.97	173.47	−17.53	162.88	173.72	−17.74	162.75	175.97	−20.98
	0.8	229.18	171.81	95.75	229.14	172.03	93.44	229.01	174.23	86.94
	Avg.	**174.89**	**173.17**	-	**174.82**	**173.42**	-	**174.71**	**175.66**	-
MRWS 1-1	0.2	144.70	171.86	−45.86	144.60	173.09	−46.93	144.52	174.16	−47.65
	0.3	150.33	171.67	−36.04	150.23	172.90	−37.34	150.14	173.96	−38.30
	0.4	146.39	171.79	−42.90	146.28	173.02	−44.05	146.18	174.09	−44.87
	0.5	147.18	171.76	−41.52	147.07	172.99	−42.70	146.96	174.06	−43.56
	0.6	147.72	171.74	−40.57	147.60	172.97	−41.78	147.50	174.04	−42.66
	0.7	156.11	171.47	−25.95	155.99	172.70	−27.53	155.89	173.76	−28.73
	0.8	170.98	171.03	−0.09	170.82	172.24	−2.35	170.76	173.29	−4.07
	Avg.	**151.91**	**171.62**	-	**151.80**	**172.85**	-	**151.71**	**173.91**	-

Table 6. *Cont.*

Samples	α	10 °C·min^{-1}			20 °C·min^{-1}			40 °C·min^{-1}		
		ΔH/ kJ·mol^{-1}	ΔG/ kJ·mol^{-1}	ΔS/ J·mol^{-1}·K^{-1}	ΔH/ kJ·mol^{-1}	ΔG/ kJ·mol^{-1}	ΔS/ J·mol^{-1}·K^{-1}	ΔH/ kJ·mol^{-1}	ΔG/ kJ·mol^{-1}	ΔS/ J·mol^{-1}·K^{-1}
MRWS 3-7	0.2	152.58	171.60	−32.13	152.50	172.21	−32.58	152.41	172.98	−33.23
	0.3	156.88	171.47	−24.64	156.79	172.07	−25.25	156.69	172.84	−26.07
	0.4	151.18	171.64	−34.55	151.09	172.25	−34.96	150.98	173.02	−35.59
	0.5	151.28	171.63	−34.37	151.18	172.24	−34.80	151.07	173.01	−35.43
	0.6	157.15	171.45	−24.15	157.05	172.05	−24.79	156.94	172.82	−25.64
	0.7	181.47	170.76	18.09	181.39	171.35	16.59	181.28	172.09	14.83
	0.8	281.90	168.63	191.28	281.84	169.17	186.18	281.76	169.87	180.71
	Avg.	**176.06**	**171.03**	-	**175.98**	**171.62**	-	**175.88**	**172.38**	-
WS	0.2	147.81	170.81	−39.04	147.75	171.44	−39.33	147.64	172.23	−39.91
	0.3	154.52	170.60	−27.28	154.45	171.22	−27.85	154.34	172.00	−28.66
	0.4	154.44	170.59	−27.43	154.36	171.22	−27.99	154.24	172.00	−28.82
	0.5	146.92	170.83	−40.58	146.84	171.46	−40.88	146.72	172.24	−41.43
	0.6	152.87	170.64	−30.16	152.78	171.26	−30.70	152.66	172.05	−31.46
	0.7	155.98	170.54	−24.72	155.90	171.16	−25.34	155.77	171.94	−26.25
	0.8	196.82	169.42	46.51	196.72	170.02	44.33	196.63	170.78	41.97
	Avg.	**158.48**	**170.49**	-	**158.40**	**171.11**	-	**158.29**	**171.89**	-

To comprehend the pyrolysis process, the variation in the thermodynamic parameters versus the conversion for MRPS blends at 20 °C·min^{-1} is plotted in Figure 8. For the pyrolysis reaction system, the ΔH corresponded to the difference in total energy between the pyrolysis products and reactants. That is to say, ΔH represented the total energy expended for raw material decomposing into various solid, liquid, and gaseous products. As observed from Figure 8a, the ΔH presented positive values, demonstrating that the pyrolysis reactions of MR, PS, and their blends were endothermic. The variation in the ΔH values over conversion was because of the enhancement or abatement in endothermicity. The MR had a higher average ΔH value than that of the PS, indicating a higher energy requirement for the breakage of chemical bonds in reactants, which was probably due to the higher mineral content in MR. The minerals in feedstocks may obstruct the diffusion of heat and released volatiles during the pyrolysis process leading to more energy needed for the formation of products [39]. For the conversion varying from 0.2 to 0.8, the MRPS blends showed lower average ΔH values (160.67, 146.58, and 157.33 kJ·mol^{-1} for MRPS7-3, MRPS1-1, and MRPS3-7, respectively) than that of the MR (202.83 kJ·mol^{-1}), indicating the lower endothermicity for the co-pyrolysis of MRPS blends. The energy differences between reactants and products for the MRPS blends were smaller compared to MR. It was noteworthy that the change trends of ΔH with conversion were the same as those of the activation energies (Figure 7a). Similar trends have also been observed for the pyrolysis of wood sawdust and rice husk [27,39].

Figure 8b illustrates the intermediate energy barriers (*EB*, deviation between E_α and ΔH_α) versus conversion for MRPS blends. The *EB* represented the additional energy required for the generation of products from the reactants. For all the samples, the values of *EB* varied from 4.65 to 5.82 kJ·mol^{-1}. The *EB* values showed a slight increasing trend with the increment in conversion, indicating the increase in additional energy requirement with the pyrolysis reaction proceeding. Furthermore, the pyrolysis process of MR, PS, and MRPS blends showed relatively small values of *EB* (avg. *EB* = 5.13–5.18 kJ·mol^{-1}), implying that the formation of the activated complex required relatively low additional energy, and the generation of pyrolysis products from these feedstocks was easily impacted by energy [49].

The Gibbs free energy versus conversion for MRPS blends is presented in Figure 8c. The ΔG referred to the total energy increase in a system needed for the generation of products [50]. Positive values of ΔG suggested non-spontaneous reactions that could happen with the input of external energy. For MR, PS, and their blends, the ΔG showed positive values and remained almost unchanged throughout the entire conversion process. These values implied that the pyrolysis reaction was a non-spontaneous process, which can be promoted by providing sufficient energy [49]. Furthermore, the degree of disorder of the pyrolysis system was evaluated by entropy change (ΔS). Negative changes in entropy demonstrated the reduction in the disorder degree of the reaction system. Small or negative values of ΔS implied the achievement of thermodynamic equilibrium of the pyrolysis reaction system, whereas positive values of ΔS indicated that the reaction system did not reach its thermodynamic equilibrium. The ΔS of MR presented negative values exclusive of the initial ($0.2 \leq \alpha \leq 0.3$) and final ($\alpha = 0.8$) stages of the pyrolysis reaction. The ΔS values of MRPS7-3 were negative except at the conversion of 0.8. However, the ΔS always showed negative values with the increase in conversion for PS, MRPS3-7, and MRPS1-1. These random changes in ΔS signified the complicacy of reactions that happened during the pyrolysis of biomass waste into various products. For the pyrolysis system with a negative ΔS, the ΔG demonstrated larger values than the ΔH, which signified that partial heat energy fed to the system was superfluous. The ΔS values of MRPS blends were more negative than positive, demonstrating that the degree of disorder of the pyrolysis system declined compared to the initial state of the system during the co-pyrolysis process [41].

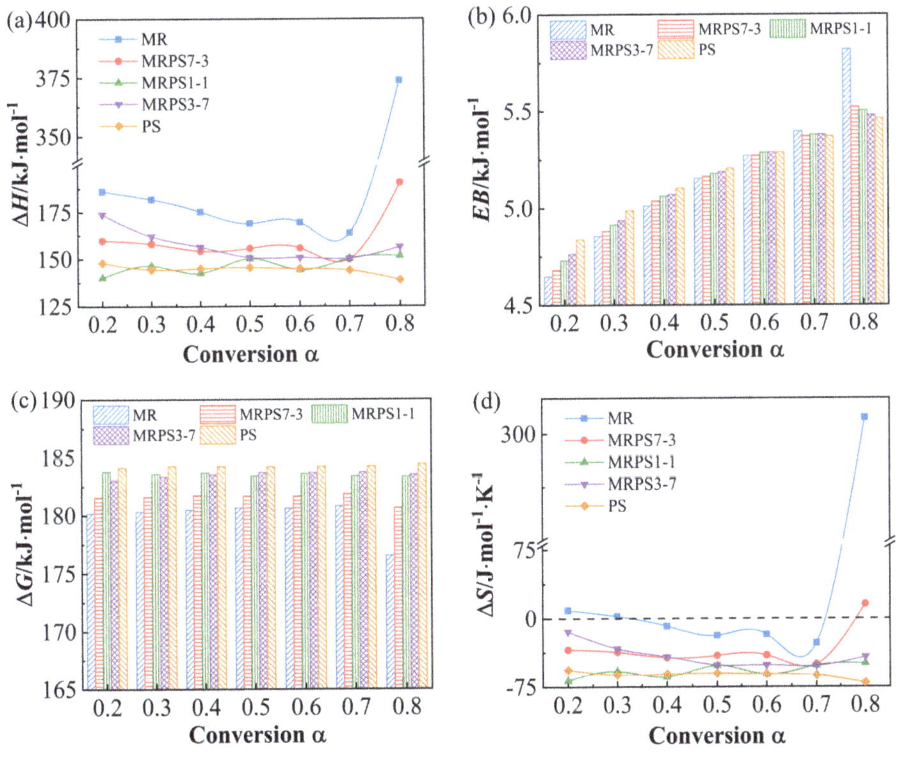

Figure 8. Variation in (**a**) ΔH versus conversion, (**b**) energy barrier versus conversion, (**c**) ΔG versus conversion, and (**d**) ΔS versus conversion for co-pyrolysis process of MRPS blends at 20 °C·min^{-1}.

The variation in the thermodynamic parameters versus conversion for the MRWS blends at 20 °C·min^{-1} is illustrated in Figure 9. The ΔH, *EB*, and ΔG presented similar change patterns to those of MRPS blends, whereas the variation in ΔS was slightly different. The ΔS showed positive values at specific conversion for MRWS7-3 ($\alpha = 0.1$ and 0.8) and MRWS3-7 ($\alpha = 0.7$ and 0.8), yet the ΔS values always showed negative values for MRWS1-1. The ΔS values of MRWS blends were also more negative than positive, signifying the reduction in the disorder degree of the system after the pyrolysis reaction.

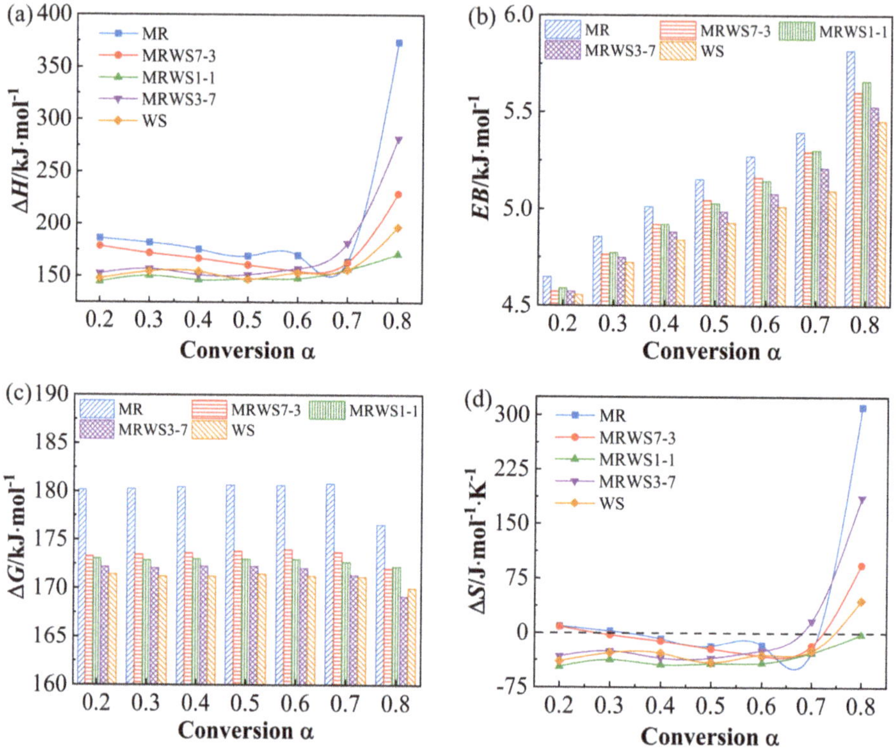

Figure 9. Variation in (**a**) ΔH versus conversion, (**b**) energy barrier versus conversion, (**c**) ΔG versus conversion, and (**d**) ΔS versus conversion for co-pyrolysis process of MRWS blends at 20 °C·min^{-1}.

3.5. Structure Properties of Co-Pyrolytic Chars

3.5.1. Carbon Microcrystalline Structure

The Raman spectrums of pyrolytic chars produced at different blending ratios are presented in Figure 10. The spectrums of all the char samples illustrated two evident peaks at about 1590 cm^{-1} and 1360 cm^{-1}. The total Raman area, the integral area of the spectrum between 800 and 1800 cm^{-1}, was used to express the Raman peak intensity of pyrolytic chars. The Raman intensity of pyrolytic chars from carbon-based materials was primarily impacted by the Raman scattering ability, the light absorptivity, and oxygen-containing structures [32]. The total Raman intensity can be enhanced by oxygen-containing structures owing to the resonance effect between oxygen and aromatic rings. On the contrary, the condensed aromatic rings in chars resulted in the reduction in the total Raman intensity owing to the high light absorption ability.

Figure 11 illustrates the total Raman area versus blending ratio for MRPS chars and MRWS chars. The total Raman area of MRPS chars increased as the PS blending ratio increased from 30% to 70%. The enlargement in the total Raman area for MRPS chars was probably caused by the increment in oxygen-containing structures or the decreased light absorption ability. As seen from Figure 11c, the area ratio of small aromatic rings to larger ones for MRPS7-3 char was smaller, suggesting a higher amount of condensed aromatic rings, which reduced the Raman intensity. As the PS mass ratio increased, the reduction in the condensed aromatic rings decreased the light absorptivity, resulting in the increase in Raman intensity. Consequently, the change in the total Raman area for MRPS chars was primarily influenced by the amount of condensed aromatic rings. It can be observed from Figure 11b that the total Raman area of MRWS chars illustrated a reduction trend as the

WS blending ratio was increased from 30% to 70%. As presented in Figure 11d, the area ratio of small aromatic rings to larger ones for MRWS chars declined with the increase in the WS mass ratio, signifying that the increase in large aromatic rings would reduce the Raman intensity, consistent with the reduction in Figure 11b. Therefore, the increased light absorption ability caused by the enlargement of large aromatic rings was also probably the principal reason leading to the decrease in the total Raman area for MRWS chars. To sum up, due to the similar oxygen content of these three feedstocks, the difference in the small and large aromatic ring structures was the primary reason for the change in the Raman intensity of co-pyrolytic chars in this study.

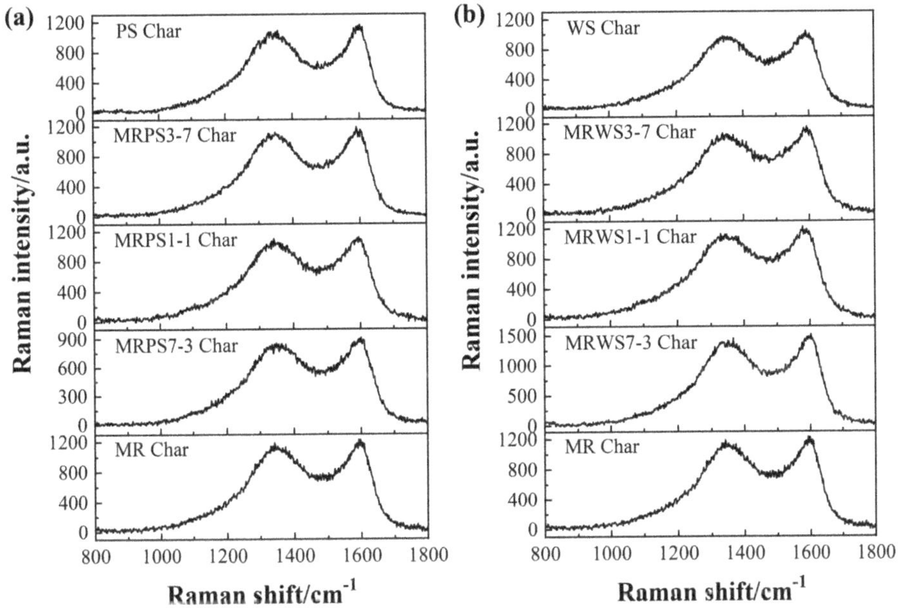

Figure 10. Raman spectrum of co-pyrolytic chars: (**a**) MRPS char and (**b**) MRWS char.

The changes in the area ratio between major Raman bands are also displayed in Figure 11. The three bands (G_R, V_L, and V_R) represented the small aromatic structures with <6 fused benzene rings, and D band mainly corresponded to the large aromatic structures with not less than 6 fused benzene rings. The area ratio of small aromatic rings to larger ones ($A_{(GR+VL+VR)}/A_D$) can be employed to describe the evolution of aromatic structures. Analyzing the changes in the relative content of small and large rings was significant for understanding the reaction pathways of the co-pyrolysis process.

As observed from Figure 11c, MRPS chars with different blending ratios showed different $A_{(GR+VL+VR)}/A_D$ values, indicating that the pyrolysis reaction pathways were different. For MRPS chars, the ratios illustrated a growing trend with the increase in the PS mass ratio. It was suggested that PS blending reduced the large aromatic rings and/or increased the small rings. A possible reason was that PS might enhance the release of oxygen-containing structures. The decomposition of oxygen-containing groups would probably promote the large oxygen-containing aromatic rings into small rings [32]. In addition, PS pyrolysis would generate more volatiles, which enhanced the volatile–char interactions. Thus, more small aromatic rings were generated in MRPS chars attributed to the accumulation of small molecule volatiles on the char surface [35]. These two impacts interpreted the increment in small aromatic rings during co-pyrolysis with PS. Furthermore, the change in small and large aromatic rings would influence the observed Raman intensity from different aromatic rings. The A_T in Figure 11c represented the total Raman area, as previously described. The $A_{(GR+VL+VR)}/A_T$ and A_D/A_T, respectively, referred to the relative proportion of the observed Raman intensity from small aromatic rings and large aromatic rings. It can be observed that the $A_{(GR+VL+VR)}/A_T$ illustrated an increasing trend with the increase in the PS blending ratio, supporting the increase in small aromatic rings at a high PS blending ratio. The $A_{(GR+VL+VR)}/A_T$ was higher than A_D/A_T, indicating that the small aromatic rings contributed more to the observed total Raman area of MRPS chars.

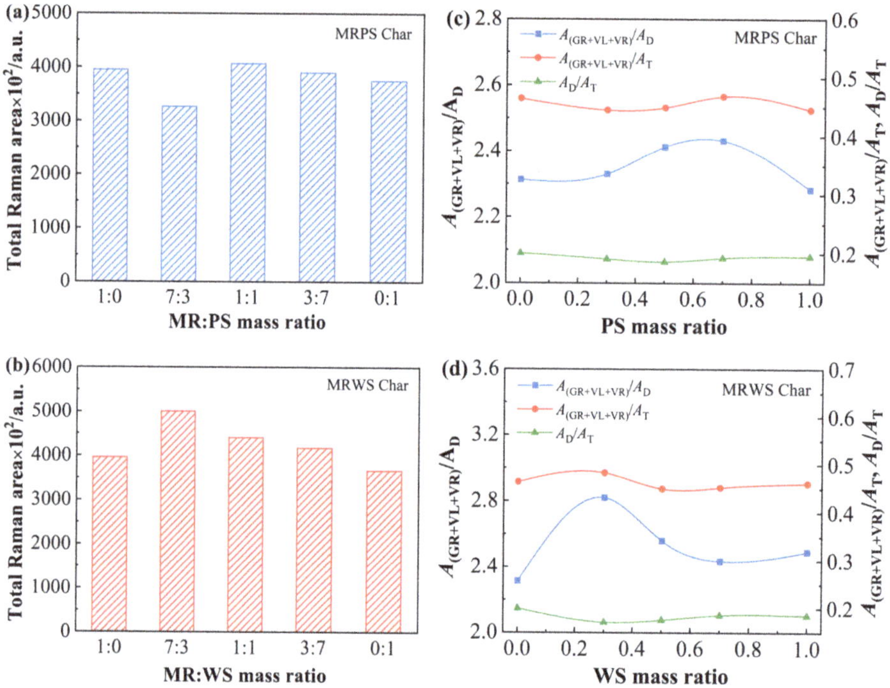

Figure 11. Total Raman area and major band area ratios versus blending ratio for co-pyrolytic chars: (**a**,**c**) MRPS char and (**b**,**d**) MRWS char.

As seen from Figure 11d, the $A_{(GR+VL+VR)}/A_D$ of MRWS chars declined obviously with the increment in the WS blending ratio. It was implied that WS blending reduced the small aromatic rings and/or increased the larger ones, which was probably due to the high mineral content of the WS. During the pyrolysis process, the alkali and alkaline-earth metal (AAEM) species in the WS can accelerate the breakage of weak bonds and enhance the generation of stronger bonds, thus increasing the large aromatic ring structures [51]. As shown in Figure 11d, the $A_{(GR+VL+VR)}/A_T$ decreased and the A_D/A_T increased with the increase in the WS mass ratio, confirming the conversion of small aromatic rings to larger rings. It was also found that the small aromatic rings contributed more to the observed total Raman area of MRWS chars. Generally speaking, the conclusion can be drawn that the formation of small or large aromatic rings during the co-pyrolysis process was impacted by the feedstock types and blending ratios.

3.5.2. Surface Texture

The surface textures of pyrolytic chars from MR, PS, WS, and their blends are shown in Figure 12. No obvious cavities were observed on the surface of MR char. The pyrolytic char of coprinus comatus fungi residue also showed similar surface textures [12]. Compared to the MR char, the PS char and WS char contained more porous and fibrous structures, which were attributed to the larger releasing rate of volatiles. A previous investigation has also reported similar morphological features for residual chars from wood biomass [45]. The surface of the MR and WS char showed more bright and agglomerated white substances than that of PS, which was because the inorganic substance content of the MR and WS was higher. As the PS mass ratio was improved, the surface morphology of MRPS chars became porous, and the fibrous structural substances increased significantly. The surface characteristics of MRWS chars became more fragmented with the WS addition. The addition of PS or WS made the char surface coarser, and a similar surface morphology was observed for the co-pyrolytic char from spent mushroom substrate and marine macroalgae [30]. Various pyrolytic chars with diverse surface characteristics can be acquired from the co-pyrolysis of MR with different biomasses by adjusting the blending ratios.

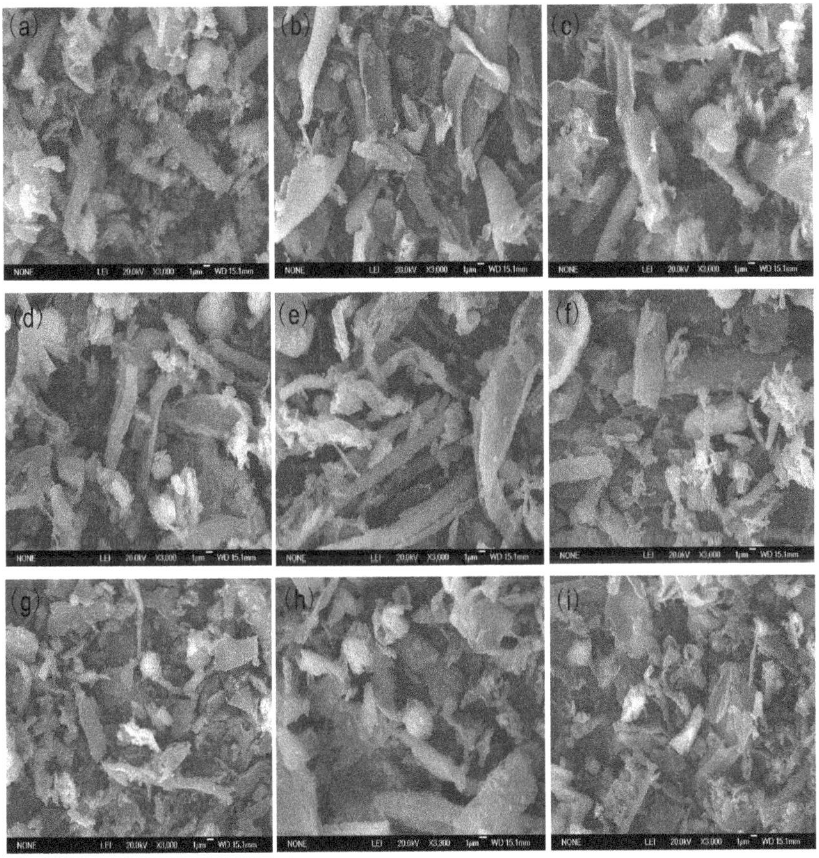

Figure 12. SEM micrographs of the pyrolytic chars: (**a**) MR char, (**b**) PS char, (**c**) WS char, (**d**) MRPS7-3 char, (**e**) MRPS1-1 char, (**f**) MRPS3-7 char, (**g**) MRWS7-3 char, (**h**) MRWS1-1 char, and (**i**) MRWS3-7 char.

The surface texture of pyrolytic chars was quantitatively explored by the fractal dimension (D_S) on the basis of fractal theory. The value of D_s can be calculated according to the following formula:

$$D_s = \frac{\ln(N_r)}{\ln(1/r)} \quad (10)$$

where r is the selected ruler, and N_r is the measurement results from the determinate ruler. Detailed information of the computational method was described in previous research [34]. The higher the D_s was, the more irregular and coarser the char surface was. Table 7 presents the fractal dimensions of surface morphology for pyrolytic chars from individual and blended samples. The correlation coefficients were higher than 0.99, suggesting the high reliability of the results. The fractal dimensions of the MR char, PS char, and WS char were 1.68, 1.65, and 1.67, respectively. It was observed that the fractal dimensions of MRPS chars and MRWS chars were higher than those of isolated chars, demonstrating that the surface textures of co-pyrolytic chars were obviously affected by the interactions between MR and PS or WS. The increase in the fractal dimension suggested that PS or WS blending promoted the char surface towards more irregular and rough structures, which would probably affect the reactivity of residual chars. The surface morphology of co-pyrolytic char from spent mushroom substrate and textile dyeing sludge was also reported to become rough and uneven due to the interactions [29].

Table 7. Fractal dimension of pyrolytic chars from individual and blended samples.

MR Blending Ratio	MRPS Char		MRWS Char	
	D_s	R^2	D_s	R^2
1	1.68	0.9921	1.68	0.9921
0.7	1.69	0.9921	1.83	0.9915
0.5	1.69	0.9921	1.63	0.9924
0.3	1.74	0.9918	1.76	0.9918
0	1.65	0.9923	1.67	0.9921

It can be concluded that the carbon structure of MRPS chars and MRWS chars was mainly composed of small aromatic rings. The pyrolytic chars with a high content of small aromatic rings were easier to combust or gasify due to their higher reactivity [35]. In addition, the co-pyrolytic chars were porous, which was conducive to the diffusion and adsorption of substances. Therefore, co-pyrolytic chars from MRPS blends or MRWS blends may be used as clean solid fuels, adsorption materials, catalysts, soil conditioners, etc. Further study should emphasize the application performance of co-pyrolytic chars from the blends of MR and wood/straw biomass.

4. Conclusions

Thermal characteristics and kinetic and thermodynamic analysis during the co-pyrolysis of mushroom residue (MR) and pine sawdust (PS) or wheat straw (WS), as well as the carbon microcrystalline structure and surface textures of co-pyrolytic chars, were investigated. The pyrolytic process of the blends consisted of three phases: the dehydration phase, volatile releasing phase, and carbonization phase. With an increase in the PS or WS blending ratio from 30% to 70%, the devolatilization index increased from 13.09 to 15.30 for MRPS blends and from 11.29 to 21.03 for MRWS blends, indicating that the volatile release was promoted and concentrated. Weak interactions were observed for the blends, which were affected by pyrolysis temperatures and blending ratios. The interactions mainly affected the thermal decomposition temperatures and rates in the stage of volatile release between 250 °C and 400 °C but had no prominent influence on the volatile yields for the co-pyrolysis process. The average activation energies of the blends (151.73–180.98 kJ·mol^{-1}) were smaller than that of MR (207.99 kJ·mol^{-1}) and presented non-additive performance owing to the interactions between individual components. The lowest average activation energy was obtained when the PS or WS mass ratio was 50%. The lower average pre-exponential factor of the blends indicated the reduced complicacy of the pyrolysis reaction. All the blends presented positive average ΔH and ΔG, and the relatively small deviation between E_α and ΔH_α (4.94–5.18 kJ·mol^{-1}) signified the energy sensitivity of product formation from the pyrolysis of these feedstocks. The observed trends of Raman intensity of co-pyrolytic chars were dependent on the difference in the small and large aromatic ring structures. As the PS or WS blending ratio increased, the

PS promoted the generation of small aromatic rings, whereas the WS favored the production of larger rings. The surface textures of co-pyrolytic chars became porous, and the increased fractal dimensions (1.69–1.83) indicated the promotion of surface irregularity and roughness. The results of this investigation can provide some valuable knowledge for developing the co-pyrolysis technology of edible fungi residue with wood/straw biomass.

Author Contributions: Conceptualization, H.M. and Z.W. (Zhiqiang Wu); Formal analysis, H.Y.; Funding acquisition, H.M., Z.W. (Zhiqiang Wu), D.W., H.W. and J.L.; Investigation, H.Y. and D.L.; Methodology, H.M.; Supervision, Z.W. (Zhiqiang Wu), H.L. and J.L.; Writing—original draft, H.Y.; Writing—review and editing, H.M., Z.W. (Zhiqiang Wu), Z.W. (Zhe Wang), D.W., H.W. and J.L. All authors have read and agreed to the published version of the manuscript.

Funding: This work was financially supported by the Natural Science Basic Research Plan in Shaanxi Province of China (Grant No: 2018JQ5101), the National Natural Science Foundation of China (Grant Nos: 51976168, 42107030, 52070156), and the "Scientists+Engineers" Team Construction Based on QinChuangYuan Platform, Shaanxi Province (Grant No. 2022KXJ-115).

Institutional Review Board Statement: Not applicable.

Informed Consent Statement: Not applicable.

Data Availability Statement: The article data is not archived anywhere and all study data can be seen in the article.

Conflicts of Interest: The authors declare that they have no known competing financial interests or personal relationships that could have appeared to influence the work reported in this paper.

References

1. Wu, M.G.; Wang, Z.W.; Chen, G.F.; Zhang, M.J.; Sun, T.L.; Wang, Q.; Zhu, H.N.; Guo, S.H.; Chen, Y.; Zhu, Y.J.; et al. Synergistic effects and products distribution during Co-pyrolysis of biomass and plastics. *J. Energy Inst.* **2023**, *111*, 101392. [CrossRef]
2. Kar, T.; Kaygusuz, Ö.; Güney, M.S.; Cuce, E.; Keles, S.; Shaik, S.; Owolabi, A.B.; Nsafon, B.E.K.; Ogunsua, J.M.; Huh, J.S. Fast Pyrolysis of Tea Bush, Walnut Shell, and Pine Cone Mixture: Effect of Pyrolysis Parameters on Pyrolysis Crop Yields. *Sustainability* **2023**, *15*, 13718. [CrossRef]
3. Xu, Z.X.; Dou, R.; Gao, F.; Chen, Y.X.; Leng, L.J.; Osman, S.M.; Luque, R. Bio-adhesives derived from sewage sludge via hydrothermal carbonization: Influence of aqueous phase recycling. *Chem. Eng. J.* **2024**, *490*, 151685. [CrossRef]
4. Jiang, Y.C.; Li, C.; Zhang, L.J.; Lin, H.S.; Zhang, S.; Wang, Y.; Hu, S.; Xiang, J.; Hu, X. Influence of torrefaction with microwave and furnace heating on pyrolysis of poplar sawdust. *Fuel Process. Technol.* **2023**, *245*, 107696. [CrossRef]
5. Thangaraj, B.; Wongyao, N.; Solomon, P.R.; Gupta, V.; Abdullah, A.; Abedrabbo, S.; Hassan, J. Synthesis of reduced graphene oxide from onion peel waste by single-stage pyrolysis, characterization and evaluation of its antibacterial activity. *J. Environ. Chem. Eng.* **2024**, *12*, 113474. [CrossRef]
6. Kong, X.; Zhu, Y.F.; Lei, H.W.; Wang, C.X.; Zhao, Y.F.; Huo, E.G.; Lin, X.N.; Zhang, Q.F.; Qian, M.; Mateo, W.D.; et al. Synthesis of graphene-like carbon from biomass pyrolysis and its applications. *Chem. Eng. J.* **2020**, *399*, 125808. [CrossRef]
7. Finney, K.N.; Ryu, C.; Sharifi, V.N.; Swithenbank, J. The reuse of spent mushroom compost and coal tailings for energy recovery: Comparison of thermal treatment technologies. *Bioresour. Technol.* **2009**, *100*, 310–315. [CrossRef]
8. Atallah, E.; Zeaiter, J.; Ahmad, M.N.; Leahy, J.J.; Kwapinski, W. Hydrothermal carbonization of spent mushroom compost waste compared against torrefaction and pyrolysis. *Fuel Process. Technol.* **2021**, *216*, 106795. [CrossRef]
9. Qin, P.F.; Li, T.Y.; Liu, C.; Liang, Y.S.; Sun, H.B.; Chai, Y.Z.; Yang, T.Y.; Gong, X.M.; Wu, Z.B. Extraction and utilization of active substances from edible fungi substrate and residue: A review. *Food Chem.* **2023**, *398*, 133872. [CrossRef] [PubMed]
10. Leong, Y.K.; Ma, T.W.; Chang, J.S.; Yang, F.C. Recent advances and future directions on the valorization of spent mushroom substrate (SMS): A review. *Bioresour. Technol.* **2022**, *344*, 126157. [CrossRef] [PubMed]
11. Zhang, B.; Tan, G.W.; Zhong, Z.P.; Ruan, R. Microwave-assisted catalytic fast pyrolysis of spent edible mushroom substrate for bio-oil production using surface modified zeolite catalyst. *J. Anal. Appl. Pyrolysis* **2017**, *123*, 92–98. [CrossRef]
12. Chen, G.Y.; Yu, H.D.; Lin, F.W.; Zhang, Z.M.; Yan, B.B.; Song, Y.J. Utilization of edible fungi residues towards synthesis of high-performance porous carbon for effective sorption of Cl-VOCs. *Sci. Total Environ.* **2020**, *727*, 138475. [CrossRef] [PubMed]
13. Perez-Chavez, A.M.; Mayer, L.; Alberto, E. Mushroom cultivation and biogas production: A sustainable reuse of organic resources. *Energy Sustain. Dev.* **2019**, *50*, 50–60. [CrossRef]
14. Li, J.L.; Pu, T.; Wang, Z.H.; Liu, T.Z. Thermal Behavior and Pyrolysis Kinetics of Mushroom Residue with the Introduction of Waste Plastics. *Polymers* **2023**, *15*, 3824. [CrossRef] [PubMed]
15. Slezak, R.; Krzystek, L.; Ledakowicz, S. Steam gasification of pyrolysis char from spent mushroom substrate. *Biomass Bioenerg.* **2019**, *122*, 336–342. [CrossRef]

16. Sun, Z.S.; Yang, X.T.; Zhang, B.; Yang, B.L.; Shang, J.X.; Wu, Z.Q. Product-oriented thermodynamic function construction of wheat straw and thermodynamic analysis for rapid pyrolysis. *Fuel* **2024**, *358*, 130285. [CrossRef]
17. Huang, J.L.; Zhang, J.H.; Liu, J.Y.; Xie, W.M.; Kuo, J.H.; Chang, K.L.; Buyukada, M.; Evrendilek, F.; Sun, S.Y. Thermal conversion behaviors and products of spent mushroom substrate in CO2 and N2 atmospheres: Kinetic, thermodynamic, TG and Py-GC/MS analyses. *J. Anal. Appl. Pyrolysis* **2019**, *139*, 177–186. [CrossRef]
18. Slezak, R.; Krzystek, L.; Ledakowicz, S. Thermogravimetric analysis coupled with mass spectrometry of spent mushroom substrate and its fractions. *J. Anal. Appl. Pyrolysis* **2018**, *133*, 1–8. [CrossRef]
19. Wang, G.X.; Guo, B.W.; Song, Z.L.; Wang, W.L.; Mao, Y.P.; Sun, J.; Zhao, X.Q. Pyrolysis characteristics and kinetic analysis of edible fungus chaff. *Trans. Chin. Soc. Agric. Eng.* **2020**, *36*, 301–307.
20. Koido, K.; Ogura, T.; Matsumoto, R.; Endo, K.; Sato, M. Spent mushroom substrate performance for pyrolysis, steam co-gasification, and ash melting. *Biomass Bioenerg.* **2021**, *145*, 105954. [CrossRef]
21. Jiang, H.F.; Cheng, Z.Q.; Zhao, T.Q.; Liu, M.Z.; Zhang, M.Y.; Li, J.N.; Hu, M.J.; Zhang, L.; Li, J.F. Pyrolysis kinetics of spent lark mushroom substrate and characterization of bio-oil obtained from the substrate. *Energy Conv. Manag.* **2014**, *88*, 259–266. [CrossRef]
22. Sarfraz, R.; Li, S.W.; Yang, W.H.; Zhou, B.Q.; Xing, S.H. Assessment of Physicochemical and Nutritional Characteristics of Waste Mushroom Substrate Biochar under Various Pyrolysis Temperatures and Times. *Sustainability* **2019**, *11*, 277. [CrossRef]
23. Wang, J.; Zhong, Z.P.; Ding, K.; Xue, Z.Y. Catalytic fast pyrolysis of mushroom waste to upgraded bio-oil products via pre-coked modified HZSM-5 catalyst. *Bioresour. Technol.* **2016**, *212*, 6–10. [CrossRef] [PubMed]
24. Wang, J.; Zhong, Z.P.; Song, Z.W.; Ding, K.; Deng, A.D. Modification and regeneration of HZSM-5 catalyst in microwave assisted catalytic fast pyrolysis of mushroom waste. *Energy Conv. Manag.* **2016**, *123*, 29–34. [CrossRef]
25. Jiang, H.F.; Deng, S.H.; Chen, J.; Zhang, L.; Zhang, M.Y.; Li, J.N.; Li, S.; Li, J.F. Preliminary Study on Copyrolysis of Spent Mushroom Substrate as Biomass and Huadian Oil Shale. *Energy Fuels* **2016**, *30*, 6342–6349. [CrossRef]
26. Jerzak, W.; Bieniek, A.; Magdziarz, A. Multifaceted analysis of products from the intermediate co-pyrolysis of biomass with Tetra Pak waste. *Int. J. Hydrogen Energy* **2023**, *48*, 11680–11694. [CrossRef]
27. Sharma, A.; Kumar, A.A.; Mohanty, B.; Sawarkar, A.N. Critical insights into pyrolysis and co-pyrolysis of poplar and eucalyptus wood sawdust: Physico-chemical characterization, kinetic triplets, reaction mechanism, and thermodynamic analysis. *Renew. Energy* **2023**, *210*, 321–334. [CrossRef]
28. Wu, Z.Q.; Wang, S.Z.; Zhao, J.; Chen, L.; Meng, H.Y. Thermal behavior and char structure evolution of bituminous coal blends with edible fungi residue during co-pyrolysis. *Energy Fuels* **2014**, *28*, 1792–1801. [CrossRef]
29. Huang, J.L.; Liu, J.Y.; Chang, K.L.; Buyukada, M.; Evrendilek, F. (Co-)pyrolytic performances and by-products of textile dyeing sludge and spent mushroom substrate. *J. Clean. Prod.* **2020**, *261*, 121195. [CrossRef]
30. Sewu, D.D.; Boakye, P.; Jung, H.; Woo, S.H. Synergistic dye adsorption by biochar from co-pyrolysis of spent mushroom substrate and *Saccharina japonica*. *Bioresour. Technol.* **2017**, *244*, 1142–1149. [CrossRef]
31. Jiang, H.F.; Zhang, M.Y.; Chen, J.; Li, S.; Shao, Y.F.; Yang, J.Q.; Li, J.F. Characteristics of bio-oil produced by the pyrolysis of mixed oil shale semi-coke and spent mushroom substrate. *Fuel* **2017**, *200*, 218–224. [CrossRef]
32. Li, T.T.; Zhang, L.; Dong, L.; Qiu, P.H.; Wang, S.; Jiang, S.J.; Li, C.Z. Changes in char structure during the low-temperature pyrolysis in N-2 and subsequent gasification in air of Loy Yang brown coal char. *Fuel* **2018**, *212*, 187–192. [CrossRef]
33. Chen, J.Z.; Lu, Z.M.; Jian, J.; Bao, Z.Y.; Cai, J.F.; Yao, S.C. Effect of torrefaction on yield, reactivity and physicochemical properties of pyrolyzed char from three major biomass constituents. *J. Anal. Appl. Pyrolysis* **2023**, *173*, 106104. [CrossRef]
34. Wu, Z.Q.; Yang, W.C.; Yang, B.L. Thermal characteristics and surface morphology of char during co-pyrolysis of low-rank coal blended with microalgal biomass: Effects of Nannochloropsis and Chlorella. *Bioresour. Technol.* **2018**, *249*, 501–509. [CrossRef] [PubMed]
35. Meng, H.Y.; Wang, S.Z.; Chen, L.; Wu, Z.Q.; Zhao, J. Thermal behavior and the evolution of char structure during co-pyrolysis of platanus wood blends with different rank coals from northern China. *Fuel* **2015**, *158*, 602–611. [CrossRef]
36. Wang, C.A.; Du, Y.B.; Che, D.F. Reactivities of coals and synthetic model coal under oxy-fuel conditions. *Thermochim. Acta* **2013**, *553*, 8–15. [CrossRef]
37. Tariq, R.; Zaifullizan, Y.M.; Salema, A.A.; Abdulatif, A.; Ken, L.S. Co-pyrolysis and co-combustion of orange peel and biomass blends: Kinetics, thermodynamic, and ANN application. *Renew. Energy* **2022**, *198*, 399–414. [CrossRef]
38. Meng, H.Y.; Wang, M.Z.; Wu, Z.Q.; Zhao, J.; Li, J.K.; Wang, S.Z. Physico-Chemical Structure and Gasification Performance of Co-Pyrolytic Char Produced by the Pyrolysis of Polyvinyl Chloride Blends with Two Rank Coals. *ACS Omega* **2022**, *7*, 32280–32291. [CrossRef]
39. Mallick, D.; Poddar, M.K.; Mahanta, P.; Moholkar, V.S. Discernment of synergism in pyrolysis of biomass blends using thermo-gravimetric analysis. *Bioresour. Technol.* **2018**, *261*, 294–305. [CrossRef]
40. Lei, J.L.; Ye, X.F.; Wang, H.; Zhao, D.N. Insights into Pyrolysis Kinetics, Thermodynamics, and the Reaction Mechanism of Wheat Straw for Its Resource Utilization. *Sustainability* **2023**, *15*, 12536. [CrossRef]
41. Kim, Y.S.; Kim, Y.S.; Kim, S.H. Investigation of thermodynamic parameters in the thermal decomposition of plastic waste-waste lube oil compounds. *Environ. Sci. Technol.* **2010**, *44*, 5313–5317. [CrossRef] [PubMed]
42. Wang, S.; Wu, L.P.; Hu, X.; Zhang, L.; Li, T.T.; Li, C.Z. Effects of the Particle Size and Gasification Atmosphere on the Changes in the Char Structure during the Gasification of Mallee Biomass. *Energy Fuels* **2018**, *32*, 7678–7684. [CrossRef]

43. Burra, K.G.; Gupta, A.K. Kinetics of synergistic effects in co-pyrolysis of biomass with plastic wastes. *Appl. Energy* **2018**, *220*, 408–418. [CrossRef]
44. Nyambura, S.M.; Wang, J.F.; Li, H.; Feng, X.B.; Pan, X.J.; Li, B.H.; Ahmad, R.; Xu, J.L.; Bertrand, G.V.; Ndiithi, J.; et al. Microwave co-pyrolysis of kitchen food waste and rice straw for waste reduction and sustainable biohydrogen production: Thermo-kinetic analysis and evolved gas analysis. *Sustain. Energy Technol. Assess.* **2022**, *52*, 102072. [CrossRef]
45. Zheng, Y.; Zhang, Y.M.; Xu, J.N.; Li, X.Y.; Xu, C.B. Co-pyrolysis behavior of fermentation residues with woody sawdust by thermogravimetric analysis and a vacuum reactor. *Bioresour. Technol.* **2017**, *245*, 449–455. [CrossRef] [PubMed]
46. Zhang, T.T.; Yuchi, W.; Bai, Z.Q.; Hou, R.R.; Feng, Z.H.; Guo, Z.X.; Kong, L.X.; Bai, J.; Meyer, B.; Li, W. Insight into the charging methods effects during clean recycling of plastic by co-pyrolysis with low-rank coal. *J. Clean. Prod.* **2022**, *333*, 130168. [CrossRef]
47. Yuan, X.S.; He, T.; Cao, H.L.; Yuan, Q.X. Cattle manure pyrolysis process: Kinetic and thermodynamic analysis with isoconversional methods. *Renew. Energy* **2017**, *107*, 489–496. [CrossRef]
48. Jerzak, W.; Gajek, M.; Magdziarz, A. Oat straw pyrolysis with ammonium chloride doping: Analysis of evolved gases, kinetic triplet, and thermodynamic parameters. *Bioresour. Technol.* **2023**, *388*, 129784. [CrossRef] [PubMed]
49. Ming, X.; Xu, F.F.; Jiang, Y.; Zong, P.J.; Wang, B.; Li, J.; Qiao, Y.Y.; Tian, Y.Y. Thermal degradation of food waste by TG-FTIR and Py-GC/MS: Pyrolysis behaviors, products, kinetic and thermodynamic analysis. *J. Clean. Prod.* **2020**, *244*, 118713. [CrossRef]
50. Xu, Y.L.; Chen, B.L. Investigation of thermodynamic parameters in the pyrolysis conversion of biomass and manure to biochars using thermogravimetric analysis. *Bioresour. Technol.* **2013**, *146*, 485–493. [CrossRef]
51. Keown, D.M.; Li, X.J.; Hayashi, J.I.; Li, C.Z. Characterization of the structural features of char from the pyrolysis of cane trash using Fourier transform—Raman spectroscopy. *Energy Fuels* **2007**, *21*, 1816–1821. [CrossRef]

sustainable chemistry

MDPI

Article

Valorization of *Agave angustifolia* Bagasse Biomass from the Bacanora Industry in Sonora, Mexico as a Biochar Material: Preparation, Characterization, and Potential Application in Ibuprofen Removal

Hylse Aurora Ruiz-Velducea [1,2], María de Jesús Moreno-Vásquez [2,*], Héctor Guzmán [1,*], Javier Esquer [1], Francisco Rodríguez-Félix [3], Abril Zoraida Graciano-Verdugo [2], Irela Santos-Sauceda [4], Idania Emedith Quintero-Reyes [5], Carlos Gregorio Barreras-Urbina [6], Claudia Vásquez-López [7], Silvia Elena Burruel-Ibarra [4], Karla Hazel Ozuna-Valencia [3] and José Agustín Tapia-Hernández [3,*]

[1] Programa de Posgrado en Sustentabilidad, Departamento de Ingeniería Industrial (DII), Universidad de Sonora, Blvd. Luis Encinas y Rosales, S/N, Colonia Centro, Hermosillo 83000, Sonora, Mexico; a221230205@unison.mx (H.A.R.-V.); javier.esquer@unison.mx (J.E.)

[2] Departamento de Ciencias Químico-Biológicas (DCQB), Universidad de Sonora, Blvd. Luis Encinas y Rosales, S/N, Colonia Centro, Hermosillo 83000, Sonora, Mexico; abril.graciano@unison.mx

[3] Departamento de Investigación y Posgrado en Alimentos (DIPA), Universidad de Sonora, Blvd. Luis Encinas y Rosales, S/N, Colonia Centro, Hermosillo 83000, Sonora, Mexico; francisco.rodriguezfelix@unison.mx (F.R.-F.); a215208659@unison.mx (K.H.O.-V.)

[4] Departamento de Investigación en Polímeros y Materiales (DIPM), Universidad de Sonora, Blvd. Luis Encinas y Rosales, S/N, Colonia Centro, Hermosillo 83000, Sonora, Mexico; irela.santos@unison.mx (I.S.-S.); silvia.burruel@unison.mx (S.E.B.-I.)

[5] Departamento de Ciencias de la Salud (DCS), Universidad de Sonora, Campus Cajeme, Blvd. Bordo Nuevo N/N, Antiguo Ejido Providencia, Ciudad Obregón 85010, Sonora, Mexico; idania.quintero@unison.mx

[6] Centro de Investigación en Alimentación y Desarrollo (CIAD), A. C., Coordinación de Tecnología de Alimentos de Origen Vegetal, Carretera Gustavo Enrique Astiazarán Rosas, Núm. 46, La Victoria, Hermosillo C.P. 83304, Sonora, Mexico; carlosgbarrerasu@gmail.com

[7] Centro de Investigación Científica de Yucatán (CICY), Calle 43 # 130 x 32 y 34, Colonia Chuburná de Hidalgo, Mérida C.P. 97205, Yucatán, Mexico; claudia.vasquez@iq.uson.mx

* Correspondence: mariadejesus.moreno@unison.mx (M.d.J.M.-V.); hector.guzman@unison.mx (H.G.); joseagustin.tapia@unison.mx (J.A.T.-H.)

Abstract: The aim of this research was to separate the over-the-counter nonsteroidal anti-inflammatory drug (NSAID), ibuprofen, from an aqueous solution using the adsorption method, as this NSAID is one of the most globally consumed. An adsorbent was crafted from the *Agave angustifolia* bagasse, a byproduct of the bacanora industry (a representative alcoholic beverage of the state of Sonora, in northwestern Mexico). Three bioadsorbents (BCT1, BCT2, and BCT3) were produced via pyrolysis at a temperature of 550 °C, with slight variations in each process for every bioadsorbent. The bioadsorbents achieved material yields of 25.65%, 31.20%, and 38.28% on dry basis respectively. Characterization of the bagasse and adsorbents involved scanning electron microscopy (SEM), Fourier-transform infrared spectroscopy (FT-IR), thermogravimetric analysis (TGA), and differential scanning calorimetry (DSC). The biomass morphology exhibited a cracked surface with holes induced via the bacanora production process, while the surface of the bioadsorbents before ibuprofen adsorption was highly porous, with a substantial surface area. After adsorption, the surface of the bioadsorbents was transformed into a smoother grayish layer. The macromolecules of cellulose, hemicellulose, and lignin were present in the biomass. According to functional groups, cellulose and hemicellulose degraded to form the resulting bioadsorbents, although traces of lignin persisted after the pyrolysis process was applied to the biomass. In an adsorption study, BCT1 and BCT2 bioadsorbents successfully removed 100% of ibuprofen from aqueous solutions with an initial concentration of 62.6 mg/L. In conclusion, the biocarbon derived from *Agave angustifolia* bagasse exhibited significant potential for removing ibuprofen via adsorption from aqueous solutions.

Keywords: bioadsorbent; adsorption; bagasse; *Agave angustifolia*; removal; ibuprofen; biochar

1. Introduction

Emerging contaminants, defined as substances that were not previously identified or recognized as pollutants, have become a significant interest in science due to growing concern about the potential consequences of their occurrence, although their presence in the natural environment is not a recent phenomenon [1]. These substances have been sporadically detected in the environment, including in water sources, groundwater, and even drinking water [2]. Pharmaceuticals, such as anti-inflammatories, antibiotics, antivirals, and analgesics, represent examples of these emerging contaminants [3,4]. The European Union, the United States, and China have regulatory frameworks for monitoring such contaminants. However, in many countries, they are not regulated by any law or authority [5,6].

In recent decades, ibuprofen has become one of the most widely consumed over-the-counter nonsteroidal anti-inflammatory drugs (NSAIDs) [7]. Its physicochemical properties contribute to high mobility in aquatic environments, making it a significant contaminant in aquatic ecosystems [8]. Studies have indicated that its environmental presence poses a risk of long-term exposure and chronic toxic effects on organisms [9–11]. Ibuprofen can enter aquatic environments through various pathways originating from anthropogenic activities. Improper disposal, including flushing unused medications down the drain, contributes significantly. Manufacturing plants and wastewater treatment plants (WWTPs) are also sources, as ibuprofen is excreted by humans and animals in its unmetabolized form, along with its more toxic metabolites. Also, ibuprofen wastewater effluents from various sources such as veterinary facilities, domestic premises, hospitals, and drug-production factories; these pose environmental concerns due to their potential impact on aquatic ecosystems. Once in the environment, ibuprofen can travel through rivers, lakes, and oceans, contaminating aquatic ecosystems. It can also seep into soil and groundwater, further spreading its presence. Aquatic plants and organisms inadvertently consume ibuprofen, incorporating it into the food web [12,13].

Various methods exist for removing ibuprofen from aqueous solutions, including ozonation and electro-Fenton oxidation [14–16]. Even though these methods have proven effective, their application involves high costs. Consequently, environmentally friendly and economically viable techniques, such as adsorption, have gained interest [17–19]. The use of natural sorbents, like biochar, facilitates efficient and economically viable adsorption processes due to the nature of the adsorptive materials employed [20]. Biochar, a carbon-rich, porous, aromatic, and stable material produced from biomass carbonization under high-temperature and oxygen-limited conditions, possesses excellent physicochemical properties. Its adjustable surface area, controllable porous structures, and abundant oxygen functional groups make it applicable in different fields, including energy, environment, and agriculture [21]. Biochar, generated primarily from biomass wastes through pyrolysis at temperatures ranging from 500 to 800 °C and durations of 1 to 2 h under an inert atmosphere, serves as a versatile tool for circular waste utilization and sustainability. Its applications extend to environmental remediation, including the removal of water pollutants, amelioration of soil quality, and reduction in solid waste [22]. A highly promising application of biochar lies in its ability to effectively capture and retain various pollutants from water sources and can be used as organic fertilizer as well as soil amendment thanks to its porous structure [23]. For the removal of contaminants, the choice of precursor (biomass) is crucial for economic feasibility and efficiency. The precursor should be readily available, low-cost, nonhazardous, with a high carbon and oxygen content, a resistance to abrasion, high thermal stability, and have small pore diameters [24,25].

The current research focused on elucidating the absorption mechanisms of biocarbons, applying them to various emerging pharmaceuticals, and enhancing their surface properties. In this regard, different absorption mechanisms were elucidated, such as the biocarbon surface exhibiting flaky structures with numerous small pores, enhancing permeability and porosity and providing ample vacant sites for adsorption. Additionally, identified oxygen-containing molecules on the biocarbon surface facilitating the adsorption of emerging pharmaceuticals through significant hydrogen bonding interactions. Furthermore, the sulfonyl group present in many emerging pharmaceuticals enables π-π electron interactions with the benzene ring, which is particularly evident around neutral pH levels, while electrostatic and coordination interactions further contribute to the adsorption of these pharmaceuticals onto the biocarbon surface [26]. Hence, the ongoing research is centered on the production and application of a novel biocarbon derived from the food industry with high adsorption capacity and with the potential to remove ibuprofen.

The traditional bacanora industry in the state of Sonora, Mexico, benefits the region socially, economically, and culturally. Bacanora, a distilled alcoholic beverage, has been produced in the state for over 300 years, involving the processing and distillation of *Agave angustifolia* [27,28]. Annually, around 1.5 million tons of agave are used for alcoholic beverage production, generating approximately 14.41 tons of bagasse per harvested hectare [29,30]. This waste is considered an environmental problem due to its natural degradation by altering soil pH, causing landscape deterioration, generating odors, attracting rodents, and when burnt, releasing greenhouse gases [31]. Given its components and characteristics, the *Agave angustifolia* bagasse resulting from the bacanora production process is considered a lignocellulosic material. Authors argue that such materials have great potential for the production of biopolymers via cellulolytic bacteria and for obtaining phenolic compounds, due to their low cost and abundance [32]. The use of this waste could contribute to the region's economic growth and mitigate environmental impact by avoiding excessive use of natural resources [33]. In this regard, *Agave angustifolia* has been used for the production of biochar and activated carbon [34,35], however, it has not been tested for the removal of emerging pharmaceuticals such as ibuprofen. Thus, utilizing *Agave angustifolia* bagasse biomass from the bacanora industry in Sonora, Mexico, as biochar material may result in a biochar product with favorable properties for effective adsorption of ibuprofen, thus offering a promising approach for water remediation applications.

Therefore, this project aimed to prevent, eliminate, and/or reduce environmental contamination caused by ibuprofen by removing it from a synthetic aqueous solution. Additionally, it sought to add value to the *Agave angustifolia* bagasse by producing a bioadsorbent. Both the bagasse and the resulting biochar treatments were characterized using scanning electron microscopy (SEM), infrared spectroscopy (FTIR), thermogravimetric analysis (TGA), and differential scanning calorimetry (DSC). The potential application of these materials in ibuprofen removal was also explored.

2. Materials and Methods

2.1. Reagents

Distilled water, infant-type suspended ibuprofen, and hydrochloric acid (HCl) were employed in adsorption tests.

2.2. Biomass Acquisition and Conditioning

The *Agave angustifolia* bagasse was donated by the Agave and Bacanora Producers Association in Hermosillo, Sonora, Mexico, to the University of Sonora. The agave residue underwent a drying and grinding process using the methodology described by Robles-García et al. [36] with some adaptations. The bagasse was shredded using a forage shredder. Subsequently, the resulting product was dried in a tunnel at 80 °C for 2 h to reduce particle size. The dried bagasse fibers were further processed with a Pulvex 200 device after sieving them with a 100-mesh screen. The obtained powder underwent a second sieving process to ensure uniform particle size. Particles that did not pass through the 100-mesh screen were subjected to additional crushing until achieving a consistent particle size (Figure 1).

(A)

(B)

Figure 1. Raw material used for bioadsorbent preparation: (**A**) *Agave angustifolia* bagasse, (**B**) conditioned *Agave angustifolia* bagasse.

2.3. Proximal Analysis of Biomass

Mexican standards NOM-116-SSA1-1994, and NMX-F-607-NORMEX-2020, as well as the ASTM E872 Standard, were employed for calculating the percentage of moisture, ash and, volatile matter respectively [37–39]. Additionally, fixed carbon was obtained via percentage difference using the following formula:

$$\text{Fixed carbon } (\%) = 100 - \%\text{ Moisture} - \%\text{ Ash} - \%\text{ Volatile matter} \quad (1)$$

2.4. Bioadsorbent Preparation

Bioadsorbents BCT1, BCT2, and BCT3 were prepared (Figure 2). Before subjecting the biomass samples to each treatment, they were dried and weighed. The carbonization processes were carried out using an 18.6 L INDBERG-BLUE M muffle furnace. For bioadsorbent BCT1, four crucibles were placed in the muffle at an initial temperature of 24 °C, each containing 5 g of biomass. Once the temperature reached 550 °C, the equipment was turned off and after a 20 s wait, the samples were left inside for 5 min before being removed and placed in a desiccator. The process was repeated for bioadsorbent BCT2, with the only variation being that the initial muffle temperature was 550 °C, and the bioadsorbent was removed immediately after the 20 s and placed in a desiccator. Finally, for bioadsorbent BCT3, the same procedure as for BCT2 was followed, except that the bioadsorbent was removed after 20 s in the muffle and was reintroduced into the equipment and exposed to 550 °C for an additional 20 s before being removed again and placed in a desiccator.

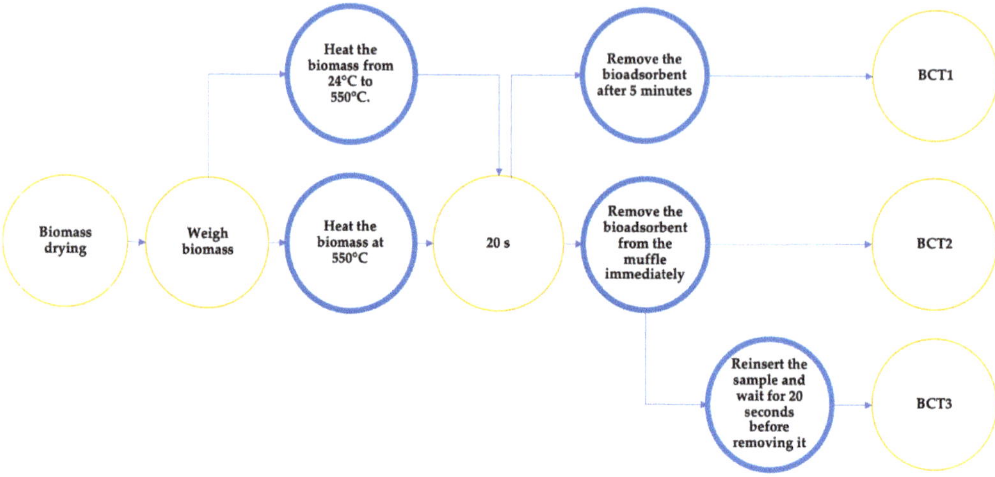

Figure 2. Preparation processes of bioadsorbents BCT1, BCT2, and BCT3.

2.5. *Characterization of Agave angustifolia Bagasse and Bioadsorbents*

2.5.1. Bioadsorbent Yield

The calculation of the bioadsorbent yield on dry basis was carried out using the following formula:

$$\text{Yield } (\%) = \frac{\text{Weight of obtained bioadsorbent}}{\text{Weight of dry biomass}} \times 100. \tag{2}$$

2.5.2. Scanning Electron Microscopy (SEM)

Both the biomass and the bioadsorbents underwent morphological characterization using scanning electron microscopy (SEM) with a scanning electron microscope (JEOL JSM-5410LV, Tokyo, Japan). This system operated at an acceleration voltage of 20 kV and a magnification of 10,000×.

2.5.3. Fourier-Transform Infrared Spectroscopy (FTIR)

The identification of different functional groups in biomass and the three bioadsorbents was carried out via Fourier-transform infrared spectroscopy (FTIR) using a PerkinElmer Frontier spectrometer with an attenuated total reflection (ATR) accessory and a transmittance scale ranging from 4000 cm^{-1} to 500 cm^{-1}.

2.6. *Thermal Analysis*

2.6.1. Thermogravimetric Analysis (TGA)

Thermogravimetric analyses (TGA) and first derivative analyses (DTG) of the starting compounds and resulting bioadsorbents were conducted to study the degradation behavior concerning temperature, ranging from 0 °C to 700 °C. The PerkinElmer TGA-8000 equipment (PerkinElmer, Shelton, CT, USA) was utilized for this purpose.

2.6.2. Differential Scanning Calorimetry (DSC)

The temperature at which specific phase transitions occurred in the bioadsorbents was determined via differential scanning calorimetry (DSC) using a conventional Perkin Elmer DSC model 8500 over a temperature range from 20 °C to 220 °C.

2.7. Preliminary Adsorption Tests

2.7.1. Preparation of Mother Solution

A quantity of 100 mL of mother solution was prepared with a concentration of lignin. The ibuprofen used originated from an infant suspension (Motrin) and was chosen to facilitate dilution in the solution and avoid the use of substances with significant environmental risks. After preparing the solution, pH was measured using a pocket pH/temperature meter, pHep®4, with a resolution of 0.1. The pH was adjusted to 2 by adding drops of hydrochloric acid (HCl). Finally, the adsorption experiments were conducted.

2.7.2. Adsorption Experiments

For the experiment setup, 10 mL of the mother solution was dispensed into a precipitate beaker containing each bioadsorbent. Subsequently, 0.041 g of bioadsorbent was added to each beaker, excluding the control beaker. These containers were covered with Parafilm and placed on plates with a constant stirring speed of 150 rpm for 48 h. At the end of this period, samples were extracted from the beakers and subjected to centrifugation at 6000 gravities at 4 °C for 30 min. Later, the resulting supernatant from each bioadsorbent was analyzed using a UV–visible spectrophotometer, Thermo Scientific Series US 6,414,753 B1 (Thermo Scientific, Waltham, MA, USA) at a maximum adsorption wavelength of 253 nm, which is the wavelength of ibuprofen.

2.7.3. Demonstration of Bioadsorbent Functionality

To exhibit the functionality of the bioadsorbents developed from *Agave angustifolia* bagasse, the samples (bioadsorbent + ibuprofen) were analyzed using scanning electron microscopy (SEM) with the aim of observing crystalline structures and reliefs derived from adsorption in the micrographs. Additionally, the samples were analyzed via Fourier-transform infrared spectroscopy (FTIR) to describe the physical interactions between the bioadsorbent and ibuprofen, thus elucidating a potential adsorption mechanism.

2.8. Statistical Analysis

Proximal analyses were conducted in quadruplicate to determine average values and standard deviations.

3. Results

3.1. Bioadsorbent Preparation

The bioadsorbents from each treatment displayed similar color and texture (Figure 3). The bioadsorbent yield of 25.65% (BTC1), 31.20% (BTC2), and 38.28% (BTC3) were relatively high compared to previous research using lignocellulosic biomass under similar conditions, were yields ranged from 1.27% to 18.48% [40]. Similar findings were reported (25% to 32%) for biomass exposed to 400 °C to 550 °C [41]. These findings align with expectations, as efficient biomass calcination typically yield 30% to 40% [42]. However, as corroborated by various studies, the yield of pyrolyzed sorbents is highly dependent on the raw material [43].

Figure 3. Bioadsorbents before and after ash removal; (A) BCT1, (A1) BCT1 without ash (B) BCT2, (B1) BCT2 without ash and (C) BCT3, (C1) BCT3 without ash.

3.2. Proximate Analysis of Biomass

The proximate analysis of biomass (Table 1), revealed moisture (1.94%), ash (9%), volatile matter (84.82%), and fixed carbon (4.24%). The low moisture content likely stems from the predrying process during biomass conditioning. Unlike some studies using nondried biomass, this approach may influence results [44]. The ash, volatile matter, and fixed carbon content fell within the average range documented for lignocellulosic materials like sugarcane bagasse, laurel sawdust, cinnamon, and eucalyptus [45]. Importantly, the high volatile matter content as noted by Velázquez's et al. [46] is advantageous for bioadsorbent production [47].

Table 1. Proximate analysis results of *Agave angustifolia* bagasse.

Proximate Analysis	Mass Percentage
Moisture	1.94 ± 0.154%
Ash	9.00 ± 0.100%
Volatile matter	84.82 ± 1.297%
Fixed carbon	4.24 ± 0.982%
Protein	4.02 ± 0.107%
Ethereal extract	0.70 ± 0.097%

3.3. Morphological Characterization (SEM) of Biomass and Bioadsorbents

SEM images (Figure 4) revealed that the bagasse exhibited fibers with perforations and a rough surface, suggesting the presence of macromolecules like cellulose, hemicellulose, and lignin [48]. Physical activation significantly altered the surface morphology of the bioadsorbents. The bagasse's holes and rough surface transformed into abundant irregular pores, indicating a highly porous structure. These porosity characteristics are crucial for contaminant removal materials as they provide a larger surface area, facilitating superior adsorption capacity [49].

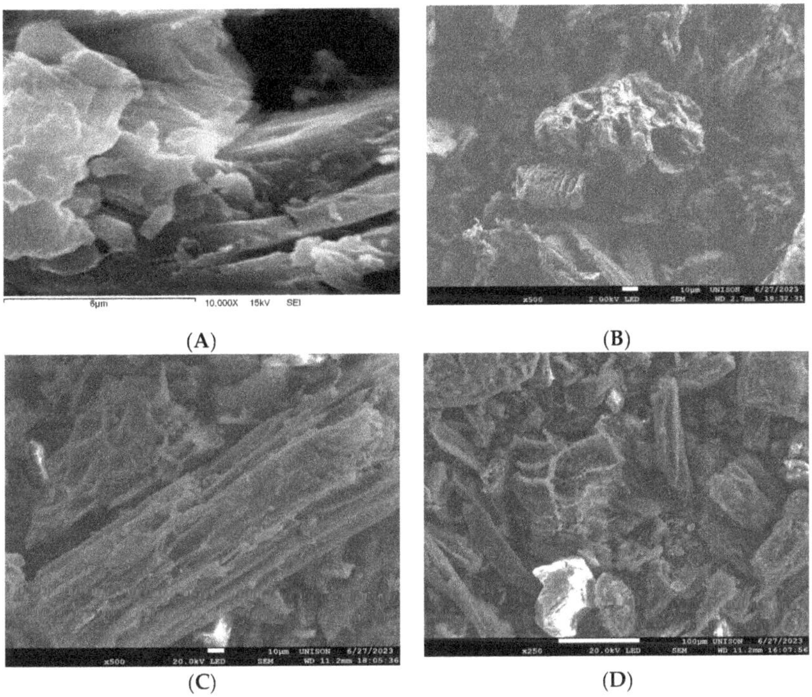

Figure 4. Scanning electron microscopy (SEM) of biomass and bioadsorbents. (**A**) *Agave angustifolia bagasse*, (**B**) BCT1, (**C**) BCT2, and (**D**) BCT3.

3.4. Biomass and Bioadsorbents Structural Characterization (FTIR)

Figure 5 presents the results of Fourier-transform infrared spectroscopy (FTIR) of biomass and bioadsorbents, highlighting key wavenumbers and their corresponding functional groups. The biomass spectrum displayed characteristic functional groups of hemicellulose (beta-glucosidic linkage of cellulose at 769 cm^{-1}) and lignin (aromatic rings at 1635 cm^{-1}). There was a strong intensity absorption peak at 3272 cm^{-1}, being attributed to the stretching of O-H bonds due to the presence of a hydroxyl group, while peaks at 2918 cm^{-1} indicated the presence of a C-H group, at 1616 cm^{-1} of absorbed O-H or carbonyl bands, at 1431 cm^{-1} of a $-CH_2$ group, at 1242 cm^{-1} of C-O bending of cellulose and hemicellulose, at 1027 cm^{-1} of a C-O bond, C-H bending bond and C-C stretching bond, at 773 cm^{-1} indicated the presence of β-glycosidic bond in cellulose or out-of-plane C-H bonds in lignin, and 595 cm^{-1} of carbon–carbon (C-C) bonds in the aromatic structures of lignin. In contrast, the bioabsorbents showed a decrease in intensity of higher wavenumbers, suggesting moisture loss due to high-temperature treatment. Furthermore, the absence of hemicellulose and cellulose bands implied their degradation at these temperatures [50]. Notably, the characteristic aromatic groups of lignin persisted in the strong waves at 1410–1414 cm^{-1} and 1608 cm^{-1} (BCT1-3, respectively) [51]. This finding indicates that the carbonization process selectively degrades cellulose and hemicellulose, while preserving some lignin. This is advantageous because abundant lignin can contribute to high adsorption yield [52].

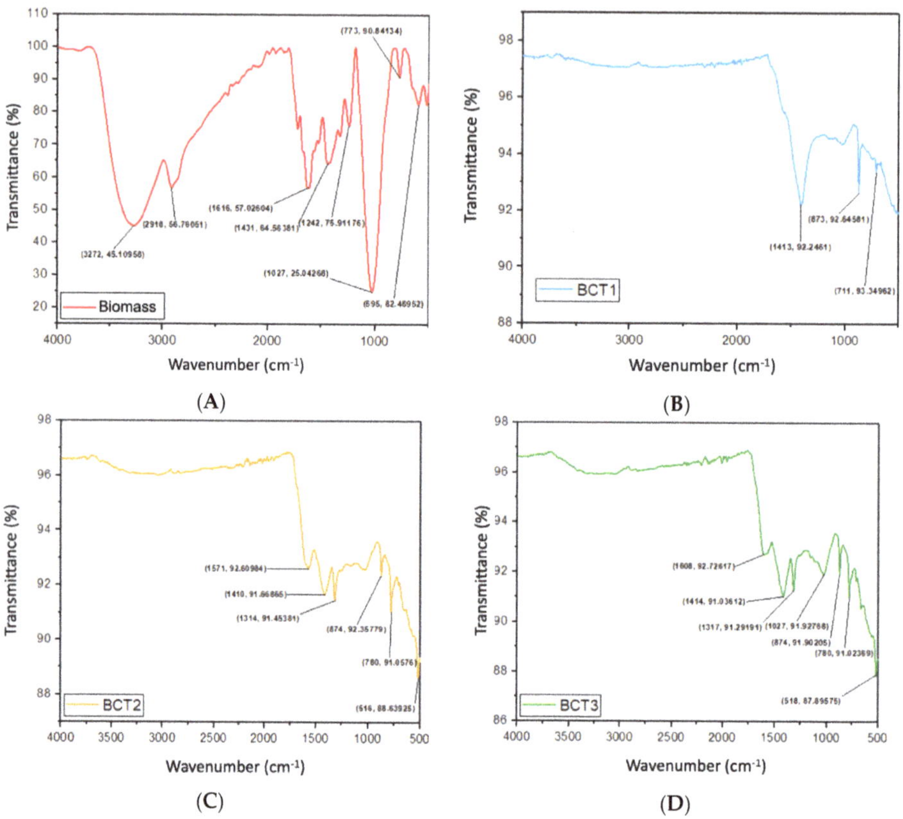

Figure 5. Fourier-transform infrared spectroscopy (FTIR) graphs of biomass and bioadsorbents. (**A**) *Agave angustifolia* bagasse, (**B**) BCT1, (**C**) BCT2, and (**D**) BCT3.

Lignin, a key component found in plant biomass such as wood and agricultural residues, plays a crucial role in the formation of the porous structure of biochar during the pyrolysis process (thermal decomposition in the absence of oxygen). Lignin is a complex polymer containing various functional groups, including hydroxyls and aromatic groups, which can interact with organic and inorganic contaminants in the environment. Due to this chemical structure, lignin exhibits high adsorption capacity, meaning it can bind to and retain contaminants on its surface. When lignin-rich biomass is used to produce biochar, the resulting product has a higher lignin content in its composition. This increased lignin content directly contributes to a higher adsorption capacity of the biochar. Therefore, in a biochar, a greater presence of lignin implies a greater potential to adsorb and retain contaminants from soil, water, or air, making it advantageous for environmental remediation and water purification purposes [52–54].

3.5. Thermal Analysis

3.5.1. Thermogravimetric Analysis (TGA) of Biomass and Bioadsorbents

TGA thermograms showed a nearly 10% mass loss between 100 °C and 200 °C, corresponding to moisture evaporation (Figure 6). A second mass decrease around 390–400 °C indicated cellulose and hemicellulose degradation [55]. Further mass loss (below 10% remaining) between 400 °C and 700 °C was attributed to lignin decomposition [56,57]. While bioadsorbents were not entirely degraded at 700 °C, BCT2 and BCT3 exhibited a weight reduction from 350 °C to 700 °C. Notably, BCT1 displayed only a 25% mass decrease, suggesting the least material degradation among the bioadsorbents. This observation aligns with the DTG curves, where BCT1 showed a single degradation peak, while BCT2 and BCT3 exhibited three and four peaks, respectively.

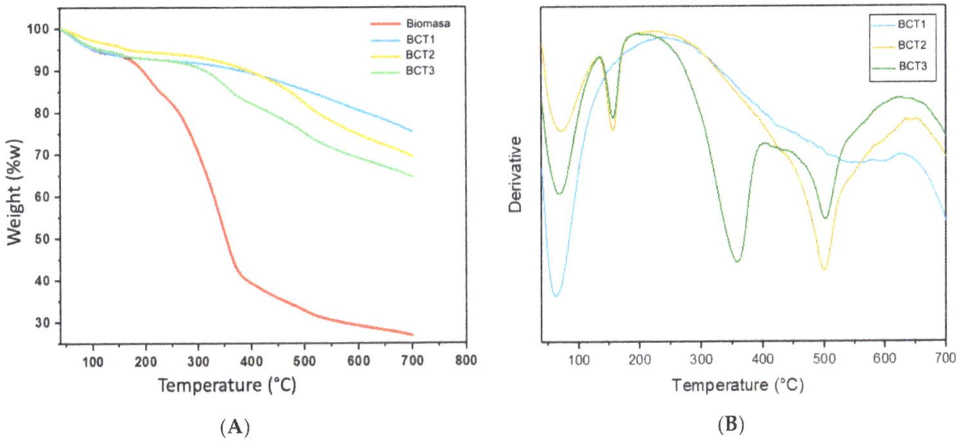

Figure 6. Thermograms of TGA, (**A**) *Agave angustifolia* Bagasse, BCT1, BCT2, and BCT3; and (**B**) derivative BCT1, BCT2, and BCT3.

The presence of a large peak in the derivative before 100 °C in a thermogravimetric analysis (TGA) of the bioadsorbent (BTC1 biochar) typically indicates the release of adsorbed or free water present in the sample. This phenomenon can be attributed to several factors. Firstly, the bioadsorbent's porous nature allows it to adsorb water from the surrounding environment, which is rapidly evaporated upon heating, manifesting as a weight loss peak in the TGA derivative. Additionally, free water trapped within the pores or structure of the bioadsorbent contributes to this initial weight loss. Furthermore, residual volatile compounds from the production process or precursor materials may also evaporate at relatively low temperatures, further adding to the observed peak in the TGA derivative [58].

3.5.2. Differential Scanning Calorimetry (DSC)

Figure 7 shows a differential scanning calorimetry (DSC) thermogram of *Agave angustifolia* bagasse biomass and bioadsorbents after the pyrolysis of lignocellulosic material; peaks observed at 40 °C, 160 °C, and 200 °C likely correspond to distinct thermal degradation phases. The peak at 40 °C (initial water desorption temperature) likely signifies the desorption of moisture or water present within the material's pores in biomass and bioadsorbents [59]. At 160 °C (decomposition temperature of organic matter) in biomass, BCT2, and BCT3, the peak suggests the onset of decomposition for organic components such as cellulose and hemicellulose, releasing volatile compounds and altering the system's specific heat capacity, similarly to the findings of De Dios Naranjo et al. [60], where fibers from *Agave salmiana* showed endotherms at 110 °C and 139 °C. A common observation across the bioadsorbents BCT2 and BCT3 was cellulose degradation at around 200 °C, aligning with findings on sugarcane bagasse by Bernardino et al. [61]. Interestingly, BCT1 did not exhibit any additional phase transitions as seen in Figure 7.

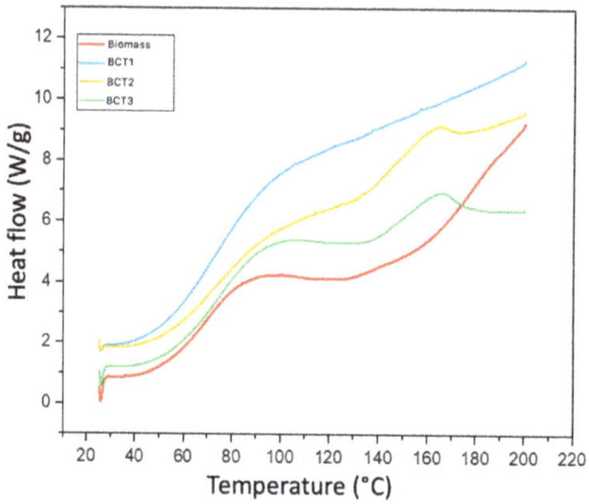

Figure 7. DSC thermograms of biomass, BCT1, BCT2, and BCT3.

3.6. Preliminary Adsorption Tests

BCT1 and BCT2 achieved complete ibuprofen removal (100%) at pH 2 after 48 h with an agitation speed of 150 rpm, suggesting surface properties compatible with ibuprofen adsorption. These bioadsorbents likely possess a combination of functional groups, such as hydroxyl, carboxyl, ether, and amide, among others, as well as surface radicals and surface charge that are compatible with ibuprofen molecules that facilitate adsorption of molecules with similar structures [62]. Figure 8 visually depicts the changes in the microstructure of these bioadsorbents before and after adsorption. While SEM and FTIR analyses of BCT3 after adsorption supported potential adsorption, further testing is necessary due to inconsistencies in specific adsorption test readings.

Figure 8. Adsorption process of ibuprofen on bioadsorbents BCT1 and BCT2.

4. Discussion

4.1. Bioadsorbent Structure

The characterization results revealed that the bioadsorbents derived from *Agave angustifolia bagasse* exhibited a porous structure with a high surface area, which is desirable for adsorption applications. The presence of functional groups on the biochar surface, as identified via FTIR analysis, suggests its potential for interacting with target contaminants like ibuprofen. Other similar studies have been carried out to remove pollutants in an aquatic environment, such as the research presented by Liu et al. [63]. The bioadsorbents prepared by Chladophora (CB) were used to adsorb the ibuprofen and ofloxacin in sewage. SEM showed that CB preserved the natural tubular structure and increased its specific surface area. According to the BET results, the specific surface area of CB was 296.831 m^2 g^{-1}. The adsorption of ibuprofen and ofloxacin on biochar conformed to the pseudo second-order kinetic model, indicating that their adsorption rate was controlled via the chemical adsorption mechanism. The bioadsorbents obtained from *Agave angustifolia* bagasse presented better ibuprofen removal results than the bioadsorbent result from Chladophora, due to the characteristics of the material and functional groups present on the surface. Nama et al. [64] produced bioadsorbent from a commercially significant microalga, such as Dunaliella biomass, through pyrolysis. FT-IR was employed to examine both the morphological and functional aspects of the biochar, as well as its thermal degradation behavior. Pyrolytic behavior was further explored using thermogravimetric analysis under an oxidizing environment. SEM analysis revealed diverse and rough-textured structures of the biochar, supporting enhanced adsorption capabilities. The biochar was used to remove naphthalene in aquatic environments, obtaining a removal of 63% of 100 mg L^{-1} of the pollutant. The structure of the bioadsorbents obtained by Nama et al. [64] was similar to that obtained in the present study for *Agave angustifolia* bioadsorbents.

4.2. Demonstration of the Functionality of Bioadsorbents

SEM images (Figure 9) revealed significant changes in the morphology of the postadsorption bioadsorbents (BCCT1, BCCT2, BCCT3) compared to their initial structure. All three displayed a grayish layer obstructing pores, with BCCT1 and BCCT3 exhibiting a reduction in the number of pores. Interestingly, BCCT2 showed a reduced surface roughness, indicating ibuprofen incorporation on its surface. These observations align with previous studies reporting similar postadsorption characteristics, supporting the efficiency of the adsorption process [65,66]. The adsorption performance of bioadsorbents depended on the reaction conditions, including the raw materials used, kinetic parameters, pH value, and specific surface area.

Figure 10 shows the infrared spectrum of ibuprofen (Figure 10A) and the infrared spectra of the bioadsorbents after removal of ibuprofen (Figure 10B–D). The analysis of infrared spectra (FTIR) provides valuable information about molecular structure and interactions present in the chemical compounds under study. Starting with the spectra of ibuprofen and the bioadsorbents derived from *Agave angustifolia* bagasse, characteristic bands were identified that reveal specific bond vibrations. In the ibuprofen spectrum, the bands at 2362 cm^{-1} and 2302 cm^{-1} are attributed to the stretching vibrations of the carboxyl (C=O) and C=C functional groups, respectively. The band at 1618 cm^{-1} corresponds to the bending vibration of the aromatic ring. The values of 1710–1708 cm^{-1} correspond to the vibration of the ester functional group (–COO–), specifically the carbonyl bond (–C=O), which is present in the structure of ibuprofen. The bands at 1314 cm^{-1} and 1160 cm^{-1} are associated with the stretching vibrations of C–O and C–C bonds, respectively, and are indicative of the presence of ester and alkyl groups in the ibuprofen molecule. The bands at 1034 cm^{-1}, 778 cm^{-1}, and 514 cm^{-1} are related to characteristic vibrations of the aromatic structure and C-H bond in the ibuprofen spectrum.

In the spectra of bioadsorbents (BCCT1, BCCT2, and BCCT3), bands suggesting the presence of different functional groups were observed. Intense bands in the 3700–2900 cm^{-1} region are typical of the stretching vibrations of –OH and –CH groups present in organic compounds. The bands at 2154 cm^{-1}, 2006 cm^{-1}, and 1948 cm^{-1} may be related to vibrations of C≡C bonds. In the three bioadsorbents, the spectra showed the band at 1710–1708 cm^{-1} (inverted) and 1610–1598 cm^{-1} corresponding to the carbonyl carbon (–C=O) of the carboxyl group of ibuprofen. Also, the band at 1598 cm^{-1} indicates the presence of double C=C bonds in carbonaceous structures. The band at 1434–1402 cm^{-1} may be related to the bending vibration of the –C–H bond in alkenes and aromatic rings of the bioadsorbent and ibuprofen. The bands at 1324 cm^{-1} and 1168 cm^{-1} correspond to the stretching vibrations of C–O and C–C, respectively, similar to those observed in the ibuprofen spectrum.

Figure 9. Scanning electron microscopy (SEM) of bioadsorbents after adsorption test. (**A**) BCCT1, (**B**) BCCT2, and (**C**) BCCT3.

The overlap of ibuprofen bands in the spectra of the bioadsorbents suggests an interaction between the drug and the adsorbent material. Adsorption mechanisms likely involve a combination of hydrophobic interactions, hydrogen bonding, and π-π interactions between the benzene ring of ibuprofen and the carbonaceous structures of the bioadsorbents. The presence of functional groups on the surface of the bioadsorbents, such as –OH and –COOH, facilitates hydrogen bonding interactions with the functional groups of ibuprofen, while π-π interactions benefit from the aromatic structure of both components. These observations indicate effective adsorption of ibuprofen molecules on the bioadsorbent surface through interactions with functional groups, as evidenced by the presence of characteristic ibuprofen bands [67].

The difference in adsorption capacity among the three bioadsorbents can be attributed to variations in chemical composition and surface morphology. For example, the presence of specific functional groups such as hydroxyl (–OH) or carboxyl (–COOH) groups may increase the bioadsorbents affinity for ibuprofen through hydrogen bonding interactions. Additionally, differences in pore structure and specific surface area of each bioadsorbent may influence the amount of available adsorption sites. Overall, the bioadsorbent exhibiting a higher quantity of functional groups and a greater surface area is expected to have a higher adsorption capacity for ibuprofen, such as BCCT1 and BBCT2 bioadsorbents.

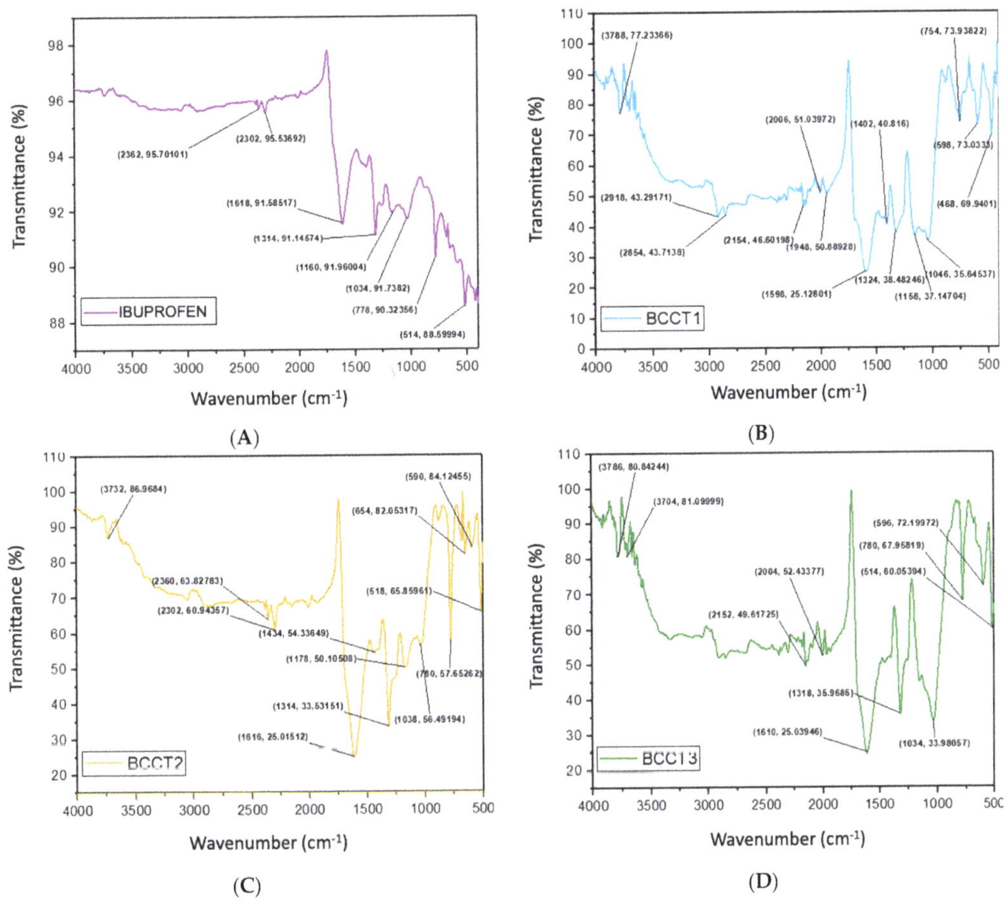

Figure 10. Fourier-transform infrared spectroscopy (FTIR) graphs of ibuprofen and bioadsorbents after adsorption test. (**A**) Ibuprofen, (**B**) BCCT1, (**C**) BCCT2, and (**D**) BCCT3.

The interaction mechanisms between biochar and pollutants in water exhibit diversity, encompassing physical adsorption, chemical adsorption, redox, and microbial activity. With numerous micropores and mesopores, biochar effectively absorbs dissolved pollutants. Moreover, its high specific surface area offers ample adsorption sites for a range of compounds, including organic pollutants, heavy metals, and nitrogen oxides. Beyond physical adsorption, biochar demonstrates chemical adsorption capabilities through surface functional groups forming covalent or hydrogen bonds with contaminants. Additionally, biochar engages in redox reactions, facilitating the conversion of pollutants into harmless substances through chemical reduction or oxidation. The efficacy of biochar in adsorbing and removing pollutants like ibuprofen varies depending on the material and its adsorption mechanism [68].

Specifically, the absorption of ibuprofen into biochar varies depending on the material used. Various studies have elucidated the mechanisms underlying ibuprofen by adsorption onto biochar. The mechanisms of absorption and removal of ibuprofen via biochar encompass a variety of processes. These include multimolecular adsorption, π-π interaction, pore filling, hydrogen bonding, and interactions between π-acceptors and electrostatics. Additionally, the application of pseudo-second-order kinetics and Langmuir–Freundlich isotherm models in adsorption was observed. Mechanisms such as pore filling-diffusion, hydrophobic interaction, and π-π electron donor-acceptor interaction also contribute to ibuprofen removal. Improved adsorption and separation were achieved through the preparation of magnetic biochar, while stability across multiple adsorption cycles was evidenced in successful biochar reuse. These findings underscore the efficacy and versatility of biocarbon derived from biomass waste in environmentally friendly and economically viable ibuprofen removal in aqueous environments [68–70].

To examine the adsorption capacity of a material, researchers primarily use batch experiments. In batch experiments, the design typically encompasses several stages. First, the material undergoes preparation, which may involve raw biomass, chemically modified biomass, or thermal treatment to produce biochars, among other methods. Next, various experimental parameters such as initial and final pH of the solution, contact time, initial concentration of the pollutant, ionic strength, particle size of the adsorbent, and dose of adsorbent are studied and optimized. Subsequently, the adsorption mechanism is investigated, followed by desorption experiments to assess the release of adsorbed pollutants. Finally, the potential for regeneration and reusability of the material is evaluated [71], these last two points being a limitation of the present study.

4.3. *Contribution of the Study to Sustainability*

The presence of pharmaceutical residues in the environment, including ibuprofen, poses a significant threat to ecosystems. According to the FAO and UNEP [72], over 4000 widely used pharmaceutical and chemical products, including drugs, contribute to environmental pollution and the emergence of drug-resistant diseases. Furthermore, soil pollution is intrinsically linked to the degradation and loss of aquatic and terrestrial resources, impacting Sustainable Development Goals (SDGs) 14 (life below water) and 15 (life on land) [72–74]. In this sense, developing effective methods for managing pharmaceutical residues is critical to achieve a more sustainable future.

Also, the study exploration of utilizing bacanora industry bagasse for biochar production represents a significant contribution to sustainability efforts. By repurposing this byproduct, the research not only addresses waste management concerns but also offers a viable solution for mitigating emerging pollution, such as pharmaceutical residues. Biochar derived from bacanora bagasse has the potential to effectively remove these pharmaceutical residues from various environmental matrices, thereby promoting cleaner ecosystems and healthier communities. This innovative approach highlights the intersection of agricultural industry byproducts and environmental stewardship, showcasing how sustainable practices can emerge from unlikely sources.

5. Conclusions

Agave angustifolia bagasse exhibits great potential as a pyrolysis-derived adsorbent (yield: 25.65–38.28%) for ibuprofen removal. Both biomass and bioadsorbents (BCT1, BCT2, and BCT3) possessed functional groups associated with cellulose, hemicellulose, and lignin. These macromolecules underwent degradation at around 350 °C (cellulose and hemicellulose), while lignin remained partially undegraded at 550 °C. While biomass displayed a fissured and porous morphology, bioadsorbents displayed a highly porous surface. After the adsorption process, these surfaces showed surface coverage consistent with ibuprofen adsorption. Treatments BCT1 and BCT2 successfully removed 100% of ibuprofen from an aqueous solution with an initial concentration of 62.5 mg/L. Supplementary adsorption tests are needed for BCT3, as SEM and FTIR results suggest successful adsorption despite inconsistent readings in specific adsorption tests. Overall, bioadsorbents derived from *Agave angustifolia* bagasse biomass demonstrate significant potential for the removal of ibuprofen from aqueous solutions.

Author Contributions: Conceptualization, M.d.J.M.-V., H.G., I.E.Q.-R. and J.A.T.-H.; methodology, H.A.R.-V., M.d.J.M.-V., I.S.-S., C.V.-L., S.E.B.-I. and J.A.T.-H.; validation, H.G. and J.A.T.-H.; formal analysis, H.A.R.-V. and J.A.T.-H.; investigation, H.A.R.-V., M.d.J.M.-V., K.H.O.-V. and J.A.T.-H.; resources, M.d.J.M.-V., F.R.-F. and J.A.T.-H.; data curation, M.d.J.M.-V. and J.A.T.-H.; writing—original draft preparation, H.A.R.-V. and J.A.T.-H.; writing—review and editing, H.A.R.-V., H.G., J.E. and J.A.T.-H.; visualization, J.E., I.E.Q.-R. and J.A.T.-H.; supervision, H.G., A.Z.G.-V., C.G.B.-U. and J.A.T.-H.; project administration, F.R.-F. and J.A.T.-H. All authors have read and agreed to the published version of the manuscript.

Funding: This research received no external funding.

Institutional Review Board Statement: Not applicable.

Informed Consent Statement: Not applicable.

Data Availability Statement: The data generated from this research are available from the authors.

Acknowledgments: The authors are grateful to the University of Sonora for their support and to CONAHCYT for the postgraduate scholarship granted to the first author.

Conflicts of Interest: The authors declare no conflicts of interest.

References

1. Nwokediegwu, Z.Q.S.; Daraojimba, O.H.; Oliha, J.S.; Obaigbena, A.; Dada, M.A.; Majemite, M.T. Review of emerging contaminants in water: USA and African perspectives. *Int. J. Sci. Res. Arch.* **2024**, *11*, 350–360. [CrossRef]
2. Thi, L.A.P.; Ly, L.T.M.; Do, H.T.; Chinh, P.M. Occurrence and risks of emerging pollutants in water bodies. In *Advanced Functional Materials and Methods for Photodegradation of Toxic Pollutants*; Elsevier: Amsterdam, The Netherlands, 2024; pp. 1–36. [CrossRef]
3. Munzhelele, E.P.; Mudzielwana, R.; Ayinde, W.B.; Gitari, W.M. Pharmaceutical Contaminants in Wastewater and Receiving Water Bodies of South Africa: A Review of Sources, Pathways, Occurrence, Effects, and Geographical Distribution. *Water* **2024**, *16*, 796. [CrossRef]
4. Fraiha, O.; Hadoudi, N.; Zaki, N.; Salhi, A.; Amhamdi, H.; Akichouh, E.H.; Mourabit, F.; Ahari, M. Comprehensive review on the adsorption of pharmaceutical products from wastewater by clay materials. *Desalination Water Treat.* **2024**, *317*, 100114. [CrossRef]
5. Puri, M.; Gandhi, K.; Kumar, M.S. Emerging environmental contaminants: A global perspective on policies and regulations. *J. Environ. Manag.* **2023**, *332*, 117344. [CrossRef]
6. Nain, P.; Anctil, A. End-of-life solar photovoltaic waste management: A comparison as per European Union and United States regulatory approaches. *Resour. Conserv. Recycl. Adv.* **2024**, *21*, 200212. [CrossRef]
7. Rainsford, K.D. Ibuprofen: Pharmacology, efficacy and safety. *Inflammopharmacology* **2009**, *17*, 275–342. [CrossRef]
8. Buser, H.-R.; Poiger, T.; Müller, M.D. Occurrence and Environmental Behavior of the Chiral Pharmaceutical Drug Ibuprofen in Surface Waters and in Wastewater. *Environ. Sci. Technol.* **1999**, *33*, 2529–2535. [CrossRef]
9. Marchlewicz, A.; Guzik, U.; Wojcieszyńska, D. Over-the-Counter Monocyclic Non-Steroidal Anti-Inflammatory Drugs in Environment—Sources, Risks, Biodegradation. *Water Air Soil Pollut.* **2015**, *226*, 355. [CrossRef]
10. Muñiz-González, A.-B. Ibuprofen as an emerging pollutant on non-target aquatic invertebrates: Effects on Chironomus riparius. *Environ. Toxicol. Pharmacol.* **2021**, *81*, 103537. [CrossRef]
11. Parolini, M.; Binelli, A.; Provini, A. Chronic effects induced by ibuprofen on the freshwater bivalve Dreissena polymorpha. *Ecotoxicol. Environ. Saf.* **2011**, *74*, 1586–1594. [CrossRef]
12. Chopra, S.; Kumar, D. Ibuprofen as an emerging organic contaminant in environment, distribution and remediation. *Heliyon* **2020**, *6*, e04087. [CrossRef] [PubMed]
13. Jan-Roblero, J.; Cruz-Maya, J.A. Ibuprofen: Toxicology and Biodegradation of an Emerging Contaminant. *Molecules* **2023**, *28*, 2097. [CrossRef] [PubMed]
14. Dey Chowdhury, S.; Surampalli, R.Y.; Bhunia, P. Advanced Oxidation Processes for Microconstituents Removal in Aquatic Environments. In *Microconstituents in the Environment: Occurrence, Fate, Removal and Management*; John Wiley & Sons, Inc.: Hoboken, NJ, USA, 2023; pp. 367–404. [CrossRef]
15. Zhao, S.; Li, H.; Zhou, J.; Sumpradit, T.; Salama, E.-S.; Li, X.; Qu, J. Simultaneous degradation of NSAIDs in aqueous and sludge stages by an electron-Fenton system derived from sediment microbial fuel cell based on a novel Fe@Mn biochar GDC. *Chem. Eng. J.* **2024**, *482*, 148979. [CrossRef]
16. Zamora, J.; Rodríguez, M.; Vera, L.; Gómez, Y. Técnicas aplicadas a la remoción de fármacos, usados en el tratamiento del COVID-19: Una revisión. *Polo Del Conoc.* **2021**, *6*, 15–36.
17. Qiu, X.; Wang, B.; Zhao, X.; Zhou, X.; Wang, R. Green and Sustainable Imprinting Technology for Removal of Heavy Metal Ions from Water via Selective Adsorption. *Sustainability* **2023**, *16*, 339. [CrossRef]
18. El-Taweel, R.M.; Mohamed, N.; Alrefaey, K.A.; Husien, S.; Abdel-Aziz, A.; Salim, A.I.; Mostafa, N.G.; Said, L.A.; Fahim, I.S.; Radwan, A.G. A review of coagulation explaining its definition, mechanism, coagulant types, and optimization models; RSM, and ANN. *Curr. Res. Green Sustain. Chem.* **2023**, *6*, 100358. [CrossRef]

19. Verma, A.; Sharma, G.; Kumar, A.; Dhiman, P.; Mola, G.T.; Shan, A.; Si, C. Microplastic pollutants in water: A comprehensive review on their remediation by adsorption using various adsorbents. *Chemosphere* **2024**, *352*, 141365. [CrossRef] [PubMed]
20. Wang, Y.; Chen, L.; Zhu, Y.; Fang, W.; Tan, Y.; He, Z.; Liao, H. Research status, trends, and mechanisms of biochar adsorption for wastewater treatment: A scientometric review. *Environ. Sci. Eur.* **2024**, *36*, 25. [CrossRef]
21. Hu, B.; Ai, Y.; Jin, J.; Hayat, T.; Alsaedi, A.; Zhuang, L.; Wang, X. Efficient elimination of organic and inorganic pollutants by biochar and biochar-based materials. *Biochar* **2020**, *2*, 47–64. [CrossRef]
22. Li, R.; Zhang, C.; Chen, W.-H.; Kwon, E.E.; Rajendran, S.; Zhang, Y. Multistage utilization of soybean straw-derived P-doped biochar for aquatic pollutant removal and biofuel usage. *Bioresour. Technol.* **2023**, *387*, 129657. [CrossRef]
23. Janiszewska, D.; Olchowski, R.; Nowicka, A.; Zborowska, M.; Marszalkiewicz, K.; Shams, M.; Giannakoudakis, D.A.; Anastopoulos, I.; Barczak, M. Activated biochars derived from wood biomass liquefaction residues for effective removal of hazardous hexavalent chromium from aquatic environments. *GCB Bioenergy* **2021**, *13*, 1247–1259. [CrossRef]
24. Ali, I.; Asim, M.; Khan, T.A. Low cost adsorbents for the removal of organic pollutants from wastewater. *J. Environ. Manag.* **2012**, *113*, 170–183. [CrossRef]
25. Homem, V.; Santos, L. Degradation and removal methods of antibiotics from aqueous matrices—A review. *J. Environ. Manag.* **2011**, *92*, 2304–2347. [CrossRef] [PubMed]
26. Sherpa, S.W.; Ponnuchamy, M.; Kapoor, A.; Jacob, M.M.; Sivaraman, P. Facile removal of sulfamethoxazole antibiotic from contaminated water using bagasse-derived pyrolytic biocarbon: Parametric assessment, mechanistic insights and scale-up analysis. *J. Water Process. Eng.* **2024**, *60*, 105110. [CrossRef]
27. Clavero, V.O.; Weber, A.; Schröder, W.; Curticapean, D. Spectral analysis of bacanora (agave-derived liquor) by using FT-Raman spectroscopy. In *Optical Sensing and Detection IV*; SPIE: Bellingham, WA, USA, 2016; Volume 9899, pp. 618–625. [CrossRef]
28. Salazar, V. La industria del bacanora: Historia y tradición de resistencia en la sierra sonorense. *Reg. Soc.* **2007**, *19*, 105–133. [CrossRef]
29. Cortés, C. Propiedades Mecánicas a Tensión de las Fibras del Bagazo del Agave Angustifolia Haw, Residuo Proveniente de la Producción Artesanal del Mezcal. Master's Thesis, Instituto Politécnico Nacional, Mexico City, Mexico, 2009.
30. Sierra, E.; Alcaraz, J.; Valdivia, A.; Rosas, A.; Hernández, M.; Vivaldo, E.; Martínez, A. Bagazo de Agave: De Desecho Agroindustrial a Materia Prima en las Biorrefinerías. Universidad Autónoma de México. 2021. Available online: http://ciencia.unam.mx/leer/1112/bagazo-de-agave-de-desecho-agroindustrial-a-materia-prima-en-las-biorrefinerias- (accessed on 3 February 2024).
31. Hidalgo-Reyes, M.; Caballero-Caballero, M.; Hernández-Gómez, L.H.; Urriolagoitia-Calderón, G. Chemical and morphological characterization of Agave angustifolia bagasse fibers. *Bot. Sci.* **2015**, *93*, 807–817. [CrossRef]
32. Jiménez-Ortega, L.A.; Valdez-Baro, O.; Bernal-Millán, M.J.; Rivera-Salas, M.M.; Heredia, J.B. Agave Byproducts: As Sources of Phytochemicals with Functional Activities and Their Management for Industrial Applications. In *Food Byproducts Management and Their Utilization*; Apple Academic Press: Palm Bay, FL, USA, 2024; pp. 385–416. [CrossRef]
33. Gómora-Hernández, J.C.; Tecante, A.; del Carmen Carreño-De-León, M.; Flores-Álamo, N.; Ventura-Cruz, S. Preparation of porous microcrystalline cellulose from mezcal industry agave bagasse by low reagent loading sequential chemical treatment. *Cellulose* **2023**, *30*, 2067–2084. [CrossRef]
34. Ayala-Cortés, A.; Arancibia-Bulnes, C.A.; Villafán-Vidales, H.I.; Lobato-Peralta, D.R.; Martínez-Casillas, D.C.; Cuentas-Gallegos, A.K. Solar pyrolysis of agave and tomato pruning wastes: Insights of the effect of pyrolysis operation parameters on the physicochemical properties of biochar. *AIP Conf. Proc.* **2019**, *2126*, 180001. [CrossRef]
35. Lobato-Peralta, D.R.; Ayala-Cortes, A.; Longoria, A.; Pacheco-Catalan, D.E.; Okoye, P.U.; Villafan-Vidales, H.I.; Arancibia-Bulnes, C.A.; Cuentas-Gallegos, A.K. Activated carbons obtained by environmentally friendly activation using solar energy for their use in neutral electrolyte supercapacitors. *J. Energy Storage* **2022**, *52*, 104888. [CrossRef]
36. Robles-García, M.; Del-Toro-Sánchez, C.L.; Márquez-Ríos, E.; Barrera-Rodríguez, A.; Aguilar, J.; Aguilar, J.A.; Reynoso-Marín, F.J.; Ceja, I.; Dórame-Miranda, R.; Rodríguez-Félix, F. Nanofibers of cellulose bagasse from Agave tequilana Weber var. azul by electrospinning: Preparation and characterization. *Carbohydr. Polym.* **2018**, *192*, 69–74. [CrossRef]
37. NORMA Oficial Mexicana. *NOM-116-SSA1-1994, Determinación de Humedad en Alimentos por Tratamiento Térmico. Método por Arena o Gasa*; Pub. L. No. 116; Diario Oficial de la Federación: Mexico City, Mexico, 1994.
38. Proyecto de Norma Mexicana PROY-NMX-F-607-NORMEX-2020. *Determinación de Cenizas en Alimentos-Método de Prueba*; Diario Oficial de la Federación: Mexico City, Mexico, 2020.
39. U.S. Department of Energy. *Standard Test Method for Volatile Matter in the Analysis of Particulate Wood Fuels*; ASTM International: West Conshohocken, PA, USA, 2019.
40. Martínez, A.; Bohórquez, L. *Evaluación de la Eficiencia de Biochar Producido a Partir de Pirólisis Lenta de Bagazo de Caña Como Medio Filtrante Para Retención de Fenoles en Matriz Acuosa [Monography]*; Universidad de La Salle: Bogotá, Colombia, 2017.
41. Williams, P.T.; Nugranad, N. Comparison of products from the pyrolysis and catalytic pyrolysis of rice husks. *Energy* **2000**, *25*, 493–513. [CrossRef]
42. Iglesias-Abad, S.; Alvarez-Vera, M.; Salas, C.; Vázquez, J. Biochar of residual biomass from eucalyptus (*Eucalyptus globulus*) by two pyrolysis methods. *Manglar* **2020**, *17*, 105–111. [CrossRef]
43. Cha, J.S.; Park, S.H.; Jung, S.-C.; Ryu, C.; Jeon, J.-K.; Shin, M.-C.; Park, Y.-K. Production and utilization of biochar: A review. *J. Ind. Eng. Chem.* **2016**, *40*, 1–15. [CrossRef]

44. Flores, P.; Robles, C.; Castañeda, E. Generación y caracterización básica de bagazos de la agroindustria del mezcal en Oaxaca. *Rev. Mex. Cienc. Agrícolas* **2020**, *11*, 1437–1445. [CrossRef]
45. Asimbaya, C.; Rosas, N.; Endara, D.; Guerrero, V. Obtención de Carbón Activado a partir de Residuos Lignocelulósicos de Canelo, Laurel y Eucalipto. *Rev. Politécnica* **2016**, *36*, 24.
46. Velázquez, A.; Bolaños, E.; Pliego, Y. Optimización de la Producción de Carbón Activado a Partir de Bambú. *Rev. Mex. Ing. Química* **2010**, *9*, 359–366.
47. Alonso, M.; Rigal, L. Caracterización y valorización del bagazo de Agave tequilana Weber de la industria del tequila. *Rev. Chapingo* **1997**, *3*, 31–39. [CrossRef]
48. Azeh, Y.; Olatunji, G.A.; Mamza, P.A. Scanning Electron Microscopy and Kinetic Studies of Ketene-Acetylated Wood/Cellulose High-Density Polyethylene Blends. *Int. J. Carbohydr. Chem.* **2012**, *2012*, 456491. [CrossRef]
49. Weber, K.; Quicker, P. Properties of biochar. *Fuel* **2018**, *217*, 240–261. [CrossRef]
50. Wang, J.; Minami, E.; Asmadi, M.; Kawamoto, H. Effect of delignification on thermal degradation reactivities of hemicellulose and cellulose in wood cell walls. *J. Wood Sci.* **2021**, *67*, 19. [CrossRef]
51. Janu, R.; Mrlik, V.; Ribitsch, D.; Hofman, J.; Sedláček, P.; Bielská, L.; Soja, G. Biochar surface functional groups as affected by biomass feedstock, biochar composition and pyrolysis temperature. *Carbon Resour. Convers.* **2021**, *4*, 36–46. [CrossRef]
52. Hu, J.; Shen, D.; Wu, S.; Zhang, H.; Xiao, R. Effect of temperature on structure evolution in char from hydrothermal degradation of lignin. *J. Anal. Appl. Pyrolysis* **2014**, *106*, 118–124. [CrossRef]
53. Gul, E.; Alrawashdeh, K.A.B.; Masek, O.; Skreiberg, Ø.; Corona, A.; Zampilli, M.; Wang, L.; Samaras, P.; Yang, Q.; Zhou, H.; et al. Production and use of biochar from lignin and lignin-rich residues (such as digestate and olive stones) for wastewater treatment. *J. Anal. Appl. Pyrolysis* **2021**, *158*, 105263. [CrossRef]
54. Li, Y.; Wang, F.; Miao, Y.; Mai, Y.; Li, H.; Chen, X.; Chen, J. A lignin-biochar with high oxygen-containing groups for adsorbing lead ion prepared by simultaneous oxidization and carbonization. *Bioresour. Technol.* **2020**, *307*, 123165. [CrossRef]
55. Tripathi, M.; Sahu, J.; Ganesan, P. Effect of process parameters on production of biochar from biomass waste through pyrolysis: A review. *Renew. Sustain. Energy Rev.* **2016**, *55*, 467–481. [CrossRef]
56. Pereira, M.E.; Varanda, L.D.; de Carvalho, N.R.; Sette, C., Jr.; de Padua, F.A.; De Conti, A.C.; Yamaji, F.M. Biochar produced from poultry litter waste. *Res. Soc. Dev.* **2021**, *10*, e351101119704. [CrossRef]
57. De la Cuesta, J.; Montoya, U.; Betancourt, S.; Álvarez, C. Effect of processing temperature on the mechanical properties of binderless fiberboards. *Prospect* **2011**, *9*, 75–80.
58. Rasapoor, M.; Young, B.; Asadov, A.; Brar, R.; Sarmah, A.K.; Zhuang, W.-Q.; Baroutian, S. Effects of biochar and activated carbon on biogas generation: A thermogravimetric and chemical analysis approach. *Energy Convers. Manag.* **2020**, *203*, 112221. [CrossRef]
59. El-Nemr, M.A.; Abdelmonem, N.M.; Ismail, I.M.A.; Ragab, S.; El Nemr, A. Ozone and Ammonium Hydroxide Modification of Biochar Prepared from *Pisum sativum* Peels Improves the Adsorption of Copper (II) from an Aqueous Medium. *Environ. Process.* **2020**, *7*, 973–1007. [CrossRef]
60. De Dios Naranjo, C.; Alamilla, L.; Gutiérrez, G.; Terres, E.; Solorza, J.; Romero, S.; Yee-Madeira, H.; Flores, A.; Mora, R. Isolation and characterization of cellulose obtained from Agave salmiana fibers using two acid alkali extraction methods. *Rev. Mex. Cienc. Agrícolas* **2016**, *7*, 31–43.
61. Bernardino, C.A.R.; Mahler, C.F.; Veloso, M.C.C.; Romeiro, G.A. Preparation of Biochar from Sugarcane By-product Filter Mud by Slow Pyrolysis and Its Use Like Adsorbent. *Waste Biomass-Valorization* **2017**, *8*, 2511–2521. [CrossRef]
62. Xiao, X.; Chen, B.; Chen, Z.; Zhu, L.; Schnoor, J.L. Insight into Multiple and Multilevel Structures of Biochars and Their Potential Environmental Applications: A Critical Review. *Environ. Sci. Technol.* **2018**, *52*, 5027–5047. [CrossRef] [PubMed]
63. Liu, L.; Shang, D.; Zhao, Y.; Zhao, Q.; Guo, Y.; Zhang, F.; Kong, Q.; Zhang, H.; Wang, Q.; Zhao, C. Preparation of Chladophora-Based Biochar and Its Adsorption Properties for Antibiotics. *ChemistrySelect* **2024**, *9*, e202303598. [CrossRef]
64. Nama, M.; Satasiya, G.; Sahoo, T.P.; Moradeeya, P.G.; Sadukha, S.; Singhal, K.; Saravaia, H.T.; Dineshkumar, R.; Kumar, M.A. Thermo-chemical behaviour of Dunaliella salina biomass and valorising their biochar for naphthalene removal from aqueous rural environment. *Chemosphere* **2024**, *353*, 141639. [CrossRef]
65. Mondal, S.; Bobde, K.; Aikat, K.; Halder, G. Biosorptive uptake of ibuprofen by steam activated biochar derived from mung bean husk: Equilibrium, kinetics, thermodynamics, modeling and eco-toxicological studies. *J. Environ. Manag.* **2016**, *182*, 581–594. [CrossRef]
66. Show, S.; Mukherjee, S.; Devi, M.S.; Karmakar, B.; Halder, G. Linear and non-linear analysis of Ibuprofen riddance efficacy by *Terminalia catappa* active biochar: Equilibrium, kinetics, safe disposal, reusability and cost estimation. *Process. Saf. Environ. Prot.* **2021**, *147*, 942–964. [CrossRef]
67. Chakraborty, P.; Show, S.; Banerjee, S.; Halder, G. Mechanistic insight into sorptive elimination of ibuprofen employing bi-directional activated biochar from sugarcane bagasse: Performance evaluation and cost estimation. *J. Environ. Chem. Eng.* **2018**, *6*, 5287–5300. [CrossRef]
68. Zhan, D.; Ye, A.; Hou, T. Research progress on biochar-based material adsorption and removal of ibuprofen. *Front. Environ. Sci.* **2023**, *11*, 1327000. [CrossRef]
69. Lekene, R.B.N.; Ntep, T.M.M.; Fetzer, M.N.A.; Strothmann, T.; Nsami, J.N.; Janiak, C. The efficient removal of ibuprofen, caffeine, and bisphenol A using engineered egusi seed shells biochar: Adsorption kinetics, equilibrium, thermodynamics, and mechanism. *Environ. Sci. Pollut. Res.* **2023**, *30*, 100095–100113. [CrossRef]

70. Osman, A.I.; Ayati, A.; Farghali, M.; Krivoshapkin, P.; Tanhaei, B.; Karimi-Maleh, H.; Krivoshapkina, E.; Taheri, P.; Tracey, C.; Al-Fatesh, A.; et al. Advanced adsorbents for ibuprofen removal from aquatic environments: A review. *Environ. Chem. Lett.* **2024**, *22*, 373–418. [CrossRef]
71. Anastopoulos, I.; Giannakoudakis, D.A.; Fronstistis, Z.; Zorpas, A.A.; Pashalidis, I.; Triantafyllidis, K. Biomass-derived adsorbents: Universal database development for their synthesis and remediation efficiency as a necessary step to move from laboratory- to pilot-scale applications. *Curr. Opin. Green Sustain. Chem.* **2024**, *47*, 100902. [CrossRef]
72. Gworek, B.; Kijeńska, M.; Wrzosek, J.; Graniewska, M. Pharmaceuticals in the Soil and Plant Environment: A Review. *Water Air Soil Pollut.* **2021**, *232*, 145. [CrossRef]
73. FAO; UNEP. *Global Assessment of Soil Pollution—Summary for Policy Makers*; FAO: Rome, Italy, 2021.
74. De Oliveira Souza, H.; dos Santos Costa, R.; Quadra, G.R.; dos Santos Fernandez, M.A. Pharmaceutical pollution and sustainable development goals: Going the right way? *Sustain. Chem. Pharm.* **2021**, *21*, 100428. [CrossRef]

sustainability

MDPI

Systematic Review

Biochars Derived from Olive Mill Byproducts: Typology, Characterization, and Eco-Efficient Application in Agriculture—A Systematic Review

José Ferreira Lustosa Filho [1,*], Ana Paula Ferreira da Silva [2,3,4], Silvana Teixeira Costa [1], Hélder Teixeira Gomes [2,3], Tomás de Figueiredo [2,3] and Zulimar Hernández [2,3]

1 Laboratorio Colaborativo Montanhas de Investigação (MORE CoLab), Edifício Brigantia Ecopark, Av. Cidade de Léon 506, 5300-358 Bragança, Portugal; scosta@morecolab.pt

2 Centro de Investigação de Montanha (CIMO), Instituto Politécnico de Bragança, Campus de Santa Apolónia, 5300-253 Bragança, Portugal; anapaula.silva@ipb.pt (A.P.F.d.S.); htgomes@ipb.pt (H.T.G.); tomasfig@ipb.pt (T.d.F.); zulimar@ipb.pt (Z.H.)

3 Laboratório Associado para a Sustentabilidade e Tecnologia em Regiões de Montanha (SusTEC), Instituto Politécnico de Bragança, Campus de Santa Apolónia, 5300-253 Bragança, Portugal

4 Centro de Química de Vila Real (CQVR), Universidade de Trás-os-Montes e Alto Douro, Quinta de Prados, 5000-801 Vila Real, Portugal

* Correspondence: filhoze04@hotmail.com

Abstract: In the context of climate change and the circular economy, biochar agricultural and environmental applications have attracted a good deal of attention. Biochar has unique characteristics like surface area, porosity, water-holding capacity, pH, surface charge, and nutrients. This study reviews the biochar production from olive pomace (OP) and olive stone (OS) byproducts, its chemical and physical characterization, and its environmental application. The current review highlights the conditions for biochar production, the effects of pyrolysis temperature, and feedstock type on the physicochemical properties of biochar. High pyrolysis temperature (>500 °C) promotes a high specific surface area, high porosity (especially for OS biochars), and pH as well as the content of ash and fixed carbon, but generates low cation exchange capacity (CEC) and electrical conductivity (EC) and high values of O/C and H/C ratio. OP biochar also presents a high C amount, and ash content, i.e., rich in nutrients and high alkalizing capacity. OP biochar serves as an important source of plant nutrients, especially potassium. After adding both types of biochar, aggregate stability and the amount of water held in soil increase, and bulk density and bioavailability of trace elements decrease. Thus, biochar from olive mill wastes can be a potential plant nutrient reservoir, a good amendment to improve soil properties and long-term carbon sequestration. Results presented in this review can be used to build designer biochars from olive mill wastes to help solve environmental issues (water purification and pollutant remediation) and are suitable for improving soil physical chemistry characteristics and crop growth.

Keywords: environmental remediation; olive pomace biochar; olive stone biochar; plant growth; soil amendments

1. Introduction

The world production of olives was 21,449 Gg in 2022, with the most producing countries being Spain (29.9%), Italy (13.5%), Greece (12.1%), Turkey (7.4%), Morocco (5.1%), Tunisia (4.4%), and Portugal (2.4%) [1]. In the last 20 years, the annual olive production has increased by 25% due, among other factors, to the increase in the olive-growing area and the intensification of olive farming practices in Mediterranean countries. In 2022 alone, more than 3010 Gg of olive oil were produced worldwide, 15% more than in previous years [2].

This increase in olive oil production has contributed to a large amount of byproducts and effluents being produced, depending on the centrifugation process. In the three-phase continuous centrifugation process, in addition to the oily phase (20%), a solid byproduct (30%), and an aqueous phase (50%) known as olive mill wastewater (OMW) is produced [3]. On the other hand, in the modern two-phase centrifugation process, only one byproduct is created, known as wet olive pomace (OP) (for every kilogram of olive oil produced 4 kg of OP). Therefore, the new effluent comprises a mixture of washes and vegetable solid residue, comprising the skin, pulp, physiological water, and olive stone (OS). In the two-phase centrifugation process, although it produces no OMW, it produces a single effluent stream in semi-solid form, OP, which doubles the amount of "solid" waste requiring disposal [3]. In Spain alone, the world's main producer, more than 6000 Gt of OP were produced in 2022 [1].

The main reason that has led to the modernization of the oil mills, from three phases to two phases, has been the elimination of OMW, which was a problem in the oil mills due to their high contaminant load. With this industrial change, the water consumption of the oil mills has also been reduced, from 1.5 L to 0.25 L per liter of oil produced [4]. The great advantage has been, without a doubt, reducing the 10–12 $\times$ 10^6 m^3 of OMW produced in the 1990s in olive-growing countries, which is compared to an equivalent contamination of 10–16 $\times$ 10^6 inhabitants [5].

However, OP is in sludge form, with about 55–70% moisture content, making the industrial recovery of the residual oil difficult and expensive. The main alternative for economic recovery of OP is the extraction of residual oil (3–4% oleic acid) in seed-oil refineries, where it is necessary to centrifuge and dry the conventional three-phase olive cake, producing a large quantity of OMW. The considerably higher energy requirements than in the three-phase process and, finally, the extraction of residual fat with oil solvents make the industrial recovery of residual oil difficult and expensive [3].

Different treatment processes can be carried out with olive cake. The stone-free OP is obtained after sieving the wet OP and de-oiling and drying [6], which is used as biomass energy given its high energy value. This process is expensive since drying is energy-consuming and highly polluting due to the CO_2 emissions generated. It is estimated that OP could provide more than 20.6 Kw kg^{-1} OP processed electric power by direct burning [7], representing high CO_2 emission rates [8]. Furthermore, this industry generates little economic profit [9].

Converting OP and OS into biochar, the solid product of thermochemical conversion under oxygen-limited conditions (pyrolysis), seems a promising approach to address the above-mentioned environmental concerns of olive oil byproduct management. The advantage of pyrolysis is that it does not require O_2 for combustion, so there are no CO_2 emissions into the atmosphere [10]. In this sense, this type of technology, in addition to its industrial scale, has the advantage of aligning with Green Europe's decarbonization objectives. The interest in producing biochar from these olive oil byproducts relates to its chemical composition. In the case of OP, the material is made up of a high proportion of C in the form of cellulose, hemicellulose, and lignin (20%); low ash content (8%); and high humidity [11]. In the case of OS, these properties are improved due to low humidity, which is of interest to avoid undesirable combustion in pyrolysis ovens.

In this sense, biochar production is a win–win strategy, considering that it reduces the volume of waste from the olive oil industry while favoring decarbonization compensation for C emissions. Due to its beneficial characteristics, e.g., surface area, porosity, water-holding capacity, pH, surface charge, functional groups, and mineral components, biochar is used as a soil conditioner, a promising alternative for agricultural applications [12,13]. Biochar's physical and chemical properties largely depend not only on the type of feedstock, but also on the pyrolysis conditions such as temperature and heating rate [14–18]. Therefore, biochar's pyrolysis conditions are essential when choosing the most suitable biochar for the intended application, which can range from application as an inorganic amendment in soils or remediation of contaminated soils [13,19,20], wastewater treatment [21,22], or carbon capture [16,19], etc.

During biochar production, pyrolysis temperature plays a key role in the thermochemical conversion of biomass [13,16–19,23,24]. In general, biomass heated to high temperatures (>500 °C) produces biochar with a porous structure, large specific surface area (SSA), aromaticity, C-C content, and high pH, while those made at low temperatures (<400 °C) have a high yield and O/C and H/C ratio, and low electrical conductivity (EC), cation exchange capacity, and surface functional groups [16–19,25]. In this sense, low-temperature biochar contains a high content of compounds that easily undergo degradation in soil, and, hence, nutrients assimilable by plants so that it could be suitable for agricultural use [26,27], while those produced at high temperatures could be used in more environmentally friendly practices [28].

Pyrolysis of olive waste has been identified as a promising management technique that detoxifies organic contaminants in these byproducts [29], increasing its agricultural value both in terms of nutrient availability and C sequestration [30]. Alotaibi et al. [31] found that the physical properties (pH, EC, SSA) of OP and sewage sludge (SS) biochars increased with increasing pyrolysis temperature, demonstrating the potential of these biochars to store recalcitrant C. Furthermore, when produced only from OP, total C values can be even higher (89.6% C) [32]. Campos and la Rosa [33] showed that biochar produced from OS has physical and chemical properties (water retention capacity greater than 70%, very high surface area (473 m^2 g^{-1}), alkaline pH (9.34), low ash content (1%), among others) which may be suitable for pollutant remediation and agricultural soil. Currently, there is a discrepancy in information between biochar types and their efficient use. In this sense, the present study aims to cross-reference existing information (studies published until 2022), to optimize the best production process for the expected purpose. The specific objectives of this review are to discuss: (i) the optimization of the OP and OS biochar production process, (ii) the analysis of factors affecting the process of biochar production, such as temperature and feedstock type, (iii) the physical and chemical characterization of the biochar produced, and finally, (iv) the advantages of the applications of OP and OS biochar to soil, for the crop growth and pollutant remediation. It is expected that this review is believed to serve as a useful tool to define the wide range of OP and OS biochar applications and to establish which pyrolysis process best fits the desired biochar properties (high contaminant retention, high C store).

2. Materials and Methods

Data Survey and Screening

A literature review was conducted using the Preferred Reporting Items for Systematic Reviews and Meta-Analyses (PRISMA) framework to provide state-of-the-art on biochar produced from both olive pomace (OP) and olive stone (OS) in topics such as production, characterization, and applications. This is accomplished through a five-phase process. The screening process and reasons for exclusion are summarized and identified in Figure 1. Detailed information about this review's search protocol and steps is presented below.

(i) We formulated the search string using different combinations of the keywords "biochar", "olive pomace", and "olive stone". On this basis, the following search string was designed: (olive pomace biochar OR olive pomace char OR olive pomace charcoal OR olive stone biochar OR olive stone char OR olive stone charcoal OR olive pit biochar OR olive pit char OR olive pit charcoal OR olive mill biochar OR olive mill char OR olive mill charcoal OR olive bagasse biochar OR olive bagasse char OR olive bagasse charcoal) AND (soil application OR soil amendment OR wastewater treatment OR plant growth OR characterization physical chemistry). The references available on the papers that resulted from the mentioned search were added to the paper's list to ensure that the obtained database comprised a complete survey on this topic.

(ii) Applying exclusion and inclusion criteria—identification: the search included all papers published before May 2023 and available from Scopus, Google Scholar, and

Web of Science databases, resulting in 245, 105, and 3 scientific papers, respectively. Results were merged, and duplicated ones were removed, totaling 270 papers. Only peer-reviewed articles in English were considered, and "other" publication types, such as press releases, book chapters, and newsletters, were removed.

(iii) Titles and abstracts screening: the titles and abstracts of articles were first checked by two reviewers, and for the doubtful cases, a full-text evaluation was performed. The screening process allowed for a reduction from 270 results to 140 publications.

(iv) Excluding irrelevant papers based on their full text: the selected papers were carefully reviewed, and studies not directly related to this research aim were excluded. A total of 78 papers were excluded in this step. The selected papers thus addressed the characterization of biochar from OP and OS and/or its applications for environmental remediation and soil amendments, which constitute the sample included and analyzed in this systematic review.

(v) Snowballing: we applied backward snowballing (search in the paper's reference list). This step has the advantage of identifying important papers in the literature review that could not be identified via the systematic review method. Four papers were added by considering this step, resulting in a total of 62 papers.

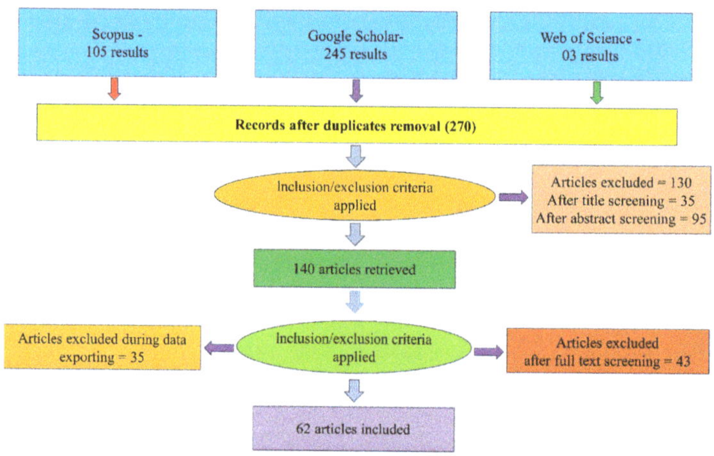

Figure 1. Flowchart data screening methods according to PRISMA principles.

Data used in this paper were exported directly from the text and tables of selected papers. When the papers' results were displayed in graph form, the corresponding numerical data were collected by using GIMP software (version 2.10.34). For data interpretations on the effect of increasing temperature on the physicochemical characteristics of biochar, we use specific subsets of each feedstock, along with SigmaPlot 13.0, for graphical interpretations within SigmaPlot 13.0. We utilized regression fitting functions (linear or quadratic) that best fit the presented data (e.g., best R^2 values and greatest p values). We avoided data interpretations that lacked logical sense (such as sigmoidal or sinusoidal equations), even if they had higher R^2 values compared to regression equations that were logical and interpretable.

3. Results

3.1. Statistical Data

From the selected articles, it was observed that the research institutions responsible for publishing the results on this topic were predominantly from Spain (46.8% of all articles, n = 29), followed by Tunisia (14.5%, n = 9); Italy (9.8%, n = 6); Greece, Saudi Arabia, and Turkey (4.8%, n = 3); and Qatar (3.3%, n = 2) (Figure 2), which is directly related to the fact that Spain is the main world producer of olive oil, contributing with 45% of the world's olive oil production and 25% of the world's crop area [34].

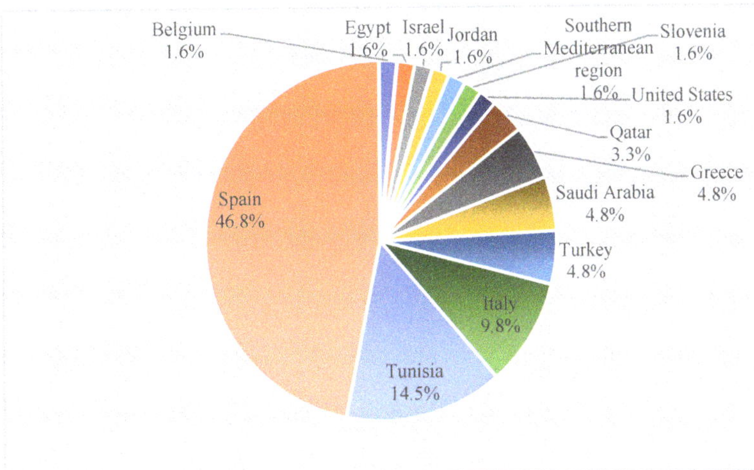

Figure 2. Global distribution of articles related to OP and OS biochar.

Complete information regarding the pyrolysis process, feedstock type, chemical and physical characterization of biochar, and pyrolysis conditions was considered in 62 papers (Table 1). Papers regarding biochar application on crop/plant productivity, pollutant remediation, and agricultural soil amendments were assessed. The literature review showed that the primary feedstocks for biochar production were OP (56%) or OS (42%). Among the 62 articles selected, 48 experiments reported biochars produced at low temperatures (from 301 °C to 500 °C), 30 trials (from the same article or not) were developed at high temperatures (from 501 °C to 800 °C), and six studies were found in other ranges (up to 300 °C and above 800 °C) (Table 1). In 60% of articles, carbons were synthesized in an N_2 atmosphere, 39% in a CO_2 atmosphere, while only 1% used He as carrier gas. Most biochars were produced through slow pyrolysis, without activation (74% of the considered data), while the remaining 26% were created either by carbonization and/or physical activation (Tables 1 and 2).

Table 1. Summary of biochar from OP and OS research conducted.

References	Feedstock	Temperature	Gas	Activation	Physico-Chemical Properties	Ultimate Analysis	Proximate Analysis	SSA	Pores	Soil	Plant	Adsorption	Nutrient
[35]	OP and OS	340, 450	CO_2	×	×	×	×	✓	×	×	×	✓	×
[36]	OS	550–600	CO_2	×	✓	×	✓	×	×	×	×	×	✓
[37]	OP	500, 600, 700	N_2	×	×	✓	✓	×	×	×	×	×	✓
[38]	OP	300, 400, 500, 600, 700	CO_2	×	✓	×	✓	✓	✓	✓	✓	×	×
[39]	OS	507	CO_2	×	✓	×	✓	×	×	✓	✓	×	✓
[40]	OS	350, 450, 550	CO_2	×	✓	×	✓	✓	×	×	✓	×	✓
[31]	OP	300, 500, 700	CO_2	×	✓	×	×	✓	×	✓	×	×	✓
[41]	OP	800	N_2	✓	×	✓	✓	✓	✓	×	×	×	×
[29]	OP	400, 500, 600	N_2	✓	×	✓	✓	×	×	×	×	×	✓
[42]	OP	500	CO_2	×	✓	×	×	×	×	✓	✓	×	×
[43]	OP	500, 700	N_2	×	×	✓	✓	×	×	×	×	×	×
[44]	OS	500, 800, 900	N_2	✓	×	×	×	✓	✓	×	×	×	×
[33]	OS	500	N_2	×	✓	✓	✓	✓	×	✓	✓	✓	×
[45]	OS	350, 400, 500, 600	N_2	×	✓	✓	✓	×	×	×	×	×	×
[46]	OS	500	N_2	×	✓	✓	✓	✓	×	✓	✓	✓	×
[47]	OS	400, 500	N_2	×	✓	✓	✓	×	×	✓	✓	×	×
[48]	OS	500	N_2	×	✓	✓	✓	×	×	✓	×	×	×
[49]	OS	500	N_2	×	✓	✓	✓	✓	×	✓	×	×	×
[50]	OP	1100–1200	CO_2	×	✓	×	×	×	×	✓	✓	×	✓
[51]	OP	500	N_2	×	✓	×	✓	×	×	✓	✓	×	×

Table 1. *Cont.*

References	Feedstock	Temperature	Gas	Activation	Physico-Chemical Properties	Ultimate Analysis	Proximate Analysis	SSA	Pores	Soil	Plant	Adsorption	Nutrient
[52]	OP	400	N_2	×	✔	✔	✔	×	×	×	×	×	×
[32]	OS	300, 500	N_2	×	✔	×	×	✔	✔	×	×	✔	✔
[53]	OP	350, 550	CO_2	✔	✔	×	✔	✔	×	×	×	✔	×
[54]	OS	500	CO_2	✔	×	×	×	✔	✔	×	×	✔	×
[55]	OP	500	N_2	×	✔	✔	✔	×	×	✔	✔	×	✔
[56]	OS	500	CO_2	✔	×	×	×	✔	✔	×	×	✔	×
[57]	OP	500	CO_2	×	✔	×	×	×	×	✔	✔	×	✔
[58]	OP	500	CO_2	×	×	×	×	×	×	✔	✔	×	×
[59]	OP	500	CO_2	✔	×	×	×	×	×	×	✔	×	✔
[60]	OP	500	CO_2	×	×	×	×	×	×	✔	✔	×	×
[61]	OP	400	N_2	×	✔	×	×	×	×	✔	✔	✔	×
[62]	OP	400	N_2	×	✔	×	×	×	×	✔	×	×	×
[63]	OS	750	N_2	✔	×	×	×	✔	✔	×	×	✔	×
[64]	OS	600	N_2	×	×	×	×	✔	✔	×	×	×	×
[65]	OS	350, 450, 550	N_2	×	×	✔	✔	×	×	×	×	×	×
[66]	OS	500, 800	N_2	✔	×	×	×	✔	✔	×	×	✔	×
[67]	OP	430, 480, 530	CO_2	×	✔	✔	✔	×	×	×	✔	×	×
[68]	OP	400–450	CO_2	×	✔	✔	✔	×	×	✔	✔	×	×
[69]	OP	350, 700	N_2	×	✔	×	✔	✔	✔	✔	×	×	✔
[70]	OP	350, 700	N_2	×	✔	×	✔	✔	✔	✔	×	×	✔
[71]	OS	400	N_2	✔	×	×	×	×	×	×	×	✔	×
[72]	OP	400	N_2	×	✔	✔	✔	✔	×	✔	✔	×	✔
[73]	OP	400	N_2	×	✔	×	×	×	×	✔	×	×	×
[74]	OP	300, 500, 800, 1000	CO_2	×	✔	✔	×	×	×	×	✔	×	×
[75]	OS	850	N_2	✔	×	×	×	×	✔	×	×	×	×
[76]	OS	750	CO_2	×	×	✔	×	✔	✔	×	×	✔	×
[77]	OS	400	N_2	✔	×	×	×	×	×	×	×	✔	×
[78]	OP	400, 700	CO_2	×	✔	✔	✔	✔	×	✔	×	×	✔
[79]	OP	400, 600	N_2	×	✔	✔	✔	✔	×	✔	✔	×	✔
[80]	OP	450	N_2	×	✔	✔	✔	×	×	✔	×	×	×
[81]	OS	450	N_2	×	✔	✔	✔	×	×	×	✔	×	×
[82]	OS	459	N_2	×	×	×	×	✔	✔	×	×	✔	×
[83]	OP	700	CO_2	×	×	✔	✔	✔	×	✔	×	×	×
[84]	OS	800, 850, 900	N_2	✔	×	×	×	✔	✔	×	×	×	×
[85]	OS	200, 300, 400, 500, 700, 900	N_2	×	×	✔	✔	✔	✔	×	×	✔	×
[86]	OP	280	CO_2	×	✔	✔	×	×	×	✔	×	×	✔
[87]	OP	450	CO_2	×	×	✔	×	×	×	×	×	×	×
[88]	OP	500	CO_2	✔	✔	×	✔	×	✔	✔	×	×	×
[89]	OS	800	N_2	✔	×	✔	✔	✔	✔	×	×	✔	×
[90]	OP	325, 400, 500, 650	He	×	×	✔	✔	×	-	×	×	×	×
[91]	OP	600	N_2	×	✔	✔	✔	✔	✔	×	×	×	✔
[92]	OS	600	N_2	✔	×	×	✔	✔	✔	×	×	×	×

OP: Olive Pomace; OS: Olive Stone; SSA: Brunauer–Emmett–Teller surface area; "✔" means studied; "×" means not studied.

Table 2. Biochar and AC production summary from OP and OS research conducted.

Reference	FD	T (°C)	Carbonization	Treatment
[41]	OP	800	Slow Pyrolysis	Superheated steam Superheated steam + H_2SO_4
[44]	OS	400 500 600 800 900 800	Slow Pyrolysis	Olive mill wastewater Gasification CO_2 NaOH $Ba(OH)_2$ $Ca(CH_3COO)_2 \cdot H_2O$
[53]	OP	350 550 350 550	-	Pre-treatment with $FeCl_3$ Post-treatment with $FeCl_3$
[54]	OS	500	-	Fe_3O_4
[56]	OS	500	Fast Pyrolysis	Co_3O_4

Table 2. *Cont.*

Reference	FD	T (°C)	Carbonization	Treatment
[59]	OP	500	Slow Pyrolysis	HNO_3
[63]	OS	750	-	Water vapor
[66]	OS	500 800 250	-	Impregnation H_3PO_4 3:1 Gasification CO_2 NH_4VO_3 $NH_4VO_3 + Zr(OCH_2CH_2CH_3)_4$ $NH_4VO_3 + Zr(OCH_2CH_2CH_3)_4$ $Zr(OCH_2CH_2CH_3)_4$ NH_4VO_3 $NH_4VO_3 + Zr(OCH_2CH_2CH_3)_4$ $Zr(OCH_2CH_2CH_3)_4$
[71]	OS	400	Slow and Mild Pyrolysis	$Ni(NO_3)_2{\cdot}6H_2O$ $Cu(NO_3)_2{\cdot}3H_2O$ $Cu(NO_3)_2{\cdot}3H_2O + Ni(NO_3)_2{\cdot}6H_2O$
[75] [77]	OS OS	850 400	Slow Pyrolysis Slow Pyrolysis	Supercritical water Citric Acid Citri Acid + $Ni(NO_3)_2{\cdot}6H_2O$ Citri Acid + $Cu(NO_3)_2{\cdot}3H_2O + Ni(NO_3)_2{\cdot}6H_2O$ Citri Acid + $Cu(NO_3)_2{\cdot}3H_2O + Ni(NO_3)_2{\cdot}6H_2O$ $Cu(NO_3)_2{\cdot}3H_2O$ +
[84]	OS	850 900 800 850 900 850	-	CO_2 Steam CO_2/Steam
[88]	OP	500	-	Na_2SiO_3 KH_2PO_4 and KNO_3
[89]	OS	800	-	CO_2 CO_2/H_3PO_4 H_3PO_4
[92]	OS	600	Slow Pyrolysis	Demineralized Ball-milled Demineralized and ball-milled

OP: Olive Pomace; OS: Olive Stone; FD: Feedstock; T: Temperature.

We found that 35%, 29%, and 35% of the studies characterized biochar from a physicochemical, ultimate, and proximate analysis. However, only 16 studies (26%) analyzed all these physicochemical properties simultaneously. In 48% of studies, they measured specific surface area, while 32% of articles reported porous structure results. Finally, 34% of the studies approached crop growth, 45% soil amendments, 29% the amount of nutrients, and 26% pollutant remediation.

3.2. Olive Pomace and Olive Stone Biochar Production Process

The biochar characteristics from olive pomace (OP) and stone (OS) are summarized in Tables S1 and S2. Data indicate highly variable results in activation conditions and associated physicochemical properties. It is confirmed that the biochar quality (which is determined by its composition and properties and is dependent on the purpose-specific applications) and its potential application are greatly affected by the pyrolysis process [13–18], as well as the raw material used [14,16,18].

Feedstock plays a critical role in the yield. OS biochar yield is substantially higher than OP biochar (39% versus 33%, means, respectively), which the biomass type may explain. A higher lignin content (50%), in comparison with cellulose (12%) or hemicellulose (24%), probably promotes a higher C-C link formation [91,93,94]. In addition to the variation of content in lignin, cellulose, and hemicellulose, mineral salts in the feedstock will also influence pyrolysis yields [45]. The OS has a lower ash content (Table S3).

Moreover, biochar yield also decreases with increasing pyrolysis temperature (Figure 3A,B). Both OP biochars (Figure 3A) and OS biochars (Figure 3B) presented a significant correlation between pyrolysis temperature and yield ($r^2 = 0.64$ vs. 0.60, OP vs. OS, respectively). The effect of temperature in combustion has been widely documented in the literature, e.g., at higher temperatures, a decrease in biochar yield is realized [17,18,95,96]. A high and fast pyrolysis temperature intensifies the carbonization of the biomass through rapid dehydrogenation and dehydration of hydroxyl groups, thus reducing the yield [97].

Production of Biochar and Activated Carbon

A novel methodology to maximize biochar production involves its transformation into activated carbon (AC) through chemical and physicochemical activations or functionalization.

Functionalizations focus on introducing specific functional groups or chemical moieties onto the OP or OS biochar surface. This process is more targeted and aims to tailor the chemical nature of a material. For example, adding functional groups like -OH, -COOH, or $-NH_2$ to a material's surface can give specific chemical or biological functionalities. On the other hand, activation generally refers to increasing a material's reactivity or surface area, making it more porous or chemically reactive. In the context of carbon materials, it often involves exposing the material to gases like steam or carbon dioxide. This process creates pores and increases the surface area, enhancing the material's adsorption properties [44,63,75,84].

Of the multiple papers reviewed on the activation and functionalization of OP and OS biochar (Table 2), the authors prefer slow pyrolysis processes at different temperatures up to 800–900 °C [41,44,66,75,84,89]. Slow pyrolysis conducted at elevated temperatures is observed to yield optimal results for activating and functionalizing the olive mill byproducts. It is the preferred method to produce activated carbons (ACs), although pyrolysis is still carried out [53,56,71].

Biochar activation processes can be used either before or after slow pyrolysis, with post-treatment being preferred to produce ACs. In El Hanandeh et al. [53], the post-treated method consisted of soaking the pristine biochar in 0.1 M $FeCl_3$ for 1 h, then drying it at 105 °C for 24 h. Due to the acidic digestion by $FeCl_3$, iron oxide was deposited on the biochar's surface, and the biomass structures opened. This process enhanced the biochar´s porous structure, and the increase in surface area was likely due to the washing of ashes caused by the acidic nature of the solution. This post-treatment can be considered a surface functionalization rather than a direct activation process.

Pre-treatment aims to introduce specific chemical groups or enhance the surface properties of the feedstock, paving the way for a more tailored and controlled pyrolytic process. Omiri et al. [77] incorporated citric acid with copper and nickel nitrates before performing slow pyrolysis to examine the influence of pre-treatment on the nanophysical properties of the biochar, which acts as a support matrix for the deposition of metal nanoparticles. Using citric acid induced the formation of bimetallic copper/nickel nanoparticles exhibiting an agglomerated or even raspberry-shaped morphology. This unexpected outcome can be considered functionalization and activation.

On the other hand, functionalizations can be performed after slow pyrolysis as an additional step to modify and optimize the properties of ACs. This flexibility allows for a comprehensive and customizable approach to meet specific application requirements. The mentioned functionalizations generally involve metallic oxides (e.g., Co_3O_4) [56], acids (e.g., HNO_3, H_2SO_4, H_3PO_4) [41,59,66], and metallic salts (e.g., $Ni(NO_3)_2{\cdot}6H_2O$, $Cu(NO_3)_2{\cdot}3H_2O$), or a mixture of chemical compounds, water vapor, and gases [71,77,84].

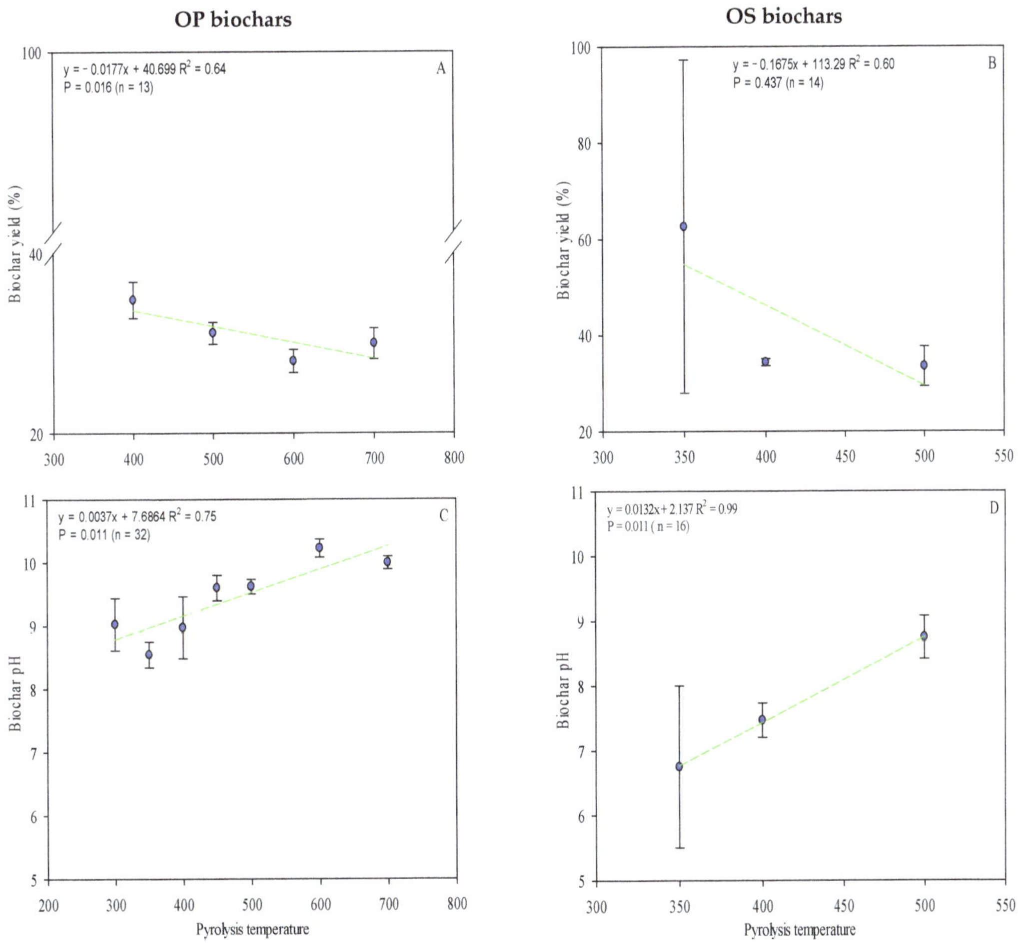

Figure 3. A linear relationship between pyrolysis temperature and yield (**A**,**B**) and pH (**C**,**D**) in OP biochar and OS biochar. Complete data for the 62 articles reviewed are provided in Tables S1 and S2. Green lines are mean and the symbol (⫱) represents a standard error.

3.3. Olive Pomace and Olive Stone Biochar Characteristics

3.3.1. Physical and Chemical Properties of Biochar

OP biochar and OS biochar show pH values varying from 5.33 to 10.7 and 5.50 to 9.80, respectively (Table S2), increasing with pyrolysis temperatures (Figure 3C,D). As expected, a linear correlation exists between pH and biochar pyrolysis temperature (r^2 = 0.75 vs. 0.99, OP vs. OS, respectively) (Figure 3C,D). Most of the biochars from olive byproducts have an alkaline pH. In general, high pyrolytic temperatures, high metal oxide, ash content, and the development of heterocyclic nitrogen compounds influence the production of high-pH biochar [98]. Biochar often can have an initially high pH (alkaline), which is desirable when activated with acidic, degraded soils; nonetheless, if soil pH becomes too alkaline, plants may suffer nutrient deficiencies [99].

Although biochar exhibits a considerable cation exchange capacity (CEC), which refers to the capacity to exchange positively charged species, such as Ca^{2+}, Mg^{2+}, K^{+}, and Na^{+} [18], only five reviewed studies show CEC data.

The highest CECs are observed in OS and OP biochar produced at low temperatures (25.5 and 20.3 $cmol_c\ kg^{-1}$, respectively), in contrast with biochar produced at relatively high temperatures (>400 °C), ranging from 11.9 to 18.6 $cmol_c\ kg^{-1}$ [40,79], which may be linked to the loss of some acidic/carboxylic functional groups at high temperatures [100,101]. Moreover, higher CEC biochar may indicate an excellent capacity to adsorb/remove potentially toxic elements [102].

Electrical conductivity (EC) in biochar is mainly influenced by feedstock (Table S2), registering the highest value for OP biochar (3.65 $dS\ m^{-1}$, mean value) and the lowest for OS biochar (0.98 $dS\ m^{-1}$, mean value). There was no significant correlation between EC and biochar pyrolysis temperature (Figure S1A,B). This could be an important biochar characteristic, limiting its final application in agronomic use. For example, for agronomic applications, a high EC value indicates the presence of cations (e.g., K^{+}, Ca^{2+}, and Mg^{2+}) and anions (SO_4^{2-} and PO_4^{3-}), nutrients essential for crop growth. This fact highlights the importance of in-depth biochar production and characterization knowledge, allowing the creation of more versatile biochars with potential socioeconomic impacts and vegetable nutrition.

3.3.2. Proximate Analysis

The proximate analysis includes the moisture content, ash content, volatile matter, and fixed carbon. As referred to above, ash content varies according to the type of feedstock used (Table S1) and pyrolysis temperature (Figure 4C,D). OP biochar has a higher ash content, which can be correlated with higher mineral content in olive fruit, mainly potassium (Table S2). Moreover, ash content positively correlated with temperature pyrolysis (Figure 4C,D). It could play an important role in agriculture, where, in addition to reducing soil acidity, it could improve productivity in acid soils. In other cases, it may be more advisable to use OS biochar (with low ash content and pH, when processed under the same conditions [103]) to avoid the concentration of soil mineral salts.

Fixed carbon contents for the biochars ranged from 25.5% to 89.0% for OP and 4.60 to 90.2% for OS (Table S1). When pyrolysis temperature increases, the fixed carbon content increases while the volatile matter decreases (Figure 4). The C content fixed in OP and OS biochar increased by 7 and 8% with the temperature increase and, at the same time, the volatile matter decreased by 31–37%, considering the same thermal conditions, resulting in a strong correlation (r^2 = >0.7) between volatile matter and pyrolysis temperature (Figure 4). Considering the mentioned data, it is clear that the biochars' fixed carbon content is affected by both pyrolysis temperature and feedstock. Despite the higher carbon content of OP biochar, the volatile matter level was higher in OS (OP—mean value: 27.7%; O—mean value: 41.7%). Even though previous studies have demonstrated that biochar with greater fixed carbon is obtained from biomasses with greater volatile content [104], this review recorded poor results, which may be explained due to variations in moisture and residence time of pyrolysis. The moisture content significantly reduces the mass loss and release of volatile matter, while extended residence times lead to a significant part of volatile matter being recombined and polymerized within the sample's structure [105].

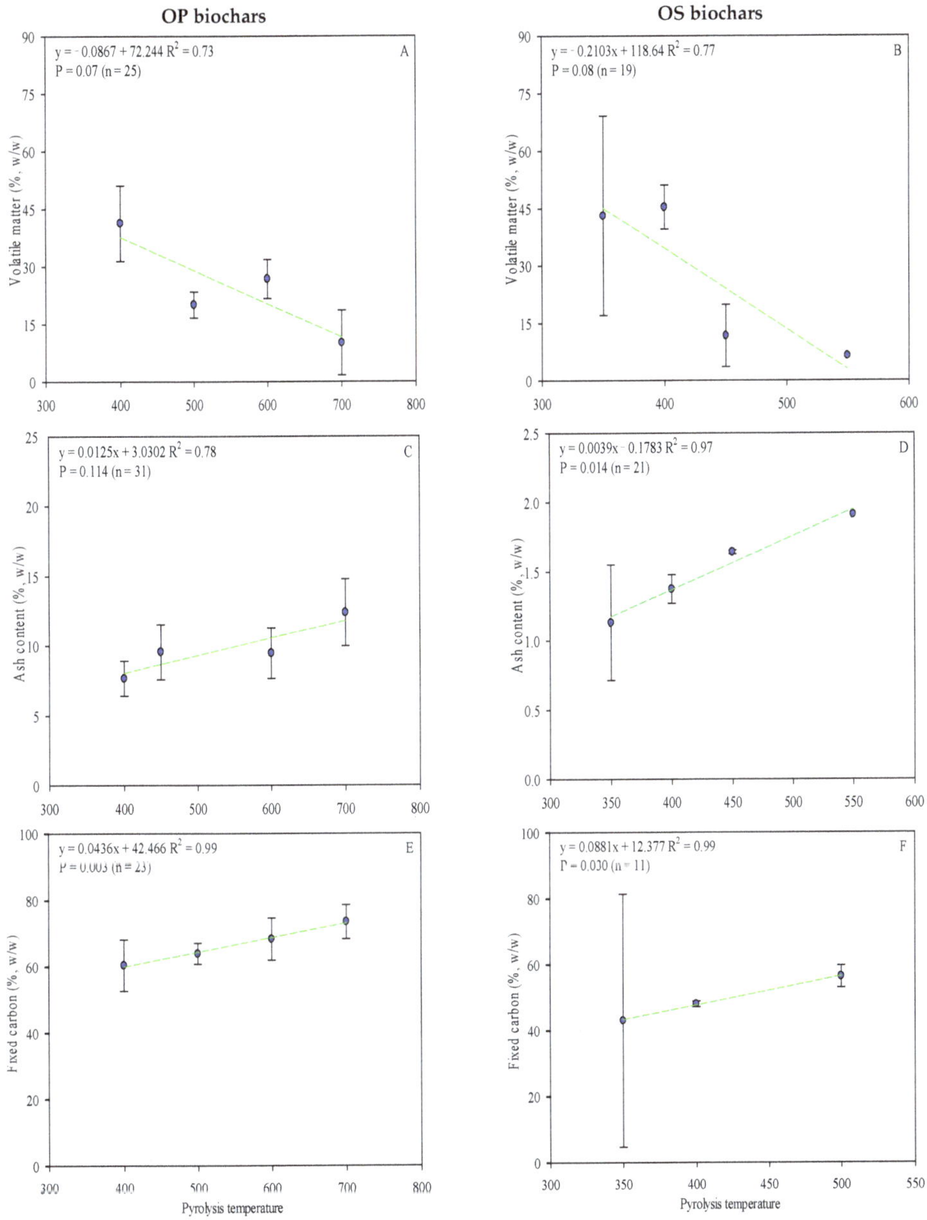

Figure 4. Linear correlation between pyrolysis temperature and OP and OS biochars' volatile matter (**A**,**B**), ash (**C**,**D**), and fixed carbon content (**E**,**F**) (*w*/*w*). Complete data for the 62 articles reviewed are provided in Table S1. Green lines are mean and the symbol (⫫) represents a standard error.

The volatile matter is a good indicator of biochar stability and nitrogen availability when applied as a soil amender, considering its potential impact on short-term microbial activity and plant growth [106]. Thus, biochar may constitute a suitable and ecological material for soil amendment, nutrient source, and carbon sequestration.

3.3.3. Ultimate Analysis

The ultimate analysis indicates the elemental constituents of biochar: carbon (C), hydrogen (H), nitrogen (N), sulfur (S), and oxygen (O) [107]. Elemental compositions for biochar obtained with different temperatures are listed in Table S1. OS biochar contains the highest C content (77.2%), while OP biochar contains the highest N content (1.79%). The contents of H and O were similar for the two biochar types. When the pyrolysis temperature increased, C content increased, whereas the H and O decreased, regardless of the type of biomass, indicating an increasing degree of biochars' carbonization (Table 3).

Table 3. Elemental composition of OP and OS biochars at different temperatures.

OP biochars							
Temperature	C (%)	H (%)	N (%)	O (%)	H/C	O/C	C/N
400 °C	65.3 ± 4.6	3.82 ± 1.0	1.72 ± 0.5	25.5 ± 7.6	0.35 ± 0.14	0.37 ± 0.13	35.2 ± 11
500 °C	73.3 ± 3.7	2.37 ± 0.6	1.79 ± 0.6	11.5 ± 4.3	0.06 ± 0.03	0.17 ± 0.06	80.0 ± 10
600 °C	77.6 ± 4.7	2.38 ± 0.3	3.11 ± 1.1	6.73 ± 3.1	0.04 ± 0.00	0.17 ± 0.07	24.0 ± 7.4
700 °C	76.9 ± 6.5	1.29 ± 1.1	2.69 ± 0.1	-	0.02 ± 0.02	-	39.0 ± 13
OS biochars							
Temperature	C (%)	H (%)	N (%)	O (%)	H/C	O/C	C/N
350 °C	66.4 ± 18	4.43 ± 1.3	0.24 ± 0.08	28.8 ± 17	0.72 ± 0.7	0.42 ± 0.28	291 ± 20
400 °C	78.7 ± 1.8	4.27 ± 0.2	0.31 ± 0.05	15.8 ±1.9	0.52 ± 0.2	0.17 ± 0.03	284 ± 54
500 °C	82.7 ± 2.9	3.50 ± 0.3	0.30 ± 0.03	12.8 ± 2.7	0.46 ± 0.1	0.13 ± 0.04	319 ± 47

An increase in C concentrations may occur when pyrolysis temperature increases due to cleavage and polymerization reactions that cause easily degradable C compounds to be restructured. At the same time, other elements may be lost to volatilization, and thus, overall total C content increases [19,108]. The great majority of the reviewed literature classifies the OP and OS biochars as class 1 on the International Biochar Initiative (IBI), i.e., C content higher than 60% [109]. In this study, N content in both biochars presented an increasing trend, reaching a peak for an intermediary pyrolysis temperature followed by a slight reduction of N content when subjected to higher temperatures (Table 3), which could be justified by the formation of N compounds, which do not readily volatilize at lower pyrolysis temperatures [110].

The pyrolysis temperature and feedstocks affected the H/C and O/C atomic ratios (Table 3), tending to decrease with pyrolysis temperature and consequently increasing aromaticity [110]. In addition, it was observed that OP-derived biochars generally presented lower H/C and O/C ratios, indicating higher aromaticity and resistance to biodegradation.

Results showed that the C/N ratio increases with the increment of pyrolysis temperature up to 500 °C. It was observed that there is a reduction in the C/N ratio in OP biochar at higher temperatures. Only OP biochar produced at 600 °C may increase N availability in the soil [111]. OS biochar with high C/N is practically inert for microbial activity but could increase the recalcitrant soil organic C pool.

3.3.4. Surface Area and Porosity

Comparisons of the specific surface area (SSA) and pore volume (PV) of the different biochars indicated that these properties were closely associated with the feedstock type. The reviewed papers recorded that SSA ranged from 0.30 $m^2\ g^{-1}$ to 507 $m^2\ g^{-1}$ (Table S2), with OS biochars registering much higher values than those produced from OP. As the pyrolysis temperature increases, a gradual increase in the SSA for both types of biochar (OP and OS) was recorded (Figure 5A,B) due to biochar experiencing crystallization and recombination during the carbonization process, leading to an expansion of the total pore volume and, consequently, on the SSA [25,112]. A strong correlation between biochar SSA and pyrolysis temperature was observed (r^2 = 0.70 and 0.94 for OS and OP biochars, respectively) (Figure 5A,B).

The European Biochar Certificate (EBC) states specified that SSA should preferably be higher than 150 m^2 g^{-1} [113], as with the OS biochars produced over 300 °C. A higher surface area is preferable due to its significant role in soil fertility, leading to improvement of the structure of the soil and provision of ecological niches for soil microorganisms [114]. In addition, a huge SSA of OS biochar makes it an ideal adsorption medium for trace metal elements, organic pollutants, and pesticides.

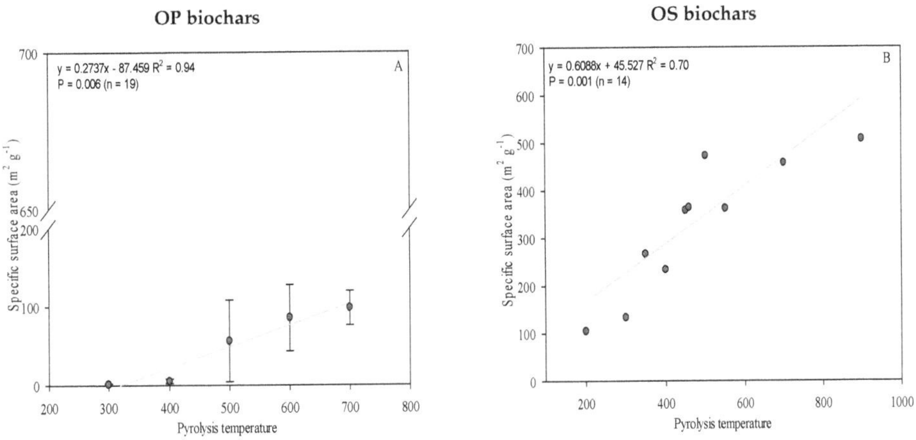

Figure 5. The linear relationship between pyrolysis temperature and the biochars' specific surface area (**A**,**B**). The complete data is given in Table S2. Green lines are mean and the symbol (∓) represents a standard error.

Alazzaz et al. [38] and Sanginés et al. [85] showed that OP and OS biochars produced under low temperatures (300 °C and 200 °C, respectively) resulted in lower values of total pore volume (0.002 cm^3 g^{-1} and 0.04 cm^3 g^{-1}, respectively); as expected, higher pyrolysis temperatures (700 °C and 900 °C, respectively) led to high values of total pore volume (0.083 cm^3 g^{-1} and 0.193 cm^3 g^{-1}, OP and OS biochars, respectively) (Table S2). The pores in the C-skeleton of the biochar developed due to the decomposition of lignin, the quick degradation and the release of volatiles from feedstock material, and the reaction of aromatic condensation as the temperature increases [115]. Both the porosity and SSA play key roles in biochar anion and cation exchange capacities, therefore influencing their adsorption abilities [116].

3.3.5. Nutrient Content

Besides C and N, biochar also contains some essential plant nutrients such as P, K, Ca, Mg, and micronutrients, as shown in Table S2. Therefore, biochar incorporation into agricultural soils may reduce the need for chemical fertilizers. In general, OP biochar has a higher nutrient content than OS biochar, reflecting the composition of the feedstocks (Table S3).

Pyrolysis temperature also influenced total nutrient contents, but to a much smaller extent; for example, Delgado-Moreno et al. [32] observed that upon further heating from 300 to 500 °C, OS biochars showed slight enrichment for P, K, Ca, Mg, and Fe (Table S2). On the other hand, Pellera and Gidarakos [78] showed that with the increase in pyrolysis temperature, OP biochars showed a pronounced enrichment for P, K, Ca, Mg, and Fe. It is essential to highlight that the nutrient content of OP may vary between regions, as reported by Fornes et al. [57]. On the other hand, the concentration of trace metals in all biochars was below the maximum concentrations allowed for biochars, according to IBI guidelines [109]. The relatively high nutrient concentrations and low trace metals concentrations in OP biochar show the versatility of its applicability for agricultural use.

3.3.6. Effect of Activation and Functionalization on Activated Carbon Characteristics

Activation focuses on developing the biochar´s surface area, making it more porous, with the aim of application in gas or liquid pollutant adsorption processes. These materials are usually activated by CO_2 gasification, steam, demineralization, and others (Table 4).

Table 4. Correlation of activation and functionalization of biochars with surface area, pore, and micropore volume.

Reference	AC	Treatment	SSA_T	ESA	Vt	V_{micro}
-	-	-	$m^2\ g^{-1}$		$cm^3\ g^{-1}$	
[41]	OPAC	Superheated steam	345.31	-	0.24	0.106
	OPACS	Superheated steam + H_2SO_4	618.18	-	0.32	0.203
[44]	ACG800	Gasification CO_2	1355	77		0.59
	ACG900	Gasification CO_2	2360	172		0.802
	NaC	NaOH	528	49		0.217
	BaC	$Ba(OH)_2$	123	2		0.048
	3CaO/ACG800	$Ca(CH_3COO)_2{\cdot}H_2O$ solution	1235	146		0.499
	6CaO/ACG900	$Ca(CH_3COO)_2{\cdot}H_2O$ solution	1960	225		0.693
[53]	BC350_pre	Pre-treatment with $FeCl_3$	28.19	-		
	BC550_pre	Pre-treatment with $FeCl_3$	31.95	-		
	BC350_post	Post-treatment with $FeCl_3$	27.6	-		
	BC550_post	Post-treatment with $FeCl_3$	29.44	-		
[54]	MAG–OSBC	Fe_3O_4	33.82	-	0.166	
[56]	Co–OSBC	Co_3O_4	39.85	-	0.168	
[63]	OSAC_PW360	Water vapor	807	-		0.3
[66]	ACP	Impregnation H_3PO_4 3:1	1300	594		0.389
	AC	Gasification CO_2	1355	50		0.59
	1V/ACP	NH_4VO_3	722	290		0.233
	0.9V–0.1Zr/ACP	NH_4VO_3 + $Zr(OCH_2CH_2CH_3)_4$	746	275		0.245
	0.8V–0.2Zr/ACP	NH_4VO_3 + $Zr(OCH_2CH_2CH_3)_4$	830	267		0.291
	1Zr/ACP	$Zr(OCH_2CH_2CH_3)_4$	818	268		0.282
	1V/AC	NH_4VO_3	1000	40		0.43
	0.8V–0.2Zr/AC	NH_4VO_3 + $Zr(OCH_2CH_2CH_3)_4$	940	39		0.403
	1Zr/AC	$Zr(OCH_2CH_2CH_3)_4$	906	33		0.389
[84]	C-850-30	CO_2	446	56		0.243
	C-850-60	CO_2	572	90		0.319
	C-850-90	CO_2	761	124		0.406
	C-850-120	CO_2	778	177		0.406
	C-900-60	CO_2	603	139		0.3
	S-800-30 (2.4)	Steam	808	166		0.426
	S-850-30 (2.4)	Steam	813	174		0.427
	S-850-20 (1.6)	Steam	589	101		0.312
	S-850-20-F (3)	Steam	713	233		0.358
	S-900-30 (2.4)	Steam	437	155		0.216
	S-850-60 (4.8)	Steam	1074	417		0.525
	CS-850-30	CO_2/Steam	674	125		0.357
	CS-850-60	CO_2/Steam	1187	469		0.553
[89]	ACG800	CO_2	1355	-		0.57
	ACG800-P2	CO_2/H_3PO_4	1506	-		0.591
	ACP2800	H_3PO_4	1380	-		0.514
[92]	DOS-BC	Demineralized	268	-	0.115	0.163
	BM-OS-BC	Ball-milled	438	-	0.193	0.152
	BM-DOS-BC	Demineralized and ball-milled	435	-	0.194	0.154

SSA: Brunauer–Emmett–Teller surface area; ESA: External surface area; Vt: Total pore volume; V_{micro}: micropore volume.

The surface area (SSA) of biochar can vary greatly depending on the selected activation technique, reaching values of up to 2360 $m^2\ g^{-1}$, which is an exceptionally large SSA. In this case, the procedure involved carbonizing OS under continuous N_2 flow at 500 °C to achieve this elevated surface area. Subsequently, the resulting biochar underwent physical activation with CO_2. The gasification temperature was set at 800 °C, and the gas feed was switched to CO_2 with a flow rate of 150 cm^3 STP min^{-1} for 7 h, resulting in an AC with high SSA [44]. Guerrero-Pérez et al. [66] employed a methodology similar to Calzado et al. [44], but the material underwent a two-hour CO_2 gasification process, obtaining an AC with 1355 $m^2\ g^{-1}$ of SSA. In these processes, in addition to transforming a byproduct into a high-added value material and a broad spectrum, it also promotes decarbonization by using greenhouse gases such as CO_2.

Other studies examine the combination of water vapor, with or without gases, achieving noteworthy values for SSA [84]. Combining CO_2 and steam as activating agents significantly influences the development of AC porous structure. For example, the production process for CS-850-60, which involves CO_2 and steam as activating agents at 850 °C for 60 min, exhibits distinctive characteristics from creating micropores (V_{micro} = 0.553 $cm^3\ g^{-1}$ and SSA = 1187 $m^2\ g^{-1}$) and mesopores.

Among the described activation methods, demineralization stands out as a pretreatment methodology that eliminates minerals from OS biochar, enhancing the resulting biochar's porosity and SSA. Ball milling, a post-treatment approach, also mechanically crushes and grinds biochar particles, contributing to increased porosity and SSA. These treatments are activation methods as they alter biochar's physicochemical properties, achieving SSA of up to 438 $m^2\ g^{-1}$.

Finally, the wide range of activations can be combined with functionalization. Fourier Transform Infrared analysis (FT-IR) can accurately identify functional groups on the surface of adsorbents (Table S4). The main structures encountered in AC from olive mill residues include hydroxyl (OH), carbonyl (C=O), alkyl, and hydrocarbons. These functional groups exert significant influence over the adsorptive performance of ACs, playing pivotal roles in surface interactions and engagement with reactants [29,41,53,56,66,88].

Moreover, aromatic structures in ACs add an extra layer of significance, contributing to the material's resilience against decomposition under heat exposure and enhancing its durability when exposed to acidic or alkaline environments [56,63,92].

On the other hand, introducing specific groups on the chemical surface, such as sulfonic acid (SO_3H) and sulfates (SO_4^{2-}), enhances catalytic activity. The catalytic activity of AC is complemented by carbonyl groups, which also serve as active sites for catalytic reactions. Their involvement is attributed to the ability of carbonyl groups to participate in redox reactions and adsorb species on their surface, amplifying the catalytic efficacy of the materials [41].

3.3.7. Elemental Composition and Thermogravimetric Analysis of Activated Carbons from OP and OS

OP is a valuable raw material for synthesizing AC because of its high oxygen concentration and fixed carbon content (41.51% of total carbon in the final product material).

Ayadi et al. [41] carried out an elemental analysis on the ACs and biochar before and after pyrolysis, seeking changes in the material's composition during the activation and sulfonation processes. Authors identified that the C content increased from 77.95% to 85%, whereas oxygen, nitrogen, and hydrogen contents were reduced by half due to decarboxylation or depolymerization and dehydration reactions.

Azzaz et al. [29] observed that materials that underwent hydrothermal carbonization had higher volatile matter and lower C content compared to un-treatment biochar, which showed higher levels of nutrients (magnesium, potassium, sodium) due to combustion at high temperatures.

Valero-Romero et al. [89] showed that the phosphoric acid activation process increased ash content and oxygen content, accompanied by the integration of phosphorus on the surface. Materials showed an ash value of 0.3–3.7%, a C content of 93.2–87.9%, an oxygen content of 4.7–6.7%, and no detectable phosphorus (Table S5). The authors analyzed the re-oxidation process after a thermal treatment at 900 °C. The authors observed that the carbon surface treated at high temperatures in N_2 is unstable and may re-adsorb oxygen even at room temperature. Since no loss of phosphorus was observed on the surface after several reduction/oxidation cycles, it was deduced that the re-oxidation of the phosphoric-acid-ACs at room temperature occurs due to the large number of highly reactive free sites susceptible to re-oxidation, where oxidation from C-P to C-O-P occurs, as this redox reaction is reversible at high temperatures.

On the other hand, Guerrero-Pérez et al. [66] concluded that the presence of phosphorus in the AC material ACP significantly enhanced its thermal stability and oxidation resistance compared to the AC without phosphorus. The TGA results showed that the ACP material remained resistant in air up to 500 °C, while the AC sample began to gasify around 400 °C. This difference in thermal stability indicated that the phosphorus-containing ACP material was more resistant to oxidation at high temperatures and in oxidant atmospheres.

Valero-Romero et al. [89] analyzed that both biochars were thermally stable in the 100–450 °C range. The weight loss observed at other temperature ranges indicated the instability of OP/OS. Additionally, the presence of magnetite on the surface contributed to its enhanced thermal stability, particularly after 550 °C. This finding suggests that incorporating magnetic nanoparticles improved the thermal properties of the biochar, making AC more resistant to high temperatures.

El-Shafie et al. [56] demonstrated that both OS biochar and Co–OSBC (Cobalt-loaded OS biochar) exhibited thermal stability within the 100–450 °C temperature range. The weight loss observed between 50 and 100 °C and 550–800 °C, respectively, was attributed to the vaporization of free water in the first case and to the carbonization of polymeric material in the second.

Zouari et al. [92] showed that the samples subjected to demineralization exhibited higher weight loss values (10%), indicating a reduction in the material's thermal resistance. It was attributed to the removal of minerals and hemicelluloses during the demineralization. The presence of ashes in the OP biomass was found to have a catalytic effect on the thermal degradation behavior. Ash minerals enhanced biochar formation during thermal decomposition. Demineralization decreased yield, indicating that the high ash content in the OP biomass promoted biochar formation by thermal catalysis.

3.4. *Applications of OP and OS Biochars*

3.4.1. As Soil Amendments or Conditioners

Due to the physical and chemical characteristics of OP and OS biochar, the soil properties were moderately affected after biochar application, as listed in Table 5.

Biochar's alkalinity allows it to function as a soil amendment for acidic soils, improving soil health and productivity. Alazzaz et al. [38] reported a slight increase in pH after application of 1% and 3% OP biochar in calcareous soils due to biochar-induced alkalinity as a result of the oxides and carbonates accumulated during the pyrolytic process. Alburquerque et al. [39] attributed the notable increases in soil pH to the soil's low buffering capacity. The application of OS-derived biochar led to a pH increase of approximately one unit (from 6.5 to 7.5) with a 7% application rate. Moreover, ash-rich biochar led to even higher soil pH values (8.5) for the same biochar application rate. This liming effect is considered positive in acidic soils but can have adverse impacts in non-acidic soils [39].

After applying OP biochars, even more pronounced impacts in soil EC were recorded [38,42,51,55,67]. Hmid et al. [67] reported that the OP biochar application with 15% in sandy soil increased the EC from 0.06 dS m^{-1} to 0.19 dS m^{-1}, making it unlikely that amended soils have a detrimental effect on microbial activity and crop growth [117]. Consequently, OP or OS biochars could be safely added to soils with relatively high buffering capacity without seriously harming soil salinization and alkalinity.

Table 5. Summary of papers on the effect of OP- and OS-derived biochar on soil properties.

References	Biomass/ Pyrolysis Temperature	Country	Experiment Type	Biochar Dose	Soil Type	Impact on Soil Properties
[38]	OP/300,500 and 700 °C	Saudi Arabia	Greenhouse	1% and 3%	Calcareous loamy sand soil	The biochar application, especially with increasing pyrolytic temperature and/or application rate, significantly increased soil pH and EC and extractable K, Na, Ca, Mg, Fe, and Zn.
[39]	OS/507 °C	Spain	Greenhouse	0.5%, 1%, 2.5%, 5%, and 7.5%	Haplic Luvisol	Biochar application significantly increased field capacity, soil pH, soil EC, and resin-nitrate and decreased soil bulk density, soil nitrate content, and resin-ammonium.
[31]	OP/300, 500 and 700 °C	Saudi Arabia	Incubation	3%	Sandy loam texture	The Resin-Pi, $NaHCO_3$-Pi, $NaHCO_3$-Po, and HCl-Pi fractions content was decreased after biochar addition.
[42]	OP/600 °C	Spain	Incubation	5%, 15% and 30%	Calcaric Arenosol	The bulk density and particle density were decreased, and total pore space, water-holding capacity, air volume, soil pH, EC, and organic matter were increased after biochar addition, especially at the largest doses (15–30%).
[33]	OS/500 °C	Spain	Greenhouse	2% and 5%	Sandy loam *Typic Xerofluvent*	The addition of biochar to acidic and moderately acidic soils increased soil pH. Furthermore, EC showed a slight increase in applying 2% of biochar. However, the application of biochar has not modified the water-holding capacity.
[46]	OS/500 °C	Spain	Field	8 t ha^{-1}	Fluvisol	Enzymatic activities and soil respiration rates were not modified with biochar application but increased the total carbon content of soils.
[47]	OS/400 and 500 °C	Spain	Incubation and greenhouse	2%, 5% and 10%	Fluvisol	Biochar addition increased the soil pH (dose 5 and 10%), C content, and C/N ratio. The application of biochar reduced the β-glucosidase activity. Furthermore, the concentrations of $CaCl_2$-extractable heavy metals significantly decreased in biochar-amended pots.
[49]	OS/500 °C	Spain	Field	10%	Fluvisols	Biochar addition caused an increase in the total carbon content, but the N content was not modified.

Table 5. *Cont.*

References	Biomass/ Pyrolysis Temperature	Country	Experiment Type	Biochar Dose	Soil Type	Impact on Soil Properties
[48]	OS/500 °C	Spain	Incubation	8 t ha^{-1}	Fluvisols	Biochar amendment increased the C content of the soils after two years. In general, biochar application did not reduce the low bioavailable contents of trace elements found (As, Cu, Pb, Ba, and Zn) compared to non-amended soils.
[50]	OP/1100–1200 °C	Italy	Field	5 t ha^{-1}	Lithic Ruptic-Inceptic Haploxeralfs	Soil chemical characteristics at the end of the experiment (3-month duration) did not show significant differences for application biochar.
[51]	OP/500 °C	Spain	Field	6.67 t ha^{-1}	Sandy loam (Xerochrept)	Adding biochar increased pH and EC up to 10 cm deep and water content for the soil samples below 5 cm depth. In Situ, applying biochar increased soil moisture of the topsoil (0–5 cm depth) and reduced penetration resistance measured in the field.
[55]	OP/500 °C	Tunisia	Greenhouse	1%, 2.5% and 5%	Agricultural soil	Applying biochar considerably increased the pH and EC of the resulting soil–biochar mixture, an effect that increased with the biochar application rate.
[68]	OP/400–450 °C	Belgium	Greenhouse	5%, 10%, and 15 %	Sandy soil	The increasing biochar rates increased the soil's pH and EC and reduced from 37% for Pb, 37% for Cd, and 49% for Zn, compared to the non-amended soil (0% biochar).
[69]	OP/350 and 700 °C	Turkey	Incubation	1%	Hypereutric Fluvisols	Biochar application did not affect NO_3^-, Br^-, and SO_4^{2-} leaching from sandy loam soils in a short time (46 days).
[70]	OP/350 and 700 °C	Turkey	Incubation	4% and 8%	Mollisol	Applying biochar improved soil aggregate stability and soil total carbon while the proportion of >4 mm aggregates and mean weight diameter of aggregates decreased. Biochar produced at high temperatures significantly increased soil aggregate stability value compared to low temperatures.
[72]	OP/400 °C	Greece	Field	10 and 25 t ha^{-1}	-	Biochar application also positively impacted C and P sequestration in the soil, improved soil structure, and improved soil nutrient content at deeper soil levels.

Table 5. *Cont.*

References	Biomass/ Pyrolysis Temperature	Country	Experiment Type	Biochar Dose	Soil Type	Impact on Soil Properties
[78]	OP/400 and 700 °C	Greece	Incubation	5%, 10%, 15% and 20%	Loamy sand	It was noticed that the addition of both biochars caused an increase in soil pH and organic matter content. Both biochars enhanced the ability of soil to immobilize cadmium and nickel; however, in most cases, the biochar produced at 700 °C was proven to be more efficient.
[80]	OP/450 °C	Italy	Italy	~8%	Clay loam	The biochar did not interfere with pyrene bioavailability or affect microbial pyrene-degrading activity.
[86]	OP/280 °C	Germany	Incubation	4 t ha^{-1}	Alkaline clay soil and acidic sandy soil	Applying biochar to the acidic sandy soil significantly increased soil pH, CO_2, and N_2O fluxes. Additionally, there was a clear variation in bacterial community structure in the acidic sandy soil (phyla *Acidobacteria*, *Nitrospirare*, and *Arthrobacter*) with the biochar addition.
[88]	OP/500 °C	Saudi Arabia	Incubation	1% and 2%	Sandy loam soil	The modification of biochar with silica and nutrients showed an excellent ability to decrease metal (Pb) contents and bioavailability in soil and also increase essential nutrient availability in soil (P, NO_3^-, NH_4^+).

Nutrients released from biochar into the soil solution vary depending on the element's specific sorption affinity with either the biochar or the soil [118]. Alazzaz et al. [38] recorded that the application of 3% biochar increased the soil's exchangeable K content from 40.5 mg kg^{-1} (control) to 324 mg kg^{-1}. Additionally, the application of 3% biochar exhibited significant increases in the soil's availability of Na, Ca, Mg, and some micronutrients (Fe and Zn). This suggests that the availability of soil nutrients amended with OP biochar could likely depend on biochar type and application rate [38].

The application of biochar has been proposed as a promising long-term tool to increase soil C sequestration (Figure 6). This is due to biochar's high stability and aromaticity, whose average residence time is longer than plant biomass, contributing alternatives to promote carbon sequestration [119]. Some studies showed that applying OS and OP biochar significantly increased the total C content on the soil's top layers [42,47–49,72].

Olive-mill-derived biochars may reduce trace metals in contaminated soils, influencing their mobility and toxicity through diverse physicochemical mechanisms [120]. Despite the few existing OP and OS biochar application studies in contaminated soils, Hmid et al. [67] observed a significant reduction of Cd, Pb, and Zn with increasing rates of OS biochar added to contaminated soil. Contrastingly, Campos et al. [49] did not find that biochar application reduces the bioavailability of trace elements (As, Cu, Pb, Ba, and Zn). However, biochar has a rich pore structure and a huge SSA that can improve soil's physical characteristics, such as bulk density, surface area, water-holding capacity, and permeability [121–123].

In several studies, OS and OP biochar application in different doses (2–30%) decreased soil bulk density and particle density and increased the total pore space and soil water-holding capacity, hence positively impacting plant growth [39,42]. Kavdır et al. [70] found that the application of OP biochar at high temperatures significantly improved soil aggregate stability. In contrast, the proportion of macroaggregates (>4 mm) and mean weight diameter of aggregates decreased. On the other hand, Lilli et al. [72] found that the distribution of water-stable aggregates was significantly increased by the addition of OP biochar (an increase in macroaggregates by 16.4% and a decrease in microaggregates by 3.3%).

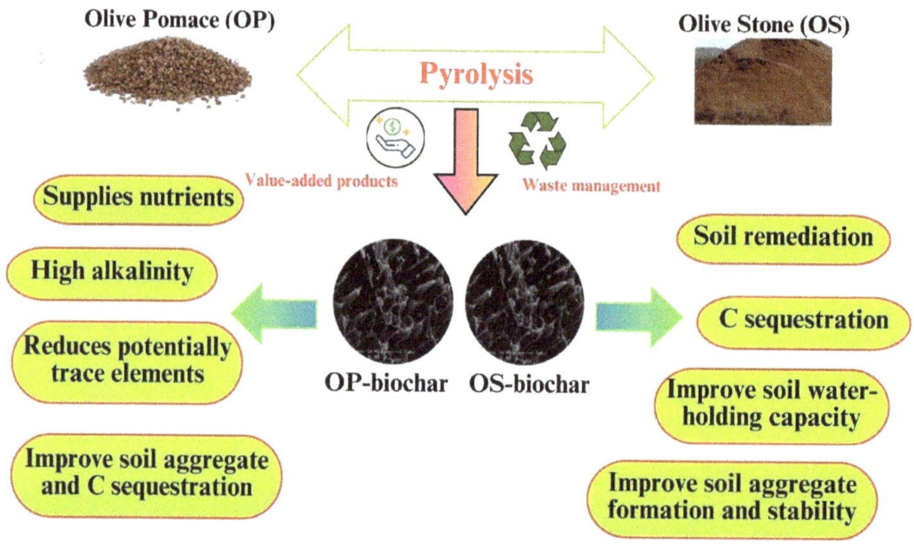

Figure 6. Positive effects of OP- and OS-derived biochar on soil.

Based on previous studies, the variables affecting the microbial response to biochar application were mainly related to biochar properties, application load, and soil properties [124]. However, our current knowledge of the impact of applying OP and OS biochar on the soil's microbial status remains scarce. Lilli et al. [72] found that in the rhizosphere, the OP biochar-treated plants showed a higher number of unique taxa. On the other hand, Campos et al. [45] found that applying biochar on contaminated soils changed the bacterial diversity, reducing the presence of extremophilic bacteria and increasing the presence of bacteria typical of non-contaminated soils. These authors underlined that changes in the soil bacterial richness and diversity after biochar application were correlated with the increment of soil pH, bacterial diversity, and plant growth.

3.4.2. Plant Growth

Several impacts of biochar application on germination and seedling growth range have been reported. Hormesis is regularly detected, meaning that high biochar rates can harm plants, while low rates can be stimulatory. Both positive [39,40,46,47,51,55,68,72,79,81] and negative [33,42,57,59,60,74] responses have been found in field trials to enhance yield crop as a result of biochar application to soils (Table 6).

Plant growth improvements and crop yields due to biochar application are the result of the modification of soil's physical, chemical, and biological properties. In the research conducted by El-Bassi et al. [55], OP biochar application rates of 2.5% and 5% used as a soil amendment promoted tomato seedling growth. Therefore, the produced biochar had great potential to be used as a biofertilizer in agriculture. Lilli et al. [72] found that the yield of tomato plants treated with biochar was 29% higher than the control, which could be attributed to the minerals contained in the biochar, especially the K, and a direct contribution of C fixed in the biochar [38,55].

Table 6. Summary of research on the effect of biochar from OP and OS on plant growth.

References	Biomass/ Pyrolysis Temperature	Biochar Dose	Soil Type	Plant	Effects
[39]	OS/507 °C	0.5%, 1%, 2.5%, 5%, and 7.5%	Haplic Luvisol	Sunflower	Adding ash-poor biochar (produced from OS) positively affected seed germination; however, it did not increase total plant dry biomass.
[40]	OS/350, 450 and 550 °C	-	-	Cress and lettuce	All the biochars (350, 450, and 550 °C) showed germination indices > 82% for both cress and lettuce, typical of non-phytotoxic materials.
[42]	OP/600 °C	5%, 15% and 30%	Calcaric Arenosol	Mastic and myrtle	The biochar of olive mill waste proved phytotoxic for myrtle in mixes with coir in 25% and 50% doses. However, mastic showed relatively good growth in the biochar mixes, the mix only being adverse at the 50% dose.
[33]	OS/500 °C	2% and 5%	Sandy loam *Typic Xerofluvent*	Chinese cabbage	The plant survival rates and height were significantly lower for the pots amended with biochar. In addition, the root and shoot dry weight of the Brassica rapa plant was not higher in plots amended with biochar compared to the unamended plots.
[46]	OS/500 °C	8 t ha^{-1}	Fluvisol	"native plants"	Applying biochar in highly polluted soil plots significantly increased the area of vegetation cover in comparison with the control plots. In contrast, although an increase in vegetation cover was observed for moderately polluted soil amended with biochar, it was not statistically significant.
[47]	OS/400 and 500 °C	2%, 5% and 10%	Fluvisol	Chinese cabbage	Biochar addition significantly increased the root-to-shoot ratio compared to the control soil. In particular, 10% of amendment increased this ratio to the greatest extent.
[50]	OP/1100–1200 °C	5 t ha^{-1}	Lithic Ruptic-Inceptic Haploxeralfs	Tomato	Biochar application alone did not affect any of the parameters of tomato production and quality compared with the control, and in combination with organic fertilizers, no synergistic effect was observed.
[51]	OP/500 °C	6.67 t ha^{-1}	Sandy loam (Xerochrept)	Olive	The biochar led to a relative olive production increase of about 15% compared with the control. Furthermore, an improvement was observed in the physiological parameters of the plants from the biochar-amended plots.
[55]	OP/500 °C	1%, 2.5% and 5%	Agricultural soil	Tomato	Biochar amendment promoted tomato seedling growth after ten weeks, which was highest in the 2.5% and 5% application rates for both biochar OP and biochar from OP impregnation with olive mill wastewater.

Table 6. *Cont.*

References	Biomass/ Pyrolysis Temperature	Biochar Dose	Soil Type	Plant	Effects
[57]	OP/500 °C	-	-	Cress and lettuce	Biochar was highly phytotoxic, with an average reduction of germination index of 82%.
[59]	OP/500 °C	-	-	Cress, lettuce, pea and Tomato	The biochar from OP showed considerable phytotoxicity with a reduction of seed germination by 80% and early seedling death.
[58]	OP/500 °C	0%, 10%, 25%, 50%, 75%, and 100%	"substrates"	Tomato	Biochar affected plant growth negatively only at doses larger than 25%. The effect of biochar on fruit quality was very limited.
[60]	OP/500 °C	10%, 25% and 50%	"substrates"	Petunia and calendula	Biochar at 10% markedly reduced the germination percentage, shoot and root growth, and the percentage of flowering plants in these species compared to the control. Furthermore, the plants showed an unbalanced nutrient uptake and low chlorophyll content.
[68]	OP/400–450 °C	5%, 10%, and 15%	Sandy soil	Bean	With the increasing biochar application rate, the growth of bean plants improved; leaf metal contents were reduced, the activities of antioxidative stress enzymes decreased, and soluble protein contents increased.
[72]	OP/400 °C	25 t ha^{-1}	-	Tomato	The yield of tomato plants treated with biochar was 29% higher than the control treatment.
[74]	OP/300, 500, 800 and 1000 °C	-	In vitro	Garden cress and field mustard	The root growth of *Lepidium sativum* and *Brassica rapa*, as well as the survival of the nematode *Meloidogyne incognita*, were inhibited by biochar produced at 300 °C, but toxicity decreased at higher pyrolysis temperatures.
[79]	OP/400 and 600 °C	-	1%	Cress	OP-derived biochars exerted a phytostimulant effect on Lepidium sativum seeds because their corresponding germination indexes were greater than 100%.
[81]	OP/450 °C	-	-	Cress	Germination indexes of biochar water extracts at two different dilution ratios (1:9 and 3:7) were 107 and 58%, respectively.

An excessive application rate of biochar addition may have an inhibitory effect on seed germination and crop yield. This is due to biochar's high alkalinity and salinity, mainly caused by the large content of soluble K in the biochar [57,59]. Therefore, a clear understanding of these parameters is essential before the biochar can be used in a plant growth substrate. In this sense, Fornes et al. [57] used seed germination bioassays to prove the produced biochar's phytotoxic character due to its high salt content. Similarly, Fornes et al. [58] evaluated that the application of 10 and 25% of OP biochars as substrates in soilless cultivation of tomato plants did not affect shoot and root growth; however, higher doses had a negative impact on the production and number of tomato fruits, without consequences on the fruit quality. This negative impact was attributed to one of the biochar-containing substrates being saline. Nevertheless, Fornes and Belda [59] demonstrated that salt excess and alkalinity were easily reduced by treating biochar with diluted nitric acid, thus removing the phytotoxic factors.

Plant stress is one of the major problems encountered in agricultural environments. There are both biotic and abiotic types of plant stresses [125]. Literature has confirmed that adding biochar was generally effective in reducing the hazard of polluted soils and restoring balanced plant mineral nutrition [126]. Campos et al. [47] found that Cd, Cu, and Zn bioavailable contents decreased in soils amended with OS biochars. In addition, biochar application enhanced the germination in acidic, polluted soil, where germination was impossible. Likewise, applying OS biochar to metal-contaminated soils provided environmental benefits, including increased diversity and plant growth [46].

Some studies have shown that biochar might improve the photosynthetic capacity of crop leaves and the activities of antioxidant enzymes [127–129]. In this regard, Hmid et al. [68] observed that increasing OP biochar increased 15-day-old bean growth and soluble protein contents and decreased the activities of antioxidant enzymes and leaf Cd, Pb, and Zn contents. Besides, De la Rosa et al. [51] found that olive trees amended with OP biochar had better water and photosynthetic conditions.

The effective application of biochar as a practical soil amendment in agriculture might face limitations due to economic, sustainability, and environmental factors [130]. Collection of waste material, processing of feedstocks (e.g., drying), and pyrolysis operations (i.e., pyrolysis reactor and operating cost), and costs associated with the handling and application of biochar, one must consider and weigh them against the environmental benefits and enhancements in crop production [131]. However, in Europe, just using quantities available of OS and OP as feedstock for the production of biochar could sequester CO_2 in the range of 0.93–2.32 $MtCO_2eq\ yr^{-1}$ and 2.14–2.97 $MtCO_2eq\ yr^{-1}$, respectively [132].

3.4.3. Biochar and Activated Carbon Applications

The application of AC represents a dynamic and innovative approach in the field of environmental and agricultural sciences. The use of biochar and AC has increasingly attracted the attention of researchers, given their multiple applications, ranging from soil improvement and carbon sequestration to water purification and pollutant remediation. As researchers delve deeper into the potential of AC, their diverse and sustainable applications continue to unfold, promising solutions to contemporary challenges.

In catalysts, Ayadi et al. [41] demonstrated the potential of different solid AC acid catalysts produced from OP and coconut husks through sulfonation for sustainable and efficient biodiesel production from OP. The solid AC acid catalysts were employed in the esterification of olive pomace oil (OPO), facilitating the conversion of free fatty acids into esters. The results indicated that, after 5 h of reaction, both AC catalysts resulted in acid values lower than 2 mg KOH g^{-1} and met the requirements of EN14214 standard [133], with a content higher than 97%, indicating the successful esterification process facilitated by the catalysts, highlighting their catalytic properties and applicability in biodiesel production processes. In Guerrero-Pérez et al. [69], the aim was to evaluate the behavior of the vanadium-containing activated carbon catalysts for the partial oxidation of propylene to oxygenates. Furthermore, the study discussed how adding zirconium as a dopant could enhance the catalytic performance. The higher propylene conversion percentages for the vanadium-zirconium-doped catalysts indicate the synergistic effect of incorporating both elements in enhancing the AC catalytic activity, highlighting their catalytic activity and selectivity towards oxygenate products. At last, Omiri et al. [77] deployed biochar as catalysts for methyl orange (MO) degradation.

Among the benefits of AC in adsorption processes, El-Azazy et al. [54] used OS biochar and magnetic OS biochar as adsorbents in the removal of antibiotic ciprofloxacin in aqueous solution, highlighting the strong adsorption capabilities (174.03 mg g^{-1}) of the magnetic OS biochar.

Ghouma et al. [63] analyzed the AC performance in NO_2 adsorption under various conditions, finding that the AC, produced through steam activation of OS, proved to be a highly effective adsorbent for NO_2 removal. The study highlighted a temperature-dependent variation in the NO_2 adsorption capacity of the AC, suffering a substantial reduction in adsorption capacity (39%) when temperature increased from 20 °C to 50 °C.

Finally, Usama et al. [88] aimed to evaluate the effectiveness of biochars in immobilizing lead (Pb) and enhancing nutrient availability in Pb-contaminated sandy loam soil. The silica-embedded (BCSi) and nutrient-loaded (BCSiNP) biochars showed the highest adsorption capacity for Pb (57.78 mg g^{-1}) in BCSiNP followed by BCSi (54.01 mg g^{-1}), indicating their excellent potential to immobilize Pb in the soil. The modified biochars also increased the bioavailability of essential nutrients (P, NO_3^-, NH_4^+) in the soil. Overall, the study demonstrated that the modified biochars effectively immobilized Pb and improved nutrient availability in the contaminated soil.

3.5. Challenges and Future Perspectives

The olive industry's waste constitutes an environmental hazard due to its acidity and high phenol content. Converting these low-value feedstocks into cost-effective materials such as biochar is a sustainable method of waste valorization with several emerging applications. A literature review on biochar production from olive mill wastes revealed academia's extensive efforts on biochar characterization, property optimization, and applications during the past few decades. Nevertheless, there are several gaps on this topic, requiring deeper knowledge and substantial efforts to understand the biochar production process and how to manipulate certain features in order to obtain functional biochar, answering the needs and challenges presented by the several targets of application, as well as soil and water.

Future studies should focus on the mechanisms involved in agriculture and environmental applications of OP and OS biochars. Although several studies have proven that OP biochar may promote desired effects in both soil and plants, the mechanisms involved and how to control those processes are not well understood.

Since characteristics of biochar are highly controlled by types of feedstocks and pyrolysis conditions (e.g., temperature), it is essential to select appropriate feedstock, i.e., OP or OS, according to the specific physicochemical biochar properties aiming at previously established purposes.

Furthermore, studies related to the use of OP and OS biochar in optimizing the removal of polyphenols from olive mill wastewater can be a cost-effective and profitable alternative, which can be included in conventional wastewater treatment plants. Thus, biochar produced from olive mill byproducts can serve as an ecological alternative and a promising material for future wastewater treatment and residue management strategies, fostering sustainability and a circular economy policy.

When employed in agriculture, biochar is typically applied in significant amounts, which makes it difficult to posteriorly remove it from the soil in case of any unfavorable outcome. Future research should focus on biochar's effectiveness for different methodologies and time of application, considering as well the cropping system and plant growth stage.

Modified biochar (e.g., acid, base, salt, or steam/air activation) may result in different outcomes imposed by its distinct elemental and chemical composition (e.g., SSA, porosity, CEC, surface functional groups, and pH), which should also be evaluated when considering modified biochar application, not only on soil but also in other environmental compartments.

Regulatory entities, from local to national and European bodies, must provide legal support and financial incentives to produce and use biochar, especially from local waste, such as olive oil byproducts.

4. Conclusions

This systematic review presented a summary of current research developed to produce and characterize OP and other environmental applications. Without a doubt, the pyrolysis temperature and feedstock type have a massive impact on the characteristics of the resultant biochar. Therefore, it is important to have in-depth knowledge of the relation of these two variables in order to optimize the productive process and maximize biochar's efficiency and its intended uses. OS biochar has advantageous properties such as large surface area, high porosity, low ash content, and high C amount, making it an efficient tool to improve the surface adsorption process. These properties make it an efficient and suitable tool for various applications such as soil remediation, and wastewater treatment, but not to crop growth. However, the OP biochar presents a high C amount and ash content that makes it suitable for amendment in agricultural soils and long-term C sequestration. Biochar alters not only the amended soil's physical, chemical, and biological properties but also influences the nutrient dynamics, immobilizing trace metals, and thereby improving soil quality and crop yield. However, the application rate of OP biochar may have some inhibitory effect on seed germination and crop yield, which requires further study. Possibly, the phenomenon is related to an overdose or inappropriate production of biochar (toxic polycyclic aromatic hydrocarbons or too aromatic, C-C) that must be previously tested with the quality indicators presented (pH, CEC, SSA, porosity, H/C vs. O/C, ashes). Overall, the use of byproducts from olive oil is a promising feedstock to produce sustainable biochar, though the productive process needs to be further improved and tailored to maximize its characteristics and applications.

Supplementary Materials: The following supporting information can be downloaded at: https://www.mdpi.com/article/10.3390/su16125004/s1, Table S1: Summary of yield and ultimate and proximate analyses of biochars from OP and OS from different studies; Table S2: Summary of physical and chemical characteristics and nutrients of biochars from OP and OS from different studies; Table S3: Physical and chemical characteristics, elemental composition, and nutrients of the feedstocks (OP and OS) used in this review; Table S4: Correlation of activation and functionalization of biochars and AC with functional groups; Table S5. Elemental analysis of CNHS-O. Figure S1: Linear relationship between pyrolysis temperature and electrical conductivity in OP biochar and OS biochar. Com-plete data for the 62 articles reviewed are provided in Tables S1 and S2. PRISMA 2020 Checklist.

Author Contributions: Conceptualization, J.F.L.F. and Z.H.; methodology, J.F.L.F. and Z.H.; software, J.F.L.F.; formal analysis, J.F.L.F., S.T.C. and Z.H.; investigation, J.F.L.F. and Z.H.; resources, Z.H. and T.d.F.; writing—original draft preparation, J.F.L.F., S.T.C., A.P.F.d.S. and Z.H.; writing—review and editing, J.F.L.F., H.T.G., T.d.F., A.P.F.d.S. and Z.H.; supervision, Z.H.; project administration, Z.H. and T.d.F.; funding acquisition, Z.H. and T.d.F. All authors have read and agreed to the published version of the manuscript.

Funding: The authors are grateful for the financial support of the research carried out by project "EEAGRANT_CALL#5—Innovative nature-based solutions for restoring ecosystem services in areas degraded by the great fire of Picões, Portugal—SOILING" funded by European Economic Area (EEA) Financial Mechanism 2014–2021 Environment, Climate Change and Low Carbon Economy Programme. 'Environment Programme (*); and Programme 13/REACT-EU/2021 "ForestWaterUp-Nature-based solutions for the ecological restoration of degraded soils in the Sabor Lakes" (POCI-07-62G4-FEDER-181557), funded by COMPETE 2020 Operational Programme. The authors are also grateful to the Foundation for Science and Technology (FCT, Portugal) for the financial support through national funds by FCT/MCTES (PIDDAC): CIMO, UIDB/00690/2020 (DOI: 10.54499/UIDB/00690/2020) and UIDP/00690/2020 (DOI: 10.54499/UIDP/00690/2020); and SusTEC, LA/P/0007/2020 (DOI:10.54499/LA/P/0007/2020), and by the institutional scientific employment program-contract with Zulimar Hernández. Ana Paula Ferreira thanks her doctoral Grant with reference PRT/BD/153090/2021, financed by FCT, with funds from NORTE2020, under Program MIT Portugal. (*) Through the Agreement on the European Economic Area (EEA), Iceland, Liechtenstein, and Norway are partners in the internal market with the Member States of the European Union. The Secretary-General of the Environment operated EEA Grants support to Portugal to promote a continuous and balanced strengthening of economic and trade relations, reduce social and economic disparities in Europe, and strengthen bilateral relations between these three countries and the beneficiary countries.

Institutional Review Board Statement: Not applicable.

Informed Consent Statement: Not applicable.

Data Availability Statement: Data are contained within the article.

Acknowledgments: The authors would like to acknowledge 11 Call#5-SOILING for funding a full-time Pos-Doctoral Grant (MORE/SOILING/1) for the first author (J.F.L.F.), which allowed the publication of the present research.

Conflicts of Interest: The authors declare no conflicts of interest.

References

1. FAO Food and Agriculture Organization of the United Nations: FAOSTAT Statistical Database. Available online: https://www.fao.org/statistics/en (accessed on 20 January 2023).
2. International Olive Council Prices & Balances. Available online: https://www.internationaloliveoil.org/ (accessed on 20 January 2023).
3. Borja, R.; Raposo, F.; Rincón, B. Treatment Technologies of Liquid and Solid Wastes from Two-Phase Olive Oil Mills. *Grasas Aceites* **2006**, *57*, 32–46. [CrossRef]
4. Alba, J.; Hidalgo, F.; Martínez, F.; Ruíz, M.A.; Moyano, M.J.; Borja, R. Evolución Medioambiental de Los Sistemas de Elaboración de Aceite de Oliva En Andalucía. *Mercacei* **1995**, *2*, 20–22.
5. Cabrera, F.; López, R.; Martinez-Bordiú, A.; Dupuy de Lome, E.; Murillo, J.M. Land Treatment of Olive Oil Mill Wastewater. *Int. Biodeterior. Biodegrad.* **1996**, *38*, 215–225. [CrossRef]
6. Giannoutsou, E.; Lambraki, M.; Karagouni, A.D. Microbial Treatment of Olive Mill Effluents. In Proceedings of the Symposium Olive Wastes, Kalamata, Greece, 5–8 November 1997; pp. 89–96.
7. Vera, D.; Jurado, F.; Carpio, J.; Kamel, S. Biomass Gasification Coupled to an EFGT-ORC Combined System to Maximize the Electrical Energy Generation: A Case Applied to the Olive Oil Industry. *Energy* **2018**, *144*, 41–53. [CrossRef]
8. EEA EMEP/EEA Air Pollutant Emission Inventory Guidebook 2023. Technical Guidance to Prepare National Emission Inventories. Available online: https://www.eea.europa.eu/publications/emep-eea-guidebook-2023 (accessed on 20 January 2023).
9. Ministry of Economy Estudios Cadena de Valor Alimentos Frescos. Available online: https://www.mapa.gob.es/es/alimentacion/temas/observatorio-cadena/estudios-e-informes/default.aspx (accessed on 20 May 2023).
10. Huang, Y.; Li, B.; Liu, D.; Xie, X.; Zhang, H.; Sun, H.; Hu, X.; Zhang, S. Fundamental Advances in Biomass Autothermal/Oxidative Pyrolysis: A Review. *ACS Sustain. Chem. Eng.* **2020**, *8*, 11888–11905. [CrossRef]
11. Alburquerque, J. Agrochemical Characterisation of "Alperujo", a Solid by-Product of the Two-Phase Centrifugation Method for Olive Oil Extraction. *Bioresour. Technol.* **2004**, *91*, 195–200. [CrossRef]
12. Kamali, M.; Sweygers, N.; Al-Salem, S.; Appels, L.; Aminabhavi, T.M.; Dewil, R. Biochar for Soil Applications-Sustainability Aspects, Challenges and Future Prospects. *Chem. Eng. J.* **2022**, *428*, 131189. [CrossRef]
13. Rodriguez, J.A.; Lustosa Filho, J.F.; Melo, L.C.A.; de Assis, I.R.; de Oliveira, T.S. Classifying the Potential of Biochars from Agricultural and Industrial Waste for the Recovery of Fe and Mg Mining Tailings. *J. Anal. Appl. Pyrolysis* **2022**, *161*, 105383. [CrossRef]
14. Li, S.; Harris, S.; Anandhi, A.; Chen, G. Predicting Biochar Properties and Functions Based on Feedstock and Pyrolysis Temperature: A Review and Data Syntheses. *J. Clean. Prod.* **2019**, *215*, 890–902. [CrossRef]
15. Lu, H.; Zhang, W.; Wang, S.; Zhuang, L.; Yang, Y.; Qiu, R. Characterization of Sewage Sludge-Derived Biochars from Different Feedstocks and Pyrolysis Temperatures. *J. Anal. Appl. Pyrolysis* **2013**, *102*, 137–143. [CrossRef]
16. Pariyar, P.; Kumari, K.; Jain, M.K.; Jadhao, P.S. Evaluation of Change in Biochar Properties Derived from Different Feedstock and Pyrolysis Temperature for Environmental and Agricultural Application. *Sci. Total Environ.* **2020**, *713*, 136433. [CrossRef]
17. Rodriguez, J.A.; Ferreira, J.; Filho, L.; Carrijo, L.; Melo, A.; Rodrigues De Assis, I.; Senna De Oliveira, T. Influence of Pyrolysis Temperature and Feedstock on the Properties of Biochars Produced from Agricultural and Industrial Wastes. *J. Anal. Appl. Pyrolysis* **2020**, *149*, 104839. [CrossRef]
18. Rodriguez, J.A.; Lustosa Filho, J.F.; Melo, L.C.A.; de Assis, I.R.; de Oliveira, T.S. Co-Pyrolysis of Agricultural and Industrial Wastes Changes the Composition and Stability of Biochars and Can Improve Their Agricultural and Environmental Benefits. *J. Anal. Appl. Pyrolysis* **2021**, *155*, 105036. [CrossRef]
19. Domingues, R.R.; Trugilho, P.F.; Silva, C.A.; De Melo, I.C.N.A.; Melo, L.C.A.; Magriotis, Z.M.; Sánchez-Monedero, M.A. Properties of Biochar Derived from Wood and High-Nutrient Biomasses with the Aim of Agronomic and Environmental Benefits. *PLoS ONE* **2017**, *12*, e0176884. [CrossRef]
20. Garcia-Ibañez, P.; Sanchez-Garcia, M.; Sánchez-Monedero, M.A.; Cayuela, M.L.; Moreno, D.A. Olive Tree Pruning Derived Biochar Increases Glucosinolate Concentrations in Broccoli. *Sci. Hortic.* **2020**, *267*, 109329. [CrossRef]

21. Enaime, G.; Baçaoui, A.; Yaacoubi, A.; Lübken, M. Biochar for Wastewater Treatment-Conversion Technologies and Applications. *Appl. Sci.* **2020**, *10*, 3492. [CrossRef]
22. Leng, L.; Xiong, Q.; Yang, L.; Li, H.; Zhou, Y.; Zhang, W.; Jiang, S.; Li, H.; Huang, H. An Overview on Engineering the Surface Area and Porosity of Biochar. *Sci. Total Environ.* **2021**, *763*, 144204. [CrossRef]
23. Danesh, P.; Niaparast, P.; Ghorbannezhad, P.; Ali, I. Biochar Production: Recent Developments, Applications, and Challenges. *Fuel* **2022**, *337*, 126889. [CrossRef]
24. Vilas-Boas, A.C.M.; Tarelho, L.A.C.; Oliveira, H.S.M.; Silva, F.G.C.S.; Pio, D.T.; Matos, M.A.A. Valorisation of Residual Biomass by Pyrolysis: Influence of Process Conditions on Products. *Sustain. Energy Fuels* **2024**, *8*, 379–396. [CrossRef]
25. Figueiredo, C.; Lopes, H.; Coser, T.; Vale, A.; Busato, J.; Aguiar, N.; Novotny, E.; Canellas, L. Influence of Pyrolysis Temperature on Chemical and Physical Properties of Biochar from Sewage Sludge. *Arch. Agron. Soil Sci.* **2018**, *64*, 881–889. [CrossRef]
26. Sarfaraz, Q.; Silva, L.; Drescher, G.; Zafar, M.; Severo, F.; Kokkonen, A.; Molin, G.; Shafi, M.; Shafique, Q.; Solaiman, Z. Characterization and Carbon Mineralization of Biochars Produced from Different Animal Manures and Plant Residues. *Sci. Rep.* **2020**, *10*, 955. [CrossRef]
27. Singh, S.; Kumar, V.; Dhanjal, D.S.; Datta, S.; Bhatia, D.; Dhiman, J.; Samuel, J.; Prasad, R.; Singh, J. A Sustainable Paradigm of Sewage Sludge Biochar: Valorization, Opportunities, Challenges and Future Prospects. *J. Clean. Prod.* **2020**, *269*, 122259. [CrossRef]
28. Ahmad, Z.; Gao, B.; Mosa, A.; Yu, H.; Yin, X.; Bashir, A.; Ghoveisi, H.; Wang, S. Removal of Cu(II), Cd(II) and Pb(II) Ions from Aqueous Solutions by Biochars Derived from Potassium-Rich Biomass. *J. Clean. Prod.* **2018**, *180*, 437–449. [CrossRef]
29. Azzaz, A.A.; Ghimbeu, C.M.; Jellai, S.; El-Bassi, L.; Jeguirim, M. Olive Mill By-Products Thermochemical Conversion via Hydrothermal Carbonization and Slow Pyrolysis: Detailed Comparison between the Generated Hydrochars and Biochars Characteristics. *Processes* **2022**, *10*, 231. [CrossRef]
30. Marks, E.A.N.; Kinigopoulou, V.; Akrout, H.; Azzaz, A.A.; Doulgeris, C.; Jellali, S.; Rad, C.; Zulueta, P.S.; Tziritis, E.; El-Bassi, L.; et al. Potential for Production of Biochar-Based Fertilizers from Olive Mill Waste in Mediterranean Basin Countries: An Initial Assessment for Spain, Tunisia, and Greece. *Sustainability* **2020**, *12*, 6081. [CrossRef]
31. Alotaibi, K.D.; Arcand, M.; Ziadi, N. Effect of Biochar Addition on Legacy Phosphorus Availability in Long-Term Cultivated Arid Soil. *Chem. Biol. Technol. Agric.* **2021**, *8*, 47. [CrossRef]
32. Delgado-Moreno, L.; Bazhari, S.; Gasco, G.; Méndez, A.; El Azzouzi, M.; Romero, E. New Insights into the Efficient Removal of Emerging Contaminants by Biochars and Hydrochars Derived from Olive Oil Wastes. *Sci. Total Environ.* **2021**, *752*, 141838. [CrossRef]
33. Campos, P.; De la Rosa, J.M. Assessing the Effects of Biochar on the Immobilization of Trace Elements and Plant Development in a Naturally Contaminated Soil. *Sustainability* **2020**, *12*, 6025. [CrossRef]
34. MAPA. Available online: https://www.mapa.gob.es/es/agricultura/temas/producciones-agricolas/aceite-oliva-y-aceituna-mesa (accessed on 26 January 2023).
35. Abdelhadi, S.O.; Dosoretz, C.G.; Rytwo, G.; Gerchman, Y.; Azaizeh, H. Production of Biochar from Olive Mill Solid Waste for Heavy Metal Removal. *Bioresour. Technol.* **2017**, *244*, 759–767. [CrossRef]
36. Aguirre, J.L.; González Egido, S.; González-Lucas, M.; González-Pernas, F.M. Medium-Term Effects and Economic Analysis of Biochar Application in Three Mediterranean Crops. *Energies* **2023**, *16*, 4131. [CrossRef]
37. Aissaoui, M.H.; Trabelsi, A.B.H.; Bensidhom, G.; Ceylan, S.; Leahy, J.J.; Kwapinski, W. Insights into Olive Pomace Pyrolysis Conversion to Biofuels and Biochars: Characterization and Techno-Economic Evaluation. *Sustain. Chem. Pharm.* **2023**, *32*, 101022. [CrossRef]
38. Alazzaz, A.; Usman, A.R.A.; Ahmad, M.; Ibrahim, H.M.; Elfaki, J.; Sallam, A.S.; Akanji, M.A.; Al-Wabel, M.I. Potential Short-Term Negative versus Positive Effects of Olive Mill-Derived Biochar on Nutrient Availability in a Calcareous Loamy Sand Soil. *PLoS ONE* **2020**, *15*, e0232811. [CrossRef]
39. Alburquerque, J.A.; Calero, J.M.; Barrón, V.; Torrent, J.; del Campillo, M.C.; Gallardo, A.; Villar, R. Effects of Biochars Produced from Different Feedstocks on Soil Properties and Sunflower Growth. *J. Plant Nutr. Soil Sci.* **2014**, *177*, 16–25. [CrossRef]
40. Alburquerque, J.A.; Sánchez, M.E.; Mora, M.; Barrón, V. Slow Pyrolysis of Relevant Biomasses in the Mediterranean Basin. Part 2. Char Characterisation for Carbon Sequestration and Agricultural Uses. *J. Clean. Prod.* **2016**, *120*, 191–197. [CrossRef]
41. Ayadi, M.; Awad, S.; Villot, A.; Abderrabba, M.; Tazerout, M. Heterogeneous Acid Catalyst Preparation from Olive Pomace and Its Use for Olive Pomace Oil Esterification. *Renew. Energy* **2021**, *165*, 1–13. [CrossRef]
42. Belda, R.M.; Lidón, A.; Fornes, F. Biochars and Hydrochars as Substrate Constituents for Soilless Growth of Myrtle and Mastic. *Ind. Crops Prod.* **2016**, *94*, 132–142. [CrossRef]
43. Caballero, B.M.; López-Urionabarrenechea, A.; Pérez, B.; Solar, J.; Acha, E.; de Marco, I. Potentiality of "Orujillo"(Olive Oil Solid Waste) to Produce Hydrogen by Means of Pyrolysis. *Int. J. Hydrogen Energy* **2020**, *45*, 20549–20557. [CrossRef]
44. Calzado, M.; Valero-Romero, M.J.; Garriga, P.; Chica, A.; Guerrero-Pérez, M.O.; Rodríguez-Mirasol, J.; Cordero, T. Lignocellulosic Waste-Derived Basic Solids and Their Catalytic Applications for the Transformation of Biomass Waste. *Catal. Today* **2015**, *257*, 229–236. [CrossRef]
45. Campos, P.; Miller, A.Z.; Knicker, H.; Costa-Pereira, M.F.; Merino, A.; De la Rosa, J.M. Chemical, Physical and Morphological Properties of Biochars Produced from Agricultural Residues: Implications for Their Use as Soil Amendment. *Waste Manag.* **2020**, *105*, 256–267. [CrossRef]

46. Campos, P.; Miller, A.Z.; Prats, S.A.; Knicker, H.; Hagemann, N.; De la Rosa, J.M. Biochar Amendment Increases Bacterial Diversity and Vegetation Cover in Trace Element-Polluted Soils: A Long-Term Field Experiment. *Soil Biol. Biochem.* **2020**, *150*, 108014. [CrossRef]
47. Campos, P.; Knicker, H.; López, R.; De la Rosa, J.M. Application of Biochar Produced from Crop Residues on Trace Elements Contaminated Soils: Effects on Soil Properties, Enzymatic Activities and Brassica Rapa Growth. *Agronomy* **2021**, *11*, 1394. [CrossRef]
48. Campos, P.; Knicker, H.; Velasco-Molina, M.; De la Rosa, J.M. Assessment of the Biochemical Degradability of Crop Derived Biochars in Trace Elements Polluted Soils. *J. Anal. Appl. Pyrolysis* **2021**, *157*, 105186. [CrossRef]
49. Campos, P.; Knicker, H.; Miller, A.Z.; Velasco-Molina, M.; De la Rosa, J.M. Biochar Ageing in Polluted Soils and Trace Elements Immobilisation in a 2-Year Field Experiment. *Environ. Pollut.* **2021**, *290*, 118025. [CrossRef] [PubMed]
50. Cavoski, I.; Al Chami, Z.; Jarrar, M.; Mondelli, D. Solutions for Soil Fertility Management to Overcome the Challenges of the Mediterranean Organic Agriculture: Tomato Plant Case Study. *Soil Res.* **2016**, *54*, 125–133. [CrossRef]
51. De la Rosa, J.M.; Campos, P.; Diaz-Espejo, A.; Rosa Arranz, J.M.; Campos Díaz de Mayorga, P.; Díaz-Espejo, A.; De la Rosa, J.M.; Campos, P.; Diaz-Espejo, A.; Rosa Arranz, J.M.; et al. Soil Biochar Application: Assessment of the Effects on Soil Water Properties, Plant Physiological Status, and Yield of Super-Intensive Olive Groves under Controlled Irrigation Conditions. *Agronomy* **2022**, *12*, 2321. [CrossRef]
52. del Pozo, C.; Rego, F.; Puy, N.; Bartrolí, J.; Fàbregas, E.; Yang, Y.; Bridgwater, A.V. The Effect of Reactor Scale on Biochars and Pyrolysis Liquids from Slow Pyrolysis of Coffee Silverskin, Grape Pomace and Olive Mill Waste, in Auger Reactors. *Waste Manag.* **2022**, *148*, 106–116. [CrossRef]
53. El Hanandeh, A.; Albalasmeh, A.; Gharaibeh, M.; Alajlouni, M. Modification of Biochar Prepared from Olive Oil Processing Waste to Enhance Phenol Removal from Synthetic and Olive Mill Wastewater. *Sep. Sci. Technol.* **2021**, *56*, 1659–1671. [CrossRef]
54. El-Azazy, M.; Nabil, I.; Hassan, S.S.; El-Shafie, A.S. Adsorption Characteristics of Pristine and Magnetic Olive Stones Biochar with Respect to Clofazimine. *Nanomaterials* **2021**, *11*, 963. [CrossRef]
55. El-Bassi, L.; Azzaz, A.A.; Jellali, S.; Akrout, H.; Marks, E.A.N.; Ghimbeu, C.M.; Jeguirim, M. Application of Olive Mill Waste-Based Biochars in Agriculture: Impact on Soil Properties, Enzymatic Activities and Tomato Growth. *Sci. Total Environ.* **2021**, *755*, 142531. [CrossRef]
56. El-Shafie, A.S.; Ahsan, I.; Radhwani, M.; Al-Khangi, M.A.; El-Azazy, M. Synthesis and Application of Cobalt Oxide (Co_3O_4)-Impregnated Olive Stones Biochar for the Removal of Rifampicin and Tigecycline: Multivariate Controlled Performance. *Nanomaterials* **2022**, *12*, 379. [CrossRef]
57. Fornes, F.; Belda, R.M.; Lidón, A. Analysis of Two Biochars and One Hydrochar from Different Feedstock: Focus Set on Environmental, Nutritional and Horticultural Considerations. *J. Clean. Prod.* **2015**, *86*, 40–48. [CrossRef]
58. Fornes, F.; Belda, R.M.; Fernández de Córdova, P.; Cebolla-Cornejo, J. Assessment of Biochar and Hydrochar as Minor to Major Constituents of Growing Media for Containerized Tomato Production. *J. Sci. Food Agric.* **2017**, *97*, 3675–3684. [CrossRef] [PubMed]
59. Fornes, F.; Belda, R.M. Acidification with Nitric Acid Improves Chemical Characteristics and Reduces Phytotoxicity of Alkaline Chars. *J. Environ. Manag.* **2017**, *191*, 237–243. [CrossRef] [PubMed]
60. Fornes, F.; Belda, R.M. Biochar versus Hydrochar as Growth Media Constituents for Ornamental Plant Cultivation. *Sci. Agric.* **2018**, *75*, 304–312. [CrossRef]
61. Galán-Pérez, J.A.; Gámiz, B.; Celis, R. Soil Modification with Organic Amendments and Organo-Clays: Effects on Sorption, Degradation, and Bioactivity of the Allelochemical Scopoletin. *J. Environ. Manag.* **2022**, *302*, 114102. [CrossRef] [PubMed]
62. Gámiz, B.; Pignatello, J.J.; Cox, L.; Hermosín, M.C.; Celis, R. Environmental Fate of the Fungicide Metalaxyl in Soil Amended with Composted Olive-Mill Waste and Its Biochar: An Enantioselective Study. *Sci. Total Environ.* **2016**, *541*, 776–783. [CrossRef] [PubMed]
63. Ghouma, I.; Jeguirim, M.; Dorge, S.; Limousy, L.; Ghimbeu, C.M.; Ouederni, A. Activated Carbon Prepared by Physical Activation of Olive Stones for the Removal of NO_2 at Ambient Temperature. *C. R. Chim.* **2015**, *18*, 63–74. [CrossRef]
64. Ghouma, I.; Jeguirim, M.; Guizani, C.; Ouederni, A.; Limousy, L. Pyrolysis of Olive Pomace: Degradation Kinetics, Gaseous Analysis and Char Characterization. *Waste Biomass Valorization* **2017**, *8*, 1689–1697. [CrossRef]
65. Gómez, N.; Rosas, J.G.; Cara, J.; Martínez, O.; Alburquerque, J.A.; Sánchez, M.E. Slow Pyrolysis of Relevant Biomasses in the Mediterranean Basin. Part 1. Effect of Temperature on Process Performance on a Pilot Scale. *J. Clean. Prod.* **2016**, *120*, 181–190. [CrossRef]
66. Guerrero-Pérez, M.O.; Valero-Romero, M.J.; Hernández, S.; Nieto, J.M.L.; Rodríguez-Mirasol, J.; Cordero, T. Lignocellulosic-Derived Mesoporous Materials: An Answer to Manufacturing Non-Expensive Catalysts Useful for the Biorefinery Processes. *Catal. Today* **2012**, *195*, 155–161. [CrossRef]
67. Hmid, A.; Mondelli, D.; Fiore, S.; Fanizzi, F.P.; Al Chami, Z.; Dumontet, S. Production and Characterization of Biochar from Three-Phase Olive Mill Waste through Slow Pyrolysis. *Biomass Bioenergy* **2014**, *71*, 330–339. [CrossRef]
68. Hmid, A.; Al Chami, Z.; Sillen, W.; De Vocht, A.; Vangronsveld, J. Olive Mill Waste Biochar: A Promising Soil Amendment for Metal Immobilization in Contaminated Soils. *Environ. Sci. Pollut. Res.* **2015**, *22*, 1444–1456. [CrossRef]
69. İlay, R. Short-Lived Effects of Olive Pomace Biochar Produced at Different Temperatures on Nitrate (NO_3^-), Bromide (Br^-), Sulfate (SO_4^{2-}) and Phosphate (PO_4^{3-}) Leaching from Sandy Loam Soils. *Commun. Soil Sci. Plant Anal.* **2020**, *51*, 2223–2243. [CrossRef]

70. Kavdır, Y.; İlay, R.; Güven, O.B.; Sungur, A. Characterization of Olive Pomace Biochar Produced at Different Temperatures and Their Temporal Effects on Soil Aggregation and Carbon Content. *Biomass Convers. Biorefinery.* **2023**. [CrossRef]
71. Khalil, A.M.; Michely, L.; Pires, R.; Bastide, S.; Jlassi, K.; Ammar, S.; Jaziri, M.; Chehimi, M.M. Copper/Nickel-Decorated Olive Pit Biochar: One Pot Solid State Synthesis for Environmental Remediation. *Appl. Sci.* **2021**, *11*, 8513. [CrossRef]
72. Lilli, M.A.; Paranychianakis, N.V.; Lionoudakis, K.; Kritikaki, A.; Voutsadaki, S.; Saru, M.L.; Komnitsas, K.; Nikolaidis, N.P. The Impact of Sewage-Sludge- and Olive-Mill-Waste-Derived Biochar Amendments to Tomato Cultivation. *Sustainability* **2023**, *15*, 3879. [CrossRef]
73. López-Cabeza, R.; Gámiz, B.; Cornejo, J.; Celis, R. Behavior of the Enantiomers of the Herbicide Imazaquin in Agricultural Soils under Different Application Regimes. *Geoderma* **2017**, *293*, 64–72. [CrossRef]
74. Marra, R.; Vinale, F.; Cesarano, G.; Lombardi, N.; D'Errico, G.; Crasto, A.; Mazzei, P.; Piccolo, A.; Incerti, G.; Woo, S.L.; et al. Biochars from Olive Mill Waste Have Contrasting Effects on Plants, Fungi and Phytoparasitic Nematodes. *PLoS ONE* **2018**, *13*, e0198728. [CrossRef] [PubMed]
75. Molina-Sabio, M.; Sánchez-Montero, M.J.; Juarez-Galan, J.M.; Salvador, F.; Rodríguez-Reinoso, F.; Salvador, A. Development of Porosity in a Char during Reaction with Steam or Supercritical Water. *J. Phys. Chem. B* **2006**, *110*, 12360–12364. [CrossRef]
76. Monteagudo, J.M.; Durán, A.; Mänttäri, M.; López, S. Insights into the Adsorption of CO_2 Generated from Synthetic Urban Wastewater Treatment on Olive Pomace Biochar. *J. Environ. Manag.* **2023**, *339*, 117951. [CrossRef]
77. Omiri, J.; Snoussi, Y.; Bhakta, A.K.; Truong, S.; Ammar, S.; Khalil, A.M.; Jouini, M.; Chehimi, M.M. Citric-Acid-Assisted Preparation of Biochar Loaded with Copper/Nickel Bimetallic Nanoparticles for Dye Degradation. *Colloids Interfaces* **2022**, *6*, 18. [CrossRef]
78. Pellera, F.-M.; Gidarakos, E. Effect of Dried Olive Pomace—Derived Biochar on the Mobility of Cadmium and Nickel in Soil. *J. Environ. Chem. Eng.* **2015**, *3*, 1163–1176. [CrossRef]
79. Pellera, F.-M.; Regkouzas, P.; Manolikaki, I.; Diamadopoulos, E. Biochar Production from Waste Biomass: Characterization and Evaluation for Agronomic and Environmental Applications. *Detritus* **2021**, *17*, 15–29. [CrossRef]
80. Piscitelli, L.; Malerba, A.D.; Mezzapesa, G.N.; Dumontet, S.; Mondelli, D.; Miano, T.; Bruno, G.L. Potential Microbial Remediation of Pyrene Polluted Soil: The Role of Biochar. *Soil Res.* **2019**, *57*, 807–813. [CrossRef]
81. Piscitelli, L.; Rasse, D.P.; Malerba, A.D.; Miano, T.; Mondelli, D. Slow Pyrolysis of Olive Mill Solid Residues as a Sustainable Valorization Strategy for Waste Biomass. *J. Mater. Cycles Waste Manag.* **2023**, *25*, 1688–1698. [CrossRef]
82. Reda, R.M.; El Gaafary, N.M.; Rashwan, A.A.; Mahsoub, F.; El-Gazzar, N. Evaluation of Olive Stone Biochar as Valuable and Inexpensive Agro-waste Adsorbent for the Adsorption and Removal of Inorganic Mercury from Nile Tilapia Aquaculture Systems. *Aquac. Res.* **2022**, *53*, 1676–1692. [CrossRef]
83. Rittenhouse, J.L.; Rice, P.J.; Spokas, K.A.; Koskinen, W.C. Assessing Biochar's Ability to Reduce Bioavailability of Aminocyclopyrachlor in Soils. *Environ. Pollut.* **2014**, *189*, 92–97. [CrossRef] [PubMed]
84. Román, S.; González, J.F.; González-García, C.M.; Zamora, F. Control of Pore Development during CO_2 and Steam Activation of Olive Stones. *Fuel Process. Technol.* **2008**, *89*, 715–720. [CrossRef]
85. Sanginés, P.; Domínguez, M.P.; Sánchez, F.; San Miguel, G. Slow Pyrolysis of Olive Stones in a Rotary Kiln: Chemical and Energy Characterization of Solid, Gas, and Condensable Products. *J. Renew. Sustain. Energy* **2015**, *7*, 043103. [CrossRef]
86. Senbayram, M.; Saygan, E.P.; Chen, R.; Aydemir, S.; Kaya, C.; Wu, D.; Bladogatskaya, E. Effect of Biochar Origin and Soil Type on the Greenhouse Gas Emission and the Bacterial Community Structure in N Fertilised Acidic Sandy and Alkaline Clay Soil. *Sci. Total Environ.* **2019**, *660*, 69–79. [CrossRef]
87. Tamošiūnas, A.; Chouchène, A.; Valatkevičius, P.; Gimžauskaitė, D.; Aikas, M.; Uscila, R.; Ghorbel, M.; Jeguirim, M. The Potential of Thermal Plasma Gasification of Olive Pomace Charcoal. *Energies* **2017**, *10*, 710. [CrossRef]
88. Usama, M.; Rafique, M.I.; Ahmad, J.; Ahmad, M.; Al-Wabel, M.I.; Al-Farraj, A.S.F. A Sustainable Approach towards the Restoration of Lead-Contaminated Soils through Nutrient-Doped Olive Waste-Derived Biochar Application. *Sustainability* **2023**, *15*, 2606. [CrossRef]
89. Valero-Romero, M.J.; García-Mateos, F.J.; Rodríguez-Mirasol, J.; Cordero, T. Role of Surface Phosphorus Complexes on the Oxidation of Porous Carbons. *Fuel Process. Technol.* **2017**, *157*, 116–126. [CrossRef]
90. Volpe, M.; D'Anna, C.; Messineo, S.; Volpe, R.; Messineo, A. Sustainable Production of Bio-Combustibles from Pyrolysis of Agro-Industrial Wastes. *Sustainability* **2014**, *6*, 7866–7882. [CrossRef]
91. Windeatt, J.H.; Ross, A.B.; Williams, P.T.; Forster, P.M.; Nahil, M.A.; Singh, S. Characteristics of Biochars from Crop Residues: Potential for Carbon Sequestration and Soil Amendment. *J. Environ. Manag.* **2014**, *146*, 189–197. [CrossRef]
92. Zouari, M.; Marrot, L.; DeVallance, D.B. Effect of Demineralization and Ball Milling Treatments on the Properties of Arundo Donax and Olive Stone-Derived Biochar. *Int. J. Environ. Sci. Technol.* **2023**, *21*, 101–114. [CrossRef]
93. Blanco López, M.; Blanco, C.; Martínez-Alonso, A.; Tascón, J.M. Composition of Gases Released during Olive Stones Pyrolysis. *J. Anal. Appl. Pyrolysis* **2002**, *65*, 313–322. [CrossRef]
94. Stefanidis, S.D.; Kalogiannis, K.G.; Iliopoulou, E.F.; Michailof, C.M.; Pilavachi, P.A.; Lappas, A.A. A Study of Lignocellulosic Biomass Pyrolysis via the Pyrolysis of Cellulose, Hemicellulose and Lignin. *J. Anal. Appl. Pyrolysis* **2014**, *105*, 143–150. [CrossRef]
95. Kan, T.; Strezov, V.; Evans, T.J. Lignocellulosic Biomass Pyrolysis: A Review of Product Properties and Effects of Pyrolysis Parameters. *Renew. Sustain. Energy Rev.* **2016**, *57*, 1126–1140. [CrossRef]

96. dos Santos, S.R.; Lustosa Filho, J.F.; Vergütz, L.; Melo, L.C.A. Biochar Association with Phosphate Fertilizer and Its Influence on Phosphorus Use Efficiency by Maize. *Ciênc. Agrotecnol.* **2019**, *43*, e025718. [CrossRef]
97. Li, S.; Chen, G. Thermogravimetric, Thermochemical, and Infrared Spectral Characterization of Feedstocks and Biochar Derived at Different Pyrolysis Temperatures. *Waste Manag.* **2018**, *78*, 198–207. [CrossRef] [PubMed]
98. Bolan, N.; Hoang, S.A.; Beiyuan, J.; Gupta, S.; Hou, D.; Karakoti, A.; Joseph, S.; Jung, S.; Kim, K.H.; Kirkham, M.B.; et al. Multifunctional Applications of Biochar beyond Carbon Storage. *Int. Mater. Rev.* **2022**, *67*, 150–200. [CrossRef]
99. Xu, G.; Lv, Y.; Sun, J.; Shao, H.; Wei, L. Recent Advances in Biochar Applications in Agricultural Soils: Benefits and Environmental Implications. *Clean.-Soil Air Water* **2012**, *40*, 1093–1098. [CrossRef]
100. Laghari, M.; Naidu, R.; Xiao, B.; Hu, Z.; Mirjat, M.S.; Hu, M.; Kandhro, M.N.; Chen, Z.; Guo, D.; Jogi, Q.; et al. Recent Developments in Biochar as an Effective Tool for Agricultural Soil Management: A Review. *J. Sci. Food Agric.* **2016**, *96*, 4840–4849. [CrossRef] [PubMed]
101. Zhao, L.; Cao, X.; Mašek, O.; Zimmerman, A. Heterogeneity of Biochar Properties as a Function of Feedstock Sources and Production Temperatures. *J. Hazard. Mater.* **2013**, *256–257*, 1–9. [CrossRef] [PubMed]
102. Inyang, M.I.; Gao, B.; Yao, Y.; Xue, Y.; Zimmerman, A.; Mosa, A.; Pullammanappallil, P.; Ok, Y.S.; Cao, X. A Review of Biochar as a Low-Cost Adsorbent for Aqueous Heavy Metal Removal. *Crit. Rev. Environ. Sci. Technol.* **2016**, *46*, 406–433. [CrossRef]
103. Singh, B.; Dolk, M.M.; Shen, Q.; Camps-Arbestain, M. Conductivity and Liming Potential. In *Biochar: A Guide to Analytical Methods*; Singh, B., Camps-Arbestain, M., Lehmann, J., Eds.; CRC Press: New York, NY, USA, 2017; pp. 23–38.
104. Chaturvedi, S.; Singh, S.V.; Dhyani, V.C.; Govindaraju, K.; Vinu, R.; Mandal, S. Characterization, Bioenergy Value, and Thermal Stability of Biochars Derived from Diverse Agriculture and Forestry Lignocellulosic Wastes. *Biomass Convers. Biorefinery* **2023**, *13*, 879–892. [CrossRef]
105. Zhang, B.; Heidari, M.; Regmi, B.; Salaudeen, S.; Arku, P.; Thimmannagari, M.; Dutta, A. Hydrothermal Carbonization of Fruit Wastes: A Promising Technique for Generating Hydrochar. *Energies* **2018**, *11*, 2022. [CrossRef]
106. Nagel, K.; Hoilett, N.O.; Mottaleb, M.A.; Meziani, M.J.; Wistrom, J.; Bellamy, M. Physicochemical Characteristics of Biochars Derived From Corn, Hardwood, Miscanthus, and Horse Manure Biomasses. *Commun. Soil Sci. Plant Anal.* **2019**, *50*, 987–1002. [CrossRef]
107. Fawzy, S.; Osman, A.I.; Yang, H.; Doran, J.; Rooney, D.W. *Industrial Biochar Systems for Atmospheric Carbon Removal: A Review*; Springer International Publishing: Berlin/Heidelberg, Germany, 2021; Volume 19, ISBN 0123456789.
108. Weber, K.; Quicker, P. Properties of Biochar. *Fuel* **2018**, *217*, 240–261. [CrossRef]
109. International Biochar Initiative (IBI). Standardized Product Definition and Product Testing Guidelines for Biochar That Is Used in Soil. Available online: https://biochar-international.org/wp-content/uploads/2023/01/IBI_Biochar_Standards_V2.1_Final2.pdf (accessed on 22 October 2023).
110. Wang, W.; Bai, J.; Lu, Q.; Zhang, G.; Wang, D.; Jia, J.; Guan, Y.; Yu, L. Pyrolysis Temperature and Feedstock Alter the Functional Groups and Carbon Sequestration Potential of Phragmites Australis- and Spartina Alterniflora-Derived Biochars. *GCB Bioenergy* **2021**, *13*, 493–506. [CrossRef]
111. Wiedner, K.; Rumpel, C.; Steiner, C.; Pozzi, A.; Maas, R.; Glaser, B. Chemical Evaluation of Chars Produced by Thermochemical Conversion (Gasification, Pyrolysis and Hydrothermal Carbonization) of Agro-Industrial Biomass on a Commercial Scale. *Biomass Bioenergy* **2013**, *59*, 264–278. [CrossRef]
112. Yang, G.; Wu, L.; Xian, Q.; Shen, F.; Wu, J.; Zhang, Y. Removal of Congo Red and Methylene Blue from Aqueous Solutions by Vermicompost-Derived Biochars. *PLoS ONE* **2016**, *11*, e0154562. [CrossRef]
113. EBC. European Biochar Certificate—Guidelines for a Sustainable Production of Biochar'. 2012. Available online: http://www.european-biochar.org/en/download (accessed on 9 October 2023).
114. Jaafar, N.M.; Clode, P.L.; Abbott, L.K. Soil Microbial Responses to Biochars Varying in Particle Size, Surface and Pore Properties. *Pedosphere* **2015**, *25*, 770–780. [CrossRef]
115. Zhao, S.X.; Ta, N.; Wang, X.D. Effect of Temperature on the Structural and Physicochemical Properties of Biochar with Apple Tree Branches as Feedstock Material. *Energies* **2017**, *10*, 1293. [CrossRef]
116. Nkoh, J.N.; Ajibade, F.O.; Atakpa, E.O.; Baquy, M.A.-A.; Mia, S.; Odii, E.C.; Xu, R. Reduction of Heavy Metal Uptake from Polluted Soils and Associated Health Risks through Biochar Amendment: A Critical Synthesis. *J. Hazard. Mater. Adv.* **2022**, *6*, 100086. [CrossRef]
117. Khadem, A.; Raiesi, F.; Besharati, H.; Khalaj, M.A. The Effects of Biochar on Soil Nutrients Status, Microbial Activity and Carbon Sequestration Potential in Two Calcareous Soils. *Biochar* **2021**, *3*, 105–116. [CrossRef]
118. El-Naggar, A.; El-Naggar, A.H.; Shaheen, S.M.; Sarkar, B.; Chang, S.X.; Tsang, D.C.W.; Rinklebe, J.; Ok, Y.S. Biochar Composition-Dependent Impacts on Soil Nutrient Release, Carbon Mineralization, and Potential Environmental Risk: A Review. *J. Environ. Manag.* **2019**, *241*, 458–467. [CrossRef] [PubMed]
119. Song, B.; Almatrafi, E.; Tan, X.; Luo, S.; Xiong, W.; Zhou, C.; Qin, M.; Liu, Y.; Cheng, M.; Zeng, G.; et al. Biochar-Based Agricultural Soil Management: An Application-Dependent Strategy for Contributing to Carbon Neutrality. *Renew. Sustain. Energy Rev.* **2022**, *164*, 112529. [CrossRef]
120. Ghosh, D.; Maiti, S.K. Biochar Assisted Phytoremediation and Biomass Disposal in Heavy Metal Contaminated Mine Soils: A Review. *Int. J. Phytoremediat.* **2021**, *23*, 559–576. [CrossRef]

121. Igaz, D.; Šimanský, V.; Horák, J.; Kondrlová, E.; Domanová, J.; Rodný, M.; Buchkina, N.P. Can a Single Dose of Biochar Affect Selected Soil Physical and Chemical Characteristics? *J. Hydrol. Hydromech.* **2018**, *66*, 421–428. [CrossRef]
122. Hussain, R.; Garg, A.; Ravi, K. Soil-Biochar-Plant Interaction: Differences from the Perspective of Engineered and Agricultural Soils. *Bull. Eng. Geol. Environ.* **2020**, *79*, 4461–4481. [CrossRef]
123. Toková, L.; Igaz, D.; Horák, J.; Aydin, E. Effect of Biochar Application and Re-application on Soil Bulk Density, Porosity, Saturated Hydraulic Conductivity, Water Content and Soil Water Availability in a Silty Loam Haplic Luvisol. *Agronomy* **2020**, *10*, 1005. [CrossRef]
124. Xu, W.; Whitman, W.B.; Gundale, M.J.; Chien, C.; Chiu, C. Functional Response of the Soil Microbial Community to Biochar Applications. *GCB Bioenergy* **2021**, *13*, 269–281. [CrossRef]
125. Kavitha, B.; Reddy, P.V.L.; Kim, B.; Lee, S.S.; Pandey, S.K.; Kim, K.-H. Benefits and Limitations of Biochar Amendment in Agricultural Soils: A Review. *J. Environ. Manag.* **2018**, *227*, 146–154. [CrossRef] [PubMed]
126. Hasnain, M.; Munir, N.; Abideen, Z.; Zulfiqar, F.; Koyro, H.W.; El-Naggar, A.; Caçador, I.; Duarte, B.; Rinklebe, J.; Yong, J.W.H. Biochar-Plant Interaction and Detoxification Strategies under Abiotic Stresses for Achieving Agricultural Resilience: A Critical Review. *Ecotoxicol. Environ. Saf.* **2023**, *249*, 114408. [CrossRef] [PubMed]
127. Abbas, T.; Rizwan, M.; Ali, S.; Adrees, M.; Zia-ur-Rehman, M.; Qayyum, M.F.; Ok, Y.S.; Murtaza, G. Effect of Biochar on Alleviation of Cadmium Toxicity in Wheat (*Triticum aestivum* L.) Grown on Cd-Contaminated Saline Soil. *Environ. Sci. Pollut. Res.* **2018**, *25*, 25668–25680. [CrossRef]
128. Ibrahim, M.; Li, G.; Chan, F.K.S.; Kay, P.; Liu, X.X.; Firbank, L.; Xu, Y.Y. Biochars Effects Potentially Toxic Elements and Antioxidant Enzymes in *Lactuca sativa* L. Grown in Multi-Metals Contaminated Soil. *Environ. Technol. Innov.* **2019**, *15*, 100427. [CrossRef]
129. Rehman, M.; Liu, L.; Bashir, S.; Saleem, M.H.; Chen, C.; Peng, D.; Siddique, K.H. Influence of Rice Straw Biochar on Growth, Antioxidant Capacity and Copper Uptake in Ramie (*Boehmeria nivea* L.) Grown as Forage in Aged Copper-Contaminated Soil. *Plant Physiol. Biochem.* **2019**, *138*, 121–129. [CrossRef]
130. Palansooriya, K.N.; Ok, Y.S.; Awad, Y.M.; Lee, S.S.; Sung, J.-K.; Koutsospyros, A.; Moon, D.H. Impacts of Biochar Application on Upland Agriculture: A Review. *J. Environ. Manag.* **2019**, *234*, 52–64. [CrossRef]
131. Lehmann, J.; Joseph, S. (Eds.) *Biochar for Environmental Management*; Routledge: London, UK, 2015; ISBN 9780203762264.
132. Galán-Martín, Á.; Contreras, M.d.M.; Romero, I.; Ruiz, E.; Bueno-Rodríguez, S.; Eliche-Quesada, D.; Castro-Galiano, E. The Potential Role of Olive Groves to Deliver Carbon Dioxide Removal in a Carbon-Neutral Europe: Opportunities and Challenges. *Renew. Sustain. Energy Rev.* **2022**, *165*, 112609. [CrossRef]
133. *EN 14214:2008*; Automotive Fuels—Fatty Acid Methyl Esters (FAME) for Diesel Engines—Requirements and Test Methods. European Committee for Standardization: Brussels, Belgium, 2008.

sustainable chemistry

MDPI

Article

Sustainable Production of Chitin Nanowhiskers from Crustacean Biomass Using Cost-Effective Ionic Liquids: Strategies to Avoid Byproduct Formation

Alexander S. Shkuratov [1,†], Reshma Panackal Shibu [1,†], Obste Therasme [2], Paula Berton [3,*] and Julia L. Shamshina [1,*]

1 Fiber and Biopolymer Research Institute, Department of Plant and Soil Science, Texas Tech University, Lubbock, TX 79409, USA; alexshku@ttu.edu (A.S.S.); rpanacka@ttu.edu (R.P.S.)
2 Department of Sustainable Resources Management, State University of New York College of Environmental Science and Forestry, Syracuse, NY 13210, USA; otherasm@esf.edu
3 Chemical and Petroleum Engineering Department, University of Calgary, Calgary, AB T2N 1N4, Canada
* Correspondence: paula.berton@ucalgary.ca (P.B.); jshamshi@ttu.edu (J.L.S.)
† These authors contributed equally to this work.

Abstract: Nanochitin, especially in the form of chitin nanowhiskers (ChNWs), represents a significant advance in biopolymer technology due to its high specific surface area, superior tensile strength, and excellent thermal stability. Derived from crustacean waste, which contains 15–40% of chitin, these materials provide a sustainable option that diverts waste from landfills and contributes to environmental conservation. Traditional methods of isolating nanochitin are energy-intensive and generate substantial waste. This study introduces a more sustainable method using inexpensive ionic liquids (ILs) such as [Hmim][HSO_4] and [HN_{222}][HSO_4], which bypass the costly and destructive steps of traditional procedures. This study also identified the byproduct in IL-mediated chitin hydrolysis reaction as calcium sulfate dihydrate and presented a solution to circumvent the byproduct formation. The effectiveness of the [HN_{222}][HSO_4] IL in producing ChNWs from both purified chitin and crustacean biomass was assessed, showing a high yield and maintaining the purity and structural integrity of chitin, thereby demonstrating a significant reduction in the environmental footprint of ChNW production.

Keywords: chitin nanowhiskers; crustacean biomass; ionic liquids; gypsum formation

1. Introduction

Nanochitin, i.e., chitin in its nanostructure form, is considered to be one of the important biopolymeric nanomaterials due to its high specific surface area, high tensile strength, and thermal stability, as well as renewability and abundance [1]. Chitin is an abundant biopolymer [2] made of repeating 2-(acetylamino)-2-deoxy-D-glucose β-linked units [3], and is found in the shells of arthropods (crabs, lobster, and shrimps), the cell walls of fungi, insect cuticles (e.g., fly larvae), some mushrooms, and yeasts [4].

Nanochitin particles, also called chitin nanowhiskers (ChNWs), possess a specific stiffness (105 GPa g^{-1} cm^{-3}) that is larger than that of ceramics, metals, and even Kevlar®, suggesting their use for fabricating exceptionally strong materials. In addition, these particles are lightweight (with a density of 1.425 g cm^{-3}) [5], transparent, and have high longitudinal elastic modulus (theoretical value > 150 GPa) [5–7]. Nanochitin is mostly used as a reinforcement filler to enhance the strength and durability of synthetic plastics [8–12]. Its reinforcing ability is attributed to the formation of a hydrogen-bonded percolated network within the synthetic polymer matrix. Other applications of nanochitin include packaging replacements [13–15], cement additives [16,17], personal care, hygiene, and cosmetics [5], medical devices [18–21], filtration membranes [22–24], and flexible electronics [25–28].

Crustacean waste biomass is, by far, the most abundant source of chitin and contains 15–40% of this polymer [29,30]. Crustacean waste is considered an environmental hazard [31] since it is either dumped back into the ocean or disposed of in landfills. Using chitinous waste to manufacture chitinous products keeps waste out of landfills, benefits the environment, conserves resources, and provides additional revenue to agricultural and aquacultural businesses. However, the recovery of "bulk" chitin is challenging due to the chitinous matrix structure, which contains strong covalent and hydrogen bonds between layers of proteins (20–40%) and chitin. In addition, crustacean waste is also rich in minerals (20–50%).

To obtain nanochitin from biomass, conventional methods include two steps. The first step aims to recover chitin as pure as possible from the shell matrix. However, this initial step is destructive, wasteful, and energy-demanding since it involves deproteinization to remove proteins present in a shell matrix (using bases, e.g., NaOH) and, in the case of crustacean biomass, demineralization to remove minerals (using acids, e.g., HCl). Bleaching/discoloration is often required [32,33]. The process is conducted at relatively high temperatures (60–100 °C), and for a long time [34,35], utilizing ~1.2 kWh of electricity and generating > 500 L acidic and basic aqueous waste per 1 kg of "bulk" chitin.

In the second step, the hydrolysis of glycosidic bonds releases nanochitin particles from the "bulk" chitin [36–39]. Hydrochloric acid [36,37,39,40], sulfuric acid [41,42], 2,2,6,6-tetramethylpiperidine-1-oxyl radical (TEMPO)-mediated oxidation [43–45], periodate-mediated oxidation [46], ammonium persulfate-mediated oxidation [47], and Deep Eutectic Solvent systems [48] have been proposed for hydrolysis. For the overall 'traditional' process of obtaining chitin nanowhiskers, 74.6% of CO_2 footprint (677 kg CO_2 equiv.) comes from the chitin isolation itself, while only a quarter of its CO_2 footprint (229.8 kg CO_2 equiv.) comes from the nanowhisker production step [49]. Bypassing the first step could lower the environmental impact by 75% and address a critical sustainability gap in the preparation of chitin nanomaterials.

A dual catalyst–solvent treatment of biomass with 1-butyl-3-methylimidazolium hydrogen sulfate ([C_4mim][HSO_4]) recently showed the potential of this IL to bypass the first step of "bulk" chitin isolation in producing nanowhiskers [50,51]. In this process, biomass was pretreated with the IL and hydrolyzed in an IL–water (3:1) mixture. After quenching the reaction, the thick white gel-looking suspension was ultrasonicated, centrifuged, washed, and lyophilized. The overall mass yield of nanofillers was 79%, much higher than that reported for HCl hydrolysis (45–55%). No deacetylation took place during the treatment, and the degree of acetylation (%DA) was the same as that of commercial chitin.

Despite all its advantages and potential applications in biomass processing, [C_4mim][HSO_4] is costly, mainly due to its manufacturing method. The method involves the formation of 1-butyl-3-methylimidazolium bromide ([C_4mim]Br) by adding butyl bromide to methylimidazole, followed by an anion exchange process with sodium hydrogen sulfate [52–55]. Factors that add to the high cost include the limited sizes of ion-exchange columns and high resource (i.e., resin and solvent) consumption. In addition, this is a traditional batch process and not a continuous manufacturing.

Since IL cost, biomass load, and IL recovery are three parameters likely to exert a significant influence on the final cost of chitin nanowhiskers [56], the manufacturing scale synthesis of the ILs must not only utilize inexpensive reagents but also be simple, straightforward, and easily scalable. The first study that evaluated the costs of hydrogen sulfate-based ILs on a bulk scale was conducted by Chen et al. [57]. The study considered factors such as raw material cost, other operating costs (salary/wages), utility costs (e.g., electricity and cooling water), and wastewater treatment [58,59]. It was found that the ILs triethylammonium hydrogen sulfate ([HN_{222}][HSO_4]) and methylimidazolium hydrogen sulfate ([Hmim][HSO_4]) are significantly cheaper than [C_4mim][HSO_4], within a price range for typical organic solvents (USD 1.44–1.56 kg^{-1}) [57]. In this work, we assess the effectiveness of these cost-effective ILs (Figure 1) for isolating chitin nanowhiskers from both purified chitin and crustacean biomass and evaluate the properties of the resultant materials.

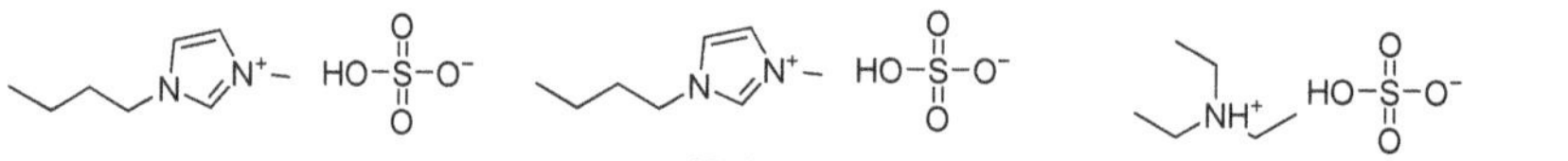

1-butyl-3-methylimidazolium hydrogen sulfate ($[C_4mim][HSO_4]$) (a); Methylimidazolium hydrogen sulfate ($[Hmim][HSO_4]$) (b); Triethylammonium hydrogen sulfate ($[HN_{222}][HSO_4]$) (c)

Figure 1. Ionic liquids used in this study: (**a**) 1-butyl-3-methylimidazolium hydrogen sulfate ($[C_4mim][HSO_4]$); (**b**) methylimidazolium hydrogen sulfate ($[Hmim][HSO_4]$); and (**c**) triethylammonium hydrogen sulfate ($[HN_{222}][HSO_4]$).

2. Materials and Methods

2.1. Materials

Commercial grade (purified) chitin, provided by Sigma-Aldrich (cat no: C7170, St. Louis, MO, USA) and Fisher Scientific (cat no: AAJ6120622, Hampton, NH, USA), was dried in an oven at 50 °C for 12 h before use. Crustacean (shrimp shells, SS, USA) biomass containing 24% chitin was obtained from Mari Signum, LLC (Richmond, VA, USA). To facilitate the extraction, the shells were washed, dried (50 °C for 12 h), ground, and sifted to achieve a particle size of less than 125 μm. The IL $[Hmim][HSO_4]$ was purchased from Sigma-Aldrich (cat no: 59760, St. Louis, MO, USA) and $[C_4mim][HSO_4]$ was purchased from Proionic (Graz, Austria), while $[HN_{222}][HSO_4]$ was synthesized as previously reported [60,61]. Deionized water was bought from Aqua One (Amarillo, TX, USA). The calibration curve for the determination of the degree of acetylation (%DA) was constructed using chitosan from shrimp shells obtained from Sigma Aldrich (cat. No: C3646, St. Louis, MO, USA) with %DA = 30.0%, chitosan from shrimp shells obtained from Vanson HaloSource (cat. No: VNS-389, Redmond, WA, USA) with %DA = 23.9%, chitosan from Alfa Aesar (cat. No: J64143, Haverhill, MA, USA) with %DA = 15.0%, and chitin from Spectrum (cat. No: C2020, Gardena, CA, USA) with %DA = 86.0%.

2.2. Determination of Chitin Content in Biomass

The Black and Schwartz method was used to determine the amount of chitin in biomass [62]. Ground and sieved crustacean biomass (1 g) was weighed into a 50 mL round bottom flask equipped with a Teflon-coated magnetic stir bar and a condenser. Then, 24 mL of 1 M HCl was added and the mixture was heated under reflux with stirring at 750 rpm for 1 h. After 1 h, the flask was removed from heat and cooled to room temperature. The reaction mixture was transferred to a centrifuge tube and centrifuged at 3750 rpm for 10 min in a Beckman Coulter Allegra X-15R Refrigerated Centrifuge (Beckman, Brea, CA, USA). Then, the supernatant was carefully separated from the solids, 25 mL of fresh DI water was added, and the resultant suspension was centrifuged again. Centrifugation, aqueous phase decantation, and fresh DI water addition steps were consecutively repeated until the washings were no longer acidic. The remaining solid was transferred to its initial (water-rinsed) round bottom flask now with 24 mL of 1.25 M aqueous NaOH, and the mixture was heated under reflux with stirring at 750 rpm for 1 h. After 1 h, the reaction was once again removed from heat, cooled down and its content transferred to new centrifuge tubes, and centrifuged for 10 min at 3800 rpm. The supernatant was decanted and discarded, and the remaining solid was washed with the addition of fresh DI water, followed by centrifugation, again at 3750 rpm, for 10 min. The solid was repeatedly washed and centrifuged with fresh DI water (8 more times) until the supernatant was neutral. When the supernatant was neutral, it was decanted and discarded, and the remaining solid was transferred onto a Petri dish and dried overnight in an 80 °C Cole-Parmer StableTemp gravimetric convection oven (Charleston, SC, USA), and weighed to obtain the percent of chitin from biomass. The determination of chitin content was carried out in triplicate.

2.3. Preparation of Chitin Nanowhiskers

Chitin nanowhiskers were obtained by following the reported method (Figure S1, ESI) [50,51].

1. **Pretreatment with ILs**—For chitin pretreatment, a suspension of chitin in IL (1:10 *w/w*) was prepared by thoroughly mixing chitin in the IL (for example, 0.76 g chitin was added and mixed in 7.64 g $[HSO_4]^-$-based IL), resulting in a well-blended paste. The prepared paste was then properly sealed, encased with a parafilm, and kept in the oven (45–65 °C) for 24 h.
   For biomass pretreatment, a 3 wt% crushed crustacean biomass was prepared by thoroughly mixing the biomass in the IL (for example, 1.5 g crushed crustacean biomass was added and mixed in 48.5 g of the $[HSO_4]^-$-based IL), resulting in a paste. The properly blended paste was subsequently sealed, covered with a parafilm, and kept in the oven (45–65 °C) for 24 h.
2. **Isolation of Chitin Nanowhiskers**—After letting the mixture in the containers settle for 24 h, 30% DI water with respect to IL (i.e., 16.7 g of water per 48.5 g of the IL) was added to each flask. Each flask was equipped with a stirring bar and a condenser, and then heated to 110 °C in an oil bath, for 6–48 h. In the case of biomass, this resulted in a significant production of foam. After heating, 50 mL DI water was added to each flask. Following that, the mixtures were moved to 15 mL centrifuge tubes, and centrifuged (Eppendorf 5430 R, Enfield, CT; rotor CE 11017, 7830 rpm). The liquid was decanted, fresh water was added, the precipitate was stirred with a spatula, and the suspension was centrifuged again. This process was repeated 10 more times until a neutral pH was obtained. After washing, the suspensions were ultrasonicated (VibraCell Ultrasonicator, model CV 33, Newtown, CT, USA) for 10 min with 30 s cycles for the uniform distribution of particles. The suspension was then split into 2 parts. One part was diluted and analyzed using Transmission Electron Microscopy (TEM, Hitachi H-9500, Tokyo, Japan). The second part was frozen at −20 °C, and then subjected to a lyophilization (Labconco FreeZone Plus Cascade Benchtop Freeze Dryer System, Kansas City, MO, USA) process for approximately 72 h, resulting in the production of nanocrystals with precise dimensions.

2.4. Characterization

2.4.1. Estimation of Crystallinity

The crystallinity of the nanocrystals generated directly from crustacean biomass was measured using a powder X-ray diffractometer (HD 2711N, Rigaku, Tokyo, Japan) with Ni-filtered Cu Kα radiation and λ = 1.542 Å, at 44 mA and 40 kV. Diffractograms were collected under a scanning rate of 1° min^{-1} over the 2θ region ranges from 5 to 50°. Origin software (OriginLab Corporation, Northampton, MA, USA) was used for removing the background, and then for the smoothening and baseline correction of the recorded data. The Peak Analyzer tool was employed to automatically identify and analyze a wide range of multiple peaks, and the crystallinity of the product was finally computed using the equation:

$$CrI\ (\%) = \frac{\sum area\ of\ crystalline\ peaks}{\sum area\ of\ crystalline\ peaks\ and\ amorphous\ scattering} \times 100 \quad (1)$$

2.4.2. Crystallite Size Determination

The crystallite size was calculated using the Scherrer equation [63], previously employed for chitin [64]:

$$\beta = (k \times \lambda)/(L \times \cos(\theta)) \quad (2)$$

where β represents the crystallite size perpendicular to the lattice plane represented by (020) and (110) peaks, k is the Scherrer constant for a given crystal shape (k = 0.91), λ represents the wavelength of the incident X-rays (1.54 Å), L represents the width of the peak at half of its maximum in radians (FWHM), and θ is the position of the peak (half of the plotted 2θ value).

2.4.3. Transmission Electron Microscopy

The morphology, including the distribution of nanocrystals (diluted 30-fold before analysis), size, and shape, was evaluated using a Transmission Electron Microscope (Hitachi H-9500, Tokyo, Japan).

2.4.4. Field Emission Scanning Electron Microscopy

A field emission scanning electron microscope (FE-SEM, S/N 4300, Hitachi, Chiyoda, Tokyo, Japan) was used for the visualization of gypsum crystals. For this, the samples were mounted on the sample stage with adhered carbon tape (Ted Pella Inc., Redding, CA, USA) and visualized at 500–5000 magnification, with an accelerating voltage of 15,000 V. Quartz PCI Imaging software (Version 8, Quartz Imaging Corp., Vancouver, BC, Canada) (https://www.quartzimaging.com/pci-microscope-imaging-software.html, accessed on 1 April 2024) was used to analyze the recorded images.

2.4.5. Thermal Gravimetric Analysis

The thermal stability of the generated nanocrystals was examined using a thermogravimetric analyzer (TGA, Pyris1, PerkinElmer Inc., Shelton, CT, USA). The samples were kept under controlled conditions at a temperature of 21 ± 1 °C with a relative humidity (RH) of 65 ± 2% before analysis. An inert atmosphere was maintained by flowing nitrogen at a rate of 20 mL min^{-1}. The thermal analysis was carried out at a heating rate of 10 °C min^{-1}, spanning the temperature range of 40 to 600 °C. The data generated were analyzed with the assistance of Pyris software (PerkinElmer Inc., Shelton, CT, USA).

2.4.6. Attenuated Total Reflectance (ATR) Fourier-Transform Infrared Spectroscopy (FTIR)

The structural characterization of the isolated nanocrystals was confirmed using an ATR FTIR Spectrum-400 instrument (PerkinElmer, Waltham, MA, USA). The samples were kept under controlled conditions at a temperature of 21 ± 1 °C with an RH of 65 ± 2% and then placed on a chemically inert Zn-Se-diamond crystal stage for analysis. A total of 64 scans were conducted, covering the spectral range from 650 to 4000 cm^{-1}, and the spectral resolution was set at 4 cm^{-1}. The results were recorded and interpreted with the aid of OPUS Bruker software (Bruker, Billerica, MA, USA).

2.4.7. Degree of Acetylation

FTIR spectroscopy was employed to assess the degree of acetylation (%DA) of the chitin nanocrystals, following the same procedure detailed by Beil et al. [65]. First, the calibration curve was constructed as follows. The ATR FTIR spectra (64 scans) were recorded for chitin and chitosan with known %DA (15, 24, 30, and 86%). The spectra were converted to Transmittance, then smoothed employing a 9-point Savitzky–Golay algorithm, and the first derivatives of the spectra were calculated. The first derivative ATR FTIR values at 1383 cm^{-1} (MB1), 1327 cm^{-1} (MB2), and 1163 cm^{-1} (RB) were determined and the ratios (MB1 + MB2)/RB were found. The data were then plotted as %DA vs. (MB1 + MB2)/RB. This calibration curve (Figures S2 and S3, ESI) was used to determine the %DA for prepared samples.

3. Results and Discussion

3.1. Preparation and Characterization of Chitin Nanowhiskers from Pure Chitin

Chitin nanowhiskers were prepared using cheaper [Hmim][HSO_4] and [HN_{222}][HSO_4] and compared side-by-side with those prepared using the more expensive IL [C_4mim][HSO_4] [50,51]. In addition to the price difference, [Hmim][HSO_4] and [HN_{222}][HSO_4] are defined as "protic" ILs, whose pKa values for cations are 6.95 and 10.72, respectively [66,67]. The IL [Hmim][HSO_4] was commercially available, whereas

$[HN_{222}][HSO_4]$ was prepared by mixing triethylamine with sulfuric acid through a simple Brønsted acid-base proton transfer reaction [60,61].

As an initial step, the ILs were tested for the preparation of ChNWs from pure chitin, to determine their ability to hydrolyze the glycosidic linkages (Figure S1, ESI). In brief, chitin was pretreated with $[C_4mim][HSO_4]$ (used as a benchmark) [50,51], $[Hmim][HSO_4]$, or $[HN_{222}][HSO_4]$, for 24 h. Since the three ILs are solids at room temperature (with melting temperatures of 28.7 [68], 40.6 [69], and 85 °C [70] for $[C_4mim][HSO_4]$, $[Hmim][HSO_4]$, and $[HN_{222}][HSO_4]$, respectively), the pretreatment temperature was kept higher than their melting points. After pretreatment, DI water was added and the solutions were heated to 110 °C for 48 h under reflux. The precipitate was centrifuged, washed with water, and ultrasonicated. The suspension was then split into two parts. One part was diluted and used for Transmission Electron Microscopy (TEM) imaging while the second one was dehydrated using freeze-drying and used for further characterization. Based on the IL used to prepare the ChNWs from chitin, the products were denoted as ChNWs/Ch/$[C_4mim][HSO_4]$, ChNWs/Ch/$[Hmim][HSO_4]$, and ChNWs/Ch/$[HN_{222}][HSO_4]$. All the dried products appeared as white powders and similar yields, i.e., 54, 55, and 50%, were obtained for ChNWs/Ch/$[C_4mim][HSO_4]$, ChNWs/Ch/$[Hmim][HSO_4]$, and ChNWs/Ch/$[HN_{222}][HSO_4]$, respectively. Such moderate yield was reported in the literature in other acidic systems, where the acidic hydrolysis of chitin leads to the formation of not only ChNWs but also chitin oligomers [71,72]. Decreasing the time to 24 h during subsequent studies produced the nanomaterial in the same yields.

The dimensional parameters of the ChNWs processed using each IL were obtained from TEM. A 10-times dilution of the original suspensions, obtained after ultrasonication, with a concentration of ~0.1 mg mL^{-1} was the most suitable for TEM imaging. As shown in the TEM images (Figure 2), the morphology of the nanowhiskers was highly regular. The ChNWs from $[Hmim][HSO_4]$ and $[HN_{222}][HSO_4]$ presented elongated crystalline rod-like nanoparticles, characterized by pointed ends, with diameters ranging between 20 and 22 nm, and lengths ranging between 230 and 246 nm (Table 1). Given the obtained results, we estimated the average aspect ratio (L/d) to be ~12 to 12.5. These findings correlate with those previously reported when using $[C_4mim][HSO_4]$ for hydrolysis, with diameters of 21.7 ± 5.5 nm and lengths of 216 ± 78 nm. Overall, the ChNWs obtained showed a similar length and diameter, i.e., avg. 230 ± 12 nm and 21.3 ± 0.9 nm, respectively. For comparison, the aspect ratio of ChNWs obtained through hydrolysis with HCl can range from 15 to 120, depending on the chitin source used [73]. The size of ChNWs is directly dependent on the biomass it was isolated from. For example, the hydrolysis of α-chitin from shrimp shells using 3 N HCl results in nanowhiskers with an average length and width of 216 ± 91 nm and 16.4 ± 5.8 nm [74]. When chitin is sourced from *Riftia* tubes, the resulting ChNWs have a length range from 500 nm to a few μm [75]. The size of the nanowhiskers from Oyster and Button mushrooms ranging from 50 to 200 nm in length and ~11 nm in diameter [76].

Table 1. Dimensional parameters of the ChNWs isolated from chitin using different ILs.

ChNW Prepared from Chitin Hydrolyzed with	Yield %	Dimensions	Size (nm) Mean ± STD	Sample Size	Aspect Ratio Mean ± STD	Sample Size
$[C_4mim][HSO_4]$ *	54	Length	216 ± 78	72	11.0 ± 4.9	38
		Diameter	21.7 ± 5.5	38		
$[Hmim][HSO_4]$	55	Length	230 ± 84	65	12.5 ± 6.0	65
		Diameter	20.0 ± 7.7	79		
$[HN_{222}][HSO_4]$	50	Length	246 ± 89	94	12.1 ± 5.9	72
		Diameter	22.2 ± 5.9	72		

* Original study. All nanowhiskers prepared from pure chitin have approximately the same dimensions, i.e., 230 ± 12 nm in length and 21.3 ± 0.9 nm in diameter.

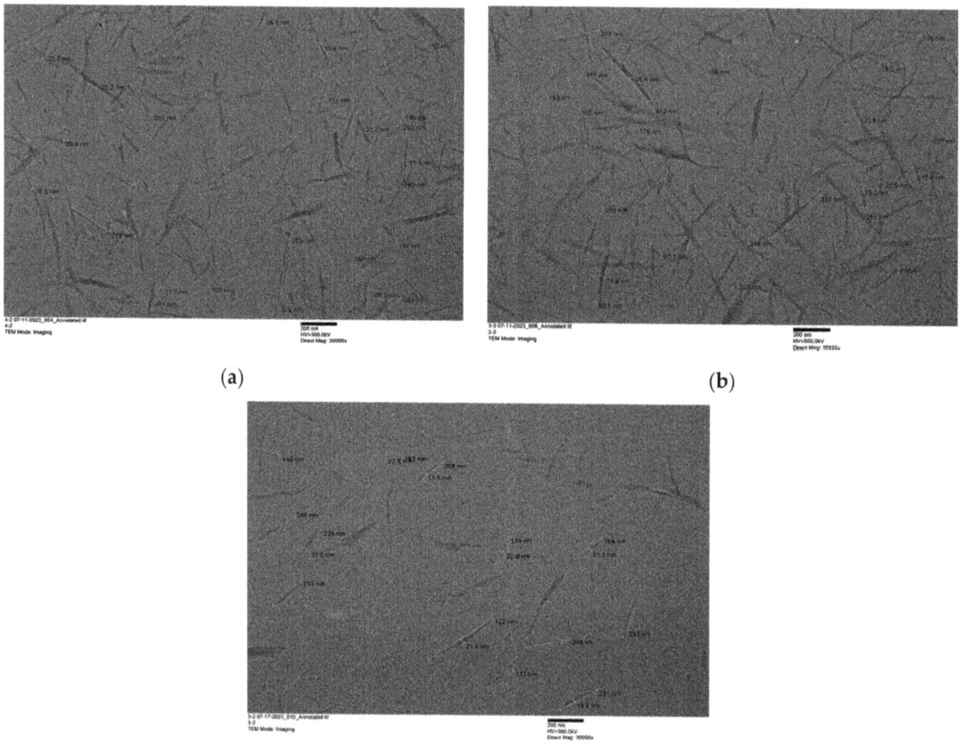

(**a**) (**b**) (**c**)

Figure 2. TEM micrograph of a dilute suspension of chitin nanowhiskers derived from purified chitin using (**a**) $[C_4mim][HSO_4]$, (**b**) $[Hmim][HSO_4]$, and (**c**) $[HN_{222}][HSO_4]$. HV 300 kV. Magnification: 30,000×; scale bar 200 nm.

Attenuated Total Reflectance (ATR) Fourier-Transform Infrared Spectroscopy (FTIR) was then used to detect chemical modifications to the chitin due to the IL treatment. The spectral peaks were compared with those identified in standard chitin reference upon the normalization of the FTIR spectra (Figure 3; expanded spectra of different FTIR regions are shown in Figures S4–S6, ESI). Independent of the IL used for hydrolysis, the obtained ChNW/Ch/ILs showed almost identical spectrum profiles (Figure 3). The spectra of ChNW/Ch/ILs were not only similar to each other but also to the spectrum of pure chitin that was used as the starting material and to those reported in the literature (summarized in Table S1, ESI). Thus, the peak attributed to O(3)H$\cdots$O(5) from the ring in the α-chitin was located at the same position in all ChNW/Ch/ILs and chitin standard, at 3438 cm^{-1}. The peaks were also of the same intensity, indicating no hydrolysis-induced changes in –O–H stretching. The same observation (i.e., peaks of close intensity situated at the same wavenumber for both ChNW/Ch/ILs products and pure chitin standard) is valid for the N–H stretching and bending, at 3103 cm^{-1} and 1556 cm^{-1}, respectively. Similarly, no differences were found in the location or intensity of Amide III peaks (namely, C–N and N–H) at 1311 cm^{-1} and N–H peak at 3261 cm^{-1}. Both the intensity and the wavenumber of the vibration modes of carbonyl (Amide I), namely, the C=O$\cdots$H–N vibration at 1654 cm^{-1} and the –$CH_2OH\cdots$O=C band at 1620 cm^{-1}, were also retained, indicating no changes in amide-type of hydrogen bonding. Concerning C–O, the stretching of glycosidic linkage (the asymmetric bridge oxygen stretching) is indicated by a peak at 1154 cm^{-1}, and pyranose ring stretching was found at 895 cm^{-1}.

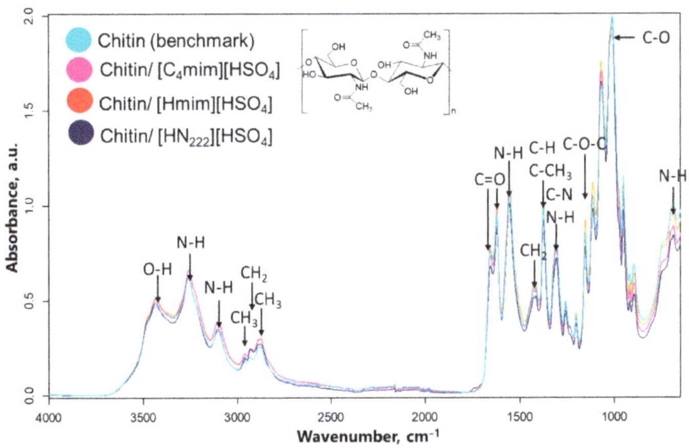

Figure 3. ATR FTIR spectra (4000–650 cm^{-1}) of chitin (aqua) and chitin nanowhiskers (ChNWs): ChNWs/Ch/[Hmim][HSO_4] (red), ChNWs/Ch/[HN_{222}][HSO_4] (blue), and ChNWs/Ch/[Hmim][HSO_4] (pink).

As for CH bonds, a symmetric stretching vibration of the CH_3 group from –NHC(O)CH_3 amide bond, a asymmetric stretching vibration of the CH_2 group from CH_2OH, and stretching vibrations of CH in the pyranose ring appeared at 2964, 2930, and 2886 cm^{-1}, respectively. The CH_3 deformation appeared at 1378 cm^{-1} in both standard and nanowhiskers, whereas –CH_2 ending in CH_2OH showed up at 1425 cm^{-1}. The bands at 970 cm^{-1} and 873 cm^{-1} correspond to vibrations of CH_3 wagging and the C–H vibrations of CH_3 in the ring structure. Overall, the obtained ChNW/Ch/ILs produced in this study did not indicate any significant change in chemical functional groups compared to the native chitin. This finding differs from what was shown before for acid hydrolysis [77], which reported that the use of an acidic environment leads to partial chitin deacetylation.

Powder X-ray diffraction (pXRD) was used to elucidate the degree of crystallinity of all products (Figure 4). For that, the diffractograms obtained for ChNW/Ch/ILs samples were compared with the diffractogram of pure chitin. In all diffractograms, the peaks were found at 9.24°, 12.69°, 19.23°, 20.58°, 23.18°, and 26.27°, corresponding to (0 2 0), (0 2 1), (1 1 0), (1 2 0), (1 3 0), and (0 1 3) reflections, respectively [78,79]. The crystallinity index (CrI) was determined using the peak height method based on a comparison of the intensity of a crystalline reflection 19.23° (I_{110}) and the height of amorphous scattering (16°, I_{am}) [80,81]. However, the reliability of this method is based on the precision of background subtraction and the presence of both anhydrous and hydrated polymorphs [82]. Therefore, we have also employed the peak deconvolution method that typically provides higher accuracy [83,84]. This method is based on calculating the ratio between the area of crystalline peaks and the total area of a pXRD pattern. Figures S8–S13 and Tables S2–S7 (ESI) provide the calculated peak fitting of the diffraction profiles, the calculated peak locations, areas under the peak, and full width at half maximum (FWHM) values.

While starting chitin was found to be 84% crystalline, the crystallinity values of ChNWs from [HN_{222}][HSO_4] was similar (84.5%), slightly lower for [Hmim][HSO_4] (79.8%), and significantly higher for the [C_4mim][HSO_4] IL (92.6%, Table 2). The crystallite size values (Tables 2, S8 and S9, ESI) were calculated using Scherrer's equation [63]. Specifically, the crystallite size (a measure of the coherent volume for chitin in the diffraction peak) of each type of nanowhiskers was determined for two of the most prominent crystalline peaks of chitin, at the (0 2 0) and (1 1 0) planes, and was found to be between 6.9 and 8.6 nm (at the (0 2 0) plane) and between 5.2 and 7.5 nm (at the (1 1 0) plane), consistent with those observed previously. Thus, nanowhiskers obtained using HCl-hydrolysis from crab chitin demonstrated a crystallinity index of 86%, whereas the crystallite dimensions at the (0 2

0) and (1 1 0) planes for CtNWs were 9.2 nm and 6.5 nm, respectively [85]. Likewise, the acid hydrolysis of α-chitin from shrimp shells produced nanowhiskers with a crystallinity index of 93.6% and crystallite dimensions of 6.3 nm at the (1 1 0) plane [86]. However, since the peaks were not fitted most appropriately for detailed structural reporting (chi-squared in the range of 10^{-4}), it would be advisable that researchers use the crystallite size data with appropriate caution.

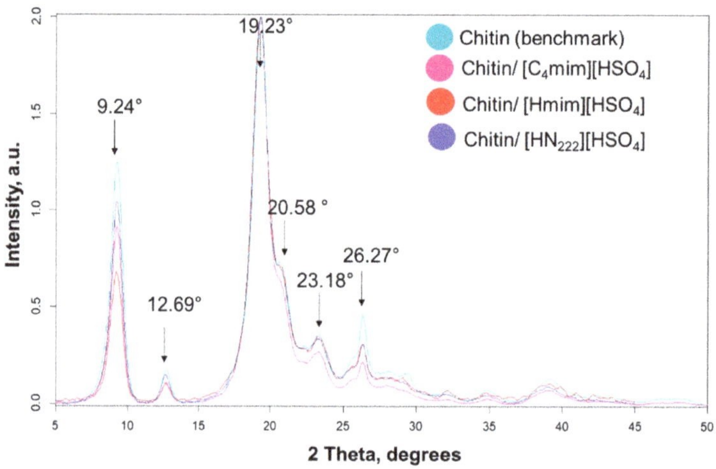

Figure 4. Full diffractograms (5–50° 2 Theta) of chitin (aqua) and chitin nanowhiskers (ChNWs): ChNWs/Ch/[Hmim][HSO_4] (red), ChNWs/Ch/[HN_{222}][HSO_4] (blue), and ChNWs/Ch/[Hmim][HSO_4] (pink).

Table 2. Crystallinity index (CrI, %) determined via peak height and peak deconvolution methods and crystallite size (nm) at the (0 2 0) and (1 1 0) planes.

	Crystallinity Index via Peak Height (CrI), %	Crystallinity Index via Deconvolution (CrI), %	Crystallite Size at the (0 2 0) Plane, nm	Crystallite Size at the (1 1 0) Plane, nm
Chitin Standard	98.0	84.0	6.9	5.2
ChNWs/Ch/[HN_{222}][HSO_4]	98.0	84.5	8.5	7.5
ChNWs/Ch/[Hmim][HSO_4]	98.0	79.8	8.1	5.8
ChNWs/Ch/[C_4mim][HSO_4]	98.1	92.6	8.6	6.3

The degree of acetylation (%DA) was also determined for pure chitin and all ChNWs/Ch/ILs, using the first derivative of FTIR spectra (Figure S13, Table S10, ESI), as reported by Beil et al. [65]. The technique compares the heights of the C–H deformation band (1383 cm^{-1}, MB1), the amide III band (1327 cm^{-1}, MB2), and the bridge oxygen stretching band (1163 cm^{-1}, RB) in the first derivative ATR FTIR spectra with those obtained for several external standards (chitins and chitosans with a known value of %DA). A calibration curve (also called a standard curve) was generated by plotting the value [(MB1 + MB2)/RB] against %DA for a series of reference standards. Back calculations using this calibration curve (Figure S2, ESI) revealed the %DA in the range of 97–100% in all cases.

3.2. Preparation and Characterization of Chitin Nanowhiskers from Crustaceous Biomass

As demonstrated in the previous section, all three ILs, i.e., [C_4mim][HSO_4], [Hmim][HSO_4], and [HN_{222}][HSO_4], were able to hydrolyze chitin and produced ChNWs with similar dimensions. These same ILs were then used to simultaneously pulp and hydrolyze crustacean biomass (shrimp shells, SS) using a two-step procedure, similar to the one described for commercial chitin. Briefly, SS was first pretreated with $[C_4mim][HSO_4]$ [50,51], $[Hmim][HSO_4]$, or $[HN_{222}][HSO_4]$ for 48 h, followed by the addition of DI water and 110 °C heating under reflux for 48 h. In comparing the reactions between the ILs with the two starting materials (chitin or SS), it is worth noting that the reaction with SS proceeded much more vigorously than the same reaction with chitin, and created foam in all cases, likely due to the release of carbon dioxide from calcium carbonate ($CaCO_3$, present in the SS) under acidic conditions. After dilution with water, centrifugation, washing, and lyophilization, the product yield was 75–80% based on chitin available in the SS (Table 3), with yields significantly higher than those obtained from pure chitin.

Table 3. Dimensional parameters of the ChNWs isolated from shrimp shells (SS) using different ILs.

ChNW Prepared from SS Hydrolyzed with	Yield %	Dimensions	Size, nm Mean ± STD	Sample Size	Aspect Ratio Mean ± STD	Sample Size
$[C_4mim][HSO_4]$ *	75	Length	561 ± 157	36	55.5 ± 22.1	36
		Diameter	10.0 ± 3.6	90		
$[Hmim][HSO_4]$	80	Length	576 ± 168	21	55.2 ± 28.1	21
		Diameter	11.7 ± 2.9	62		
$[HN_{222}][HSO_4]$	79	Length	612 ± 198	41	34.8 ± 18.8	41
		Diameter	17.3 ± 3.9	127		

* Original study [50,51]. For comparison, all nanowhiskers prepared from biomass have approximately the same dimensions, i.e., 583 ± 21 nm in length and 21.3 ± 0.9 nm in diameter.

The analysis of TEM micrographs revealed that ChNWs from SS (sample ID ChNWs/SS/ILs, Figure 5) had a length ranging from 560 to 612 nm, i.e., roughly 2.5–3 times longer compared to ChNWs/Ch/ILs, and diameters ~10–20 nm (Table 3). When considering aspect ratios, calculated as the ratio of length to diameter (L/d), ChNWs/SS/ILs showed ~3–5 times larger aspect ratio than ChNWs/Ch/ILs. Interestingly, while ChNWs prepared using imidazolium ILs were ~2 times narrower in diameter than ChNWs/Ch/imidazolium ILs, ChNWs/SS/$[HN_{222}][HSO_4]$ showed a similar diameter as ChNWs/Ch/$[HN_{222}][HSO_4]$.

Upon examining the FTIR spectra (Figure 6) of the obtained ChNWs, it became evident that while the location and intensity of the peaks of ChNWs/SS/$[C_4mim][HSO_4]$ closely resembled those of commercially available chitin [50,51], there were significant differences in the spectra of ChNWs/SS produced using "inexpensive" ILs, i.e., ChNWs/SS/$[Hmim][HSO_4]$ or ChNWs/SS/$[HN_{222}][HSO_4]$. The spectrum of ChNWs/SS/$[Hmim][HSO_4]$ exhibited peaks typical for chitin; however, the intensities of many peaks were found to be somewhat greater than those typically associated with the biopolymer. This was particularly true for weak vibrations at 3483 and 3437 cm^{-1}, normally associated with water molecules, weak peaks in the region 820–650 cm^{-1}, and medium/strong vibrations at 1155 and 1116 cm^{-1}, characteristic of sulfate ion [87–89]. The FTIR spectrum of ChNWs/SS/$[HN_{222}][HSO_4]$ also showed peaks of enhanced intensities similar to the ones described for ChNWs/SS/$[Hmim][HSO_4]$. In this sample, additional peaks at 3152, 1555, 1401, 1170, and 974 cm^{-1} were identified, which did not match either chitin or sulfate (indicated in Figure 6a–d in blue color). The peaks did not belong to ionic liquids (Figure S7). We suggest that the reader consult the SI file, where the expansion of the FTIR regions is provided. Byproduct formation was reproducible in both cases.

Powder X-ray diffractograms (Figure 7) of ChNWs/SS/$[Hmim][HSO_4]$ and ChNWs/SS/$[HN_{222}][HSO_4]$ confirmed the presence of chitin (at 9.24°, 12.69°, 19.23°, 20.58°, 23.18°, and 26.27°, respectively [78,79]), but also peaks that did not belong to the biopolymer, at 11.56, 20.68, 23.33, and 29.07° 2θ. The amount of chitin was significantly greater in ChNWs/SS/$[HN_{222}][HSO_4]$, whereas ChNWs/SS/$[Hmim][HSO_4]$ showed a significant amount of byproduct. Interestingly, no byproduct formation was detected in ChNWs/SS/

[C_4mim][HSO_4]. It is also worth noting that the same byproduct was formed regardless of the "inexpensive" IL's identity, excluding any compound that could be derived from the ILs themselves. Also important is that the byproduct formed only in the case of IL reactions with biomass but was not seen when pure chitin polymer was hydrolyzed using these ILs, despite identical reaction conditions. This suggests that the formation of the byproduct had something to do with minerals present in biomass.

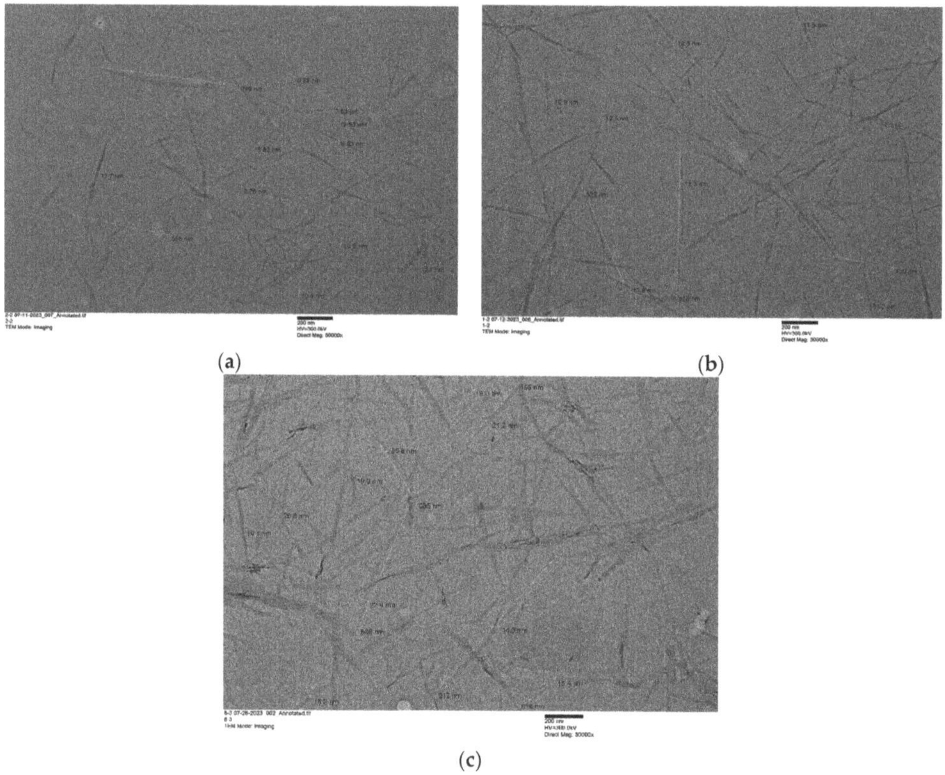

Figure 5. TEM micrograph of a dilute suspension of chitin nanowhiskers derived from purified chitin using (**a**) [C_4mim][HSO_4], (**b**) [Hmim][HSO_4], and (**c**) [HN_{222}][HSO_4]. HV 300 kV. Magnification: 30,000×; scale bar 200 nm.

Since one of the goals of this work was to generate high-purity chitin, identifying the new material was of great importance. An analysis of the literature identified the peaks at 11.56, 20.68, 23.33, and 29.07° 2θ as calcium sulfate dihydrate (also known as gypsum, $Ca(SO_4){\cdot}2H_2O$) [90]. TEM images did not show the presence of gypsum, likely because of the mineral being dissolved in water during the preparation of the sample for TEM analysis. On the other hand, lyophilized samples were expected to contain both chitin nanocrystals and gypsum after water removal. When the ChNWs/SS/[HN_{222}][HSO_4] lyophilized sample was dispersed in a small amount of water and placed onto a carbon tape, followed by slow water evaporation, Scanning Electron Microscopy (SEM) imaging revealed the presence of gypsum crystals (Figure 8).

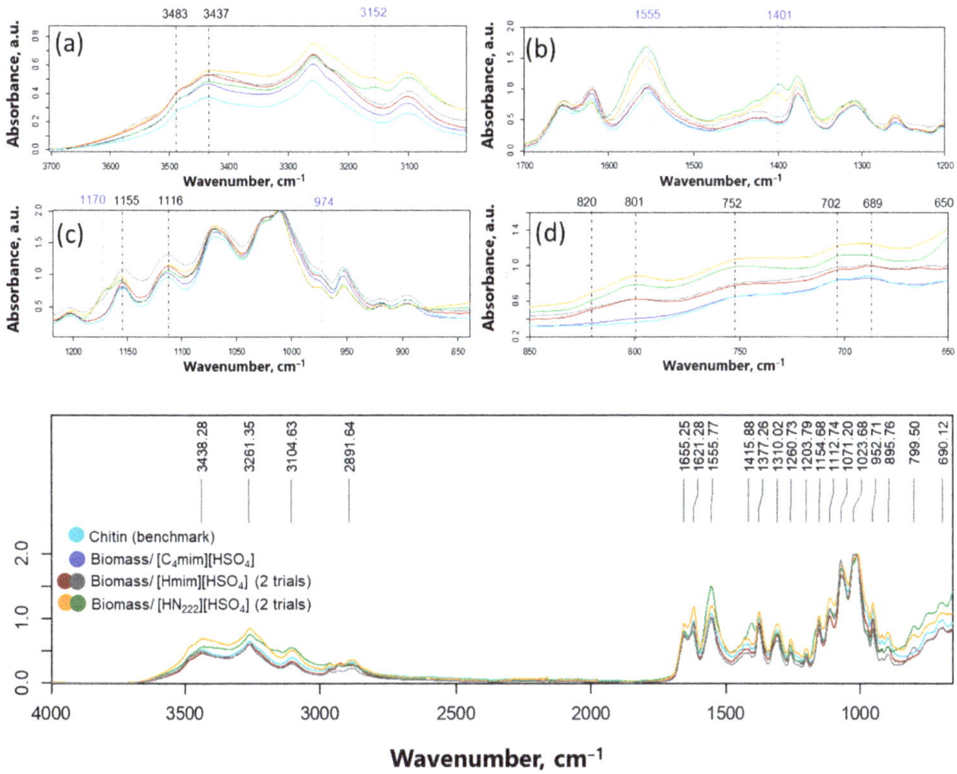

Figure 6. FTIR spectra of pure chitin (aqua) and chitin nanowhiskers (ChNWs) obtained from SS: ChNWs/SS/[C_4mim][HSO_4] (blue), ChNWs/SS/[Hmim][HSO_4] (brown and grey, two trials), and ChNWs/SS/[HN_{222}][HSO_4] (orange and green, two trials).

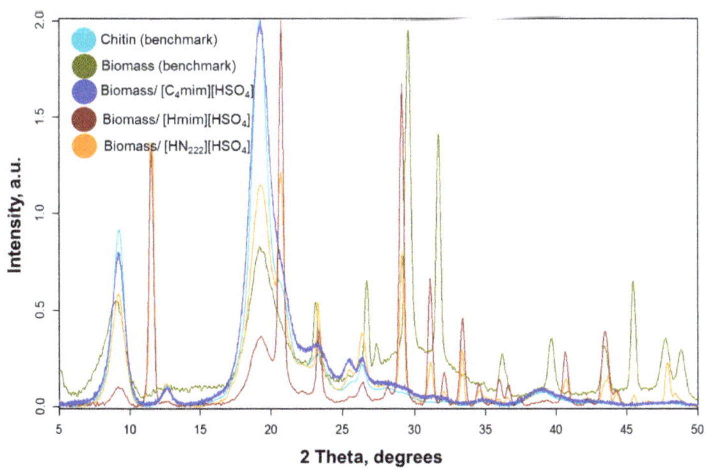

Figure 7. Diffractograms of pure chitin (aqua) and chitin nanowhiskers (ChNWs) obtained from shrimp shells: ChNWs/SS/[C_4mim][HSO_4] (blue), ChNWs/SS/[Hmim][HSO_4] (brown), and ChNWs/SS/[HN_{222}][HSO_4] (orange).

Figure 8. Gypsum crystals as obtained from biomass using [Hmim][HSO_4] and [HN_{222}][HSO_4] (left, right) looking identical to those provided in ref. [91]. No coating was used for imaging. Note that no ChNWs could be seen at this low magnification.

The reaction of acidic ILs with SS resulted in foam formation, which was attributed to CO_2 release from $CaCO_3$ in acidic conditions. This same reaction generates $CaSO_4$, which was previously reported to be ~5–10 times less soluble in the presence of 10% of alkylimidazolium hydrogen sulfate ILs than in pure water (0.055 mol/kg) [92], which would induce the coprecipitation of $CaSO_4$ together with ChNWs. However, a comparatively lesser decrease in the solubility of $CaSO_4{\cdot}2H_2O$ was observed for [C_4mim][HSO_4] when compared with [C_2mim][HSO_4], indicating an influence of the alkyl chain at the N2 atom of the imidazolium cation on the mineral's aqueous solubility [92]. With respect to the discussed reactions, no gypsum was detected when [C_4mim][HSO_4] was used to recover ChNWs from biomass, while a significant amount of gypsum was detected when [Hmim][HSO_4] was used. Considering that the difference between the ILs is only the butyl substituent on the N2 atom of imidazolium, we hypothesize that it is a protonated nitrogen atom of the cation of the IL that participates in the byproduct formation. The protonated ammonium or imidazolium could be stabilizing ${CO_3}^{2-}$ from the $CaCO_3$ mineral of SS in solution through electrostatic interactions. It could also undergo deprotonation and react with CO_2 liberated by the reaction of $[HSO_4]^-$ with $CaCO_3$ with the formation of carbamates.

Regardless, the next step was adjusting the reaction conditions to minimize the formation of gypsum. Hence, the pretreatment step was eliminated. However, when SS was treated with [Hmim][HSO_4] without pretreatment, there still was a significant amount of gypsum present. Decreasing the reaction time first to 24 h, then to 12 h, and then to 6 h did not result in the removal of gypsum. Contrarily, the treatment of SS with [HN_{222}][HSO_4] provided a pure chitin product upon excluding the pretreatment step and reducing the reaction time to 24 h. Eliminating the pretreatment step aimed to remove the need for 24 h heating at 90 °C, leading to a favorable result.

The yield of the product decreased from 79 to 72%, whereas ChNWs' dimensions remained approximately the same (Table 4), not surprising considering that TEM imaging was only suitable for chitin visualization. The crystallinity, determined using the peak fitting method, was found to be 79.4%, a value very similar to that of ChNWs/SS/[C_4mim][HSO_4] (80.6%, Tables S8 and S9, ESI). The crystallite size was determined to be 7.1 nm at the (0 2 0) plane and 6.1 nm at the (1 1 0) plane (Table S9, ESI), also very close to that obtained for ChNWs/SS/[C_4mim][HSO_4] (8.4 and 5.6 nm at the (0 2 0) and (1 1 0) planes). The %DA calculated was found to be ~100%. Thermal analysis data have shown that $T_{50\%dec}$ was 341.7 °C, compared with 347 °C reported for pure chitin [93]. Overall, the [HN_{222}][HSO_4] IL performed very similarly to the [C_4mim][HSO_4] IL for biomass treatment and required milder reaction conditions, i.e., no pretreatment and 12 h reflux.

Table 4. Dimensional parameters of the ChNWs/SS/ILs without pretreatment step.

ChNW Prepared from Biomass Hydrolyzed with	Yield %	Dimensions	Size Mean ± STD	DA %	CrI %	Crystallite Size, nm	
						(0 2 0)	(1 1 0)
$[C_4mim][HSO_4]$ *	75	Length	561 ± 157	~100 [a]	80.6	8.4	5.6
		Diameter	10.0 ± 3.6				
$[HN_{222}][HSO_4]$	72	Length	588 ± 163	~100 [a]	79.4	7.1	6.1
		Diameter	17.1 ± 4.1				

* Original study [50,51]. [a] Calculated value is >100 (empirical method).

This innovative approach to producing ChNWs from crustacean biomass will contribute to advancing the circular economy concept. Crustacean biomass (e.g., shrimp shells) which would normally be sent to a waste treatment process (landfill) can be valorized through this process. Furthermore, the environmental impact of nanochitin production is considerably diminished through this method. By reducing hazardous chemical use and waste generation, the process aligns with global sustainability goals, notably cutting CO_2 emissions by approximately 75%.

To make sure that the process is environmentally viable, future research will perform an environmental life cycle assessment (LCA) The International Standardization Organization (ISO) defines the framework and guidelines for LCA. The framework includes goal and scope definition, inventory analysis, impact assessment, and result interpretation [94,95]. The system boundary shall encompass all the activities associated with the collection and transportation of the crustacean shells to the biorefinery, processing, and conversion into nanostructured chitin.

Finally, techno-economic considerations are an integral part of defining the sustainability of a product. Process simulation models can help to estimate the material and energy balances of the process pathways producing nanostructured chitin using the process data from the experimental tasks. The economic analysis encompasses the estimation of the capital investment, operational expenses, and production costs.

4. Conclusions

This research introduces a sustainable and cost-effective method for producing nanochitin using inexpensive ionic liquids (ILs) such as $[HN_{222}][HSO_4]$, presenting a significant advancement over traditional chitin extraction processes. As mentioned in the introduction of this work, the applications to chitin nanowhiskers are broad, including packaging replacements, medical devices, and flexible electronics, to name but a few. Importantly for all the aforementioned applications, the IL reduces the overall production costs by simplifying the process, eliminating the need for initial purification steps, and reducing the consumption of chemicals and energy.

The scalability of this process promises broader implementation across sectors utilizing biopolymers. Future research should focus on further cost reductions for IL production and on evaluating the lifecycle impacts of these materials to enhance their commercial and environmental viability. This approach not only demonstrates a technological innovation in the production of chitin nanowhiskers but also exemplifies the application of green chemistry principles in materials science, setting a template for sustainable material production. The chitin nanowhiskers produced maintain high crystallinity and dimensional stability, making them suitable for various industrial applications including biomedicine, packaging, and electronics.

Supplementary Materials: The following supporting information can be downloaded at: https://www.mdpi.com/article/10.3390/suschem5020010/s1, Figure S1: General Scheme of preparation of ChNWs/Ch/ILs; Figure S2: First derivative of FTIR spectra of chitin and chitosan standards; Figure S3: Calibration curve for calculation of %DA; Figures S4–S6: Expanded FTIR spectra of chitin standard and chitin nanowhiskers (ChNWs) produced from chitin; Figures S7–S12: pXRD Peak Fitting (Origin); Figure S13: First derivative of FTIR spectra (1450–1150 cm^{-1}) of chitin standard and chitin nanowhiskers (ChNWs) produced from chitin; Table S1: Characteristic FTIR peaks identified in chitin samples reported in the literature; Tables S2–S7: Parameters of pXRD Peak Fitting (Origin); Table S8: Evaluation of crystallinity; Table S9: Evaluation of crystallite size; Table S10: MB1, MB2 and RB values elucidated from first derivative of FTIR spectra. References [96–98] are cited in Supplementary Materials.

Author Contributions: Conceptualization, J.L.S. and P.B.; methodology, J.L.S.; data collection and analysis, A.S.S., R.P.S. and J.L.S.; writing—original draft preparation, J.L.S. and P.B.; writing—review and editing, J.L.S., P.B. and O.T.; supervision, J.L.S.; funding acquisition, J.L.S. All authors have read and agreed to the published version of the manuscript.

Funding: This research was supported by NSF-I Corps 2243420 (prototype development funds).

Institutional Review Board Statement: Not applicable.

Informed Consent Statement: Not applicable.

Data Availability Statement: The original contributions presented in the study are included in the article/Supplementary Material; further inquiries can be directed to the corresponding authors.

Acknowledgments: The authors thank S.P. Kelley for the help with gypsum identification.

Conflicts of Interest: The authors declare no conflicts of interest.

References

1. Bai, L.; Liu, L.; Esquivel, M.; Tardy, B.L.; Huan, S.; Niu, X.; Liu, S.; Yang, G.; Fan, Y.; Rojas, O.J. Nanochitin: Chemistry, Structure, Assembly, and Applications. *Chem. Rev.* **2022**, *122*, 11604–11674. [CrossRef] [PubMed]
2. Li, J.; Revol, J.F.; Marchessault, R.H. Effect of degree of deacetylation of chitin on the properties of chitin crystallites. *J. Appl. Polym. Sci.* **1997**, *65*, 370–373. [CrossRef]
3. Foster, A.B.; Webber, J.M. Chitin. In *Advances in Carbohydrate Chemistry*; Melville, L., Wolfrom, R., Tipson, S., Eds.; Academic Press: Oxford, UK, 1961; pp. 371–393.
4. Visakh, P.M.; Thomas, S. Preparation of Bionanomaterials and Their Polymer Nanocomposites from Waste and Biomass. *Waste Biomass Valor.* **2010**, *1*, 121–134. [CrossRef]
5. Lee, S.; Hao, L.T.; Park, J.; Oh, D.X.; Hwang, D.S. Nanochitin and nanochitosan: Chitin nanostructure engineering with multiscale properties for biomedical and environmental applications. *Adv. Mater.* **2023**, *35*, e2203325. [CrossRef] [PubMed]
6. Vincent, J.F.V.; Wegst, U.G.K. Design and mechanical properties of insect cuticle. *Arthropod Struct. Dev.* **2004**, *33*, 187–199. [CrossRef] [PubMed]
7. Sanchez, C.; Boissiere, C.; Cassaignon, S.; Chaneac, C.; Durupthy, O.; Faustini, M.; Grosso, D.; Laberty-Robert, C.; Nicole, L.; Portehault, D.; et al. Molecular engineering of functional inorganic and hybrid materials. *Chem. Mater.* **2014**, *26*, 221–238. [CrossRef]
8. Dominic, C.D.M.; Joseph, R.; Sabura Begum, P.M.; Raghunandanan, A.; Vackkachan, N.T.; Padmanabhan, D.; Formela, K. Chitin nanowhiskers from shrimp shell waste as green filler in acrylonitrile-butadiene rubber: Processing and performance properties. *Carbohydr. Polym.* **2020**, *245*, 116505. [CrossRef] [PubMed]
9. Liu, Y.; Liu, M.; Yang, S.; Luo, B.; Zhou, C. Liquid crystalline behaviors of chitin nanocrystals and their reinforcing effect on natural rubber. *ACS Sustain. Chem. Eng.* **2018**, *6*, 325–336. [CrossRef]
10. Chi-Yan Li, S.; Sun, Y.C.; Guan, Q.; Naguib, H. Effects of chitin nanowhiskers on the thermal, barrier, mechanical, and rheological properties of polypropylene nanocomposites. *RSC Adv.* **2016**, *6*, 72086–72095. [CrossRef]
11. Wang, M.; Xue, H.; Feng, Z.; Cheng, B.; Yang, H. Increase of Tensile Strength and Toughness of Bio-Based Diglycidyl Ether of Bisphenol A with Chitin Nanowhiskers. *PLoS ONE* **2017**, *12*, e0177673. [CrossRef]
12. Rajak, D.K.; Pagar, D.D.; Kumar, R.; Pruncu, C.I. Recent progress of reinforcement materials: A comprehensive overview of composite materials. *J. Mater. Res. Technol.* **2019**, *8*, 6354–6374. [CrossRef]
13. Serventi, L.; He, Q.; Huang, J.; Mani, A.; Subhash, A.J. Advances in the preparations and applications of nanochitins. *Food Hydrocoll. Health* **2021**, *1*, 100036. [CrossRef]
14. Mushi, N.E.; Utsel, S.; Berglund, L.A. Nanostructured biocomposite films of high toughness based on native chitin nanofibers and chitosan. *Front. Chem.* **2014**, *2*, 99. [CrossRef] [PubMed]

15. Hashiguchi, T.; Yamamoto, K.; Kadokawa, J.-I. Fabrication of highly flexible nanochitin film and its composite film with anionic polysaccharide. *Carbohydr. Polym.* **2021**, *270*, 118369. [CrossRef] [PubMed]
16. Zhong, T.; Jian, G.; Chen, Z.; Wolcott, M.; Nassiri, S.; Fernandez, C.A. Interfacial interactions and reinforcing mechanisms of cellulose and chitin nanomaterials and starch derivatives for cement and concrete strength and durability enhancement: A review. *Nanotechnol. Rev.* **2022**, *11*, 2673–2713. [CrossRef]
17. Haider, M.M.; Jian, G.; Zhong, T.; Li, H.; Fernandez, C.A.; Fifield, L.S.; Wolcott, M.; Nassiri, S. Insights into setting time, rheological and mechanical properties of chitin nanocrystals- and chitin nanofibers-cement paste. *Cem. Concr. Compos.* **2022**, *132*, 104623. [CrossRef]
18. Fernandes, S.C.M.; Aguirre, G. Biopolymer micro/nanogel particles as smart drug delivery and theranostic systems. *Pharmaceutics* **2023**, *15*, 2060. [CrossRef] [PubMed]
19. Olza, S.; Salaberria, A.M.; Alonso-Varona, A.; Samanta, A.; Fernandes, S.C.M. The role of nanochitin in biologically-active matrices for tissue engineering–Where do we stand? *J. Mater. Chem. B* **2023**, *11*, 5630–5649. [CrossRef] [PubMed]
20. Zubillaga, V.; Salaberria, A.M.; Palomares, T.; Alonso-Varona, A.; Kootala, S.; Labidi, J.; Fernandes, S.C.M. Chitin nanoforms provide mechanical and topological cues to support growth of human adipose stem cells in chitosan matrices. *Biomacromolecules* **2018**, *19*, 3000–3012. [CrossRef]
21. Suneetha, M.; Kim, H.; Han, S.S. Doxorubicin-loaded fungal-carboxymethyl chitosan functionalized polydopamine nanoparticles for photothermal cancer therapy. *Pharmaceutics* **2023**, *15*, 1281. [CrossRef]
22. Dasan, Y.K.; Bhat, A.H.; Khan, I. 1. Nanocellulose and Nanochitin for Water Remediation by Adsorption of Heavy Metals. In *Nanomaterials for Water Remediation*; Kumar Mishra, A., Hussain, M.C., Mishra, S.B., Eds.; De Gruyter: Berlin, Germany; Boston, MA, USA, 2020; pp. 1–18. [CrossRef]
23. CORDIS: Functional Membranes/Filters with Anti/Low-Fouling Surfaces for Water Purification through Selective Adsorption on Biobased Nanocrystals and Fibrils. Available online: https://cordis.europa.eu/project/id/280519 (accessed on 12 February 2024).
24. Somasundaram, S.; Kumaravel, V. Application of Nanoparticles for Self-Cleaning Surfaces. In *Home Emerging Nanostructured Materials for Energy and Environmental Science*; Part of the Environmental Chemistry for a Sustainable World Book Series; Springer: Berlin/Heidelberg, Germany, 2019; Volume 23.
25. He, Y.; Lin, X.; Feng, Y.; Luo, B.; Liu, M. Carbon nanotube ink dispersed by chitin nanocrystals for thermoelectric converter for self-powering multifunctional wearable electronics. *Adv. Sci.* **2022**, *9*, e2204675. [CrossRef] [PubMed]
26. Schötz, S.; Reisbeck, F.; Schmitt, A.C.; Dimde, M.; Quaas, E.; Achazi, K.; Haag, R. Tunable polyglycerol-based redox-responsive nanogels for efficient cytochrome c delivery. *Pharmaceutics* **2021**, *13*, 1276. [CrossRef] [PubMed]
27. Andre, R.S.; dos Santos, D.M.; Mercante, L.A.; Facure, M.H.M.; Campana-Filho, S.P.; Mattoso, L.H.C.; Correa, D.S. Nanochitin-based composite films as a disposable ethanol sensor. *J. Environ. Chem. Eng.* **2020**, *8*, 104163. [CrossRef]
28. Heidarian, P.; Gharaie, S.; Yousefi, H.; Paulino, M.; Kaynak, A.; Varley, R.; Kouzani, A.Z.A. 3D printable dynamic nanocellulose/nanochitin self-healing hydrogel and soft strain sensor. *Carbohydr. Polym.* **2022**, *291*, 119545. [CrossRef] [PubMed]
29. Younes, I.; Rinaudo, M. Chitin and chitosan preparation from marine sources. Structure, properties and applications. *Mar. Drugs* **2015**, *13*, 1133–1174. [CrossRef] [PubMed]
30. Yang, N.; Xi, C. Sustainability: Don't waste seafood waste. *Nature* **2015**, *524*, 155–157. [CrossRef] [PubMed]
31. Chandrasekharan, M.; Sachindra, N.M.; Mahendrakar, N.S. Biotechnology for utilization of marine by-products. In *Fish Processing Byproducts*; Studium Press: Austin, TX, USA, 2015; pp. 43–62.
32. Poeloengasih, C.D.; Hernawan, D.; Angwarndo, M. Isolation and characterization of chitin and chitosan prepared under various processing times. *Indones. J. Chem.* **2008**, *8*, 189–192. [CrossRef]
33. Beaney, P.; Lizardi-Mendoza, J.; Healy, M. Comparison of chitins produced by chemical and bioprocessing methods. *J. Chem. Tech. Biotech.* **2005**, *80*, 145–150. [CrossRef]
34. Rinaudo, M. Chitin and chitosan: Properties and applications. *Prog. Polym. Sci.* **2006**, *31*, 603–632. [CrossRef]
35. Khoushab, F.; Yamabhai, M. Chitin Research Revisited. *Mar. Drugs* **2010**, *8*, 1988–2012. [CrossRef]
36. Joseph, B.; Sam, R.M.; Balakrishnan, P.; Maria, H.J.; Gopi, S.; Volova, T.; Fernandes, S.C.M.; Thomas, S. Extraction of nanochitin from marine resources and fabrication of polymer nanocomposites: Recent advances. *Polymers* **2020**, *12*, 1664. [CrossRef] [PubMed]
37. Druzian, S.P.; Zanatta, N.P.; Côrtes, L.N.; Fátima, A.; Streit, M. Preparation of chitin nanowhiskers and its application for crystal violet dye removal from wastewaters. *Environ. Sci. Pollut. Res.* **2019**, *26*, 28548–28557. [CrossRef] [PubMed]
38. Hong, S.; Yuan, Y.; Zhang, K.; Lian, H.; Liimatainen, H. Efficient Hydrolysis of Chitin in a Deep Eutectic Solvent Synergism for Production of Chitin Nanocrystals. *Nanomaterials* **2020**, *10*, 869. [CrossRef] [PubMed]
39. Yamamoto, Y.; Nishimura, T.; Saito, T.; Kato, T. $CaCO_3$/chitin-whisker hybrids: Formation of $CaCO_3$ crystals in chitin-based liquid-crystalline suspension. *Polym. J.* **2010**, *42*, 583–586. [CrossRef]
40. Revol, J.; Marchessaultf, R.H. In Vitro Chiral Nematic Ordering of Chitin Crystallites. *Int. J. Biol. Macromol.* **1993**, *15*, 329–335. [CrossRef] [PubMed]
41. Jin, T.; Liu, T.; Lam, E.; Moores, A. Chitin and chitosan on the nanoscale. *Nanoscale Horiz.* **2021**, *6*, 505–542. [CrossRef]
42. Li, Z.; Zhang, M.; Cheng, D.; Yang, R. Preparation of silver nano-particles immobilized onto chitin nano-crystals and their application to cellulose paper for imparting antimicrobial activity. *Carbohydr. Polym.* **2016**, *151*, 834–840. [CrossRef] [PubMed]

43. Fan, Y.; Saito, T.; Isogai, A. Chitin nanocrystals prepared by TEMPO-mediated oxidation of α-chitin. *Biomacromolecules* **2008**, *9*, 192–198. [CrossRef] [PubMed]
44. Salaberria, A.M.; Labidi, J.; Fernandes, S.C.M. Different routes to turn chitin into stunning nano-objects. *Eur. Polym. J.* **2015**, *68*, 503–515. [CrossRef]
45. Hou, C.; Zhang, J.; Zhang, X.; Wang, Y.; Li, T.; Xia, B.; Jiang, J.; Dong, W. High strength chitin nanocrystal/alginate filament prepared by wet-spinning in "green" coagulating bath. *Cellulose* **2022**, *29*, 8611–8621. [CrossRef]
46. Liu, P.; Liu, H.; Schäfer, T.; Gutmann, T.; Gibhardt, H.; Qi, H.; Tian, L.; Zhang, X.C.; Buntkowsky, G.; Zhang, K. Unexpected selective alkaline periodate oxidation of chitin for the isolation of chitin nanocrystals. *Green Chem.* **2021**, *23*, 745–751. [CrossRef]
47. Oun, A.A.; Rhim, J.W. Effect of Oxidized Chitin Nanocrystals Isolated by Ammonium Persulfate Method on the Properties of Carboxymethyl Cellulose-Based Films. *Carbohydr. Polym.* **2017**, *175*, 712–720. [CrossRef] [PubMed]
48. Li, Z.; Liu, C.; Hong, S.; Lian, H.; Mei, C.; Lee, J.; Wu, Q.; Hubbe, M.A.; Li, M. Recent Advances in Extraction and Processing of Chitin Using Deep Eutectic Solvents. *Chem. Eng. J.* **2022**, *446*, 136953. [CrossRef]
49. Berroci, M.; Vallejo, C.; Lizundia, E. Environmental impact assessment of chitin nanofibril and nanocrystal isolation from fungi, shrimp shells, and crab shells. *ACS Sustain. Chem. Eng.* **2022**, *10*, 14280–14293. [CrossRef]
50. Shamshina, J.L.; Abidi, N. Isolation of chitin nano-whiskers directly from crustacean biomass waste in a single step with acidic ionic liquid. *ACS Sustain. Chem. Eng.* **2022**, *10*, 11846–11855. [CrossRef]
51. Abidi, N.; Shamshina, J.L. Preparation of Chitin Nanocrystals and Nanowhiskers from Crustacean Biomass Using Ionic Liquid. WO2023059499A1, 29 September 2022. Available online: https://worldwide.espacenet.com/patent/search?q=PCT/US2022/045177 (accessed on 12 February 2024).
52. Singh, V.; Kaur, S.; Sapehiyia, V.; Singh, J.; Kad, G.L. Microwave accelerated preparation of [bmim][HSO_4] ionic liquid: An acid catalyst for improved synthesis of coumarins. *Catal. Commun.* **2005**, *6*, 57–60. [CrossRef]
53. Tao, D.-J.; Li, Z.-M.; Cheng, Z.; Hu, N.; Chen, X.-S. Kinetics study of the ketalization reaction of cyclohexanone with glycol using Brønsted acidic ionic liquids as catalysts. *Ind. Eng. Chem. Res.* **2012**, *51*, 16263–16269. [CrossRef]
54. Stocker, M.W.; Tsolaki, E.; Harding, M.J.; Healy, A.M.; Ferguson, S. Combining isolation-free and co-processing manufacturing approaches to access room temperature ionic liquid forms of APIs. *J. Pharm. Sci.* **2023**, *112*, 2079–2086. [CrossRef] [PubMed]
55. Alcalde, E.; Dinarès, I.; Ibáñez, A.; Mesquida, N. A Simple halide-to-anion exchange method for heteroaromatic salts and ionic liquids. *Molecules* **2012**, *17*, 4007–4027. [CrossRef]
56. Klein-Marcuschamer, D.; Simmons, B.A.; Blanch, H.W. Techno-economic analysis of a lignocellulosic ethanol biorefinery with ionic liquid pre-treatment. *Biofuels Bioprod. Bioref.* **2011**, *5*, 562–569. [CrossRef]
57. Chen, L.; Sharifzadeh, M.; Dowell, N.M.; Welton, T.; Shah, N.; Hallett, J.P. Inexpensive ionic liquids: $[HSO_4]^-$-based solvent production at bulk scale. *Green Chem.* **2014**, *16*, 3098–3106. [CrossRef]
58. George, A.; Brandt, A.; Tran, K.; Zahari, S.M.S.N.S.; Klein-Marcuschamer, D.; Sun, N.; Sathitsuksanoh, N.; Shi, J.; Stavila, V.; Parthasarathi, R.; et al. Design of low-cost ionic liquids for lignocellulosic biomass pretreatment. *Green Chem.* **2015**, *17*, 1728–1734. [CrossRef]
59. Zahari, S.M.S.N.S.; Azman, H.; Karim, L. Triethylammonium hydrogen sulfate ionic liquid as a low-cost solvent: A short review of synthesis, analysis and applications. In Proceedings of the MATEC Web conference, International Mechanical and Industrial Engineering Conference 2018 (IMIEC 2018), Malang, Indonesia, 30–31 August 2018; Volume 204. Available online: https://www.matec-conferences.org/articles/matecconf/abs/2018/63/matecconf_imiec2018_00006/matecconf_imiec2018_00006.html (accessed on 26 March 2024).
60. Brandt-Talbot, A.; Gschwend, F.J.V.; Fennell, P.S.; Lammens, T.M.; Tan, B.; Wealea, J.; Hallett, J.P. An economically viable ionic liquid for the fractionation of lignocellulosic biomass. *Green Chem.* **2017**, *19*, 3078. [CrossRef]
61. Karimi-Jaberi, Z.; Masoudi, B.; Rahmani, A.; Alborzi, K. Triethylammonium hydrogen sulfate [Et_3NH][HSO_4] as an efficient ionic liquid catalyst for the synthesis of coumarin derivatives. *Polycycl. Aromat. Compd.* **2020**, *40*, 99–107. [CrossRef]
62. Black, M.M.; Schwartz, H.M. The estimation of chitin and chitin nitrogen in crawfish waste and derived products. *Analyst* **1950**, *75*, 185–189. [CrossRef]
63. Scherrer, P. Determination of the size and internal structure of colloidal particles using X-rays. In *Kolloidchemie Ein Lehrbuch*; Springer: Berlin/Heidelberg, Germany, 1912; pp. 387–409.
64. Fan, Y.; Fukuzumi, H.; Saito, T.; Isogai, A. Comparative characterization of aqueous dispersions and cast films of different chitin nanowhiskers/nanofibers. *Int. J. Biol. Macromol.* **2012**, *50*, 69–76. [CrossRef] [PubMed]
65. Beil, S.; Schamberger, A.; Naumann, W.; Machill, S.; van Pee, K.-H. Determination of the degree of N-acetylation (DA) of chitin and chitosan in the presence of water by first derivative ATR FTIR spectroscopy. *Carbohydr. Polym.* **2012**, *87*, 117–122. [CrossRef] [PubMed]
66. PubChem. 1-Methylimidazolium (Compound). Available online: https://pubchem.ncbi.nlm.nih.gov/compound/1-Methylimidazolium#section=Synonyms (accessed on 6 April 2024).
67. Muckerman, J.T.; Skone, J.H.; Ning, M.; Wasada-Tsutsui, Y. Toward the accurate calculation of pKa values in water and acetonitrile. *Biochim. Biophys. Acta* **2013**, *1827*, 882–891. [CrossRef] [PubMed]
68. Grishina, E.P.; Ramenskaya, L.M.; Gruzdev, M.S.; Kraeva, O.V. Water effect on physicochemical properties of 1-butyl-3-methylimidazolium based ionic liquids with inorganic anions. *J. Mol. Liquids* **2013**, *177*, 267–272. Available online: https://www.sciencedirect.com/science/article/pii/S0167732212003819?via=ihub (accessed on 10 April 2024). [CrossRef]

69. Matuszek, K.; Chrobok, A.; Coleman, F.; Seddon, K.R.; Swadźba-Kwaśny, M. Tailoring ionic liquids: Structure, acidity and catalytic activity of a new class of protic ionic liquids based on oligomeric $[(HSO_4)_xH_{x-1}]$-anions. *Green Chem.* **2014**, *16*, 3463–3471. [CrossRef]
70. Belieres, J.P.; Angell, C.A. Protic ionic liquids: Preparation, characterization, and proton free energy level representation. *J. Phys. Chem. B* **2007**, *111*, 4926–4937. [CrossRef] [PubMed]
71. Zhang, X.; Mao, Y.; Schwab, N.; Briber, R. An efficient method for chitin processing using phosphoric acid: Oligomers, nanocrystals, and high polymers. In Proceedings of the APS March Meeting 2022, Chicago, IL, USA, 14–18 March 2022; abstract id.N00.094.
72. Muñoz-Núñez, C.; Fernández-García, M.; Muñoz-Bonilla, A. Chitin nanocrystals: Environmentally friendly materials for the development of bioactive films. *Coatings* **2022**, *12*, 144. [CrossRef]
73. Mincea, M.; Negrulescu, A.; Ostafe, V. Preparation, modification, and applications of chitin nanowhiskers: A review. ons of chitin nanowhiskers: A review. *Rev. Adv. Mater. Sci.* **2012**, *30*, 225–242.
74. Gómez Estaca, J.; Tovar, C.A.; Montero García, P.; Gómez Guillén, M.C. Structural, viscoelastic, and emulsifying properties of shrimp chitin nanowhisker dispersions as a function of acidic pHs. *J. Food Eng.* **2023**, *351*, 111519. [CrossRef]
75. Morin, A.; Dufresne, A. Nanocomposites of chitin whiskers from Riftia tubes and poly(caprolactone). *Macromolecules* **2002**, *35*, 22190–22199. [CrossRef]
76. Kumar, G.N.P.; Bhat, S.K. Preparation of chitin nano whiskers from mushrooms. *Int. J. Sci. Res. Publ.* **2018**, *8*, 130. [CrossRef]
77. Goodrich, J.D.; Winter, W.T. Alpha-chitin nanocrystals prepared from shrimp shells and their specific surface area measurement. *Biomacromolecules* **2007**, *8*, 252–255. [CrossRef] [PubMed]
78. Nguyen, H.V.D.; de Vries, R.; Stoyanov, S.D. Chitin nanowhiskers with improved properties obtained using natural deep eutectic solvent and mild mechanical processing. *Green Chem.* **2022**, *24*, 3834–3844. [CrossRef]
79. Abdul Haleem, M.; Parker, K.D. X-ray Diffraction Studies on the Structure of a Chitin. *Z. Naturforsch. C Biosci.* **1976**, *31*, 383–388. [CrossRef]
80. Focher, B.; Beltrame, P.L.; Naggi, A.; Torri, G. Alkaline N-deacetylation of chitin enhanced by flash treatments. Reaction kinetics and structure modifications. *Carbohydr. Polym.* **1990**, *12*, 405–418. [CrossRef]
81. Ioelovich, M. Crystallinity and hydrophility of chitin and chitosan. *Res. Rev. J. Chem.* **2014**, *3*, 7–14.
82. Osorio-Madrazo, A.; David, L.; Trombotto, S.; Lucas, J.-M.; Peniche-Covas, C.; Domard, A. Kinetics study of the solid-state acid hydrolysis of chitosan: Evolution of the crystallinity and macromolecular structure. *Biomacromolecules* **2010**, *11*, 1376–1386. [CrossRef] [PubMed]
83. Gupta, A.K.; Singhal, R.P. Carbon fibers prepared from tailored reversible-addition-fragmentation transfer copolymerization-derived poly(acrylonitrile)-co-poly(methylmethacrylate). *J. Polym. Sci. Polym. Phys. Ed.* **1983**, *21*, 2243–2262. [CrossRef]
84. Nansé, G.; Papirer, E.; Fioux, P.; Moguet, F.; Tressaud, A. Fluorination of carbon blacks: An X-ray photoelectron spectroscopy study: I. A literature review of XPS studies of fluorinated carbons. XPS investigation of some reference compounds. *Carbon* **1997**, *35*, 175–194. [CrossRef]
85. Pereira, A.G.; Muniz, E.C.; Hsieh, Y.L. Chitosan-sheath and chitin-core nanowhiskers. *Carbohydr. Polym.* **2014**, *107*, 158–166. [CrossRef] [PubMed]
86. Oun, A.A.; Rhim, J.W. Preparation of multifunctional chitin nanowhiskers/ZnO-Ag NPs and their effect on the properties of carboxymethyl cellulose-based nanocomposite film. *Carbohydr. Polym.* **2017**, *169*, 467–479. [CrossRef] [PubMed]
87. Böke, H.; Akkurt, S.; Özdemir, S.; Gökturk, E.H.; Saltik, E.N.C. Quantification of $CaCO_3$–$CaSO_3 \cdot 0.5H_2O$–$CaSO_4 \cdot 2H_2O$ mixtures by FTIR analysis and its ANN model. *Mater. Lett.* **2004**, *58*, 723–726. [CrossRef]
88. Al Dabbas, M.; Eisa, M.Y.; Kadhim, W.H. Estimation of gypsum-calcite percentages using a fourier transform infrared spectrophotometer (FTIR), in Alexandria Gypsiferous soil-Iraq. *Iraqi J. Sci.* **2014**, *55*, 1916–1926.
89. Morris, R.J. Infrared spectrophotometric analysis of calcium sulfate hydrates using internally standardized mineral oil mulls. *Anal. Chem.* **1963**, *35*, 1489–1492. [CrossRef]
90. RRUFF Data Base. Gypsum. Available online: https://rruff.info/gypsum/display=default/R040029 (accessed on 6 April 2024).
91. Kutschera, M.; Nicoleau, L.; Bräu, M. Nano-optimized Construction Materials by Nano-seeding and Crystallization Control. In *Nanotechnology in Civil Infrastructure*; Gopalakrishnan, K., Birgisson, B., Taylor, P., Attoh-Okine, N., Eds.; Springer: Berlin/Heidelberg, Germany, 2011; pp. 195–205. [CrossRef]
92. Shukla, J.; Mehta, M.J.; Kumar, A. Effect of ionic liquid additives on the solubility behavior and morphology of calcium sulfate dihydrate (gypsum) in the aqueous sodium chloride system and physicochemical solution properties at 30 °C. *J. Chem. Eng. Data* **2018**, *63*, 2743–2752. [CrossRef]
93. Moussout, H.; Ahlafi, H.; Aazza, M.; Bourakhouadar, M. Kinetics and mechanism of the thermal degradation of biopolymers chitin and chitosan using thermogravimetric analysis. *Polym. Degrad. Stab.* **2016**, *130*, 1–9. [CrossRef]
94. *ISO 14040*; Environmental Management, Life Cycle Assessment, Principles and Framework. ISO: Geneva, Switzerland, 2006. Available online: https://www.iso.org/standard/37456.html (accessed on 10 April 2024).
95. *ISO 14044*; Environmental Management, Life Cycle Assessment, Requirements and Guidelines. ISO: Geneva, Switzerland, 2006. Available online: https://www.iso.org/standard/38498.html (accessed on 10 April 2024).
96. Cárdenas, G.; Cabrera, G.; Taboada, E.; Miranda, S.P. Chitin characterization by SEM, FTIR, XRD, and ^{13}C cross polarization/mass angle spinning NMR. *J. Appl. Polym. Sci.* **2004**, 1876–1885. [CrossRef]

97. Kaya, M.; Mujtaba, M.; Ehrlich, H.; Salaberria, A.M.; Baran, T.; Amemiya, C.T.; Galli, R.; Akyuz, L.; Sargin, I.; Labidi, J. On chemistry of γ-chitin. *Carbohydr. Polym.* **2017**, *176*, 177–186. [CrossRef] [PubMed]
98. Akpan, E.I.; Gbenebor, O.P.; Adeosun, S.O. Synthesis and characterisation of chitin from periwinkle (*Tympanotonus fusatus* (L.)) and snail (*Lissachatina fulica* (Bowdich)) shells. *Int. J. Biol. Macromol.* **2018**, *106*, 1080–1088. [CrossRef]

molecules

MDPI

Article

Potential Use of Andean Tuber Waste for the Generation of Environmentally Sustainable Bioelectricity

Segundo Rojas-Flores [1,*], Magaly De La Cruz-Noriega [1], Luis Cabanillas-Chirinos [2], Nélida Milly Otiniano [1], Nancy Soto-Deza [1], Nicole Terrones-Rodriguez [1] and Mayra De La Cruz-Cerquin [1]

[1] Institutos y Centros de Investigación de la Universidad Cesar Vallejo, Universidad Cesar Vallejo, Trujillo 13001, Peru; mdelacruzn@ucv.edu.pe (M.D.L.C.-N.); notiniano@ucv.edu.pe (N.M.O.); nsoto@ucv.edu.pe (N.S.-D.); nterronesr@ucv.edu.pe (N.T.-R.); mdelacruz@ucv.edu.pe (M.D.L.C.-C.)

[2] Investigación Formativa e Integridad Científica, Universidad César Vallejo, Trujillo 13001, Peru; lcabanillas@ucv.edu.pe

* Correspondence: segundo.rojas.89@gmail.com

Abstract: The growing demand for agricultural products has increased exponentially, causing their waste to increase and become a problem for society. Searching for sustainable solutions for organic waste management is increasingly urgent. This research focuses on considering the waste of an Andean tuber, such as Olluco, as a fuel source for generating electricity and becoming a potential sustainable energy source for companies dedicated to this area. This research used Olluco waste as fuel in single-chamber microbial fuel cells using carbon and zinc electrodes. An electric current and electric potential of 6.4 ± 0.4 mA and 0.99 ± 0.09 V were generated, operating with an electrical conductivity of 142.3 ± 6.1 mS/cm and a pH of 7.1 ± 0.2. It was possible to obtain a 94% decrease in COD and an internal resistance of $24.9 \pm 2.8\ \Omega$. The power density found was 373.8 ± 28.8 mW/cm^2 and the current density was 4.96 A/cm^2. On day 14, the cells were connected in earnest, achieving a power of 2.92 V and generating enough current to light an LED light bulb, thus demonstrating the potential that Olluco waste has to be used as fuel in microbial fuel cells.

Keywords: sustainability; organic waste; Olluco; bioenergy; bioelectricity

1. Introduction

The continuous increase in the world's population has caused the demand for the production and distribution of different types of food to grow exponentially in recent years [1]. From 2000 to 2020, the generation of primary foods increased by 50%, which generated a solid economic boost for the agricultural sector [2]. This has led to increased waste generated in the distribution chain until consumption increases. In developing countries that do not have an organized system for its collection, they are heavily contaminated due to the dumping of this waste in unsuitable places [3].

On the other hand, remote places where autochthonous fruits, vegetables, or tubers are planted and harvested do not have adequate electricity supply [4]. In 2018, it was reported that 77% of the population had an energy supply; that is, 23% (1671 million people) still needed an electricity supply [5]. Due to this, many cities with this problem have decided to generate and obtain other sustainable and renewable energy [6]. For example, in 1990, no one used bioenergy as a source to obtain electricity; however, in 2022, there are reports that it produced 675.11 TWh of electrical energy according to the Energy Institute's Statistical Review of World Energy (2023) [7].

These two problems can be solved through microbial fuel cell (MFC) technology, which uses waste with organic content as fuel. It works through the conversion of chemical energy into electrical energy through the oxidation-reduction process [8]. MFCs have been investigated using waste from fruits, vegetables, and tubers; for example, Yaakop et al. (2023) used domestic organic waste as fuel, managing to generate voltage peaks and a power density of 110 mV and 0.1047 mW/m^2 [9]. Likewise, Ahmad A. (2023) also used potato waste as a substrate in his single-chamber MFCs, managing to generate 150 mV with an internal resistance of 724 Ω and a power density of 1.450 mW/m^2 [10]. Kitchen waste has also been used, generating a power density of 0.76 W/m^3 and registering an optimal operating pH of 5.74 [11]. The chemical–biological compatibility that the electrodes used must be optimal because the electrical energy is obtained from the metabolic activity of the microorganisms present in the substrate [12]; the microorganisms release electrons and these are captured by the anode electrode and travel through an external circuit to the cathode electrode. This has led to the use of different types of electrodes, the most economical being those that have obtained the best voltage, current, and power density: metallic nature electrodes [13]. Although MFCs currently have a series of difficulties scaling up, they are presented as a long-term alternative. Because the components they need for their operation are expensive, studies like these that provide alternatives to improve are vital.

In Peru, the production of Olluco (*Ullucus tuberosus*) has increased; this Andean tuber has been used for hundreds of years for gastronomy, and in recent decades due to investments in agriculture, it has begun to be produced in large quantities for export [14]. Currently, it has been reported that they have produced more than 60,000 tons in 2023, becoming one of the agricultural products with the greatest export potential. Olluco contains vitamins (A, B, and C), calcium, iron, phosphorus, a high fiber content, and a low percentage of sugar; in addition, betalains, betaxanthins, and betacyanins are Olluco's most predominant compounds [15]. The literature does not report on the use of Ullucus tuberous waste as fuel in microbial fuel cells. This research is the first report on this technology using this type of substrate.

The main objective of this research is to observe the potential of Olluco as a fuel source in single-chamber microbial fuel cells using carbon electrodes. For this, the pH, voltage, electrical conductivity, electrical current, and chemical oxygen demand (COD) values were observed for 28 days. The internal resistance of the MFC and the power density as a function of the current density were also observed, and the identification of the microorganisms on the anode electrode took place. In this way, a new and innovative way of using Olluco waste will arise, using it as fuel in MFCs for electricity generation, which could help illuminate remote places, those lacking this supply in an environmentally sustainable way.

2. Materials and Methods

2.1. Construction of the Single-Chamber Microbial Fuel Cells

The single-chamber microbial fuel cell was obtained from the Xin Tester company (Shanghai, China), which was made of silica (SiO_2). The electrodes were made of carbon (anode) and zinc (cathode) as reported by Aguero et al. (2023) [16], and a proton exchange membrane/Nafion 117 (Wilmington, DE, USA) was used. The anodic electrode's area was 30 cm^2, and the cathodic electrode's area was 15.75 cm^2; both electrodes were joined by an external circuit whose resistance was 100 Ω (see Figure 1).

2.2. Collection of Samples Used as a Substrate

The substrate used was collected from the waste of the Santo Domingo Market, Trujillo, La Libertad, Peru, whose samples were selected by the same sellers and then taken to the laboratory of the Cesar Vallejo University, collecting 1.5 kg, which was washed several times until all types of impurities acquired from the environment were eliminated. Then, it was left to dry at 25 °C for 48 h; finally, with an extractor (extractor, Labtron, LDO-B10-Orlando, FL, USA), a 550 mL liquid solution was obtained. This solution was used as a substrate and finally used in the microbial fuel cells. The initial parameters are shown in Table 1.

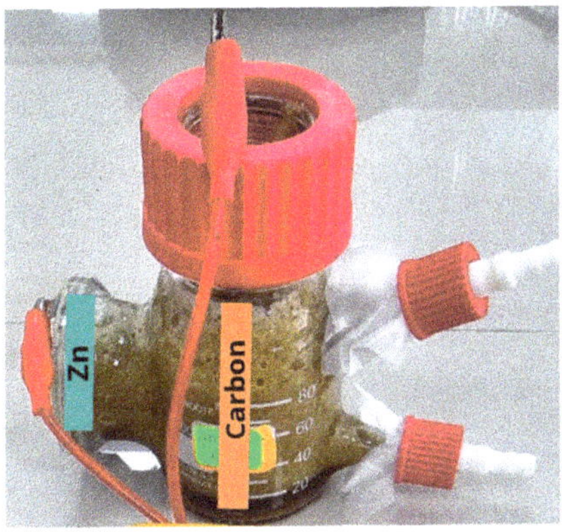

Figure 1. MFC design used.

Table 1. Initial parameters of the Olluco waste solution.

Parameters	Values
Temperature (°C)	21.5 ± 0.5
Electrical Conductivity (mS/cm)	67.8 ± 1.7
Dissolved Oxygen (mg/L)	198.2 ± 2.5
Total Dissolved Solids (mg/L)	485.7 ± 4.4
Chemical Oxygen Demand (mg/L)	987.5 ± 10.3
pH	4.2

2.3. *Characterization of the Electrochemical Parameters of the Microbial Fuel Cells*

The voltage and electrical current values were monitored for 28 days using a digital multimeter (Truper MUT—830 Digital Multimeter, Truper, Campeche, Mexico) and an external resistance of 100 Ω. The COD (chemical oxygen demand) values were measured using the closed reflux colorimetric method according to the NTP 360.502:2016 standard [16]. The internal resistance values of the MFC were carried out using an energy sensor (Vernier ±30 V and ±1000 mA). The PD (power density) and CD (current density) values were obtained with Nafion 117 using the formula PD = $V^2_{cell}/(R_{ext.}A)$ and CD = $V_{cell}/(R_{ext.}A)$ where the V_{cell} is the voltage of 118 and the MFC and Rext are 0.3 (±0.1), 3 (±0.6), 10 (±1.3), 50 (±8.7), 100 (±9.3), 220 (±13), 460 (±23.1), 119 531 (±26.8), 700 (±40.5), and 1000 (±50.6) Ω [16]. Three MFCs were carried out, where each MFC contained 100 mL of crushed Olluco waste (Olluco residues). The average value and standard deviation were calculated from the results obtained.

2.4. *Isolation and Molecular Identification of Electrogenic Bacteria*

The bacteria were isolated from the anodic chamber using the swabbing technique to recover the most significant number of bacteria. Using the streaking technique, the sample contained in the swab was sown on the surface of the Petri dishes with Nutrient Agar and Mac Conkey agar [17]. The seeded plates were incubated at 35 °C × 24 h; after the incubation time, the growth of colonies was observed. Colonies with different cultural characteristics (color, shape, and the macroscopic appearance of the colony) were then replicated in slant tubes with nutrient agar.

Before performing molecular identification, Gram staining was performed to verify the purity of the strains. The already-numbered axenic cultures were sent to the external laboratory ECOBOTECH LAB SAC for molecular identification through 16S rRNA gene sequencing.

3. Results and Analysis

Figure 2a shows the electrical potential values increase from the 1st day (0.05 ± 0.01 V) to the 14th day (0.99 ± 0.09 V) and then slowly decay until the last day (0.62 ± 0.14 V). The progressive increase in the electrical potential shown is due to the potential differential originating between the electrodes in the oxidation-reduction process that occurs within the MFC. Over time, the components necessary to initiate the oxidation-reduction process begin to decrease, which are the values shown on the monitoring days [18]. It has been reported that the use of glucose in the substrate stimulates the increase in the electrical potential; it has been possible to increase from 66.6 mV at 2.78 mM of glucose to 96.4 mV at 5.56 mM because foods with glucose concentration will consume various competitive metabolisms (methanogenesis and fermentation) by inhibiting other components of the substrate [19]. In their research, Kalagbor et al. (2020) managed to generate voltage peaks of 4.2, 3.1, and 3.0 V using tomato, banana, and pineapple waste as a substrate in single-chamber MFCs, mentioning that the decrease in potential values is due to the reduction in the organic matter available in the substrate [20]. Kebaili et al. (2021), in their double-chamber MFCs, used fruit waste leachates as a substrate, managing to generate peaks of 260 mV, mentioning that the growth of sucrose in the former allowed high voltage values, while the accumulation of metabolic waste produced the decrease in the electric potential [21]. Likewise, Rahman et al. (2021), in their single-chamber MFC, used orange, banana, and mango waste as substrates, managing to generate peaks of 259, 255, and 320 mV, mentioning that the substrates that contained a more significant amount of glucose were those that showed higher voltages [22]. The electric current values increased from the 1st day (0.40 ± 0.05 mA) to the 14th day (6.4 ± 0.4 mA) and then decreased until the 28th day (3.5 ± 0.6 mA). The microbes present in each substrate act as biocatalysts, stimulating the degradation of the organic matter present, thus generating electrons, which are captured by the anode electrode and travel through the external circuit to the cathode electrode, thus generating electricity [23]. Yahoob et al. (2022) used local waste (rambutan, langsat, and mango) from Malaysia, managing to generate 0.026 mA using mango as a substrate, attributing the decrease in electrical values to the decrease in carbon in the substrate [24]. Blueberry waste has also been used as substrates in single-chamber MFCs, managing to generate peaks of 1.130 ± 0.018 mA, attributing the rapid increase in electrical values to the high presence of dissolved oxygen and the good formation of the anodic biofilm responsible for capturing the electrons generated in the medium [25]. Naveenkumar M. and Senthilkumar K. (2021) used tannery wastewater to generate peaks of 0.43 mA, attributing the high electron release values to the low resistances shown by the MFC [26]. The existence of a relationship between the mechanisms and the degradation pathways of organic matter is revealed by electrogenic microbes that use two types of metabolisms (anaerobic and fermentative), where one uses anaerobic respiration and the other produces primary redox metabolites. The fermentative metabolism generates small organic acids and alcohols, which are used in anaerobic respiration. In contrast, the sugars that can be found in the substrate can be used up in the fermentation process as in methanogenesis [11,19,23].

The pH values are shown in Figure 3a, observing an increase in values from the first (4.2) to the last (8.4 ± 0.6) day, with the optimal operating pH of the MFC of 7.1 ± 0.2. The generation of electric current depends on the different mechanics that microbes have for the generation of electrons due to their different forms, and these microbes grow at specific pH values, which is why the standardization of an optimal pH in each MFC is important [27]. Rincon et al. (2022) standardized its optimal operating pH at 7.2, using banana waste as substrates in their single-chamber MFC, managing to generate voltage and electric current peaks of 300 mV and 0.2867 mA, attributing the pH value to the state of maturity of the banana used and its components (oxalic, citric, and malic) [28]. Molasses has been used as a substrate in single-chamber MFCs, achieving an optimal pH of 5.25 ± 0.12, and it was shown that the increase in pH value causes the performance of the MFC to decrease due to the variation in the ionization state of the amino and carboxyl functional groups that are present in the bacterial biofilm of the anode electrode [29]. Verma M. and Mishra V.

(2023) used banana peel as fuel in their MFC, obtaining an optimal pH of 7.4, showing a maximum voltage and current of 600 mV and 0.30 mA [30].

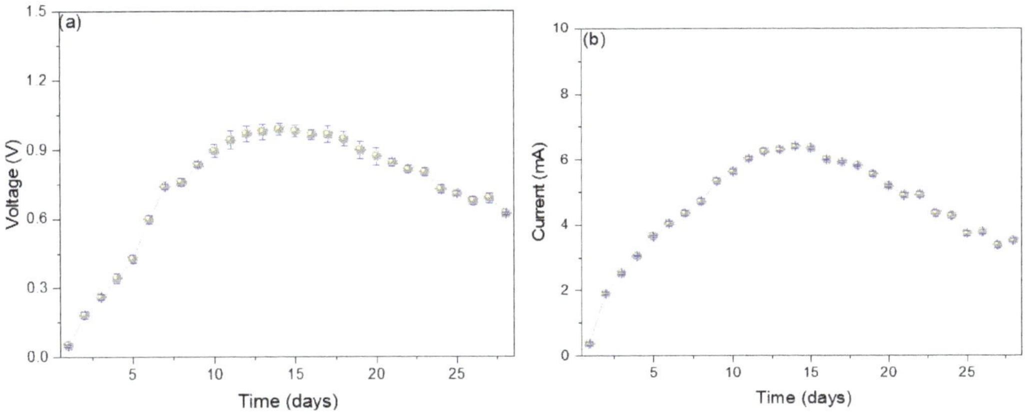

Figure 2. Monitoring values of (**a**) voltage and (**b**) electric current of reverse microbial fuel cells.

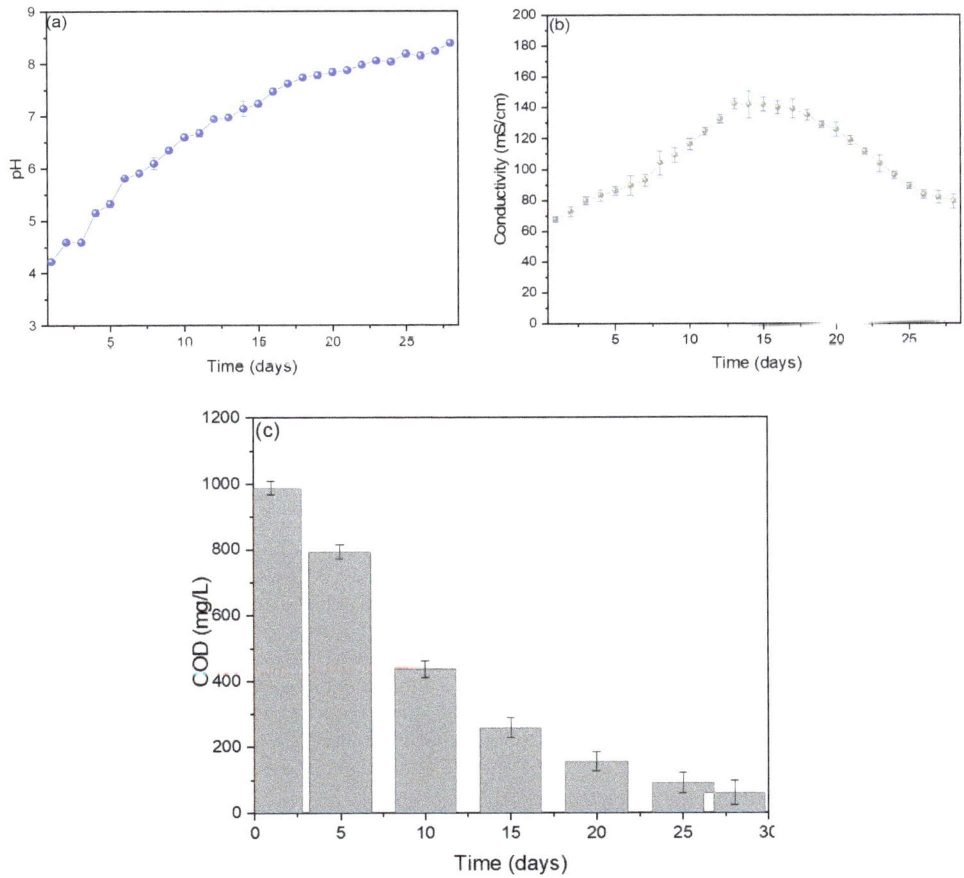

Figure 3. Monitoring of (**a**) pH, (**b**) conductivity, and (**c**) COD values of reverse microbial fuel cells.

Figure 3b shows that the electrical conductivity values increased from the 1st day (67.8 ± 1.7 mS/cm) to the 13th day (142.3 ± 6.1 mS/cm) and then decreased slightly until the 28th day (79.3 ± 8.8 mS/cm). The ions released in the first days due to oxidation-reduction reactions caused the substrate to increase its conductivity. In contrast, the sedimentation of organic compounds in the last few days is responsible for decreased electrical conductivity [31]. Similar behaviors have been reported in other investigations, mentioning that the variations in electrical conductivity depend on the internal resistance of the MFCs, which vary due to the behavior of the microorganisms in generating electric current [32]. The measured COD values are shown in Figure 3c, where a 94% decrease can be observed on the last day (60.3 ± 36.5 mg/L) from its initial value (987.5 ± 10.3 mg/L), showing a decrease of 74% for the 14th day (258.2 ± 30.6 mg/L), which was when the maximum peaks of voltage and electrical current were observed. It has been shown that there is a relationship between the decrease in COD and the values of the electric current; this is due to the activity of electron-producing microbes, which are diminished by the consumption of the organic charge in the substrate [33]. For example, Din et al. (2020) used potato waste as substrates in their single-chamber MFCs, observing that the degradation rate is higher than the hydrolysis rate, which could be due to the active oxidation of the substrate by the microorganisms present in the substrate [34]. Toczyłowska-Mamińska R. and Mamiński M. (2023) mention in their research that the aeration of the cathode chamber increases COD reduction and power density [35].

The maximum power density observed was 373.8 ± 28.8 mW/cm^2 at a current density of 4.96 A/cm^2, with a peak voltage of 901.1 ± 19.8 mV, as shown in Figure 4a. Din et al. (2023) managed to generate a PD_{max} of 0.3714 mW/cm^2 for a CD of 0.235 mA/cm^2 using potato waste as fuel, deducing that the formation of the thickness of the anodic biofilm affects the PD values; the biofilm is influenced by the temperature at which the MFC operates [36]. Du H. and Shao Z. (2022) used potato peels and activated sludge as substrates, generating PD_{max} peaks at 14.1 mW/m^2 for a CD of 320.1 mA/m^2, deducing that increasing hydrolysis rates improves the performance of the redox reactions of the soluble fractions present and promote electron transport, improving PD values [37]. Radeef A. and Ismail Z. (2021) used waste from the potato chip processing industry as a substrate in their MFCs, observing a PD_{max} of 434.8 mW/m^2 in the CD of 1165.6 mA/m, mentioning that the distance between the electrodes interferes with the results of the PD, as well as the size and material of each electrode used in the MFC [38]. The use of metallic electrodes facilitates the passage of electrons through the electrode area and between the chambers (anodic and cathodic) due to the high inherent electrical conductivity of the material [39], which explains the high power density values observed in this investigation. Similarly, Sonawane et al. (2024) developed their microbial fuel cells using organic waste, successfully generating 0.17 V and 400 mA/m$_2$ of electrical potential and maximum power density; also, a 60% reduction in COD was observed, and it was possible to reduce total solids by 82.83% [11]. Ohm's law (V = RI) is used in Figure 4b to calculate the internal resistance of the MFC; for this, the values of the electric potential in the "Y" axis and in "X" were adjusted values of the electric current, whose slope of the linear fit is the internal resistance of the MFC. The internal resistance (R_{int}) calculated was 24.9 ± 2.8 Ω; this low value would demonstrate the high voltage, electric current, electrical conductivity, and power density shown in this research. Choudhury et al. (2021) conclude that $R_{int.}$ values influence microbes' metabolic processes, biofilm formation, and attachment to the electrode [40]. Seed et al. (2022) used leachate from landfills and municipal wastewater as a substrate in their MFC, mentioning the R_{int} values. They increase when there are poor degradation kinetics of organic matter, causing ohmic resistance and anodic depletion to increase [41]. Tamilarasan et al. (2024) calculate an $R_{int.}$ of 95.257 Ω using petrochemical effluents as substrates, mentioning that the anode electrode, electrolyte, and membrane influence microbial adhesion and electron transfer and that materials with a high surface area can reduce internal resistance by promoting efficient electron transport [42]. Lawson et al. (2020) managed to calculate an $R_{int.}$ of 62 ± 4 Ω in their MFCs using ferricyanide as substrate, mentioning that any O_2 diffusion in the analyte does not affect the resistance of the anode [43]. The extraction of the genetic material and the sequencing were performed using the software MEGA (Version 4.0, Molecular Evolutionary Genetics Analysis), shown in Table 1.

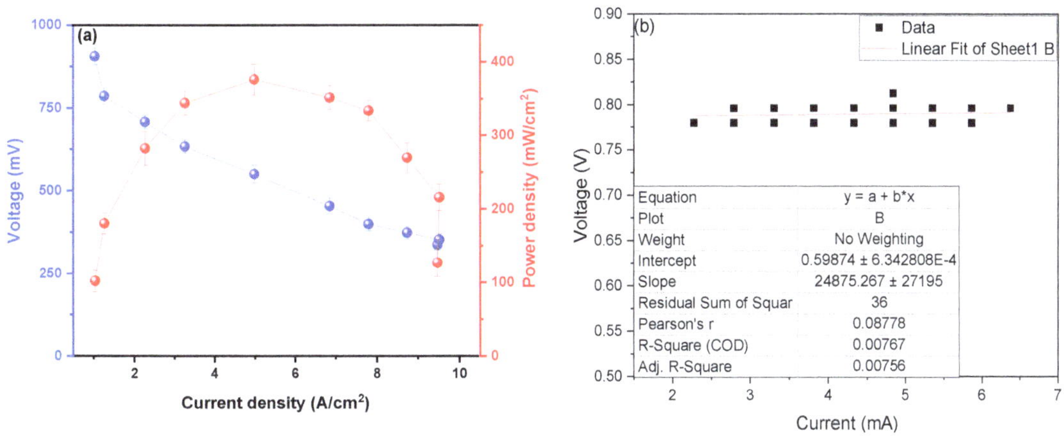

Figure 4. Values of (**a**) power density as a function of current density and (**b**) internal resistance.

In Table 2, we can see the identified microorganisms, which were two strains of *Stenotrophomonas maltophilia* with identity percentages of 99.32 and 99.59%. The results obtained are similar to those obtained by other authors, who found this microorganism as part of a microbial consortium that generated maximum peaks of 4.459 ± 0.0608 mA and 0.991 ± 0.02 V for current and voltage in onion residues [17]. In addition, *S. maltophilia* is cosmopolitan because it can be found in a wide variety of habitats, including the *rhizosphere* of crop soils [44]. On the other hand, it has been reported that the *S. maltophilia* bacteria form an anodic biofilm (above the biofilm) with a good structure of extracellular polymeric substances (EPSs), which protects antimicrobials and foreign agents whose inhibitory agents are the protein bifunctional phosphoglucomutase/phosphomannose y glucose-1-phosphate [45]. Likewise, Shu Hui Liu et al. (2022) reported that *S. maltophilia* could remove copper and generate bioelectricity in microbial fuel cells using sediments with high concentrations of copper [46]. Meanwhile, Gnanarathinam et al. (2023) found *S. maltophilia* to be part of a microbial consortium that has the property of reducing Cr^{+6} to Cr^{+3} using a dual-chamber microbial fuel cell (MFC) [47]. Alvarado-Gutierrez et al. (2020) demonstrated the biotechnological capacity of *S. maltophilia* to degrade the fungicide carbendazim, considered an ecotoxic contaminant that is frequently found in water reservoirs [48].

Table 2. Electrogenic bacteria present in the MFC anode-containing Olluco residues.

Identified Species	pb	BLAST	
		Identity (%)	Accession Number
Stenotrophomonas maltophilia	1459	99.32%	NR_112030.1
Stenotrophomonas maltophilia	1458	99.59%	NR_041577.1

The Olluco waste is seen in Figure 5, where the crushed waste was placed in single-chamber microbial fuel cells, and when connected in series, they managed to generate an electrical potential of 2.92 V on the 14th day, enough to ignite an LED spotlight, demonstrating that the waste of this tuber from the Andes can have great potential to be used as fuel and generate electricity. Although these electrical values still need to be higher to apply to larger scales, the voltage and current values found in this research are quite high, which would not give hope due to the values and costs reported by other research [49]. It has been reported that for the treatment of 113.14 m^3 of waste, there is a cost of USD 4.2 million [50]. The treatment of 1 m^3 of organic waste costs around USD 289, and the expense is more significant in the manufacture of electrodes [51]. Another way to improve the economic viability of microbial fuel cells is by using the sediments generated after waste treatment and the generation of electrical energy as fertilizer or some other type of application [52]. It has also been reported that the use of microbial fuel cells as bioremediations or for the reduction of heavy metals are other applications that can be given to this technology, and it has already been shown that the excess concentration of heavy metals reduces the growth of microbes in cells [53]. The optimal concentration level of metals would also have to be standardized to improve the performance of the systems by reducing maintenance and operating costs and time [54].

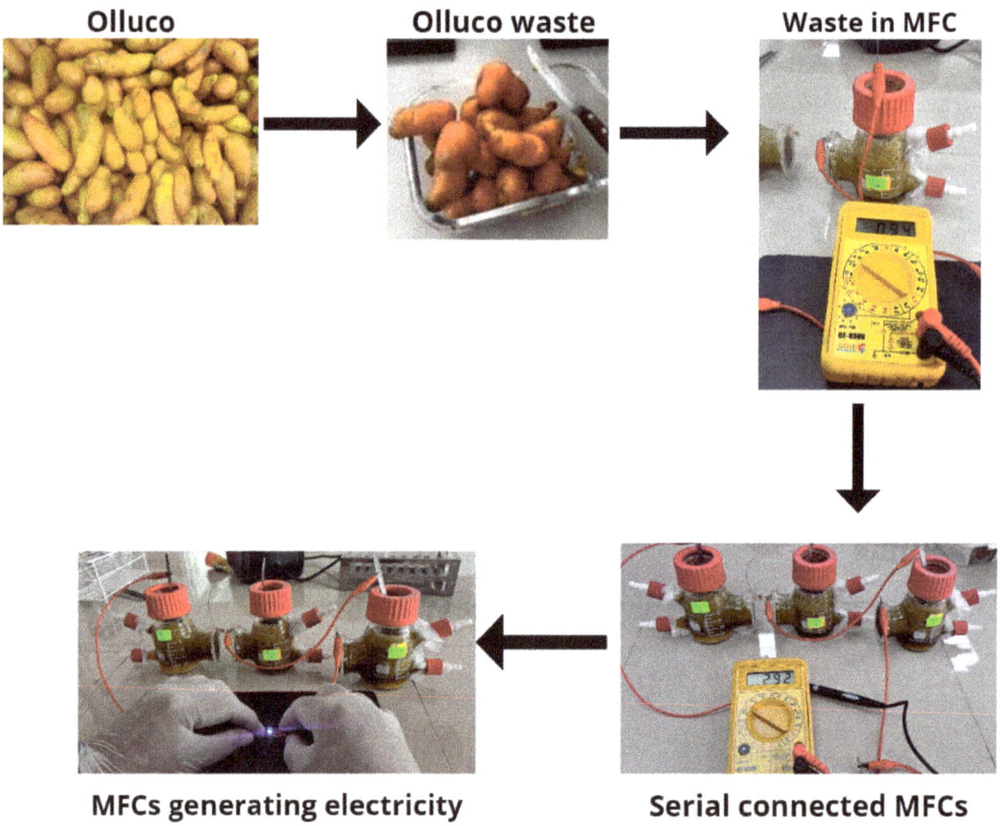

Figure 5. Schematization of the electrical energy generation process through waste.

In recent years, research on the use of different organic wastes has increased abruptly because the agroindustry has begun to produce large quantities of waste and has the need to reuse their waste efficiently and profitably. Table 3 shows recent research on waste in microbial fuel cells for bioelectricity production compared to our research. In the comparison of results, it can be observed that electrodes of a metallic nature or with metallic components significantly increase power density and current density compared to those that use only carbon, graphite, or some other derivative. This shows that our research follows the trend of using metal electrodes combined with carbon to take advantage of the porosity of the latter. Furthermore, it has been observed that the use of MFCs connected in series to increase the output voltage of the MFCs has started to gain importance; for example, Rokhim et al. (2024) used four MFCs connected in series, generating a maximum voltage of 5.53 V [55]. Likewise, using bianodes in MFCs has also been reported to improve the efficiency of the cells because by placing two anodes, the adhesion of electrons would increase, thus obtaining a greater flow of electrons. However, a problem has been observed in the reaction of these electrodes because the cathode tends to show faster wear compared to cells with a single anode [56,57]. Other researchers have reported that the coating of the metal electrode with carbon or graphite improves the adhesion of electrons because the porosity shown by these two materials adheres a more significant number of electrons than when passing to the metallic material if it repowers because the metal or non-metallic nanoparticles embedded in the carbon or graphite electrode show lower resistance to the passage of electrons [58,59].

Table 3. Comparison of electrical parameters with other types of organic waste.

Substrate Type	Types of MFC	$V_{max.}$ (V)	PD (mW/m^2)	CD (mA/m^2)	Electrodes	Ref.
Rice, vegetable, and fruit wastes	Single-chamber	5.53	12.719	2.300	Cu and Mg	[55]
Mango peel extract	Single-chamber	0.102	0.099	31.57	Graphite and carbon	[56]
Banana peel waste	Dual-chamber	602 ± 1.5	2.2 ± 0.1	129.4 ± 1.0	Stainless steel and activated carbon	[29]
Sweet lemon peels	Single-chamber	102.5	1.100	31.569	Carbon and graphite	[57]
Coriander waste	Single-chamber	882 ± 154	3 04.33 ± 16.51	506	Copper and zinc	[58]
Oily kitchen waste	Single-chamber cubic	400.011	25.156	35.213	Carbon and graphite	[59]
Olluco waste	Single-chamber	0.99 ± 0.09	373.8 ± 28.8	4.96	Cu and Carbon	This research

4. Conclusions

This research successfully demonstrated that Olluco waste can be used as fuel in single-chamber MFCs using carbon and zinc electrodes managing to generate peaks of electric potential and electric current of 0.99 ± 0.09 V and 6.4 ± 0.4 mA, which operated at an optimal pH of 7.1 ± 0.2 and whose maximum electrical conductivity was 142.3 ± 6.1 mS/cm, on the fourteenth day. In addition, the chemical oxygen demand was reduced by 94% in the 28 days of operation of the MFCs, while the internal resistance calculated was 24.9 ± 2.8 Ω. The maximum power density calculated was 373.8 ± 28.8 mW/cm^2 and the current density was 4.96 A/cm^2; in addition, the three microbial fuel cells used were connected in series on day 14, managing to generate a potential of 2.92 V, turning on an LED spotlight. This research demonstrates the potential that Olluco waste has to be used as fuel, as well as the solution for farmers who grow and harvest this tuber because the areas where this work is carried out usually lack electrical energy. For future work, it is recommended to carry out a technical study on the application of the circular economy using Olluco waste for better sustainable development from the point of view of environmental, social, and economic development to give a descriptive point of view for the future of the sustainable bioenergy area. Also, this prototype must be scaled up with a more significant amount of Olluco waste, increasing the size of the electrodes and standardizing the optimal operating pH to the value found in this research. Isolated microbes should also be used as biocatalysts to increase voltage, electrical current, and electrical potential values in microbial fuel cells.

Author Contributions: Conceptualization, S.R.-F.; methodology, L.C.-C.; validation, N.M.O. and N.S.-D.; formal analysis, S.R.-F. and M.D.L.C.-N.; investigation, S.R.-F., M.D.L.C.-C. and N.T.-R.; data curation, M.D.L.C.-N. and N.T.-R.; writing—original draft preparation, L.C.-C.; writing—review and editing, S.R.-F.; project administration, S.R.-F. and L.C.-C. All authors have read and agreed to the published version of the manuscript.

Funding: This research has been financed by the Universidad Cesar Vallejo, project code No. P-2023-113.

Institutional Review Board Statement: Not applicable.

Informed Consent Statement: Not applicable.

Data Availability Statement: Data are contained within the article.

Conflicts of Interest: The authors declare no conflicts of interest.

References

1. Lee, S.H.; Lum, W.C.; Boon, J.G.; Kristak, L.; Antov, P.; Pędzik, M.; Rogoziński, T.; Taghiyari, H.R.; Lubis, M.A.R.; Fatriasari, W.; et al. Particleboard from agricultural biomass and recycled wood waste: A review. *J. Mater. Res. Technol.* **2022**, *20*, 4630–4658. [CrossRef]
2. Phiri, R.; Rangappa, S.M.; Siengchin, S.; Oladijo, O.P.; Dhakal, H.N. Development of sustainable biopolymer-based composites for lightweight applications from agricultural waste biomass: A review. *Adv. Ind. Eng. Polym. Res.* **2023**, *6*, 436–450. [CrossRef]
3. Kober, T.; Schiffer, H.W.; Densing, M.; Panos, E. Global energy perspectives to 2060–WEC's World Energy Scenarios 2019. *Energy Strategy Rev.* **2020**, *31*, 100523. [CrossRef]
4. Li, L.; Lin, J.; Wu, N.; Xie, S.; Meng, C.; Zheng, Y.; Wang, X.; Zhao, Y. Review and outlook on the international renewable energy development. *Energy Built Environ.* **2022**, *3*, 139–157. [CrossRef]
5. Bogdanov, D.; Ram, M.; Aghahosseini, A.; Gulagi, A.; Oyewo, A.S.; Child, M.; Caldera, U.; Sadovskaia, K.; Farfan, J.; Barbosa, L.D.S.N.S.; et al. Low-cost renewable electricity as the key driver of the global energy transition towards sustainability. *Energy* **2021**, *227*, 120467. [CrossRef]
6. Liu, C.; Li, Q. Air Pollution, Global Warming and Difficulties to Replace Fossil Fuel with Renewable Energy. *Atmos. Clim. Sci.* **2023**, *13*, 526–538. [CrossRef]
7. Segundo, R.F.; Magaly DL, C.N.; Luis, C.C.; Otiniano, N.M.; Soto-Deza, N.; Terrones-Rodriguez, N.; Mayra DL, C.C. Obtaining Sustainable Electrical Energy from Pepper Waste. *Sustainability* **2024**, *16*, 3448. [CrossRef]
8. Yaakop, A.S.; Ahmad, A.; Hussain, F.; Oh, S.E.; Alshammari, M.B.; Chauhan, R. Domestic Organic Waste: A Potential Source to Produce the Energy via a Single-Chamber Microbial Fuel Cell. *Int. J. Chem. Eng.* **2023**, *2023*, 2425735. [CrossRef]
9. Ahmad, A. Conventional vegetable waste: A potential source for the high performance of benthic microbial fuel cells. *Biomass Convers. Biorefinery* **2023**, *8*, 1–13. [CrossRef]
10. Gurjar, R.; Behera, M. Optimization and modelling of volatile fatty acid generation in a leachate bed reactor for utilization in microbial fuel cells. *Water Environ. J.* **2023**, *37*, 581–593. [CrossRef]
11. Sonawane, A.V.; Rikame, S.; Sonawane, S.H.; Gaikwad, M.; Bhanvase, B.; Sonawane, S.S.; Mungray, A.K.; Gaikwad, R. A review of microbial fuel cell and its diversification in the development of green energy technology. *Chemosphere* **2024**, *350*, 141127. [CrossRef] [PubMed]
12. Christwardana, M.; Hadiyanto, H.; Motto, S.A.; Sudarno, S.; Haryani, K. Performance evaluation of yeast-assisted microalgal microbial fuel cells on bioremediation of cafeteria wastewater for electricity generation and microalgae biomass production. *Biomass Bioenergy* **2020**, *139*, 105617. [CrossRef]
13. Al-Badani, M.A.; Chong, P.L.; Lim, H.S. Enhancing microbial fuel cell performance with carbon powder electrode modifications for low-power sensors modules. *Int. J. Renew. Energy Dev.* **2024**, *13*, 80–87. [CrossRef]
14. Pico, C.; De la Vega, J.; Tubón, I.; Arancibia, M.; Casado, S. Nanoscopic Characterization of Starch Biofilms Extracted from the Andean Tubers Ullucus tuberosus, Tropaeolum tuberosum, Oxalis tuberosa, and Solanum tuberosum. *Polymers* **2022**, *14*, 4116. [CrossRef] [PubMed]
15. Paniagua-Zambrana, N.Y.; Bussmann, R.W. Ullucus tuberosus Caldas Basellaceae. In *Ethnobotany of the Andes*; Springer International Publishing: Cham, Switzerland, 2020; pp. 1853–1855.
16. Rojas-Flores, S.; De La Cruz-Noriega, M.; Benites, S.M.; Delfín-Narciso, D.; Luis, A.S.; Díaz, F.; Luis, C.C.; Moises, G.C. Electric Current Generation by Increasing Sucrose in Papaya Waste in Microbial Fuel Cells. *Molecules* **2022**, *27*, 5198. [CrossRef] [PubMed]
17. Gupta, S.; Srivastava, P.; Patil, S.A.; Yadav, A.K. A comprehensive review on emerging constructed wetland coupled microbial fuel cell technology: Potential applications and challenges. *Bioresour. Technol.* **2021**, *320*, 124376. [CrossRef] [PubMed]
18. Obileke, K.; Onyeaka, H.; Meyer, E.L.; Nwokolo, N. Microbial fuel cells, a renewable energy technology for bio-electricity generation: A mini-review. *Electrochem. Commun.* **2021**, *125*, 107003. [CrossRef]
19. Rossi, R.; Hall, D.M.; Wang, X.; Regan, J.M.; Logan, B.E. Quantifying the factors limiting performance and rates in microbial fuel cells using the electrode potential slope analysis combined with electrical impedance spectroscopy. *Electrochim. Acta* **2020**, *348*, 136330. [CrossRef]
20. Kalagbor, I.A.; Azunda, B.I.; Igwe, B.C.; Akpan, B.J. Electricity generation from waste tomatoes, banana, pineapple fruits and peels using single chamber microbial fuel cells (SMFC). *Waste Manag. Xenobiotics* **2020**, *3*, 142.
21. Kebaili, H.; Kameche, M.; Innocent, C.; Ziane, F.; Sabeur, S.; Sahraoui, T.; Ouis, M.; Zerrouki, A.; Charef, M. Treatment of fruit waste leachate using microbial fuel cell: Preservation of agricultural environment. *Acta Ecol. Sin.* **2021**, *41*, 97–105. [CrossRef]
22. Rahman, W.; Yusup, S.; Mohammad, S.N.A. Screening of fruit waste as substrate for microbial fuel cell (MFC). In *AIP Conference Proceedings*; AIP Publishing: Melville, NY, USA, 2021; Volume 2332, No. 1.
23. Malekmohammadi, S.; Ahmad Mirbagheri, S. A review of the operating parameters on the microbial fuel cell for wastewater treatment and electricity generation. *Water Sci. Technol.* **2021**, *84*, 1309–1323. [CrossRef] [PubMed]

24. Yaqoob, A.A.; Guerrero–Barajas, C.; Ibrahim, M.N.M.; Umar, K.; Yaakop, A.S. Local fruit wastes driven benthic microbial fuel cell: A sustainable approach to toxic metal removal and bioelectricity generation. *Environ. Sci. Pollut. Res.* **2022**, *29*, 32913–32928. [CrossRef] [PubMed]
25. Rojas-Flores, S.; Benites, S.M.; De La Cruz-Noriega, M.; Cabanillas-Chirinos, L.; Valdiviezo-Dominguez, F.; Álvarez, M.A.Q.; Vega-Ybañez, V.; Angelats-Silva, L. Bioelectricity production from blueberry waste. *Processes* **2021**, *9*, 1301. [CrossRef]
26. Naveenkumar, M.; Senthilkumar, K. Microbial fuel cell for harvesting bio-energy from tannery effluent using metal mixed biochar electrodes. *Biomass Bioenergy* **2021**, *149*, 106082. [CrossRef]
27. Zafar, H. Microbial Fuel Cells: A Comparative Analysis of Operational Factors, Response Metrics, and Degradation Response to Fruit Waste Degradation. Ph.D. Thesis, University of British Columbia, Vancouver, BC, Canada, 2023.
28. Rincón-Catalán, N.I.; Cruz-Salomón, A.; Sebastian, P.J.; Pérez-Fabiel, S.; Hernández-Cruz, M.d.C.; Sánchez-Albores, R.M.; Hernández-Méndez, J.M.E.; Domínguez-Espinosa, M.E.; Esquinca-Avilés, H.A.; Ríos-Valdovinos, E.I.; et al. Banana Waste-to-Energy Valorization by Microbial Fuel Cell Coupled with Anaerobic Digestion. *Processes* **2022**, *10*, 1552. [CrossRef]
29. Flores, S.R.; Pérez-Delgado, O.; Naveda-Renny, N.; Benites, S.M.; De La Cruz–Noriega, M.; Narciso, D.A.D. Generation of Bioelectricity Using Molasses as Fuel in Microbial Fuel Cells. *Environ. Res. Eng. Manag.* **2022**, *78*, 19–27. [CrossRef]
30. Verma, M.; Mishra, V. Bioelectricity generation by microbial degradation of banana peel waste biomass in a dual-chamber *S. cerevisiae*-based microbial fuel cell. *Biomass Bioenergy* **2023**, *168*, 106677. [CrossRef]
31. Jalili, P.; Ala, A.; Nazari, P.; Jalili, B.; Ganji, D.D. A comprehensive review of microbial fuel cells considering materials, methods, structures, and microorganisms. *Heliyon* **2024**, *10*, e25439. [CrossRef]
32. De La Cruz-Noriega, M.; Benites, S.M.; Rojas-Flores, S.; Otiniano, N.M.; Vargas, A.M.S.; Alfaro, R.; Cabanillas-Chirinos, L.; Rojas-Villacorta, W.; Nazario-Naveda, R.; Delfín-Narciso, D. Use of Wastewater and Electrogenic Bacteria to Generate Eco-Friendly Electricity through Microbial Fuel Cells. *Sustainability* **2023**, *15*, 10640. [CrossRef]
33. Kacmaz, G.K.; Eczacioglu, N. The mechanism of bioelectricity generation from organic wastes: Soil/plant microbial fuel cells. *Environ. Technol. Rev.* **2024**, *13*, 76–95. [CrossRef]
34. Din, M.I.; Iqbal, M.; Hussain, Z.; Khalid, R. Bioelectricity generation from waste potatoes using single chambered microbial fuel cell. *Energy Sources Part A Recovery Util. Environ. Eff.* **2020**, *1*, 1–11. [CrossRef]
35. Toczyłowska-Mamińska, R.; Mamiński, M.Ł. Application of Microbial Fuel Cell Technology in Potato Processing Industry. *Energies* **2023**, *16*, 6581. [CrossRef]
36. Din, M.I.; Ahmed, M.; Ahmad, M.; Iqbal, M.; Ahmad, Z.; Hussain, Z.; Khalid, R.; Samad, A. Investigating the Activity of Carbon Fiber Electrode for Electricity Generation from Waste Potatoes in a Single-Chambered Microbial Fuel Cell. *J. Chem.* **2023**, *2023*, 8520657. [CrossRef]
37. Du, H.; Shao, Z. Synergistic effects between solid potato waste and waste activated sludge for waste-to-power conversion in microbial fuel cells. *Appl. Energy* **2022**, *314*, 118994. [CrossRef]
38. Radeef, A.Y.; Ismail, Z.Z. Resource recovery in potato chips processing industry: Green bioelectricity production using continuous mediatorless microbial fuel cell. *Int. J. Green Energy* **2021**, *18*, 910–919. [CrossRef]
39. Zafar, H.; Peleato, N.; Roberts, D. A comparison of reactor configuration using a fruit waste fed two-stage anaerobic up-flow leachate reactor microbial fuel cell and a single-stage microbial fuel cell. *Bioresour. Technol.* **2023**, *374*, 128778. [CrossRef] [PubMed]
40. Choudhury, P.; Ray, R.N.; Bandyopadhyay, T.K.; Basak, B.; Muthuraj, M.; Bhunia, B. Process engineering for stable power recovery from dairy wastewater using microbial fuel cell. *Int. J. Hydrogen Energy* **2021**, *46*, 3171–3182. [CrossRef]
41. Saeed, T.; Yadav, A.K.; Miah, M.J. Landfill leachate and municipal wastewater co-treatment in microbial fuel cell integrated unsaturated and partially saturated tidal flow constructed wetlands. *J. Water Process Eng.* **2022**, *46*, 102633. [CrossRef]
42. Tamilarasan, K.; Shabarish, S.; Banu, J.R.; Sharmila, V.G. Sustainable power production from petrochemical industrial effluent using dual chambered microbial fuel cell. *J. Environ. Manag.* **2024**, *351*, 119777. [CrossRef]
43. Lawson, K.; Rossi, R.; Regan, J.M.; Logan, B.E. Impact of cathodic electron acceptor on microbial fuel cell internal resistance. *Bioresour. Technol.* **2020**, *316*, 123919. [CrossRef]
44. Gil-Gil, T.; Martínez, J.L.; Blanco, P. Mechanisms of antimicrobial resistance in Stenotrophomonas maltophilia: A review of current knowledge. *Expert Rev. Anti-Infect. Ther.* **2020**, *18*, 335–347. [CrossRef]
45. An, S.Q.; Berg, G. Stenotrophomonas maltophilia. *Trends Microbiol.* **2018**, *26*, 637–638. [CrossRef]
46. Liu, S.H.; Huang, W.J.; Lin, C.W.; Zhu, T.J. Enhanced copper removal and bioelectricity generation in sediment microbial fuel cells through biostimulation and bioaugmentation. *J. Clean. Prod.* **2022**, *350*, 131458. [CrossRef]
47. Gnanarathinam, V.S.; Balakrishnan, A.; Rajendran, G.; Arivarasu, A.; Dharmalingam, S.; Balasundaram, S.T.; Venkatesan, P.N.; Perumal, E.P.U. Reduction of hexavalent chromium using sulphonated polyether ether ketone as proton exchange membrane in microbial fuel cells. *Environ. Prog. Sustain. Energy* **2023**, *42*, e14158. [CrossRef]
48. Alvarado-Gutiérrez, M.L.; Ruiz-Ordaz, N.; Galíndez-Mayer, J.; Curiel-Quesada, E.; Santoyo-Tepole, F. Degradation kinetics of carbendazim by Klebsiella oxytoca, Flavobacterium johnsoniae, and Stenotrophomonas maltophilia strains. *Environ. Sci. Pollut. Res.* **2020**, *27*, 28518–28526. [CrossRef] [PubMed]
49. Yaqoob, A.A.; Ibrahim, M.N.M.; Guerrero-Barajas, C. Modern trend of anodes in microbial fuel cells (MFCs): An overview. *Environ. Technol. Innov.* **2021**, *23*, 101579. [CrossRef]
50. Maddalwar, S.; Nayak, K.K.; Kumar, M.; Singh, L. Plant microbial fuel cell: Opportunities, challenges, and prospects. *Bioresour. Technol.* **2021**, *341*, 125772. [CrossRef] [PubMed]

51. Hoang, A.T.; Nižetić, S.; Ng, K.H.; Papadopoulos, A.M.; Le, A.T.; Kumar, S.; Hadiyanto, H. Microbial fuel cells for bioelectricity production from waste as sustainable prospect of future energy sector. *Chemosphere* **2022**, *287*, 132285. [CrossRef]
52. Yaqoob, A.A.; Al-Zaqri, N.; Alamzeb, M.; Hussain, F.; Oh, S.E.; Umar, K. Bioenergy generation and phenol degradation through microbial fuel cells energized by domestic organic waste. *Molecules* **2023**, *28*, 4349. [CrossRef]
53. Hassan, H.; Zahir, R.M.; Ahmad, M.A.; Ahmad, M.M. Current Status on Microbial Fuel Cell (MFC) Technology. In *Renewable Energy from Bio-Resources in Malaysia*; Springer: Singapore, 2022; pp. 195–212.
54. Chakraborty, I.; Sathe, S.M.; Dubey, B.K.; Ghangrekar, M.M. Waste-derived biochar: Applications and future perspective in microbial fuel cells. *Bioresour. Technol.* **2020**, *312*, 123587. [CrossRef]
55. Rokhim, D.A.; Vitarisma, I.Y.; Sumari, S.; Utomo, Y.; Asrori, M.R. Optimizing Household Wastes (Rice, Vegetables, and Fruit) as an Environmentally Friendly Electricity Generator. *J. Renew. Mater.* **2024**, *12*, 275–284. [CrossRef]
56. Aleid, G.M.; Alshammari, A.S.; Alomari, A.D.; Sa'ad Abdullahi, S.; Mohammad, R.E.A.; Abdulrahman, R.M.I. Degradation of metal ions with electricity generation by using fruit waste as an organic substrate in the microbial fuel cell. *Int. J. Chem. Eng.* **2023**, *2023*, 1334279. [CrossRef]
57. Ahmad, A.; Alshammari, M.B.; Ibrahim, M.N.M. Impact of Self-Fabricated Graphene–Metal Oxide Composite Anodes on Metal Degradation and 472 Energy Generation via a Reverse microbial fuel cell. *Processes* **2023**, *11*, 163. [CrossRef]
58. Rojas-Flores, S.; De La Cruz-Noriega, M.; Cabanillas Chirinos, L.; Nazario-Naveda, R.; Gallozzo-Cardenas, M.; Diaz, F.; Murga-Torres, E. Potential Use of Coriander Waste as Fuel for the Generation of Electric Power. *Sustainability* **2023**, *15*, 896. [CrossRef]
59. Kafaei, R.; Yazdanbakhsh, A.; Sadani, M.; Alavi, N. Bioelectricity and Biohydrogen Production Using High Solid Content of Oily-Kitchen Wastes 466 in Air Cathode Reverse microbial fuel cells. *Iran. J. Chem. Chem. Eng.* **2023**, *42*, 2895–2907.

sustainability MDPI

Review

Coconut Waste: Discovering Sustainable Approaches to Advance a Circular Economy

Fabrícia Vieira [1], Hortência E. P. Santana [1], Meirielly Jesus [2,*], Joana Santos [2], Preciosa Pires [2], Manuela Vaz-Velho [2], Daniel Pereira Silva [1,2,3,4,5,*] and Denise Santos Ruzene [1,2,3,4]

1 Northeastern Biotechnology Network, Federal University of Sergipe, São Cristóvão 49100-000, Sergipe, Brazil; fabriciaviieira@academico.ufs.br (F.V.); horteenciap@academico.ufs.br (H.E.P.S.); ruzeneds@academico.ufs.br (D.S.R.)
2 CISAS-Center for Research and Development in Agrifood Systems and Sustainability, Instituto Politécnico de Viana do Castelo, Rua Escola Industrial e Comercial de Nun'Álvares, n 34, 4900-367 Viana do Castelo, Portugal; joana@estg.ipvc.pt (J.S.); ppires@estg.ipvc.pt (P.P.); mvazvelho@estg.ipvc.pt (M.V.-V.)
3 Center for Exact Sciences and Technology, Federal University of Sergipe, São Cristóvão 49100-000, Sergipe, Brazil
4 Graduate Program in Biotechnology, Federal University of Sergipe, São Cristóvão 49100-000, Sergipe, Brazil
5 Graduate Program in Intellectual Property Science, Federal University of Sergipe, São Cristóvão 49100-000, Sergipe, Brazil
* Correspondence: meiriellyjesus@ipvc.pt (M.J.); silvadp@academico.ufs.br (D.P.S.)

Abstract: The coconut tree (*Cocos nucifera*) stands as a pivotal resource in tropical regions, playing a crucial role in both subsistence and economic activities across Asia, the Pacific Islands, and South America. While the harvesting of coconut fruit is essential for producing globally utilized edible products, such as coconut oil, by small owners and large producers around the world in the food, cosmetics, and pharmaceutical industries, concerns have arisen due to the substantial amount of agro-industrial residue generated in this process, posing environmental risks if they are not properly managed. Recognizing the environmental challenges, this paper emphasizes the transformative potential inherent in coconut waste, characterized by its lignocellulosic composition rich in lignin and multifunctional groups. By delving into the historical context of coconut economic exploration and its chemical composition, this review explores the diverse applications of coconut products, focusing on the utilization and processing of residues to generate sustainable products and byproducts. Ultimately, this comprehensive review underscores the significance of repurposing coconut waste, not only to mitigate the environmental impact but also as a valuable contributor to a circular economy, promoting the use of the lignocellulosic biomass in research and bolstering its role as a raw material in the chemical and energy sectors.

Keywords: coconut waste; circular economy; sustainability; lignocellulosic biomass; bioproducts; environmental stewardship

1. Introduction

Coconut palm tree (*Cocos nucifera*), the only recorded species of the genus Cocos, is a perennial plant with multiple uses and is widely disseminated on almost all continents [1]. It can easily adapted to and can be cultivated under many soil types and climate conditions. Nonetheless, the coconut tree grows best under a humid tropical climate, being extensively found in tropical areas, wherein it is considered one of the main crops and plays a vital role in the economic and subsistence activities of the lower Pacific Islands [2]. From coconut fruit, it is possible to generate, directly or indirectly, different value-added products with applications across several sectors, whether made from its mature or immature state and *in natura* or processed [3].

The coconut fruit is a fibrous drupe comprising five parts [4]: a solid endosperm (coconut meat or kernel) and liquid endosperm (coconut water), both known as the edible parts of the coconut, and three hierarchical layers of peel enclosing the seed: the endocarp (hard inner shell), mesocarp (a thick husk), and exocarp (outer shell) (Figure 1) [5,6]. Coconut meat is the most exploited fruit component; however, every part of the plant, from the stems to husk and shells, can be used to make value-added products and byproducts [3].

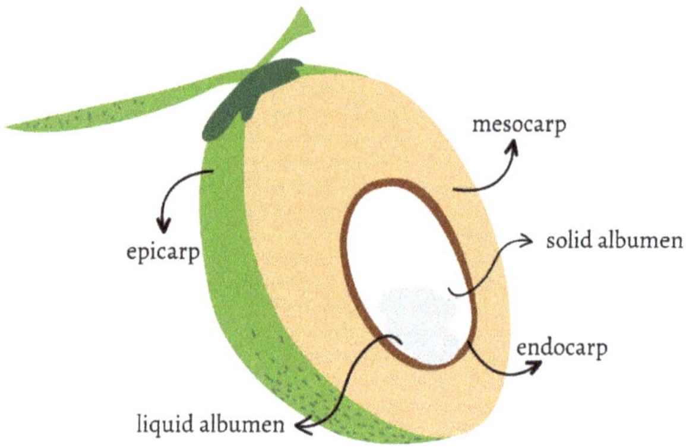

Figure 1. Constituent parts of coconut (*Cocos nucifera*).

Copra, the dried mature kernel, is the source of coconut oil, the primary marketable product of the coconut industry [6]. Coconut oil is commercialized in different forms and has wide applications in food, cosmetics, pharmaceutics, hair care products, and paint [7,8]. Other non-oil kernel applications include the obtention of desiccated coconut (grated and dehydrated coconut meat), coconut milk, coconut cream, and coconut flour. In the international coconut trade, the oil and desiccated coconut account for 60% and 17.3% of exported products and 57.4% and 15.6% of imported products, respectively [9]. The tender coconut water in fresh fruit is a nutritionally rich and refreshing liquid that has been commercialized as a natural isotonic beverage. In Brazil, the per capita consumption of coconut water is 0.74 L/habitant/year [10]. Because of the sugar content, coconut water is also used as a substrate for vinegar production [11].

A non-traditional application of coconut is the use of its residues, which have a heterogeneous chemical composition mainly composed of cellulose, hemicellulose, and lignin [4,12]. Since the mesocarp and endocarp represent about 35% and 15% of the total fruit weight [13], respectively, the processing of coconut water and meat leads to the annual generation of 1,400,980 tons of shell [14,15], which has a slow degradation rate and when mismanaged, whether by direct disposal or open burning, causes environmental pollution, which has implications for human health [16]. Therefore, a circular economy approach, involving recycling coconut residues to create valuable products, can help mitigate these negative impacts by avoiding the burning and improper disposal of these wastes [17]. By adopting this approach, coconut residues can be transformed into value-added products such as handicrafts, organic fertilizers [18], bioethanol [19], cementitious materials [20], bioadsorbents for water pollutants [21], and alternative solid fuels [22]. This not only helps reduce the amount of waste sent to landfills but it also promotes resource conservation and mitigates the effects of climate change. Thus, the circular economy offers a sustainable solution for handling coconut residues, benefiting both the environment and society as a whole [23].

Recently, to improve the sustainability and economic value of the coconut supply chain, numerous scientific studies have been conducted to develop more efficient technology pathways that allow for the exploration of the whole potential of the physicochemical properties of coconut biomass residues. The mesocarp has been evaluated as an additive to produce thermal and acoustic insulation in building construction [24], as a biosorbent to adsorb rhodamine B [25], in ethanol production [26], and to improve the mechanical properties of biocomposites [27]. Meanwhile, the endocarp, the innermost rigid layer, has been reported as a raw material for the production of porous carbon [28], the preparation of bio-additives to improve the thermal and UV-blocking properties of epoxy resins [29], and the generation of thermal energy [30].

Coconut residues play a crucial role in a circular economy, offering numerous opportunities for repurposing waste and maximizing resource efficiency. By integrating these residues into circular economy practices, such as recycling and upcycling initiatives, valuable products can be created while reducing waste generation [31]. This not only stimulates economic growth and job creation but it also contributes to environmental conservation by minimizing carbon footprints and promoting sustainable alternatives to conventional materials. Embracing coconut residues within a circular economy framework facilitates sustainable development and fosters a regenerative approach to resource management [23].

There has been an upward trend in research focusing on expanding the potential of coconut residues for new applications and improving the efficiency of a well-known conversion process. However, although remarkable advances have been achieved in this direction, many of the innovative technologies in coconut residue processing for value-added products are still at an early stage and some challenges must be overcome, such as the lack of adequate infrastructure for the collection, transportation, and processing of residues and further research to demonstrate large-scale feasibility from a technical-economic perspective. Improving and integrating the coconut lignocellulosic residue value chain into a growing economy and biorefineries are essential foe sustainability and to reduce the environmental impact of residue disposal.

This study focused on optimizing the utilization of coconut residues, highlighting the infrastructure and logistics challenges, and exploring the feasibility of developing economically viable value-added products. It also examined the integration of coconut residue valorization into the circular bioeconomy, advocating for sustainable residue management. The objective was to provide a comprehensive overview of coconut historical insights, planting characteristics, chemical compositions, and applications, with a focus on generating value-added products from residues. Recognizing the paramount importance of coconuts in the global economy and the abundance of their residues, this paper offers a concise literature review on key aspects and highlights the current status, challenges, and future prospects of integrating agro-industrial coconut residues into the circular bioeconomy. Overall, this study aims to promote sustainable practices, particularly by emphasizing the economic and industrial reintegration of the lignocellulosic biomass as a key raw material for value-added products.

2. Research Method and Approach

This study comprised a broad literature review to summarize, evaluate, and integrate the previous research on the main aspects associated with the coconut industry and its agro-industrial residues. The adopted methodology was based on the identification and in-depth analysis of original research articles, reviews, and technical notes related to the topic. First, the search for high-quality scientific publications was carried out using scientific databases such as Scopus, Web of Science, Scielo, and Google Scholar. The following Boolean searches were used: (Coconut), (Coconut fiber Or Coconut coir fiber), (Coconut endocarp), (Coconut waste Or Coconut residue) and (Coconut shells Or Coconut husk). A total of 1104 results were found for Scielo, 357 for Web of Science, 4327 on Scopus, and 967 on Google Scholar. Then, the selected articles were carefully evaluated to retrieve the relevant information and identify the chief aspects of the coconut plant and its whole value chain. The major topics evaluated were coconut history, cultivation characteristics, industrial applications, processing technologies, forms of commercialization, and the progress in the field of coconut residue valorization, which includes chemical composition analyses, current and potential applications, and future perspectives toward a sustainable bioeconomy. Thus, this methodology allowed for a complete analysis of the reuse of agro-industrial coconut residues, establishing a solid basis for boosting the coconut industry's integration into the circular economy. Figure 2 summarizes the overall and detailed aspects of coconut residue uses that were reviewed in this study.

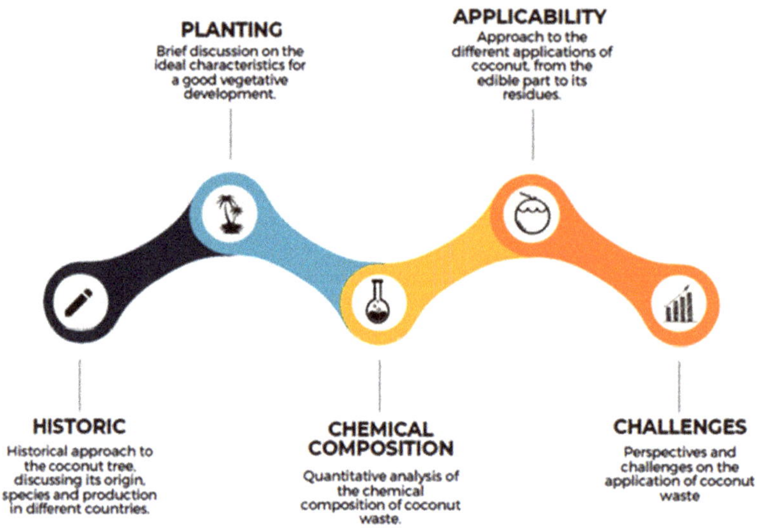

Figure 2. Aspects addressed and discussed in this study.

3. Coconut History, Cultivation, and Value Chain

3.1. Historical Records of Coconut

The origin of coconut is assigned to the Indo-Pacific region [32]. It is a native species of the Indian and Pacific ocean regions, showing a long-standing and independent evolutionary presence in both ocean basins [33]. Coconut has been cultivated in the lowland tropics of West Africa and the Neotropics, where the soil and climatic conditions favor its development [34].

Coconut was disseminated along much of the tropics, possibly due to the movement of ocean currents and navigators [1]. By the end of the XVI century, coconut palm was already present in all tropic regions, except for the tropical coast of Australia, which was still unknown to European navigators [35].

Currently, coconut is produced and commercially exploited in over 90 countries [32], accounting for an annual worldwide production of 64 million tons of fruits [36]. Indonesia, the Philippines, and India are the largest producers with 17.5 Mt, 14.7 Mt, and 11 Mt of coconut produced per year in a harvest area of 2.8, 3.6, and 2.2 million ha, respectively. Indonesia and the Philippines are also the leading suppliers of coconut fruit (fresh, dried, or dissected) and its derived products (32.1% and 24.2% of the total amount of exported products). Other important players in the global market are Sri Lanka (2.2 million tons) and Brazil (2.4 million tons) (Figure 3) [36]. In Brazil, the northeast is the central producing region, responsible for an annual production of 1.44 Mt of coconut fruits (78% total production), which is cultivated in a harvest area of 157,403 ha (82% total area) and generates a production value of BRL 1.12 billion (70% total value). The cultivation is predominant in the coastal areas of the Ceará, Bahia, Pernambuco, and Sergipe states [37].

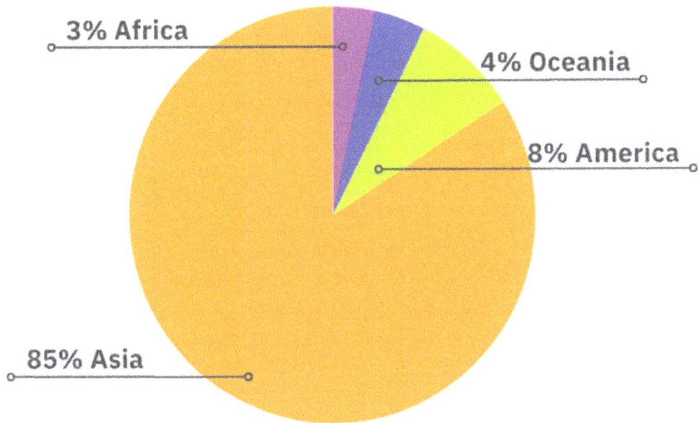

Figure 3. Average percentage contribution of coconut-producing continents from 2000 to 2021, according to FAO [15].

The contribution of each continent to global coconut production, on average between 2000 and 2021, is distributed according to the information present in Figure 4. Notably, coconut production is heavily concentrated on the Asian continent, particularly in the southeastern and southern subregions. Europe does not participate significantly in this industry, while the Antarctic is not ranked, and neither appear in the graph.

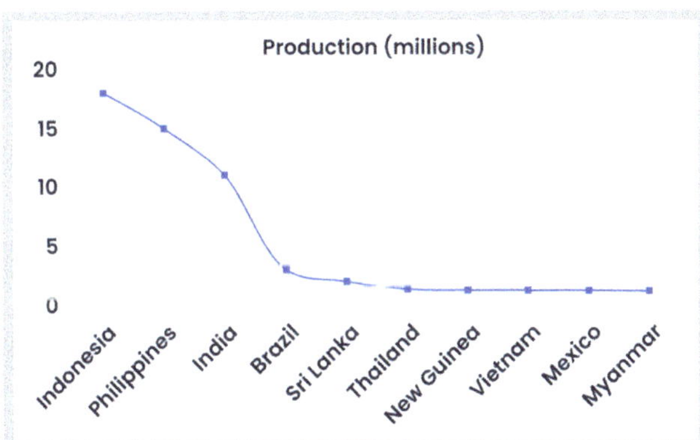

Figure 4. Average coconut production by country from 2000 to 2021, according to FAO [15].

3.2. The Main Aspects of Coconut Cultivation

The monotypic species *Cocos nucifera* belongs to the palm tree family *Arecaceae* [34] and its species can be broadly classified into *Typica*, the tall type, and *Nana*, the dwarf type. Crossing these two types result in hybrid varieties [38]. As presented in Table 1, the palm plant has different characteristics according to its variety, in which, height and lifespan are the most relevant and can reach up to 10 m and 20 years.

Table 1. Characteristics of coconut tree (*Cocos nucifera*) varieties.

Characteristic	*Typica*	*Nana*	Hybrid
Height	30 m	10–12 m	20 m
Life cycle	60–80 years	30 years	50 years
Growth	Fast	Slow	Intermediate
Fluorescence time	5–7 years	2–3 years	3–4 years
Applications	Agro-industry/food preparation	Water consumption	Water consumption/agro-industry/ food preparation

Source: adapted from Niral and Jerard [39]; Donadio et al. [40]; and Embrapa [14].

The successful development of coconuts, mostly in the world's tropical areas, requires warm conditions and high humidity for their proper growth and yield [32]. Optimal coconut production occurs under a well-distributed rainfall of 1500–2500 mm/year, mean annual temperatures of 27–30 °C, and relative humidity above 60% [3]. Weak to moderate winds also favor coconut development since they increase transpiration and, consequently, the roots' absorption of water and nutrients [41].

On the other hand, long dry spells and extreme weather events with high temperatures adversely affect coconut productivity [42], primarily when they occur during the initial stage of inflorescence. In their work, Hebbar et al. [43] identified stress during the pollen germination phase due to the influence of temperatures higher than 33.7 °C. Also, coconut is known as a cold-intolerant plant, as it can only support short and mild cold periods and should not face seasons of cool temperatures (below 21 °C) longer than three months [44].

Solar radiation is also considered a climate variable that significantly influences coconut yields. These palms require abundant sunshine and do not develop well under poor luminosity conditions [45]. Apart from climate factors, the coconut palm tree grows best in soils that are well-draining and rich in chloride [3,46], but it is adaptable to a wide range of soil textures such as laterite, alluvial, sandy loam, coastal sandy, and clayey, and it can tolerate acidic and alkaline soils (pH 5–8) and the presence of different levels of nitrogen, potassium, and phosphorous [44].

Besides being determinants in the survival and growth of coconut, all the mentioned variables and properties regarding genotypes, maturity level, and cultivation conditions significantly influence the chemical composition of coconut fibers and seeds [47,48]. In the earlier stages, coconut fruit develops a husk and shell with a cavity that reaches the maximum water amount and sugar content at 6–8 months. Water gradually reduces in volume with aging, and the husk and shell turn brown and more fibrous [39]. During nut maturation, the content of extractives in the pericarp tends to decrease while the content of lignin and carbohydrates increases, wherein the measurement of a gradual increase in the glucose level is due to the formation of cellulose [49]. Kernel formation starts 5–6 months after fertilization as a thin jelly-like material that, with development, turns white and becomes a thicker solid (10–15 mm). The formation of the copra is complete at 11–12 months when the oil content has increased from traces to 40–70% of the fruit [50,51].

Regarding genotype, Corradini et al. [52] studied the characteristics of green coconut fibers from five different cultivars of the dwarf variety and identified differences in the chemical compositions of cellulose and lignin. Although all samples were richer in lignin than cellulose, the cultivar "Brazilian Green Dwarf—Jiqui" was found to have the lowest lignin content of 37.2 ± 0.8% while "Brazilian Yellow Dwarf—Gramame" was found to have the highest lignin content of 43.9 ± 0.7. The cultivar "Malayan Red Dwarf" presented the highest level of cellulose (37.4 ± 0.5%) and "Cameroon Red Dwarf" the lowest (31.5 ± 0.1%).

In general, the selection of cultivars and maturity stage for harvest must consider the specific utilization of coconut kernels. Fruits from dwarf cultivars are smaller and have a lower copra and oil content [53]. Because of the quality of its tender nut, the young coconut of dwarf type is destined for water consumption, with harvest occurring in the eighth-month post flowering [50,54]. Dwarf palms are also more sensitive to environmental stresses. The harvest of tall and hybrid varieties happens at 11–13 months when their fruits have an optimal maturation state with thicker kernels, good-quality copra, and a high oil content (65–70%). These are mainly used for agro-industrial and domestic uses [50,54].

3.3. Coconut's Main Products

As discussed in the previous sections, coconut fruit in all its forms (water, kernel, and residues) has been widely exploited for several uses and in several industrial segments, e.g., coconut oil production, food, cosmetics, and as a water source. Figure 5 presents some of the applications of coconut.

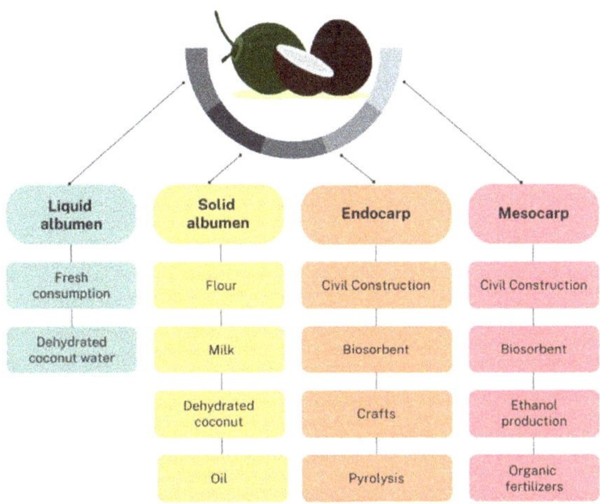

Figure 5. Various applications of the constituent parts of coconut.

The liquid endosperm or coconut water is a natural functional beverage that is widely consumed due to its significant nutritional qualities, antioxidant potential, and low carbohydrate and fat content [55]. The composition of coconut water varies depending on the maturity stage, coconut variety, and other aspects [56]. Coconut water is commercially available fresh directly from fruits, bottled, canned, frozen, or as powder, and has long been consumed in tropical countries such as Sri Lanka, India, Thailand, and Brazil, where it is cultivated in abundance. In more recent years, exportation to markets in the United States and Europe has also increased [51].

Solid albumen is the primary coconut product because it is a major commercial source of vegetable oil (coconut oil). The world demand for coconut oil is approximately 3 Mt per year [15], which is used in food industries as an alternative cooking fat [57]; in the cosmetics industry for hair products and body moisturizers, improving skin barrier function, and anti-inflammatory activity in the skin [58]; and in the pharmaceutical industry, as it has antioxidant, anti-inflammatory, and cytoprotective efficacy due to its polyphenolics [59–61].

Coconut oil is rich in saturated fatty acids (up to 91%), which mainly consist of medium-chain fatty acids (MCFAs) with small percentages of unsaturated fatty acids (UFAs) and polyunsaturated acids. The MCFAs include lauric (the major contributor in fully matured fruits—up to 52%), capric, and caprylic acids, while the UFAs comprises linoleic and oleic acids [62,63]. During fruit growth and development, there is a progressive increase in the oil percentage and a significant change in the fatty acid profile [64]. Due to this chemical profile of MCFAs, coconut oil is more resistant to oxidative rancidity during storage, more thermally stable, and more quickly metabolized than the fatty acids found in animal fat [65].

Various methods have been employed to extract coconut oil: (i) mechanical, generally by pressing the copra; (ii) traditional, by boiling with water; and (iii) leaching out with a solvent [66]. Coconut oil can present different characteristics and applications depending on the extraction method and post-treatment. Chemical means are applied to obtain refining, bleaching, and deodorizing (RBD) oil. Using solvents and high temperatures deactivates the high-value components in oil, reducing the aroma and taste and some health-promoting effects. On the other hand, virgin coconut oil (VCO) is solvent-free and extracted from fresh meat or milk through natural or mechanical techniques, with or without heat application. The absence of chemicals allows the oil to retain its natural sensory and functional characteristics [67]. Therefore, unrefined oil has increased in popularity around the world because of its versatility and health benefits, and is one of the most prominent coconut products in the in the main coconut-producing countries.

Flour, milk, and desiccated (disintegrated and dehydrated) coconut are the other coconut kernel versions that are applied in the food industry. Coconut milk is a natural oil-in-water emulsion that can be prepared domestically by manually pressing the grated solid albumen on industrial and commercial scales using a hydraulic or mechanical pressing process [68]. Coconut flour is a good source of dietary fiber and protein; it is produced from the byproducts of the coconut milk and oil industries. It is obtained after the leftovers go through bleaching, pressing, and grinding to reduce their size [69]. Desiccated coconut is produced from mature kernel following three main steps: drying to about 2.5% moisture, sterilization, and shredding. It is estimated that, to produce 1 ton of desiccated coconut, 8000 coconut nuts are needed; this product is widely used in the bakery and confectionery industries [70].

Coconut sugar is another marketable *Cocos nucifera* product with significant utility in the food and beverage industry. Instead of being obtained from the fruit, the sugar is tapped from the palm tree inflorescence sap, popularly known as 'neera'. This sap contains approximately 16% sucrose and, hence, upon heating at a high temperature, forms crystalline or amorphous sugar. Coconut sugar is brown, contains 2–3% moisture, has high nutritional value, and is rich in amino acids, vitamins, minerals, polyphenols, and antioxidants. In addition, it has a low glycemic index compared to sugar cane, and is considered a suitable sweetener for some health conditions [71,72].

The worldwide exportation and importation of coconut products total about USD 5.7 billion and USD 6.1 billion, respectively. In 2021, Brazil imported 13.3 kt (trade value of USS 29.6 million) and exported 1.2 kt (trade value of USS 29.6 million) of coconut products, changes of −33% and +24% compared to 2019. Although Brazil is among the top five coconut-producing countries, most of the production is destined for the commercialization of green fruits. In this sense, to supply the domestic demand, desiccated coconut and coconut oil are imported from Indonesia, Malaysia, the Philippines, and Uruguay [36].

4. Coconut Residues

4.1. Lignocellulosic Composition

The chemical composition is the most critical characteristic of lignocellulosic materials [73], which are mostly composed of cellulose, hemicellulose, and lignin [4,12].

Cellulose is a naturally occurring polymer and is the main component of the plant cell wall and the most abundant organic compound found in nature [74]. Structurally, it is a linear-chain polysaccharide formed by monomeric units of D-glucose linked by β 1–4 bonds. Natural cellulose chains consist of individual fibers in a parallel arrangement of crystalline (ordered) and amorphous (less-ordered) regions [75,76].

The molecular orientation and hydrogen bonding networks in crystalline regions of the cellulose chain can vary greatly and occur as different polyforms, whose structure depends on the cellulose source and extraction/treatment method used. Type I polyform cellulose has a higher degree of crystallinity and organization and is used to prepare nanocomposites with higher mechanical properties; meanwhile, type II and III cellulose presents an amorphous morphology and is applied in pharmaceuticals, as food thickening agents, and in biomedical products [77]. The high availability of cellulose in inedible parts of the plants (lignocellulosic biomass) makes it highly desired as a commercial source of chemical compounds with economic value [78].

Hemicellulose is also a polysaccharide in the plant cell wall, but contrary to cellulose, its chain is highly branched and comprises diverse sugar units (pentose and hexoses) [79]. The main sugar units are xylose (the most representative component), glucose, and galactose, which are readily hydrolyzed by dilute acids and soluble in alkaline solutions [80,81]. Thus, hemicelluloses confer to the structure of the lignocellulosic biomass properties such as increased flexibility and a high fiber bonding area, which are essential when processing lignocellulosic materials and make it a desirable resource for applications in coatings and composite materials [82].

Lignin, the third polymer found in the lignocellulosic biomass and the second most abundant organic compound on Earth, is an amorphous molecule that is gradually incorporated into the plant cell wall. It has an aromatic tridimensional structure composed of phenyl propane units synthesized by a radical route from three precursor *p*-hydroxycinnamyl alcohols: *p*-coumaryl, sinapyl, and coniferyl [79]. Lignin confers resistance and rigidity to plant cell walls, allowing for the transport of water, nutrients, and metabolites to the different regions of the plants. Further, it protects the cell against the destructive action of microorganisms and enzymes [83].

The quantity of cellulose, hemicellulose, and lignin that occurs in the lignocellulosic biomass can vary with species, as well as among samples of the same species depending on the climate and environmental cultivation conditions, due to the different physiological functions of these molecules in the plant [12,73,84,85]. These changes have a strong influence on the mechanical properties of lignocellulosic byproducts [86].

Moreover, according to Guleria et al. [87], the three components, cellulose, hemicellulose, and lignin, are connected by strong interactions due to the non-covalent forces and cross-linked covalent linkages between them, where hemicellulose acts as a linkage agent between cellulose and lignin molecules to form the so-called lignocellulosic matrix. This complex assembly is rigid and recalcitrant, making accessing and separating the fibers difficult.

Therefore, much of the work in the scientific literature on lignocellulosic material has concentrated on the characterization of their chemical composition in terms of cellulose, hemicellulose, lignin, other fractions of ashes, and moisture, and in finding efficient pre-treatments to promote the recovery of lignocellulosic components without compromising their integrity [88,89].

In particular, evaluating the lignin content in the biomass is important for optimizing the physical–chemical parameters in the pre-treatment process. Lignin and polysaccharides are isolated when macromolecules are fragmented into smaller parts, which involves the rupture of their chemical linkages and a reduction in their molecular masses. During the delignification of lignocellulosic materials, some of the solvents used promote the cleavage of the covalent bonds in lignin–carbohydrate complexes [90,91].

4.2. Coconut Residues Characterization

As mentioned, the coconut fruit weight comprises 55% to 80% non-edible parts that are rich in lignocellulosic biomass, which is characterized by a fibrous and heterogeneous structure. Most of the work on coconut residues has focused on the husk (mesocarp), whether for the characterization of chemical compositions, exploration of potential applications, or evaluation of different chemical pre-treatments to fractionate lignocellulose components [4,92,93]. This attention to the fibers is related to its easy handling and chopping procedures, since it only needs to be separated from the inner shell, and its high volume.

Table 2 summarizes the distribution of the main components of the different coconut residues; it is possible to note that the chemical compound profiles are different even for the same type of residues. Borel et al. [4] and Nascimento et al. [92] found similar values of lignin, cellulose, and hemicellulose content for coconuts collected in Brazil; however, they were lower when compared to those found in the work by Anuchi, Campbell, and Hallet [93] using samples from Costa Rica. These differences are due to several factors (genotypes, environmental and physiological conditions, and harvest techniques) and to the chosen methods and conditions of both the pre-treatment and extraction.

Table 2. Chemical composition of coconut residues.

Authors	Residue	Hemicellulose (%)	Lignin (%)	Cellulose (%)
Nascimento et al. [92]	Mesocarp	25.5	35.1	31.6
Borel et al. [4]	Mesocarp	30	32	31
Anuchi et al. [93]	Mesocarp	15	41	38
Andrade et al. [94]	Endocarp	15.2	33.7	10.4
Alharbi et al. [86]	Leaves	19	21	33
Jose and Beevi [95]	Coir pith *	14.2	41.3	34

* residue generated during coconut coir defibrillation.

Borel et al. [4] subjected coconut fibers to Soxhlet extraction using acetone as a solvent for the determination of the lignin content in moisture- and extractive-free samples, which was performed utilizing acid hydrolysis and sulfuric acid according to TAPPI standard T222 [96]. The cellulose was quantified by treating holocellulose with 5% and 24% potassium hydroxide solutions and hemicellulose by subtracting the cellulose content from the holocellulose fraction. In the work of Nascimento et al. [92], the determination of lignin also followed the TAPPI T222 [96] standard, and cellulose and hemicellulose followed TAPPI T203 [97]. Anuchi, Campbell, and Hallet [93] determined the chemical composition of fibers according to the NREL protocol NREL/TP-510-42618 [98].

Kochova et al. [99] evaluated the influence of the quantity and types of sugar units from coconut fibers on cement hydration; they concluded that the presence of arabinose, galactose, glucose, xylose, and mannoses (the main hydrolysis products of cellulose and hemicellulose and also components of the plant) corresponds well with the total heat released during cement hydration, as measured with calorimetry. Due to its rich sugar content, Mariano, Unpaprom, and Ramaraj [100] utilized coconut to produce bioethanol. In that study, they demonstrated that subjecting coconut pulp to sequential hydrothermal and acid pre-treatments can improve saccharification and maximize the release of sugar for fermentation into ethanol.

The use of agro-industrial residues is also of interest given their mechanical properties, which vary depending on the specific byproduct generated and its chemical properties. The mechanical properties commonly evaluated are tensile strength, modulus of elasticity, and density/specific weight. Danso et al. [101] measured the mechanical and physical properties of natural fibers and found that coconut fibers have 6.4% moisture, a tensile strength of 222 MPa, a modulus of elasticity of 2.8 GPa, and a density of 0.81 g/cm^3. The fibers were investigated for their ability to reinforce soil building blocks, showing an improvement in the compressive strength of 41% compared to unreinforced blocks.

Robert et al. [102] analyzed the thermal and mechanical properties of boards made from coconut husks to investigate their use in building projects. The results showed that composite boards containing any percentage of coconut husk could be a good alternative to conventional thermal building insulators like ceiling boards and partition boards. Treated coconut husk presented an apparent density of 0.333 g/cm^3, flexural strength of 435 MPa, and compression strength of 321 MPa. Then, they found that untreated husks produced boards with better thermal properties, while husks treated with NaOH improved the mechanical properties of boards.

Pereira et al. [103] found that coconut husk samples had a density of 0.169 g/cm^3, a value lower than that reported for other natural fibers, such as sugarcane bagasse, kenaf, and bamboo. This density can create difficulties in obtaining homogeneous mixtures, and cause the agglomeration of fibers and problems in production processes. This problem, however, can be solved by crushing the husks in mills [104].

4.3. *From Waste to Value: Byproducts from Coconut Lignocellulosic Residues*

Exploiting the potential of coconut residues has become an important matter in the coconut industry. Besides the residues' aggregate economic value to the coconut supply chain, renewable raw materials and waste management are major concerns in terms of environmental problems, and it is necessary to stimulate studies to find alternatives to avoid their incorrect disposal. Coconut residues can be used in the generation of handicraft goods for sale, such as carpets, mattresses, and upholstery, and they can undergo a process that changes their chemical compositions allowing them to be used as green alternatives to other natural resources extracted from the environment. Several studies have attempted to use coconut residues, whether *in natura*, minimally processed by physical pre-treatment, or when submitted to chemical pre-treatments that lead to chemical structural alterations.

Due to the structural complexity of cellulose, hemicellulose, and lignin, as discussed in Section 4.1, the conversion of the lignocellulosic biomass into value-added products generally involves a multi-stage process and requires pre-treatment steps [105]. However, an appropriate choice must be made regarding the nature of the separation process of the lignocellulosic components, which can be physical, chemical, biological, or a combination of them, as the degree of separation depends on the process used [106]. In this sense, the high potential and availability of coconut residues encourage the search for cost-competitive and efficient pre-treatment technologies that could promote the fragmentation, recovery, and utilization of the lignocellulosic biomass in large-scale biorefineries [107].

In the scientific literature, the studies on coconut residue applications, using as-obtained or pre-treated residues, have focused on the following strategies: (i) application in civil construction [108–111]; (ii) use as filtering and adsorbent materials [112–117]; (iii) in ethanol production [19,26,118]; (iv) obtention of pyrolysis products [22,119–122]; and (v) other applications [123–126].

4.3.1. Civil Constructions

Umoren et al. [108] reported the inhibitory effect of a coconut coir dust extract, produced using chemical and electrochemical techniques, on mild steel in a sulfuric acid solution. Coir dust was produced from shredded husks, and an extract was obtained using methanol and water as free acid extractive solvents; the experiments were carried out in solution at different temperatures and extract concentrations. As a result, the extracts of coconut coir dust exhibited a corrosion inhibition effect for mild steel in 0.5 MH_2SO_4. There was an increase in efficiency with increasing concentration of the inhibitor and decreasing temperature.

Ramasubramani and Gunasekaran [109] used coconut endocarp as a substitute for crushed stone aggregate for coarse aggregate concrete production. The microstructural characteristics and properties of conventional concrete (CC) and coconut shell concrete (CSC) were analyzed. For CC and CSC, the flexural strength was 5.38 N/mm^2 and 5.20 N/mm^2; tensile strength was 3.70 N/mm^2 and 2.92 N/mm^2; and the average impact resistance to form initial cracks was 398 Joules (20 blows) and 616 Joules (31 blows). The results showed that the mechanical properties of CSC were inferior to those of CC; however, the differences were not significant, indicating the feasible use of coconut endocarp as a sustainable alternative to coarse aggregate.

Narciso et al. [110] used coconut husk combined with *Pinus oocarpa* wood to produce medium-density particleboards. The physical and thermal properties of the particleboards were positively affected by husk addition, producing lower thermal conductivity values. Although the increase in coconut fiber concentration promoted a decrease in the mechanical properties (strength and stiffness) of the panels, all treatments met the market standards, indicating the potential of coconut waste as a sustainable raw material to produce panels.

Souza et al. [111] studied different coconut husk fiber levels (0%, 10%, 20%, 30%, 40% and 50%) to replace eucalyptus wood in producing wood–cement particleboards. From each board, the specimens were subjected to physical–mechanical tests. Regarding physical properties, the addition of coconut fibers improved the dimensional stability, decreased water absorption, and increased thickness swelling, while for mechanical properties, a decreased modulus of elasticity and modulus of rupture were observed as the percentage of coconut fibers increased. In general, the boards were considered inappropriate for structural uses but suitable for indoor uses.

It is important to highlight that, in both studies that used coconut waste to produce panels, it was observed that the physical properties of water absorption and thickness swelling, and the mechanical properties of modulus of rupture and modulus of elasticity decreased with increasing of residue content.

4.3.2. Filtering Material and Adsorbent

Hoang et al. [112] used coconut shells as raw materials to produce activated carbon loaded with cobalt ferrite composites via a single-step refluxing router method. The carbon was used to manufacture adsorbents to remove rhodamine B (RhB), a cationic xanthene fluorescent dye, from aqueous environments via adsorption. This method proved promising, and the coconut fiber-activated carbon showed a maximum efficiency capacity for the rhodamine B adsorption of 94.08 mg/g.

Padilha et al. [113] subjected green coconut fibers to an organosolv pre-treatment and used the obtained liquor for the synthesis of lignin/Fe_3O_4 nanoparticles to be used as an immobilization support for fungal β-glucosidase, which was evaluated as an adsorbent for textile dyes in aqueous solutions. Due to the physicochemical properties of the lignin from the coconut fibers, the obtained nanoparticles showed efficient support for improving β-glucosidase stability and synergism with cellulases during enzymatic hydrolysis and as a bioabsorbent with a high adsorption capacity for textile dyes: 203.66 mg/g for methylene blue, 112.36 mg/g for cibacron blue, and 96.46 mg/g for remazol red dyes. Both evaluated liquor uses proved to be sustainable pathways for the valorization of lignin as a raw material.

Esfandiar et al. [114] evaluated low-cost materials, including coconut coir fibers, as a sorbent for removing a mixture of polycyclic aromatic hydrocarbons (PAHs) from simulated stormwater. The sorption capacity was investigated, and the outcome revealed the strong potential of the coconut coir-based sorbent. The authors concluded that using widely available sorbent materials from agricultural waste can provide a low-cost option to address pollutant challenges.

Marín-Velásquez and Cóndor-Salvatierra [115] used coconut shells as a granular filtering material to remove oil and fat from water contaminated with diesel. The water samples were contaminated with diesel oil at 5, 10, and 15% concentrations. In this study, the properties analyzed were density, pH, water adsorption, and porosity by gravimetric determination. The use of coconut shells as filtering material provided a maximum oil removal of 85.10%, classifying the filtered water sample as drinking water according to the country's tolerance levels.

Thongsamer et al. [116] produced biochar pellets from coconut husk biochar (CH) to be used as a low-cost adsorbent for nutrient removal from eutrophic surface waters. The pelletized coconut husk biochar was modified with chitosan impregnation (CHC) and mixed with eggshell powder (CHEP) and experiments were carried out to evaluate and compare the performance of the three versions of biochar pellets. CH, CHC, and CHEG were able to adsorb ammonium, but CHC had the greatest potential in removing all nutrients (ammonium, nitrate, and phosphate) from eutrophic surface waters, achieving removal efficiencies of 61.70% and 54.37% for phosphate and nitrate, respectively.

Regarding filtering materials for air pollutants, Pettit et al. [117] evaluated the effectiveness of green walls consisting of differently sized coconut husk-based substrates as a filtering medium for ethyl acetate, benzene, and total volatile organic compounds (VOCs) in the atmosphere. The coconut husk used had a water content of 72.5% or 95% organic matter, a specific surface area of 0.75 m^2/g, and a water holding capacity of 5.5 g[H20]/g dry material, and differing proportions of granular activated carbon were added to the coconut shell substrate. The results showed that including activated carbon can improve the efficiency in removing certain pollutants, and a 50:50 composite medium provided the best VOC removal performance. However, the authors recommended adjusting the use of activated carbon according to the type of pollutant.

4.3.3. Ethanol Production

Cabral et al. [19] investigated the production of bioethanol from green coco husk fibers, which were pre-treated by an alkaline method with NaOH, hydrolyzed enzymatically and subjected to ethanol fermentation with commercial yeasts of *Saccharomyces cerevisiae*. Enzymatic hydrolysis converted 87% of the sugars and the ethanolic fermentation consumed 81% of the substrate in the hydrolysate, leading to a sugar-to-ethanol conversion efficiency of 59.6%. The authors pointed out that green coconut husk is strongly viable as a feedstock for ethanol production.

Ebrahimi et al. [118] examined the effect of an organosolv pre-treatment with acidified aqueous glycerol on coconut coir fibers for ethanol production. A simultaneous saccharification and fermentation procedure with *Saccharomyces cerevisiae* was performed and as a result, the glycerol solvent improved the fiber glucose digestibility from 11.8% (untreated) to 79.7–81.8% and ethanol production from 1.67 g/L to 8.97–8.81 g/L.

Padilha et al. [26] also reported using green coconut fibers pre-treated with glycerol to produce ethanol, lignin, and rhamnolipids, which belong to the class of glycolipid biosurfactants. Ethanol production was performed via semi-saccharification and simultaneous fermentation and the recovery of organosolv lignin by diluting black liquor in acidified deionized water. The washing waters from the organosolv treatment and the precipitation water from lignin were used to prepare the culture medium for producing rhamnolipids, which were obtained with promising emulsifying properties. The fermentations with untreated coconut fiber achieved high concentrations of ethanol (5.97–15.01 g/L), which may be linked to the impregnated sugars. However, the maximum concentration of ethanol (29.64 g/L) was obtained with treated fibers.

4.3.4. Pyrolysis Products

Pyrolysis is a thermochemical process that occurs under medium to high temperatures and in an inert atmosphere to promote the conversion of carbonaceous materials into value-added solid, liquid, and gaseous (rich in methane) products. The solid phase is used in biochar production while the liquid phase, known as bio-oil, has drawn attention due to its potential as a biofuel and its chemical products [127]. The samples with a high volatile matter content, low moisture, and low ash content had favorable outcomes for bio-oil conversion and the overall efficiency of the process [128–131].

Babatabar et al. [22] investigated the pyrolysis conversion of different lignocellulosic biomasses (rice husk, coconut shell, and walnut shell) while maintaining the same operating conditions in a fixed-bed reactor. Coconut shell showed the highest yield of bio-oil at 50.25 wt% (rice husk: 44.10%; walnut shell: 41.50%), which had a higher content of hydrocarbon components (paraffin, olefin, and aromatic hydrocarbons) and phenols due to the biochemical structure of the biomass rich in lignin. The high proportion of hydrocarbons in the coconut shell bio-oil indicates the biomass's potential as a precursor for biofuels, especially diesel and jet fuel.

Sahoo et al. [120] also performed coconut shell pyrolysis to obtain bio-oil to partially replace the silane coupling agent in rubber compounding, which is associated with a few drawbacks, such as the formation of ethanol as a byproduct. From 150 g of coconut shell powder, about 80 g of bio-oil was obtained with a yield of approximately 53%. The main constituents of the coconut shell oil were phenol and its derivatives (49%), followed by ammonium acetate, aniline, and ether. As a result, incorporating the obtained bio-oil into silica-filled styrene butadiene rubber enhanced the salinization reaction; however, it must be combined with a silane coupling agent (40–50%) to improve effectiveness.

Arantes et al. [119] studied the co-pyrolysis of green coconut pericarp (comprising the epicarp, mesocarp, and endocarp) and waste polystyrene for oil production. Experiments were performed in a fixed-bed reactor over an activated biochar catalyst. The results showed that co-processing the biomass and waste plastic (BPS) led to a higher yield of oil (59%) and an oil organic fraction of 39%, when compared to an experiment containing only the coconut pericarp (CP) (7% of the organic fraction). Nonetheless, the CP oil presented a high content of both phenolic (60.67% relative area) and furanic (15.78% relative area) compounds, while the BPS organic fraction mainly constituted aromatic (70.65% relative area) and polyaromatic (10.06% relative area) hydrocarbons. Although the catalyst favored gas formation, it improved the selectivity for monoaromatic hydrocarbons, toluene, and xylene, which are alternative to zeolites.

Chaos-Hernández et al. [121] evaluated the efficiency of char catalysts obtained from the coconut endocarp pyrolysis and improved the ability of KOH and $Ca(NO_3)_2$ catalysts to produce biodiesel. In this sense, a detailed analysis of the catalyst functionalization, activation condition of the catalysts, and their application in transesterification to produce fatty acid methyl esters (FAMEs) was performed. Experiments were carried out at varying pyrolysis temperatures and dwell time, finding the best catalyst preparation via the pyrolysis of coconut endocarp at 600 °C for 1 h (FAME formation of 90.8%).

Gopal et al. [122] used mature coconut husk, tender (immature or green) coconut husk, coconut leaf petiole (central section that connects leaf blades), and coir pith (a sponge-like material found between the hard internal shell and the outer shell) as pyrolysis feedstocks for the production of biochar for remediating acidic soils. Due to the alkaline pH (>7.5), the produced biochar was considered ideal for neutralizing humid tropical soils, which are mostly acidic. At the same time, the high potassium content (>2.5%) makes them a valuable alternative organic K source for plants. Also, the suitability of coconut biochar for soil amendment was established through seed germination and earthworm avoidance tests.

4.3.5. Other Applications

Tang et al. [124] evaluated the use of six lignocellulosic residues, including coconut coir fibers, as a source of ferulic acid and phenolic compounds for bioconversion into vanillic acid. Before conversion, the samples of oil palm empty fruit bunch fiber (OPEFBF), coconut coir fiber (CCF), pineapple peel (PP), pineapple crown leaves (PCLs), kenaf (from *Malvaceae* family) bast fiber (KBF), and kenaf core fiber (KCF) were pre-treated with organosolv (NaOH-glycerol) and an alkaline treatment (NaOH). It was found that organosolv was superior for phenolic compound extraction, whereas the alkaline treatment enhanced the lignin extraction. Among the tested agricultural residues, it was found that PP was the best source of ferulic acid, yielding 5.72 mg/g (CCF 0.17 mg/g), whereas CCF was the best source of phenolic compounds, yielding 47.99 mg/g. However, vanillic acid production was only detected in fermentation with the PP and PCL liquors.

Luis-Zarate et al. [125] evaluated the efficiency of the removal of benzene, toluene, and naphthalene from water using coconut shells (endocarp), coconut fibers (mesocarp), and coconut shells with fibers (endocarp and mesocarp) as biosorbents and further evaluate the residues' saturated potential for energy generation by combustion. CF showed the highest adsorption capacities of 222 mg/g, 96 mg/g, and 5.85 mg/g for benzene, toluene, and naphthalene, respectively, which was justified by its morphologic characteristics and high concentration of phenolic groups due to the lignin structure. The combustion heat analysis showed an increase in heat capacity up to 657 cal/g, which is related to its characteristics of a high lignin and low ash content.

Oliveira et al. [123] proposed using coconut husk fibers as a photoprotective and antiglycation agent by determining their phenolic chemical composition and photoprotection capacity according to the sun protector factor of an ethanol extract of coconut husk fibers. Seven phenolic compounds were identified by high-performance liquid chromatography: quercetin, catechin, epicatechin, and vanillic, caffeic, 4-hydroxybenzoic, and chlorogenic acids. The prepared formulations showed photoprotective activity with a maximum SPF value of 15.94, equivalent to the standards for benzoquinone and quercetin. Therefore, using coconut byproduct can be considered as an efficient ingredient in sunscreen formulations, and is a low-cost alternative source of active ingredients for the cosmetics and pharmaceutical industries.

In their work, Lasmini et al. [126] proposed the use of coconut husks as a liquid organic fertilizer source to obtain the maximum shallot yield under dryland conditions. CH was combined with two different mulches made from silver–black plastic and rice straw and tested at four concentration levels. The applications of straw mulch with coconut husks waste at a dose of 1000 L/ha resulted in the highest plant height, number of leaves per plant, number of tillers, and bulb yield.

5. Discussion

5.1. Circular Economy

The circular economy is a sustainable and integrated approach that aims to support the regeneration and restoration of biological resources to produce value-added products while minimizing the negative impacts of waste from the industrial sector [132]. Ensuring the efficient use of resources has been recognized as one of the United Nations' Sustainable Development Goals (SDGs), confirming the importance of biological resources for achieving a sustainable future [133]. In particular, SDG12 was proposed to ensure sustainable consumption and production patterns by 2030, which is directly related to the concepts of residue reintegration and biorefineries [134].

Support policies and innovative technologies for advanced biorefineries represent a crucial point for developing a modern bioeconomy, especially in regions whose economic system is based on agricultural production. Biorefineries, focused on processing residue biomass, such as lignocellulosic materials, offer the potential for generating a wide variety of bioproducts with different chemical compositions [135]. Thus, besides aiding in adopting sustainable agricultural practices coupled with a circular economy approach, biorefineries can contribute to the generation of bio-based alternative fuels and chemicals for mitigating the effects of climate change [136].

The conversion of coconut agricultural residues into high-value bio-based low-carbon materials is an effective residue management strategy for applying the biorefinery concept in a cutting-edge circular economy, adding value to the whole supply and processing chain, increasing a company's competitiveness, and providing consumers with sustainable products and services. Based on the analyses of recent studies, the strategies for coconut lignocellulosic biomass valorization may include the improvement of recovery and processing techniques and the exploration of the uses for this biomass in sectors such as civil construction [108–111], manufacturing of adsorbent materials [112,113] and filters [114,117], ethanol production [19,26,118], bio-oils [22,119–121], biochar [122,125], and organic fertilizers [126].

However, it is important to highlight that inserting biomass as feedstock into the current industrial and economic model for an efficient transition to renewable generation requires more specific policies, collaborations, and investments [137].

5.2. Insights from Coconut Residue Value Chain

Regarding the large-scale management of coconut residues, the major strategy that has been adopted is direct disposal rather than recycling, recovery of energy potential, or any other reuse method [23]. Therefore, based on the current volume of coconut production and the expected exponential growth, the coconut residue value chain presents a huge potential for supporting green economic development [24,138].

The lack of advanced recycling techniques for coconut residues represents a challenge to achieving a circular economy and stimulating local low-carb economic development. This gap is due to the absence of specific policies to support a greener transition and disseminate knowledge regarding the benefits of recycling this waste. Promoting this change requires the active participation of small and large producers in the entire value chain [23,139].

Furthermore, investment in the research and development of innovative technologies for processing coconut residues can improve productivity and product quality and establish supplier–customer partnerships with industries, small companies, and potential sectors that can use coconut byproducts. This collaboration creates synergies and business opportunities, which can guarantee the viability of a circular economy and boost local economic development based on the "from waste to profit" approach [17,140].

Given this, the main aspects that have been identified regarding the recovery and reuse of coconut residue valorization are as follows:

(i) Value addition through product diversification and improvement of lignocellulosic chemical and functional properties in coconut residues;
(ii) Integration of the whole coconut value chain, from main coconut products to byproduct generation;
(iii) Strategic partnerships with other coconut industries or sectors;
(iv) Boosting investments in the research and development of innovative technologies for processing coconut residues;
(v) Identification and exploration of market opportunities for products derived from coconut waste at the national and international levels.

5.3. *Coconut Industry: A Sustainable Pathway towards the 2030 Agenda Goals*

Over the decades, the coconut industry has demonstrated its versatility and economic importance for several industrial sectors, from food and cosmetics to construction products and sustainable fuels. Although it also stands out for its ability to incorporate sustainable practices in the production and processing chain, the coconut industry still has the potential to open new pathways for exploring the different parts of the fruit, from its water to kernel and lignocellulosic residues, which offer a promising perspective for achieving the goals of the 2030 Agenda for Sustainable Development [140] (Figure 6).

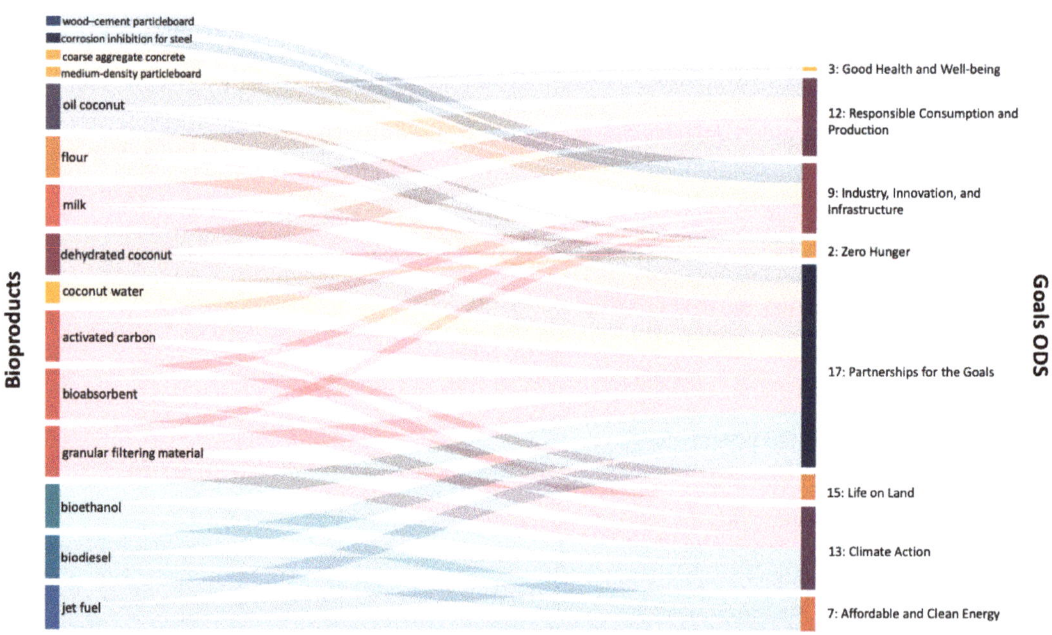

Figure 6. Interlinking coconut-derived products with the objectives of the 2030 agenda for sustainable development.

a. Goal 2: Zero Hunger

Indicator: Utilization of coconut kernel products such as coconut oil in food industries as an alternative cooking fat.

b. Goal 3: Good Health and Well-being

Indicator: Coconut oil's incorporation into cosmetics for hair products and body moisturizers, enhancing skin barrier function and performing anti-inflammatory functions.

c. Goal 7: Affordable and Clean Energy

Indicator: Production of bioethanol from green coconut husk fibers, providing a renewable energy source.

d. Goal 9: Industry, Innovation, and Infrastructure

Indicator: Use of coconut residues in civil construction, producing materials like particleboards and wood–cement particleboards.

e. Goal 12: Responsible Consumption and Production

Indicator: Development of sustainable alternatives, such as coconut flour, milk, and desiccated coconut, contributing to responsible consumption practices.

f. Goal 13: Climate Action

Indicator: Pyrolysis of coconut shells for bio-oil production, offering potential biofuels and chemical products, aligning with climate-friendly initiatives.

g. Goal 15: Life on Land

Indicator: Utilization of coconut biochar for remediating acidic soils, supporting sustainable land use and ecological health.

h. Goal 17: Partnerships for the Goals

Indicator: International trade of coconut products contributes to economic partnerships and global commerce.

The applications of coconut residues, from civil construction to filtering materials and adsorbents, align with the need for sustainable alternatives, waste management, and the development of eco-friendly technologies, promoting circular economy principles. The comprehensive utilization of coconut products contributes to addressing multiple dimensions of sustainability, fostering economic growth, and supporting environmental conservation.

6. Challenges and Future Perspectives

Given the increasing popularity of coconut products, mostly coconut oil and water, coconut has significantly increased its relevance in agribusiness and the demand for its primary derivatives is expected to grow exponentially in the following years. Thus, new technologies and strategies are being investigated to enhance the whole coconut sector and its supply and value chain by improving management, commercialization, and the quality of the cultivation system, and innovating the processing industries and techniques [141].

Consequently, coconut crop production is leading to the generation of large amounts of lignocellulosic residues and environmental pollution since such wastes are commonly considered of low value and are incorrectly disposed of [142]. Therefore, there is also a need to develop knowledge and encourage scientific investigations toward exploiting the potential of such residues for generating renewable and high-value-added products.

Agro-industrial residues are grouped into two classes based on their economic value: the first refers to residues that already have consolidated value and the second are the residues that have fewer potential applications or are chiefly restricted to scientific research, although they have shown remarkable characteristics for numerous applications [143]. Therefore, using coconut waste as an alternative and renewable raw material to generate value-added products requires the redirection of this waste to the consumer market and to be part of a sustainable circular economy [144].

Some of the challenges identified by this review are presented in Table 3, which summarizes what improvements are needed to allow the use of coconut waste in different processes according to the applications found in the literature.

Table 3. Analysis and challenges of the different biotechnological uses of coconut residues (CRs).

Application	Highlights	Challenges
Civil construction	- CR properties have a corrosion inhibition effect for mild steel in sulfuric acid solutions. - The abundance phytochemicals (tannins, flavonoids, and phenolics) make residues suitable alternatives to coarse aggregate in concrete production and as raw materials to produce panels.	- Corrosion inhibition efficiency depends on the extractive solvents used and this ability is lost at high temperatures. - Physical and mechanical properties (water absorption, swelling in thickness, modulus of rupture, and modulus of elasticity) of construction materials decrease with an increase in the level of CR inclusion. - Building materials with CRs cannot be exposed to excessive humidity. - New strategies must be developed to improve the mechanical properties of CRs.
Adsorbent material	- CRs have a high adsorption capacity to remove dyes (rhodamine B, methylene blue, cibacron blue, and remazol red) and polycyclic aromatic hydrocarbons (PAHs) from aqueous solutions. - CRs have high adsorption capacities for benzene, toluene, and naphthalene. - CRs have filtering abilities to remove oil and fat from water contaminated with diesel and hydrocarbons. - The lignin content in CRs can improve the β-glucosidase performance of crystalline cellulose in enzymatic hydrolysis.	- Dye adsorption capacity is greatly reduced with the number of reuses. - Further research is needed to investigate the mechanisms and behaviors of the adsorption reactions between the sorbents and pollutant and to determine the leaching potential and performance of a sorbent under different hydrological conditions. - Research is still lacking for some aspects concerning the energy required and economic viability on a commercial scale.
Production of organic solvents	- CRs are strongly viable substrates for ethanol production via saccharification and fermentation with *Saccharomyces cerevisiae.* - Residues can be used as a carbon source for the production of rhamnolipids with interesting emulsifying properties.	- The efficiency in producing second-generation ethanol from coconut biomasses depends on the pre-treatment steps. - More studies need to be conducted to evaluate the viability of the process concerning the loss of sugars in the pre-treatment step. - The process for producing ethanol was conducted only on a laboratory scale, making it necessary to evaluate in a pilot/at the industrial scale to prove its efficiency and good cost–benefit ratio.
Feedstock for pyrolysis process	- CRs can produce high bio-oil yields containing a high content of hydrocarbon components. - CRs demonstrate a high potential for producing phenolic compounds and biofuels. - Derived bio-oils can be used as a green alternative component to enhance the physicomechanical properties of other materials, e.g., silica-filled styrene butadiene rubber compounds. - Derived biochar can be used as a catalyst to produce biodiesel and for remediating soils.	- Pyrolysis product distribution and composition is also influenced by the process' operation conditions. - Depending on the components of interest, it is necessary to employ some catalysts to improve the selectivity. - More studies must be performed at pilot and large scales.

7. Conclusions

Coconut palm and its fruit have been cultivated and consumed in several countries around the world, with production heavily concentrated in the Asia continent and the Pacific Region, mainly in Indonesia, the Philippines, and India. Due to the high demand for coconut derivatives, residues from the coconut sector are available in large quantities, and although they have not been appropriately used, they have an interesting and promising chemical composition with a high content of lignin, cellulose, and hemicelluloses that are favorable for generating high-value-added byproducts.

Among the applications where coconut waste has been proven to be an efficient raw material for generating byproducts are in civil construction, as an adsorbent material, and in the production of ethanol, bio-oil, and organic fertilizers. Furthermore, several research studies have identified coconut residues as a rich source of phenolic compounds that are useful as antioxidants and anti-inflammatories and have potential for use in the pharmaceutical industry. Its phenolic profile is attributed to its high lignin content.

While these applications are promising, several technological challenges hinder their widespread implementation. This review not only underscores the significance of coconut waste in promoting a circular economy and sustainability but also serves as a resource for researchers seeking to enhance its applicability. By addressing the technological drawbacks, this work encourages further research and innovations, fostering the advancement of coconut waste utilization in diverse fields. To provide a broader and more effective use of these natural resources, it is essential to face the highlighted challenges by investing in innovative technologies and integrating the entire coconut value chain, from coconut production to residue recovery and processing.

Author Contributions: Conceptualization, F.V., D.P.S., M.J., J.S. and D.S.R.; methodology, F.V.; formal analysis, F.V. and H.E.P.S.; investigation, F.V., M.J. and H.E.P.S.; data curation, F.V. and H.E.P.S.; writing—original draft preparation, F.V.; writing—review and editing, F.V., D.P.S., M.J., J.S., M.V.-V., P.P. and D.S.R.; visualization, J.S., M.V.-V. and P.P.; supervision, D.P.S. and D.S.R. All authors have read and agreed to the published version of the manuscript.

Funding: This research received no external funding.

Institutional Review Board Statement: Not applicable.

Informed Consent Statement: Not applicable.

Data Availability Statement: Not applicable.

Acknowledgments: The authors acknowledge the financial assistance from Brazilian research funding agencies such as the Coordination for the Improvement of Higher Education Personnel (CAPES), under Finance Code 001, a Brazilian foundation within the Ministry of Education (MEC); National Council for Scientific and Technological Development (CNPq), a Brazilian foundation associated with the Ministry of Science, Technology, and Innovation (MCTI); Foundation of Support for Research and Technological Innovation of the State of Sergipe (FAPITEC/SE); and Federal University of Sergipe. We acknowledge the Foundation for Science and Technology (FCT, Portugal) for financial support for the CISAS UIDB/05937/2020 and UIDP/05937/2020, including a postdoc grant awarded to M.J.

Conflicts of Interest: The authors declare no conflicts of interest.

References

1. Nampoothiri, K.U.K.; Krishnakumar, V.; Thampan, P.K.; Nair, M.A. *World Coconut Economy: Sectoral Issues, Markets and Trade*; Springer: Berlin/Heidelberg, Germany, 2019; ISBN 9789811327544.
2. Subramanian, P.; Thamban, C.; Josephrajkumar, A.; Vinayaka, H.; Hebbar, K.B.; Ravi, B.; Niral, V. *Coconut Development Board*; Ministry of Agriculture & Farmers Welfare: Kerala, India, 2020.
3. Chan, E.; Elevitch, C.R. Species Profiles for Pacific Island Agroforestry: *Cocos nucifera* (Coconut). 2006. Available online: https://raskisimani.files.wordpress.com/2013/01/cocos-nucifera-coconut.pdf/ (accessed on 4 December 2023).
4. Borel, L.D.M.S.; de Lira, T.S.; Ataíde, C.H.; de Souza Barrozo, M.A. Thermochemical Conversion of Coconut Waste: Material Characterization and Identification of Pyrolysis Products. *J. Therm. Anal. Calorim.* **2021**, *143*, 637–646. [CrossRef]

5. Kalina, S.; Navaratne, S.B. Developing an Edible Food Product from Tender Coconut Mesocarp and Analyzing Its Sensory Parameters. *Pharma Innov. J.* **2018**, *7*, 62–66.
6. Siriphanich, J.; Saradhuldhat, P.; Romphophak, T.; Krisanapook, K.; Pathaveerat, S.; Tongchitpakdee, S. Coconut (*Cocos nucifera* L.). In *Postharvest Biology and Technology of Tropical and Subtropical Fruits*; Woodhead Publishing: Sawston, UK, 2011; pp. 8–33.
7. Adeloye, J.B.; Osho, H.; Idris, L.O. Defatted Coconut Flour Improved the Bioactive Components, Dietary Fibre, Antioxidant and Sensory Properties of Nixtamalized Maize Flour. *J. Agric. Food Res.* **2020**, *2*, 100042. [CrossRef]
8. Beegum, P.P.S.; Nair, J.P.; Manikantan, M.R.; Pandiselvam, R.; Shill, S.; Neenu, S.; Hebbar, K.B. Effect of Coconut Milk, Tender Coconut and Coconut Sugar on the Physico-Chemical and Sensory Attributes in Ice Cream. *J. Food Sci. Technol.* **2022**, *59*, 2605–2616. [CrossRef] [PubMed]
9. Brainer, M.S.C.P. *Coco: Produção e Mercado*; Caderno Setorial ETENE; Banco do Nordeste do Brasil: Fortaleza, Brazil, 2021; pp. 1–13.
10. ABIR—Associação Brasileira das Indústrias de Refrigerantes e de Bebidas Não Alcoólicas. Água de Coco. 2022. Available online: https://abir.org.br/o-setor/bebidas/agua-de-coco/ (accessed on 15 November 2023).
11. Xu, S.; Ma, Z.; Chen, Y.; Li, J.; Jiang, H.; Qu, T.; Zhang, W.; Li, C.; Liu, S. Characterization of the Flavor and Nutritional Value of Coconut Water Vinegar Based on Metabolomics. *Food Chem.* **2022**, *369*, 130872. [CrossRef] [PubMed]
12. Bensalah, H.; Raji, M.; Abdellaoui, H.; Essabir, H.; Bouhfid, R.; Qaiss, A.e.K. Thermo-Mechanical Properties of Low-Cost "Green" Phenolic Resin Composites Reinforced with Surface Modified Coir Fiber. *Int. J. Adv. Manuf. Technol.* **2021**, *112*, 1917–1930. [CrossRef]
13. Li, N.; Jiang, H.; Yang, J.; Wang, C.; Wu, L.; Hao, Y.; Liu, Y. Characterization of Phenolic Compounds and Anti-Acetylcholinase Activity of Coconut Shells. *Food Biosci.* **2021**, *42*, 101204. [CrossRef]
14. Nunes, M.U.C. *Coprodutos—Ccoco*; EMBRAPA: Aracaju, Brazil, 2021. Available online: https://www.embrapa.br/en/agencia-de-informacao-tecnologica/cultivos/coco/pos-producao/coprodutos (accessed on 15 November 2023).
15. Food and Agriculture Organization of the United Nations. *FAOSTAT—Food and Agriculture Data Database 2021*; FAO: Rome, Italy, 2021.
16. Schena, T.; Lazzari, E.; Primaz, C.; Canielas Krause, L.; Machado, M.E.; Bastos Caramão, E. Upgrading of Coconut Fibers Bio-Oil: An Investigation By Gc×Gc/Tofms. *J. Environ. Chem. Eng.* **2020**, *8*, 103662. [CrossRef]
17. Ighalo, J.O.; Conradie, J.; Ohoro, C.R.; Amaku, J.F.; Oyedotun, K.O.; Maxakato, N.W.; Akpomie, K.G.; Okeke, E.S.; Olisah, C.; Malloum, A.; et al. Biochar from Coconut Residues: An Overview of Production, Properties, and Applications. *Ind. Crops Prod.* **2023**, *204*, 117300. [CrossRef]
18. Li, H.; Wang, X.; Tan, L.; Li, Q.; Zhang, C.; Wei, X.; Wang, Q.; Zheng, X.; Xu, Y. Coconut Shell and Its Biochar as Fertilizer Amendment Applied with Organic Fertilizer: Efficacy and Course of Actions on Eliminating Antibiotic Resistance Genes in Agricultural Soil. *J. Hazard. Mater.* **2022**, *437*, 129322. [CrossRef]
19. Cabral, M.M.S.; Abud, A.K.d.S.; Silva, C.E.d.F.; Almeida, R.M.R.G. Bioethanol Production from Coconut Husk Fiber. *Cienc. Rural* **2016**, *46*, 1872–1877. [CrossRef]
20. Bui, H.; Boutouil, M.; Levacher, D.; Sebaibi, N. Evaluation of the Influence of Accelerated Carbonation on the Microstructure and Mechanical Characteristics of Coconut Fibre-Reinforced Cementitious Matrix. *J. Build. Eng.* **2021**, *39*, 102269. [CrossRef]
21. James, A.; Yadav, D. Valorization of Coconut Waste for Facile Treatment of Contaminated Water: A Comprehensive Review (2010–2021). *Environ. Technol. Innov.* **2021**, *24*, 102075. [CrossRef]
22. Babatabar, M.A.; Yousefian, F.; Mousavi, V.M.; Hosseini, M.; Tavasoli, A. Pyrolysis of Lignocellulosic and Algal Biomasses in a Fixed-Bed Reactor: A Comparative Study on the Composition and Application Potential of Bioproducts. *Int. J. Energy Res.* **2022**, *46*, 9836–9850. [CrossRef]
23. Doe, B.; Aboagye, P.D.; Osei-Owusu, P.K.; Amoah, T.; Aidoo, A.; Amponsah, N.Y. Towards Circular Economy and Local Economic Development in Ghana: Insights from the Coconut Waste Value Chain. *Circ. Econ. Sustain.* **2023**, *3*, 347–372. [CrossRef]
24. Ricciardi, P.; Cillari, G.; Carnevale Miino, M.; Collivignarelli, M.C. Valorization of Agro-Industry Residues in the Building and Environmental Sector: A Review. *Waste Manag. Res.* **2020**, *38*, 487–513. [CrossRef]
25. Nascimento, R.J.M.; Pereira, K.R.A.; Avelino, F. Parametric and Modelling Studies of Rhodamine-B Adsorption Using Coconut Coir-Based Materials as Eco-Friendly Adsorbents. *J. Environ. Chem. Eng.* **2021**, *9*, 105943. [CrossRef]
26. Padilha, C.E.A.; da Costa Nogueira, C.; de Santana Souza, D.F.; de Oliveira, J.A.; dos Santos, E.S. Valorization of Green Coconut Fibre: Use of the Black Liquor of Organolsolv Pretreatment for Ethanol Production and the Washing Water for Production of Rhamnolipids by *Pseudomonas aeruginosa* ATCC 27583. *Ind. Crops Prod.* **2019**, *140*, 111604. [CrossRef]
27. Hasan, K.M.F.; Horváth, P.G.; Bak, M.; Alpár, T. A State-of-the-Art Review on Coir Fiber-Reinforced Biocomposites. *RSC Adv.* **2021**, *11*, 10548–10571. [CrossRef]
28. Prasetyo, I.; Permatasari, P.R.; Laksmana, W.T.; Rochmadi, R.; Oh, W.-C.; Ariyanto, T. Lignin Refinery Using Organosolv Process for Nanoporous Carbon Synthesis. *Molecules* **2020**, *25*, 3428. [CrossRef]
29. Marciano, S.J.; Avelino, F.; da Silva, L.R.R.; Mazzetto, S.E.; Lomonaco, D. Microwave-Assisted Phosphorylation of Organosolv Lignin: New Bio-Additives for Improvement of Epoxy Resins Performance. *Biomass Convers. Biorefin.* **2022**, *12*, 619–631. [CrossRef]
30. Marafon, A.C.; Nunes, M.U.C.; Amaral, A.F.C.; dos Santos, J.P. *Aproveitamento de Cascas de Coco Para Geração de Energia Térmica: Potencialidades e Desafios*; EMBRAPA: Brasília, Brazil, 2019.

31. Elroi, H.; Zbigniew, G.; Agnieszka, W.-C.; Piotr, S. Enhancing Waste Resource Efficiency: Circular Economy for Sustainability and Energy Conversion. *Front. Environ. Sci.* **2023**, *11*, 1303792. [CrossRef]
32. Maloney, B.K. Palaeoecology and the Origin of the Coconut. *GeoJournal* **1993**, *31*, 355–362. [CrossRef]
33. Gunn, B.F.; Baudouin, L.; Olsen, K.M. Independent Origins of Cultivated Coconut (*Cocos nucifera* L.) in the Old World Tropics. *PLoS ONE* **2011**, *6*, e21143. [CrossRef] [PubMed]
34. Nayar, N.M. *The Coconut Phylogeny, Origins, and Spread*, 4th ed.; Academic Press: Cambridge, MA, USA, 2016.
35. Batugal, P.; Rao, V.R.; Oliver, J. Coconut Genetic Resources Network Its History and Achievements. *CORD* **2005**, *21*, 34. [CrossRef]
36. FAO. *FAOSTAT—Crops and Livestock Products 2021*; FAO: Rome, Italy, 2021.
37. IBGE. *Levantamento Sistemático da Produção Agrícola*; IBGE: Rio de Janeiro, Brazil, 2022.
38. Liyanage, V.D. Varieties and Forms of the Coconut Palm Grown in Ceylon. *Ceylon Coconut Q.* **1958**, *9*, 1–10.
39. Niral, V.; Jerard, B.A. *Botany, Origin and Genetic Resources of Coconut. Coconut Palm (Cocos nucifera L.)—Research and Development Perspectives*; Springer: Singapore, 2019; pp. 57–111. [CrossRef]
40. Donadio, L.C.; Lederman, I.E.; Roberto, S.R.; Stucchi, E.S. Dwarfing-Canopy and Rootstock Cultivars for Fruit Trees. *Rev. Bras. Frutic.* **2019**, *41*, e997. [CrossRef]
41. Fontes, H.R.; Ferreira, J.M.S. *A Cultura do Coco*; Área de Informação da Sede-Col Criar Plantar ABC 500p/500r Saber (INFOTECA-E); Embrapa Tabuleiros Costeiros: Aracaju, Brazil, 2006; 500p.
42. Vishweshwar, S.; Meti, S.; Champa, B.V.; Nagaraja, M.S. Climate Based Coconut Yield Model for Arsikere Taluk of Hassan District in Karnataka. *J. Plant. Crops* **2020**, *48*, 27–35.
43. Hebbar, K.B.; Neethu, P.; Sukumar, P.A.; Sujithra, M.; Santhosh, A.; Ramesh, V.S.; Niral, V.; Hareesh, G.S.; Nameer, P.O.; Prasad, V.P.V. Understanding Physiology and Impacts of High Temperature Stress on the Progamic Phase of Coconut (*Cocos nucifera* L.). *Plants* **2020**, *9*, 1651. [CrossRef]
44. Foale, M. *The Coconut Odyssey: The Bounteous Possibilities of the Tree of Life*; Australian Centre for International Agricultural Research Canberra: Canberra, Australia, 2003.
45. Das, B.; Nair, B.; Arunachalam, V.; Reddy, K.V.; Venkatesh, P.; Chakraborty, D.; Desai, S. Comparative Evaluation of Linear and Nonlinear Weather-Based Models for Coconut Yield Prediction in the West Coast of India. *Int. J. Biometeorol.* **2020**, *64*, 1111–1123. [CrossRef]
46. Naresh Kumar, S.; Aggarwal, P.K. Climate Change and Coconut Plantations in India: Impacts and Potential Adaptation Gains. *Agric. Syst.* **2013**, *117*, 45–54. [CrossRef]
47. Abad, M.; Noguera, P.; Puchades, R.; Maquieira, A.; Noguera, V. Physico-chemical and chemical properties of some coconut coir dusts for use as a peat substitute for containerised ornamental plants. *Bioresour. Technol.* **2002**, *82*, 241–245. [CrossRef] [PubMed]
48. Worku, L.A.; Bachheti, R.K.; Tadesse, M.G. Isolation and Characterization of Natural Cellulose from *Oxytenanthera abyssinica* (Lowland Ethiopian Bamboo) Using Alkali Peroxide Bleaching Stages Followed by Aqueous Chlorite in Buffer Solution. *Int. J. Polym. Sci.* **2022**, *2022*, 5155552. [CrossRef]
49. van Dam, J.E.; Oever, M.J.v.D.; Keijsers, E.R.; van der Putten, J.C.; Anayron, C.; Josol, F.; Peralta, A. Process for production of high density/high performance binderless boards from whole coconut husk: Part 2: Coconut husk morphology, composition and properties. *Ind. Crop. Prod.* **2006**, *24*, 96–104. [CrossRef]
50. Kuram, E. Degradation Effects of Completely Biodegradable Composites to Moisture Absorption and Water Aging. In *Aging Effects on Natural Fiber-Reinforced Polymer Composites: Durability and Life Prediction*; Springer Nature: Singapore, 2022; pp. 85–113. [CrossRef]
51. Rethinam, P.; Krishnakumar, V. *Coconut Water: A Promising Natural Health Drink-Distribution, Processing and Nutritional Benefits*; Springer: Berlin/Heidelberg, Germany, 2022; pp. 1–507. [CrossRef]
52. Corradini, E.; Rosa, M.d.F.; de Macedo, B.P.; Paladin, P.D.; Mattoso, L.H.C. Composição Química, Propriedades Mecânicas e Térmicas da Fibra de Frutos de Cultivares de Coco Verde. *Rev. Bras. Frutic.* **2009**, *31*, 837–846. [CrossRef]
53. EMBRAPA. *Recomendações Técnicas Para o Cultivo do Coqueiro*; EMBRAPA: Aracaju, Brazil, 1993.
54. Galvão, E.U.P. Coqueiro: Recomendações de Cultivo. Available online: http://www.infoteca.cnptia.embrapa.br/handle/doc/697390 (accessed on 19 November 2023).
55. Santos, J.L.A.; Bispo, V.S.; Filho, A.B.C.; Pinto, I.F.D.; Dantas, L.S.; Vasconcelos, D.F.; Abreu, F.F.; Melo, D.A.; Matos, I.A.; Freitas, F.P.; et al. Evaluation of Chemical Constituents and Antioxidant Activity of Coconut Water (*Cocus nucifera* L.) and Caffeic Acid in Cell Culture. *Acad. Bras. Cienc.* **2013**, *85*, 1235–1247. [CrossRef] [PubMed]
56. Naik, M.; Sunil, C.K.; Rawson, A.; Venkatachalapathy, N. Tender Coconut Water: A Review on Recent Advances in Processing and Preservation. *Food Rev. Int.* **2020**, *38*, 1215–1236. [CrossRef]
57. Kim, T.K.; Lee, M.H.; Kim, S.M.; Kim, M.J.; Jung, S.; Yong, H.I.; Choi, Y.S. Physiochemical Properties of Reduced-Fat Duck Meat Emulsion Systems: Effects of Preemulsification with Vegetable Oils and Duck Skin. *Poult. Sci.* **2021**, *100*, 1291–1298. [CrossRef] [PubMed]
58. Varma, S.R.; Sivaprakasam, T.O.; Arumugam, I.; Dilip, N.; Raghuraman, M.; Pavan, K.B.; Rafiq, M.; Paramesh, R.; Varma, S.R.; Sivaprakasam, T.O.; et al. In Vitro Anti-Inflammatory and Skin Protective Properties of Virgin Coconut Oil Author Links Open Overlay Panel. *J. Tradit. Complement. Med.* **2019**, *9*, 5–14. [CrossRef]

59. Ghani, N.A.A.; Channip, A.A.; Chok Hwee Hwa, P.; Ja'afar, F.; Yasin, H.M.; Usman, A. Physicochemical Properties, Antioxidant Capacities, and Metal Contents of Virgin Coconut Oil Produced by Wet and Dry Processes. *Food Sci. Nutr.* **2018**, *6*, 1298–1306. [CrossRef]
60. Narayanankutty, A.; Illam, S.P.; Raghavamenon, A.C. Health Impacts of Different Edible Oils Prepared from Coconut (*Cocos nucifera*): A Comprehensive Review. *Trends Food Sci. Technol.* **2018**, *80*, 1–7. [CrossRef]
61. Neto, A.S.d.S.; Silva, L.M.S.; Neto, B.M. Use of Coconut Oil in the Production of Cosmetics: A Bibliographic Review. *Res. Soc. Dev.* **2020**, *9*, e75491110397. [CrossRef]
62. Manikantan, M.R.; Pandiselvam, R.; Beegum, S.; Mathew, A.C. Harvest and Postharvest Technology. In *Coconut Palm (Cocos nucifera L.)—Research and Development Perspectives*; Springer: Singapore, 2019; pp. 635–722. [CrossRef]
63. Lekshmi Sheela, D.; Nazeem, P.A.; Narayanankutty, A.; Manalil, J.J.; Raghavamenon, A.C. In Silico and Wet Lab Studies Reveal the Cholesterol Lowering Efficacy of Lauric Acid, a Medium Chain Fat of Coconut Oil. *Plant Foods Hum. Nutr.* **2016**, *71*, 410–415. [CrossRef] [PubMed]
64. Naresh Kumar, S.; Balakrishna, A. Seasonal Variations in Fatty Acid Composition of Oil in Developing Coconut. *J. Food Qual.* **2009**, *32*, 410. [CrossRef]
65. Boateng, L.; Ansong, R.; Owusu, W.B.; Steiner-Asiedu, M. Coconut Oil and Palm Oil's Role in Nutrition, Health and National Development: A Review. *Ghana Med. J.* **2016**, *50*, 189. [CrossRef] [PubMed]
66. Pinho, A.P.S.; Souza, A.F. Extração e Caracterização do Óleo de Coco (*Cocos nucifera* L.). *Biol. Saúde* **2018**, *8*, 9–18. [CrossRef]
67. Sundrasegaran, S.; Hui Mah, S.; Taylor, J.; Jaya, S. Extraction Methods of Virgin Coconut Oil and Palm-Pressed Mesocarp Oil and Their Phytonutrients. *eFood* **2020**, *1*, 381–391. [CrossRef]
68. Patil, U.; Benjakul, S. Coconut Milk and Coconut Oil: Their Manufacture Associated with Protein Functionality. *J. Food Sci.* **2018**, *83*, 2019–2027. [CrossRef]
69. Trinidad, T.P.; Mallillin, A.C.; Valdez, D.H.; Loyola, A.S.; Askali-Mercado, F.C.; Castillo, J.C.; Encabo, R.R.; Masa, D.B.; Maglaya, A.S.; Chua, M.T. Dietary Fiber from Coconut Flour: A Functional Food. *Innov. Food Sci. Emerg. Technol.* **2006**, *7*, 309–317. [CrossRef]
70. Kumar, S.; Senanayake, G.; Visvanathan, C.; Basu, B. Desiccated Coconut Industry of Sri Lanka: Opportunities for Energy Efficiency and Environmental Protection. *Energy Convers. Manag.* **2003**, *44*, 2205–2215. [CrossRef]
71. Hebbar, K.B.; Ramesh, S.V.; Ghosh, D.K.; Beegum, P.P.S.; Pandiselvam, R.; Manikantan, M.R.; Mathew, A.C. Coconut Sugar—A Potential Storehouse of Nutritive Metabolites, Novel Bio-Products and Prospects. *Sugar Tech.* **2022**, *24*, 841–856. [CrossRef]
72. Saraiva, A.; Carrascosa, C.; Ramos, F.; Raheem, D.; Lopes, M.; Raposo, A. Coconut Sugar: Chemical Analysis and Nutritional Profile; Health Impacts; Safety and Quality Control; Food Industry Applications. *Int. J. Environ. Res. Public Health* **2023**, *20*, 3671. [CrossRef] [PubMed]
73. Rahman, A.; Ulven, C.A.; Johnson, M.A.; Durant, C.; Hossain, K.G. Pretreatment of Wheat Bran for Suitable Reinforcement in Biocomposites. *J. Renew. Mater.* **2017**, *5*, 62–73. [CrossRef] [PubMed]
74. Macdonald, R.G.; Franklin, J.N. *The Pulp of Wood (Pulp and Paper Manufacture)*; MacGraw-Hill: New York, NY, USA, 1969.
75. Magagula, L.P.; Masemola, C.M.; Ballim, M.A.; Tetana, Z.N.; Moloto, N.; Linganiso, E.C. Lignocellulosic Biomass Waste-Derived Cellulose Nanocrystals and Carbon Nanomaterials: A Review. *Int. J. Mol. Sci.* **2022**, *23*, 4310. [CrossRef] [PubMed]
76. Barros, D.; Fernandes, É.; Jesus, M.; Barros, L.; Alonso-Esteban, J.I.; Pires, P.; Vaz Velho, M. The Chemical Characterisation of the Maritime Pine Bark Cultivated in Northern Portugal. *Plants* **2023**, *12*, 3940. [CrossRef]
77. Souza, A.G.; Junqueira, M.T.; de Lima, G.F.; Rangari, V.K.; Rosa, D.S. A New Proposal of Preparation of Different Polymorphs of Nanocellulose from Eucalyptus Citriodora. *J. Polym. Environ.* **2020**, *28*, 1150–1159. [CrossRef]
78. Adeleye, A.T.; Louis, H.; Akakuru, O.U.; Joseph, I.; Enudi, O.C.; Michael, D.P. A Review on the Conversion of Levulinic Acid and Its Esters to Various Useful Chemicals. *AIMS Energy* **2019**, *7*, 165–185. [CrossRef]
79. Fengel, D.; Wegener, G. *Wood—Chemistry, Ultrastructure, Reactions*, 2nd ed.; De Gruyter: Berlin, Germany, 1989; Volume 1.
80. Borovkova, V.S.; Malyar, Y.N.; Sudakova, I.G.; Chudina, A.I.; Zimonin, D.V.; Skripnikov, A.M.; Miroshnikova, A.V.; Ionin, V.A.; Kazachenko, A.S.; Sychev, V.V.; et al. Composition and Structure of Aspen (*Pópulus trémula*) Hemicelluloses Obtained by Oxidative Delignification. *Polymer* **2022**, *14*, 4521. [CrossRef] [PubMed]
81. Santos, B.L.P.; Jesus, M.S.; Mata, F.; Prado, A.A.O.S.; Vieira, I.M.M.; Ramos, L.C.; López, J.A.; Vaz-Velho, M.; Ruzene, D.S.; Silva, D.P. Use of Agro-Industrial Waste for Biosurfactant Production: A Comparative Study of Hemicellulosic Liquors from Corncobs and Sunflower Stalks. *Sustainability* **2023**, *15*, 6341. [CrossRef]
82. Wolf, M.; Berger, F.; Hanstein, S.; Weidenkaff, A.; Endreß, H.U.; Oestreich, A.M.; Ebrahimi, M.; Czermak, P. Hot-Water Hemicellulose Extraction from Fruit Processing Residues. *ACS Omega* **2022**, *7*, 13436–13447. [CrossRef]
83. Baucher, M.; Halpin, C.; Petit-Conil, M.; Boerjan, W. Critical Reviews in Biochemistry and Molecular Biology Lignin: Genetic Engineering and Impact on Pulping Lignin: Genetic Engineering and Impact on Pulping. *Crit. Rev. Biochem. Mol. Biol.* **2003**, *38*, 305–350. [CrossRef]
84. Kumar, R.; Strezov, V.; Weldekidan, H.; He, J.; Singh, S.; Kan, T.; Dastjerdi, B. Lignocellulose Biomass Pyrolysis for Bio-Oil Production: A Review of Biomass Pre-Treatment Methods for Production of Drop-in Fuels. *Renew. Sustain. Energy Rev.* **2020**, *123*, 106763. [CrossRef]

85. Jesus, M.; Romaní, A.; Mata, F.; Domingues, L. Current Options in the Valorisation of Vine Pruning Residue for the Production of Biofuels, Biopolymers, Antioxidants, and Bio-Composites following the Concept of Biorefinery: A Review. *Polymers* **2022**, *14*, 1640. [CrossRef] [PubMed]
86. Alharbi, M.A.H.; Hirai, S.; Tuan, H.A.; Akioka, S.; Shoji, W. Effects of Chemical Composition, Mild Alkaline Pretreatment and Particle Size on Mechanical, Thermal, and Structural Properties of Binderless Lignocellulosic Biopolymers Prepared by Hot-Pressing Raw Microfibrillated Phoenix Dactylifera and *Cocos nucifera*. *Polym. Test.* **2020**, *84*, 106384. [CrossRef]
87. Guleria, A.; Kumari, G.; Lima, E.C.; Ashish, D.K.; Thakur, V.; Singh, K. Removal of Inorganic Toxic Contaminants from Wastewater Using Sustainable Biomass: A Review. *Sci. Total Environ.* **2022**, *823*, 153689. [CrossRef]
88. Marafon, A.C.; Amaral, A.F.C.; De Lemos, E.E.P. Characterization of Bamboo Species and Other Biomasses with Potential for Thermal Energy Generation. *Pesqui. Agropecu. Trop.* **2019**, *49*, e55282. [CrossRef]
89. Droepenu, E.K.; Asare, E.A.; Dampare, S.B.; Adotey, D.K.; Gyampoh, A.O.; Kumi-Arhin, E. Laboratory and Commercial Synthesized Zinc Oxide Nanoparticles Adsorption onto Coconut Husk: Characterization, Isotherm, Kinetic, and Thermodynamic Studies. *Biointerface Res. Appl. Chem.* **2021**, *11*, 7871–7889. [CrossRef]
90. Brinchi, L.; Cotana, F.; Fortunati, E.; Kenny, J.M. Production of Nanocrystalline Cellulose from Lignocellulosic Biomass: Technology and Applications. *Carbohydr. Polym.* **2013**, *94*, 154–169. [CrossRef]
91. Brown, M.E.; Chang, M.C.Y. Exploring Bacterial Lignin Degradation. *Curr. Opin. Chem. Biol.* **2014**, *19*, 1–7. [CrossRef]
92. Nascimento, D.M.; Dias, A.F.; de Araújo Junior, C.P.; de Freitas Rosa, M.; Morais, J.P.S.; de Figueirêdo, M.C.B. A Comprehensive Approach for Obtaining Cellulose Nanocrystal from Coconut Fiber. Part II: Environmental Assessment of Technological Pathways. *Ind. Crops Prod.* **2016**, *93*, 58–65. [CrossRef]
93. Anuchi, S.O.; Campbell, K.L.S.; Hallett, J.P. Effective Pretreatment of Lignin-Rich Coconut Wastes Using a Low-Cost Ionic Liquid. *Sci. Rep.* **2022**, *12*, 6108. [CrossRef]
94. Andrade, S.N.; Veloso, C.M.; Fontan, R.C.I.; Bonomo, R.C.F.; Santos, L.S.; Brito, M.J.P.; Diniz, G.A. Chemical-Activated Carbon from Coconut (*Cocos nucifera*) Endocarp Waste and Its Application in the Adsorption of β-Lactoglobulin Protein. *Rev. Mex. De. Ing. Quim.* **2018**, *17*, 463–475. [CrossRef]
95. Jose, S.; Sajeena Beevi, B. Optimization of Ultrasonication Assisted Alkaline Delignification of Coir Pith Using Response Surface Methodology. *Bioresour. Technol. Rep.* **2023**, *21*, 101330. [CrossRef]
96. TAPPI, Acid-Insoluble Lignin in Wood and Pulp, Test Method T 222 Om-21. Available online: https://imisrise.tappi.org/TAPPI/Products/01/T/0104T222.aspx (accessed on 3 April 2024).
97. TAPPI, Alpha-, Beta- and Gamma-Cellulose in Pulp, Test Method T 203 Cm-22. Available online: https://imisrise.tappi.org/TAPPI/Products/01/T/0104T203.aspx (accessed on 3 April 2024).
98. NREL; Sluiter, A.; Hames, B.; Ruiz, R.; Scarlata, C.; Sluiter, J.; Templeton, D.; Crocker, D. *Analytical Procedure—Determination of Structural Carbohydrates and Lignin in Biomass*; NREL/TP-510-42618; NREL: Golden, CO, USA, 2012.
99. Kochova, K.; Schollbach, K.; Gauvin, F.; Brouwers, H.J.H. Effect of Saccharides on the Hydration of Ordinary Portland Cement. *Constr. Build. Mater.* **2017**, *150*, 268–275. [CrossRef]
100. Mariano, A.P.B.; Unpaprom, Y.; Ramaraj, R. Hydrothermal Pretreatment and Acid Hydrolysis of Coconut Pulp Residue for Fermentable Sugar Production. *Food Bioprod. Process.* **2020**, *122*, 31–40. [CrossRef]
101. Danso, H.; Martinson, D.B.; Ali, M.; Williams, J.B. Physical, Mechanical and Durability Properties of Soil Building Blocks Reinforced with Natural Fibres. *Constr. Build. Mater.* **2015**, *101*, 797–809. [CrossRef]
102. Robert, U.W.; Etuk, S.E.; Umoren, G.P.; Agbasi, O.E. Assessment of Thermal and Mechanical Properties of Composite Board Produced from Coconut (*Cocos nucifera*) Husks, Waste Newspapers, and Cassava Starch. *Int. J. Thermophys.* **2019**, *40*, 83. [CrossRef]
103. Pereira, T.G.T.; Silva, D.W.; Eugênio, T.M.C.; Scatolino, M.V.; de Carvalho Terra, I.C.; Fonseca, C.S.; Bufalino, L.; Mendes, R.F.; Mendes, L.M. Coconut Fibers and Quartzite Wastes for Fiber-Cement Production by Extrusion. *Mater. Today Proc.* **2019**, *31*, S309–S314. [CrossRef]
104. Fonseca, C.S.; Silva, M.F.; Mendes, R.F.; Hein, P.R.G.; Zangiacomo, A.L.; Savastano, H.; Tonoli, G.H.D. Jute Fibers and Micro/Nanofibrils as Reinforcement in Extruded Fiber-Cement Composites. *Constr. Build. Mater.* **2019**, *211*, 517–527. [CrossRef]
105. Wagle, A.; Angove, M.J.; Mahara, A.; Wagle, A.; Mainali, B.; Martins, M.; Goldbeck, R.; Raj Paudel, S. Multi-Stage Pre-Treatment of Lignocellulosic Biomass for Multi-Product Biorefinery: A Review. *Sustain. Energy Technol. Assess.* **2022**, *49*, 101702. [CrossRef]
106. Kumar, A.K.; Sharma, S. Recent Updates on Different Methods of Pretreatment of Lignocellulosic Feedstocks: A Review. *Bioresour Bioprocess.* **2017**, *4*, 7. [CrossRef] [PubMed]
107. Hassan, S.S.; Williams, G.A.; Jaiswal, A.K. Emerging Technologies for the Pretreatment of Lignocellulosic Biomass. *Bioresour. Technol.* **2018**, *262*, 310–318. [CrossRef] [PubMed]
108. Umoren, S.A.; Obot, I.B.; Israel, A.U.; Asuquo, P.O.; Solomon, M.M.; Eduok, U.M.; Udoh, A.P. Inhibition of Mild Steel Corrosion in Acidic Medium Using Coconut Coir Dust Extracted from Water and Methanol as Solvents. *J. Ind. Eng. Chem.* **2014**, *20*, 3612–3622. [CrossRef]
109. Ramasubramani, R.; Gunasekaran, K. Sustainable Alternate Materials for Concrete Production from Renewable Source and Waste. *Sustainability* **2021**, *13*, 1204. [CrossRef]
110. Narciso, C.R.P.; Reis, A.H.S.; Mendes, J.F.; Nogueira, N.D.; Mendes, R.F. Potential for the Use of Coconut Husk in the Production of Medium Density Particleboard. *Waste Biomass Valoriz.* **2021**, *12*, 1647–1658. [CrossRef]

111. Souza, M.J.C.; de Melo, R.R.; Guimarães, J.B.; Carnaval, T.K.B.d.A.; Pimenta, A.S.; Mascarenhas, A.R.P. Wood–Cement Boards with Addition of Coconut Husk. *Wood Mater. Sci. Eng.* **2021**, *17*, 617–626. [CrossRef]
112. Hoang, L.P.; Van, H.T.; Hang Nguyen, T.T.; Nguyen, V.Q.; Thang, P.Q. Coconut Shell Activated Carbon/$CoFe_2O_4$ Composite for the Removal of Rhodamine B from Aqueous Solution. *J. Chem.* **2020**, *2020*, 9187960. [CrossRef]
113. Padilha, C.E.A.; da Costa Nogueira, C.; de Santana Souza, D.F.; de Oliveira, J.A.; dos Santos, E.S. Organosolv Lignin/Fe_3O_4 Nanoparticles Applied as a β-Glucosidase Immobilization Support and Adsorbent for Textile Dye Removal. *Ind. Crops Prod.* **2020**, *146*, 112167. [CrossRef]
114. Esfandiar, N.; Suri, R.; McKenzie, E.R. Simultaneous Removal of Multiple Polycyclic Aromatic Hydrocarbons (PAHs) from Urban Stormwater Using Low-Cost Agricultural/Industrial Byproducts as Sorbents. *Chemosphere* **2021**, *274*, 129812. [CrossRef]
115. Marín-Velásquez, T.D.; Cóndor-Salvatierra, E.J. Capacidad de Retención de Hidrocarburos del Endocarpio de Coco en Aguas Aceitosas. *Tecnol. Cienc. Agua* **2021**, *12*, 1–36. [CrossRef]
116. Thongsamer, T.; Vinitnantharat, S.; Pinisakul, A.; Werner, D. Fixed-Bed Biofilter for Polluted Surface Water Treatment Using Chitosan Impregnated-Coconut Husk Biochar. *Environ. Pollut.* **2023**, *334*, 122137. [CrossRef] [PubMed]
117. Pettit, T.; Irga, P.J.; Torpy, F.R. Functional Green Wall Development for Increasing Air Pollutant Phytoremediation: Substrate Development with Coconut Coir and Activated Carbon. *J. Hazard. Mater.* **2018**, *360*, 594–603. [CrossRef] [PubMed]
118. Ebrahimi, M.; Caparanga, A.R.; Ordono, E.E.; Villaflores, O.B. Evaluation of Organosolv Pretreatment on the Enzymatic Digestibility of Coconut Coir Fibers and Bioethanol Production via Simultaneous Saccharification and Fermentation. *Renew. Energy* **2017**, *109*, 41–48. [CrossRef]
119. Arantes, H.T.L.; Machado, M.A.; Santoro, M.C.; Freitas, J.C.C.; Ronconi, C.M.; Ligiero, C.B.P.; Cassini, S.T.A.; Sampaio, I.C.F.; Luz, P.P. Effect of Activated Biochar as a Low-Cost Catalyst on the Quality of Catalytic Intermediate Co-Pyrolysis Oil from Waste Polystyrene and Green Coconut Pericarp. *Fuel Process. Technol.* **2023**, *240*, 107539. [CrossRef]
120. Sahoo, S.; Basu, D.; Kumar, A.; Nawale, M.; Kadam, S.; Bhujbal, A.; Rajkumar, K.; Bhowmick, A.; Chattopadhyay, S. Bio-Based Oil Derived from Waste Coconut Shell: A Potential Additive for Enhancing Silanization in Silica Filled Styrene Butadiene Copolymer. *J. Polym. Res.* **2022**, *29*, 311. [CrossRef]
121. Chaos-Hernández, D.; Reynel-Avila, H.E.; Mendoza-Castillo, D.I.; Bonilla-Petriciolet, A.; Aguayo-Villarreal, I.A. Functionalization and Activation of Carbon-Based Catalysts with KOH and Calcium and Their Application in Transesterification to Produce Biodiesel: Optimization of Catalytic Properties and Kinetic Study. *Fuel* **2022**, *310*, 122066. [CrossRef]
122. Gopal, M.; Gupta, A.; Shahul Hameed, K.; Sathyaseelan, N.; Khadeejath Rajeela, T.H.; Thomas, V.G. Biochars Produced from Coconut Palm Biomass Residues Can Aid Regenerative Agriculture by Improving Soil Properties and Plant Yield in Humid Tropics. *Biochar* **2020**, *2*, 211–226. [CrossRef]
123. Oliveira, M.B.S.; Valentim, I.B.; Santos, T.R.; Xavier, J.A.; Ferro, J.N.S.; Barreto, E.O.; Santana, A.E.G.; Melo, V.L.; Bottoli, C.B.G.; Goulart, M.O.F. Photoprotective and Antiglycation Activities of Non-Toxic *Cocos nucifera* Linn. (Arecaceae) Husk Fiber Ethanol Extract and Its Phenol Chemical Composition. *Ind. Crops Prod.* **2021**, *162*, 113246. [CrossRef]
124. Tang, P.L.; Hassan, O.; Md-Jahim, J.; Mustapha, W.A.W.; Maskat, M.Y. Fibrous Agricultural Biomass as a Potential Source for Bioconversion to Vanillic Acid. *Int. J. Polym. Sci.* **2014**, *2014*, 509035. [CrossRef]
125. Luis-Zarate, V.H.; Rodriguez-Hernandez, M.C.; Alatriste-Mondragon, F.; Chazaro-Ruiz, L.F.; Rangel-Mendez, J.R. Coconut Endocarp and Mesocarp as Both Biosorbents of Dissolved Hydrocarbons in Fuel Spills and as a Power Source When Exhausted. *J. Environ. Manag.* **2018**, *211*, 103–111. [CrossRef]
126. Lasmini, S.A.; Rosmini, R.; Lakani, I.; Hayati, N.; Nasir, B.H. Increasing Shallot Production in Marginal Land Using Mulches and Coconut Husk Fertilizer. *Int. J. Des. Nat. Ecodyn.* **2021**, *16*, 105–110. [CrossRef]
127. Zubiolo, C.; de Santana, H.E.P.; Pereira, L.L.; Ruzene, D.S.; Silva, D.P.; Freitas, L.S. Bio-Oil Production and Characterization from Corn Cob and Sunflower Stem Pyrolysis. *Ind. Eng. Chem. Res.* **2024**, *63*, 65–77. [CrossRef]
128. Zhang, Y.; Cui, Y.; Chen, P.; Liu, S.; Zhou, N.; Ding, K.; Fan, L.; Peng, P.; Min, M.; Cheng, Y.; et al. *Gasification Technologies and Their Energy Potentials*; Elsevier: Amsterdam, The Netherlands, 2019; ISBN 9780444642004.
129. Gupta, S.; Patel, P.; Mondal, P. Biofuels Production from Pine Needles via Pyrolysis: Process Parameters Modeling and Optimization through Combined RSM and ANN Based Approach. *Fuel* **2022**, *310*, 122230. [CrossRef]
130. Grams, J.; Jankowska, A.; Goscianska, J. Advances in Design of Heterogeneous Catalysts for Pyrolysis of Lignocellulosic Biomass and Bio-Oil Upgrading. *Microporous Mesoporous Mater.* **2023**, *362*, 112761. [CrossRef]
131. Sørmo, E.; Krahn, K.M.; Flatabø, G.Ø.; Hartnik, T.; Arp, H.P.H.; Cornelissen, G. Distribution of PAHs, PCBs, and PCDD/Fs in Products from Full-Scale Relevant Pyrolysis of Diverse Contaminated Organic Waste. *J. Hazard. Mater.* **2023**, *461*, 132546. [CrossRef] [PubMed]
132. Velasco-Muñoz, J.F.; Aznar-Sánchez, J.A.; López-Felices, B.; Román-Sánchez, I.M. Circular Economy in Agriculture. An Analysis of the State of Research Based on the Life Cycle. *Sustain. Prod. Consum.* **2022**, *34*, 257–270. [CrossRef]
133. Scarano, P.; Sciarrillo, R.; Tartaglia, M.; Zuzolo, D.; Guarino, C. Circular Economy and Secondary Raw Materials from Fruits as Sustainable Source for Recovery and Reuse. A Review. *Trends Food Sci. Technol.* **2022**, *122*, 157–170. [CrossRef]
134. United Nations. *Sustainable Development Goals 2024. United Nations SDG12: Ensure Sustainable Consumption and Production Patterns*; United Nations: New York, NY, USA, 2024. Available online: https://sdgs.un.org/goals/goal12 (accessed on 3 January 2024).
135. Hayes, D.J.M. Biomass Composition and Its Relevance to Biorefining. In *The Role of Catalysis for the Sustainable Production of Bio-Fuels and Bio-Chemicals*; Elsevier: Amsterdam, The Netherlands, 2013; pp. 27–65.

136. Wagh, M.S.; Chandra Nath, P.; Chakraborty, A.; Amrit, R.; Mishra, B.; Kumar Mishra, A.; Kishore Mohanta, Y. Valorisation of Agro-Industrial Wastes: Circular Bioeconomy and Biorefinery Process—A Sustainable Symphony. *Process Saf. Environ. Prot.* **2024**, *183*, 708–725. [CrossRef]
137. Haque, F.; Fan, C.; Lee, Y.Y. From Waste to Value: Addressing the Relevance of Waste Recovery to Agricultural Sector in Line with Circular Economy. *J. Clean. Prod.* **2023**, *415*, 137873. [CrossRef]
138. Ramírez, J.F.G.; Muñoz, R.C.; Sossa, J.W.Z. Innovations and Trends in the coconut Agroindustry Supply: A Technological and Foresight Analysis. *Front. Sustain. Food Syst.* **2023**, *7*, 1048450.
139. Alan, H.; Köker, A.R. Analyzing and Mapping Agricultural Waste Recycling Research: An Integrative Review for Conceptual Framework and Future Directions. *Resour. Policy* **2023**, *85*, 103987. [CrossRef]
140. United Nations. *United Nations Transforming Our World: The 2030 Agenda for Sustainable Development*; United Nations Sustainable Knowledge Platform; Division for Sustainable Development Goals: New York, NY, USA, 2015.
141. Ignacio, I.F.; Miguel, T.S. Research Opportunities on the Coconut (*Cocos nucifera* L.) Using New Technologies. *S. Afr. J. Bot.* **2021**, *141*, 414–420. [CrossRef]
142. Krishnan, S.; Zulkapli, N.S.; Kamyab, H.; Taib, S.M.; Din, M.F.B.M.; Majid, Z.A.; Chaiprapat, S.; Kenzo, I.; Ichikawa, Y.; Nasrullah, M.; et al. Current Technologies for Recovery of Metals from Industrial Wastes: An Overview. *Environ. Technol. Innov.* **2021**, *22*, 101525. [CrossRef]
143. Azevedo, A.R.G.; Amin, M.; Hadzima-Nyarko, M.; Agwa, I.S.; Zeyad, A.M.; Tayeh, B.A.; Adesina, A. Possibilities for the Application of Agro-Industrial Wastes in Cementitious Materials: A Brief Review of the Brazilian Perspective. *Clean. Mater.* **2022**, *3*, 100040. [CrossRef]
144. Diwan, B.; Mukhopadhyay, D.; Gupta, P. *Recent Trends in Biorefinery-Based Valorisation of Lignocellulosic Biomass*; Elsevier: Amsterdam, The Netherlands, 2020; ISBN 9780128179512.

sustainability

Article

Liquid Hot Water (LHW) and Hydrothermal Carbonization (HTC) of Coffee Berry Waste: Kinetics, Catalysis, and Optimization for the Synthesis of Platform Chemicals

Alejandra Sophia Lozano-Pérez * and Carlos Alberto Guerrero-Fajardo

Departamento de Química, Universidad Nacional de Colombia, Bogotá 111321, Colombia; caguerrerofa@unal.edu.co
* Correspondence: aslozanop@unal.edu.co

Abstract: Colombia is the world's leading producer of mildly washed arabica coffee and produces 12.6 million bags of green coffee, but at the same time, 784,000 tons of waste biomass are dumped in open fields, of which only 5% is recovered or used. The objective of this project was to evaluate the production of platform chemicals from these coffee wastes for sustainable resource management. To achieve this, biomass characterization was carried out using proximate analysis, ultimate analysis, and structural analysis. Hydrothermal valorization was carried out at a temperature range of 120–180 °C (LHW) and 180–260 °C (HTC) for one hour. The platform chemicals obtained were quantified by HPLC-RI and monitored by pH and conductivity, and the solid fraction was characterized by monitoring the functional groups in IR spectroscopy and elemental analysis. Hydrolysis processes were obtained at 150 °C, production of platform chemicals at 180 °C, and maximum concentration at 180 °C-4 h; over 200 °C, degradation of the products in the liquid fraction starts to take place. Homogeneous basic and acid catalysts were used to improve the yields of the reaction. The kinetics of the hydrolysis of lignocellulosic structures to sugars were also analyzed and described, and reaction orders of 1 (LHW), 3 (HTC), and their respective reaction rate equations were reported.

Keywords: coffee berry waste; valorization; hydrothermal process; kinetic analysis; platform chemicals

1. Introduction

Coffee is a beverage that has been integrated into modern society, becoming an integral part of the daily routine of millions of people around the world. There are approximately 1 billion coffee drinkers worldwide, and over 2 billion cups of coffee are consumed daily [1]. Coffee occupies a significant position in today's industry and has a far-reaching impact on economies around the world. With an estimated annual production of more than 169.50 million bags (10 million Tons) [2], coffee has become one of the most traded commodities on the planet [3], supporting the livelihoods of countless farmers, traders, and businesses along its complex supply chain. But its importance transcends mere economic value; coffee represents culture, community, and connection, fostering social interactions and driving innovation [4,5].

Even though coffee is such an important commodity, its industry produces over 23 million tons of trash every year, from the pulp of fresh coffee cherries to the packing of roasted beans, according to sustainability researcher Gunter Pauli. Brazil, Vietnam, and Colombia are the main producers of coffee worldwide [6]. According to the FNC, in 2021, the top producer of mild-washed Arabica coffee in the world, Colombia, produced 12.6 million bags of green coffee weighing 60 kg [7]. The Colombian coffee agroindustry uses only 9.5% of the fruit for beverage preparation, with 90.5 percent being by-products dumped into water bodies. This pollution reduces ecosystem life and soil quality. Around two million tons of pulp and 420,000 tons of mucilage are dumped in open fields, affecting the value chain and environment. Efforts are being made to utilize these materials in the pig and livestock industries, beverages, vinegar, biogas, caffeine, pectins, peptic enzymes, proteins, and fertilizers [8–11]. There are several ways of coffee waste management, such as oxidation ponds [12], modular anaerobic treatment systems [13], anaerobic upflow filters [14], worm farming [15], composting [16], and hydrothermal valorization [17–21].

This paper will mainly focus on hydrothermal carbonization (HTC) and liquid hot water extraction (LHW). Hydrothermal carbonization (HTC) is a promising thermal conversion technology that transforms biomass into hydrochar through the application of heat and pressure in a water-rich environment. This process mimics the natural coal formation process, ranging from 180–260 °C, resulting in a carbon-rich product with improved fuel properties and potential applications in bioenergy, carbon sequestration, and soil improvement [22]. HTC is a sustainable and efficient method for converting waste biomass into valuable products, such as biofuels, chemicals, and other commodities, and it has the potential to improve energy recovery and economic performance in biomass waste treatment systems. Liquid hot water extraction (LHW) involves increasing the water temperature under high pressure. This enhances the solubility and extraction efficiency of target compounds, making LHW a valuable technique for obtaining bioactive compounds, essential oils, and other valuable substances from various natural sources [23,24]. LHW (120–180 °C) is considered an environmentally friendly and cost-effective alternative to traditional solvent extraction methods, as it eliminates the need for harmful organic solvents. The extracted bioactive compounds find applications in the food, pharmaceutical, and nutraceutical industries due to their potential health benefits.

Previous research has been performed regarding the use of coffee (mainly SGC (spent coffee grounds) and silverskin) (Scheme 1) and different hydrothermal processes to obtain platform molecules, such as biochar [25], polysaccharides (Cellobiose, Glucose, Xylose, Galactose, Arabinose, and Maltose) [26], and activated carbon [20] using HTC; via LHW polyphenols, hydroxymethylfurfural, feruloylquinic acid, epicatechin [27], and antioxidants (DPPH, ABTS, and HPLC) [28] were obtained. It is worth noticing that little to no information was found on the hydrothermal valorization of coffee berry waste (made up of husk, mucilage, parchment, and silverskin) in order to present a possibility for sustainable management of coffee wastes.

Scheme 1. Composition of coffee waste. Data from [29].

Hydrothermal valorization offers a promising avenue for the sustainable utilization of coffee waste for the production of valuable platform chemicals. This innovative process involves subjecting coffee grounds to high-temperature, high-pressure conditions in water, breaking down the organic components into simpler molecules. This approach not only mitigates the environmental burden posed by coffee waste but also harnesses its potential as a renewable resource. By utilizing hydrothermal valorization, various platform chemicals, such as phenolic compounds, sugars, and organic acids, can be derived from coffee waste. These chemicals serve as building blocks for numerous industrial applications, including the production of biofuels, bioplastics, and pharmaceuticals. Embracing this method contributes to a circular economy model by reducing waste generation, lowering reliance on fossil fuels, and promoting the sustainable use of resources, ultimately advancing toward a more environmentally friendly and economically viable solution.

The scope of this paper is to analyze the different platform chemicals that are produced through LHW and HTC using residual waste from coffee cherries, their yields, and the best conditions in order to make the process sustainable. This was achieved via a set of hydrothermal experiments modifying time, temperature, and the implementation of homogeneous catalysts. The quantification and characterization were performed through analytic methods (HPLC-RI, IR-spectroscopy, elemental analysis). Purification, kinetic analysis, and optimization of variables based on each platform chemical are some of the goals of this research in order to assess the sustainability of the extraction of platform chemicals from agroindustrial waste as an alternative way of managing waste.

2. Materials and Methods

2.1. Biomass Characterization

The coffee cherry biomass was dried, ground, and sieved according to the technical report NREL/TP 510-42620 using an ASTM 18 mesh (>1 mm) sieve.

2.1.1. Proximate Assay

The assays consisted of moisture (NREL/TP-510-42621 [30]), ash (NREL/TP-510-42622 [31]), volatile matter (ISO-18123-2015 [32]), and fixed carbon (ASTM D 3172-89 [33]), and the procedure is shown in the Supplementary Material.

2.1.2. Ultimate Assay

A Thermo Scientific (Waltham, MA, USA) FLASH 2000 Series Organic Elemental Analyzer was used, and the Technical Report NREL/TP-510-42620 was followed.

2.1.3. Chemical Composition

The lignin, cellulose, and hemicellulose were measured as Neutral Detergent Fiber, Acid Detergent Fiber, and lignin. The procedure presented in the Journal of Dairy Science called "Methods for Dietary Fiber, Neutral Detergent Fiber, and Nonstarch Polysaccharides in Relation to Animal Nutrition" was carried out.

2.2. Hydrothermal Experiments

Experiments were performed in a 100 mL batch reactor with a 1:9 biomass-to-water ratio (3 g of biomass and 27 g of water), which was then placed in a Memmert oven at different temperatures (120–180 °C) at autogenous pressure for 1 h (measured after reaching the desired temperature). To analyze the kinetics of the processes, a set of experiments was conducted at 180 °C, modifying the reaction time from 1 to 5 h. To test another hydrothermal range, reactions were performed at different temperatures from 200 to 260 °C (HTC) at autogenous pressure for 1 h with a 1:9 biomass–water ratio (10 g biomass, 90 g water) in a 500 mL batch reactor. To analyze the kinetics of these processes, a set of experiments was performed at 200 °C, modifying the reaction time from 1 to 5 h. The heating was performed through a heating jacket that surrounded the reactor. The reactors are shown in the Supplementary Material.

2.3. Characterization of Solid and Liquid Fractions

The solid and liquid fractions obtained via hydrothermal processes were separated through vacuum filtration, and they were measured to keep a mass balance. The solid was recovered and washed with water, ethanol, and acetone and dried at 105 °C until constant weight. In order to characterize the solid, Infrared spectroscopy was conducted with a Nicolet iS10 Spectrometer Thermo Fisher Scientific, and elemental analysis was taken with a Thermo Flash 2000 following the NREL/TP-510-42620 as well as SEM images (TESCAN VEGA3). To monitor the liquid fraction, pH and conductivity were measured, and the platform chemicals (PC) were quantified using HPLC-RI. The yields were measured using Equation (1):

$$\text{Yield} = \frac{(grams\ of\ PC)}{(grams\ of\ lignocelullosic\ structure\ in\ biomass)} * 100\% \quad (1)$$

2.4. Analytical Methods

For the quantification of PC, a method was developed for HPLC-RI using standards of xylose, glucose, formic acid, levulinic acid, HMF, and furfural. Said method was obtained with a Hitachi Elite LaChrom (Tokyo, Japan) with a Hitachi L-2490 refraction index detector at 40 °C, a SHODEX Sugar SH1821 column at 60 °C, a mobile phase of 0.005 M H_2SO_4, and a flow rate of 0.5 mL/min. The validation of this method is presented in the Supplementary Material.

2.5. Homogeneous Catalysts

In order to improve the yields obtained in the hydrothermal reactions previously made, a set of acid and basic homogeneous catalysts were used to make the reactions more efficient and specific. H_2SO_4 [0.02 M], CH_3COOH [0.2 M], KOH [0.02 M], and $NaHCO_3$ [0.2 M] solutions were prepared in order to replace the water in the biomass/water (3 g biomass/27 g catalysts solution). Reactions at 180 °C were performed for 1, 2, and 3 h. The obtained fractions were characterized as previously carried out in the other experiments.

2.6. Purification

After optimizing the conditions and obtaining the best yielding conditions, tests of purification were performed using techniques such as distillation (at 100 °C), liquid–liquid partition (octanol–water), C-18 column chromatography (water and ethanol as mobile phases), and ionic interchange chromatography (DOWEX-50W, AMBERLITE IRA 400, and DOWEX* MARATHON* MR-3 with Buffers of pH 4.3, 7 and 12 as mobile phases). All of the obtained extracts were characterized via HPLC-RI in order to test the efficiency of the separation.

3. Results

3.1. Biomass Characterization

3.1.1. Proximal Assay

Table 1 presents the characterization of coffee biomass obtained in this study and other types of coffee biomass from the previous articles. The chemical composition of the biomass is influenced by moisture content, ash content, fixed carbon, volatile matter, and geographical location. Moisture content is crucial for accurate and repeatable results as it helps to start with conditions with similar moisture content. Ash content is influenced by the mineral content of the biomass source and corresponds to inorganic products specific to the biomass. Fixed carbon is related to the organic matter in the biomass and can be influenced by the type of biomass, maturity stage, and carbon/hydrogen ratio. Volatile matter, consisting of combustible gases and compounds, is affected by factors such as biomass decomposition, volatile organic compounds, and processing methods. Geographical location, harvesting and processing methods, contaminants, age and maturity of the biomass source, and species and variety within a biomass type also affect the results of proximate analysis. These factors are important for comparing the results with reports found in the literature, as they may not be consistent across different types of coffee residues used in valorization processes.

Table 1. Proximal assay of coffee waste.

Essay	Coffee Cherry (%) [a]	Pulp (%) [b]	Husk (%) [c]	SCG * (%) [d]
Moisture (Initial Biomass)	80.79	73.85	8.88	55.2
Moisture (BHP) **	10.94	-	-	-
Ashes	7.79	6.29	0.79	15.3
Volatile matter	10.06	9.80	75.85	83.3
Fixed carbon	1.36	-	14.48	1.5

[a] Obtained results. [b] [34], [c] [35], [d] [36]. * Spent Coffee Grounds ** Biomass for hydrothermal processes.

The biomass of coffee cherries is similar to that of coffee pulp, as it contains pulp, husk, and mucilage. The moisture of the biomass after kiln drying and grinding is calculated as BHP moisture. The initial biomass moisture corresponds to the characterization of coffee cherries without pre-treatment processes. Reducing the moisture from 80.79 to 10.94 (eight times) was possible. However, due to its high moisture percentage, hydrothermal techniques can be used by calculating the biomass-to-water ratio without having to add external water. The biomass has a low amount of volatile matter compared to husks and SCG, indicating a low amount of organic components that release gases when heated. It also has a low, but relevant, amount of volatile matter when compared to husks and SCG, implying that it consists of a low amount of organic components that release gases when the biomass is heated, but that will also be the organic groups that will later be extracted in the aqueous phase, so that, perhaps for recovery processes, parts of the coffee waste with a higher percentage of volatiles could be of interest.

3.1.2. Ultimate Assay

The ultimate assay is of great importance as it provides accurate information on the elemental composition of biomass, which is essential for understanding its potential and applications. It provides information on the amount of carbon, which is directly related to the energy content of the biomass and its suitability as a renewable energy source. The most relevant factors are the O/C, H/C, and C/N ratios, where said ratios classify the biomass within the van Krevelen diagram (Figure 1) and have been widely used to describe the change in biomass composition during thermochemical decomposition [37]. On the other hand, the C/N ratio shows how relevant and feasible a biomass valorization process by means of microbial decomposition can be. When the C/N ratio is over 24/1, the rate of relative decomposition is slower; when it is equal to 24/1, it is the ideal microbial diet, and when it is under 24/1, the rate is faster.

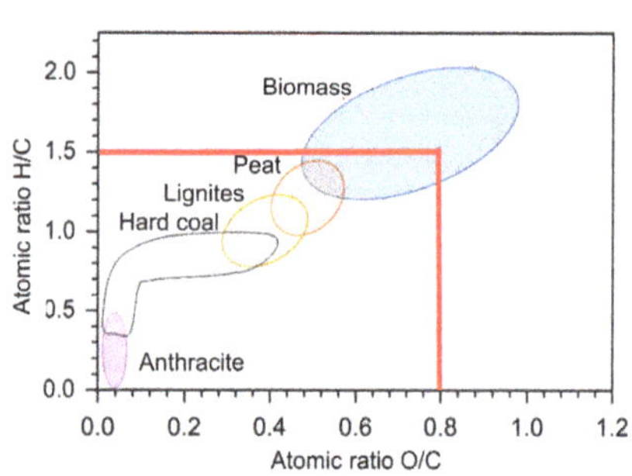

Figure 1. Van Krevelen diagram for coffee cherry waste. The red line shows the characterization of the initial biomass.

Regarding the characterization of coffee cherries by ultimate assay, the percentages of the different elements are quite similar to those reported for coffee husks (Table 2) since, again, this type of residue is included in coffee cherries. No information was reported on the characterization of either the pulp or the cherry. Considering the O/C and H/C ratios and their location in the Van Krevelen diagram (Figure 1), the sample obtained can be effectively classified as a biomass-type fuel with a higher amount of oxygen and hydrogen than traditional fossil fuels and, therefore, has acceptable conditions to be implemented as biofuel.

Table 2. Ultimate assay of coffee waste.

Element	Coffee Cherry (%) [a]	Pulp (%) [b]	Husk (%) [c]	SCG (%) [d]
Carbon	45.27		44.95	68.52
Hydrogen	4.862		5.34	11.04
Nitrogen	1.471	1.5	0.60	1.40
Sulfur	0.138		0.03	-
Elemental ratio				
Ratio O/C		0.792		
Ratio H/C		1.498		
Ratio C/N		19.98		

[a] Obtained results. [b] [38], [c] [35], [d] [39].

In the same way, the C/N ratio corresponding to 19.98 allows for the biomass to be classified according to its possibility of being valorized or treated with microorganisms. In the present case, the ratio favors the rapid decomposition of the biomass by the microorganisms (20:1), which is why one of the most common uses of coffee residues is as a soil fertilizer.

3.1.3. Structural Essay

Depending on the plant species and developmental circumstances, the structures and chemical composition of these biopolymers differ significantly. Hemicellulose and lignin are linked to cellulose by covalent bonds, while cellulose is linked to hemicellulose and lignin by hydrogen bonds. Their direct utilization is a challenge due to the complicated and dense spatial structure, so treatment, in this case hydrothermal, is required to enable efficient conversion. It is important to identify the composition of the biomass in order to know the possible platform chemicals that will be produced after valorization. Considering the results obtained in Table 3, a relevant percentage of all fractions (lignin, cellulose, and hemicellulose) and, therefore, different types of reactions and products will be obtained. For this work, the main focus was on monitoring the valorization of hemicellulose and cellulose (hydrolyzable fraction) in order to focus on the production of sugars (glucose and xylose), organic acids (levulinic, formic, and acetic), and furfural derivatives (furfural and hydroxymethylfurfural (HMF)). It is worth stressing the importance of this characterization process as it allows us to know the physicochemical routes and processes that the biomass and its possible valorization products will take.

Table 3. Structural essay on coffee waste.

Essay	Coffee Cherry (%) [a]	Pulp (%) [b]	Husk (%) [c]	SCG (%) [d]
Dry matter (%)	95.5			
Hemicellulose	12.5	3.60	7	12.1
Lignin	13.7	20.07	9	17.8
Cellulose	27.6	25.88	43	23.6

[a,b] [34], [c] [40], [d] [39].

This study compares cherries and other types of coffee waste, revealing greater similarity to SCG but a decrease in hemicellulose compared to pulp and husk. Understanding the percentages of each structure in the biomass can help predict the type of platform chemicals and hydrothermal process to be implemented. In pulp, low hemicellulose makes it unlikely to obtain xylose and derivatives, requiring the use of lignin through an HTC process. In husk, cellulose is abundant but contains little hemicellulose and lignin, making it suitable for glucose extraction through LHW and fermentation processes. Lignin (13.7%), acts as a structural component in the biomass cell wall, enveloping and protecting hemicellulose. This hinders the breaking process with water, decreasing its decomposition and valorization efficiency. Pre-treatment at scaling up or initial fractionation of cellulose, hemicellulose, and lignin could be beneficial but would require more time, energy, and expenses. It is important to note that the sum of lignin, hemicellulose, and cellulose does not equal 100% due to the non-consideration of proteins, fats, fibers, and carbohydrates in the characterization process.

3.2. *Hydrothermal Valorization via LHW*

Hydrothermal processes were carried out from 120 to 180 °C for one hour and at 180 °C for 5 h, starting in all cases with 3 g of biomass and 27 g of water. The solid, liquid, and gas phase results were obtained, as shown in Table 4. Solids and liquids were measured after filtration, and the gas phase was obtained by subtracting the values of the solids and gases from the initial weight, considering the law of conservation of mass.

Table 4. Fractions obtained via LHW.

Temperature (°C)	Time (Hours)	Solid Fraction (g)	Liquid Fraction (g)	Gas Fraction (g)
120	1	2.784	22.920	4.296
130	1	2.689	22.918	4.393
140	1	2.650	22.830	4.520
150	1	2.633	22.835	4.532
160	1	2.590	22.873	4.537
170	1	2.546	22.904	4.550
180	1	2.498	22.890	4.502
180	2	2.148	20.310	7.852
180	3	1.995	16.871	12.005
180	4	1.895	14.281	14.105
180	5	1.390	15.120	13.489

It can then be observed that the ratios of solid fraction, liquid fraction, and gas fraction remain constant in all tests performed within one hour, indicating that there is not a large number of reactions taking place and that the physical characteristics of the solid phase remain relatively constant. On the other hand, with increasing time at a constant temperature, a variation in both the solid and liquid phases and an increase in the gaseous fraction are observed. This may be due to dehydration processes in the medium leading to the removal of water in gaseous form, decarboxylation leading to the removal of CO_2, and changes in the structures in both the aqueous and solid phases.

3.2.1. Preliminary Follow-Up of Liquid Fraction

A preliminary follow-up was conducted on the obtaining species of interest by measuring pH and conductivity in liquid fractions obtained after hydrothermal experiments. Conductivity in acids is related to the capacity of an acid to conduct electric current, and it is directly proportional to the concentration of ions in the acid solution. When an acid dissolves in water, some covalent bonds in the acid molecules are broken, releasing hydrogen ions (H^+) into the solution. These ions are electric charge carriers, moving toward the opposite electrode when an electric potential is applied across the solution. pH is a logarithmic scale that measures the concentration of hydrogen ions (H^+) in an aqueous solution. In acidic solutions, the concentrations of H^+ ions are higher, making the pH lower (less than 7 on the pH scale). A neutral solution with a pH of 7 and equal concentrations of H^+ and OH- ions is the case in sugars. The variation in pH in the aqueous fraction may indicate the presence of acidic or basic species being produced in the medium.

Figure 2 shows the change in pH (decrease) and conductivity (increase) in the liquid fraction at different temperatures (120–180 °C) with a constant time (1 h). A change of 3.336 mS/cm of conductivity is observed, in general, and when applying the criterion of the first derivative, it is observed that the point of greatest change is at 140 °C, so it can be concluded that from this temperature onward, charged species start to be produced and not only the presence of sugars. As for the pH, there is a change of 0.51 pH units, and the greatest change in the acidity of the medium occurs at 150 °C; so it can be concluded that below 150 °C there are few acidic species in the system and there are no relevant changes in the liquid fraction.

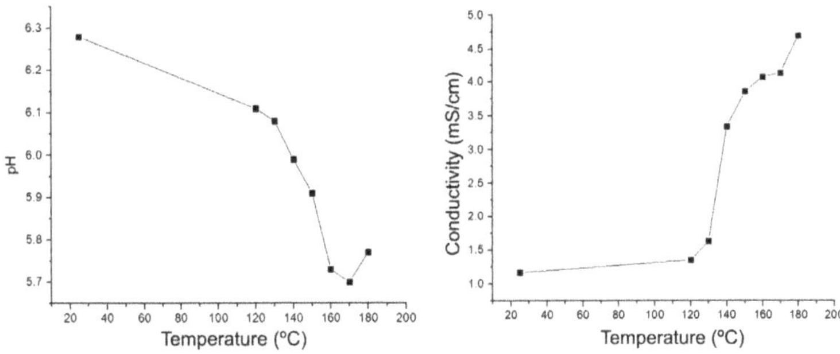

Figure 2. Preliminary follow-up on experiments from 120–180 °C for 1 h.

The same procedure was repeated by monitoring the pH and conductivity of the liquid fraction at 180 for 5 h, obtaining Figure 3. In both cases, in both conductivity and pH, it is observed that after 4 h, both values decay, which can be explained by a possible degradation of the products previously obtained and, therefore, the decrease in the species of importance for this project. Analyzing the variations from 1–4 h, there is a variation of 1.74 pH units and 5.002 conductivity units, values higher than those obtained only by modifying the temperature. Therefore, it can be preliminarily concluded that the LHW process carried out at 180 °C for 4 h will be the one with the highest production of platform chemicals of interest or, at least, the one with the best transformation of biomass to the aqueous phase in the form of organics.

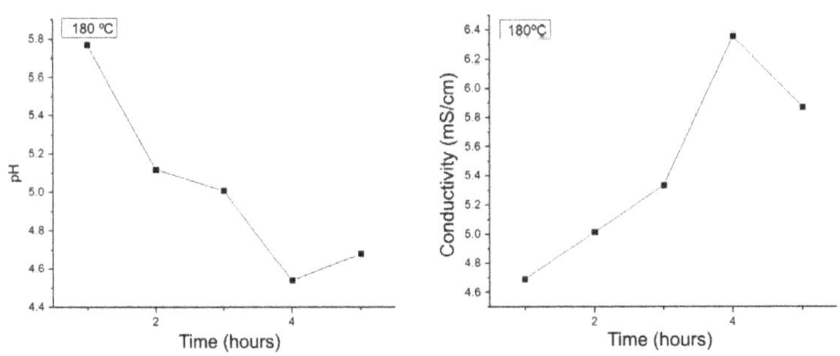

Figure 3. Preliminary follow-up of experiments from 1 to 5 h at 180 °C.

3.2.2. Characterization of Solid Fraction

Figure 4 shows the infrared spectra obtained for the solid fraction before the hydrothermal treatment and with the hydrothermal processes at 120, 150, 170, and 180 °C. It is important to highlight that the main bands (Table 5) are those of O-H stretching (3284–3308 cm^{-1}), C-H stretching (2919–2923 cm^{-1}), C=C stretching (1608–1622 cm^{-1}), and those present around 1500 cm^{-1} (C=C stretching of lignin and aromatic C), 1200 cm^{-1} (C-O-C in cellulose and hemicellulose), and finally, around 1000 cm^{-1} (corresponding to cellulose and hemicellulose as well). With the last three signals, it is possible to follow the transformation of the initial biomass and the breakdown of the primary structures (lignin, cellulose, and hemicellulose) and, thus, the disappearance of their functional groups in the solid fraction by hydrolysis to proceed to be extracted in the aqueous fraction.

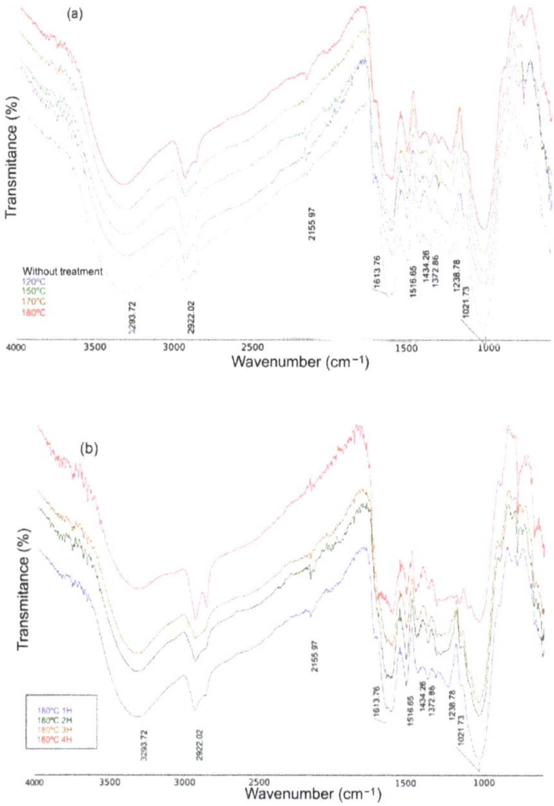

Figure 4. Infrared spectra of biomass (**a**) from 120 °C to 180 °C for 1 h and (**b**) from 1 h to 4 h at 180 °C.

Table 5. Characteristic vibrations and functionality of biomass.

Wavenumber (cm^{-1})	Characteristic Vibrations and Functionality
3284–3308	O-H vibrations of H-bonded hydroxyl groups.
2919–2923 and 853	C-H symmetric vibrations of aliphatic CHx. Methyl/Methylene.
1705	C=O vibrations of ketone, aldehyde, and acid groups.
1608–1622	C=C vibrations of aromatic compounds, C=O vibrations of conjugated ketones and quinones.
1516	C=C-vibrations, indicative of lignin and aromatic C.
1500–1350	Amorphous polysaccharides.
1300–1150	Symmetrical C-O-C, C-OH, and C-O vibrations of cellulose, hemicellulose, and lignin.
1050–1000	C-O, C=C, C-C-O vibrations in Cellulose, hemicellulose, and lignin.
900	O-H vibrations of H-bonded hydroxyl groups.
800–600	C-H symmetric vibrations of aliphatic CHx. Methyl/Methylene.

In the case of Figure 4a, there is no relevant variation in the previously mentioned signals, so no transformation of biomass into biochar is taking place yet, nor are processes of breakdown of the primary lignocellulosic structures. Figure 4b shows a decrease in the

OH stretching band, as well as a decrease in the intensity of the hemicellulose, cellulose, and lignin bands in the zone between 1000 and 1600 cm^{-1}, which allows us to conclude that an effective process of hydrolysis and obtaining organic species in the liquid fraction of the hydrothermal process has been carried out, an aspect that goes hand in hand with the results previously obtained from the monitoring of pH and conductivity. It is also important to note the appearance of a band at around 780 cm^{-1}, typical of alcohols and phenols that may have been produced after the hydrothermal reactions. On the other hand, in order to follow up and partially characterize the solid fraction, the elemental analysis of the solids obtained in each of the hydrothermal valorization processes was carried out (Table 6). It can be observed that there are no major changes in the H/C ratio under the temperature of 180 °C, where the ratio starts to decrease due to an increase in the percentage of carbon in the solid fraction. The percentage of oxygen decreases in a more relevant way in the experiments at 180 °C with a longer time, causing the O/C ratio to decrease as well.

Table 6. Elemental analysis of solid fractions.

Essay	Time (Hour)	C (%)	H (%)	N (%)	O (%)	O/C	H/C	C/N
120 °C 1 h	1	44.65	5.57	2.61	47.16	0.79	1.50	19.98
130 °C 1 h	1	46.74	5.84	1.98	45.42	0.73	1.50	27.49
140 °C 1 h	1	46.13	5.75	2.19	45.92	0.75	1.50	24.52
150 °C 1 h	1	47.81	5.81	2.19	44.16	0.69	1.46	25.39
160 °C 1 h	1	44.40	5.57	2.22	47.80	0.81	1.50	23.37
170 °C 1 h	1	48.17	5.98	2.07	43.77	0.68	1.49	27.11
180 °C 1 h	1	48.10	5.94	2.36	43.59	0.68	1.48	23.76
180 °C 2 h	2	46.71	5.60	2.63	45.05	0.72	1.44	20.71
180 °C 3 h	3	49.61	5.83	2.35	42.19	0.64	1.41	24.57
180 °C 4 h	4	51.31	5.99	2.49	40.21	0.59	1.40	24.04
180 °C 5 h	5	52.21	5.38	2.44	39.96	0.57	1.24	24.93

Considering these changes, the different experiments were placed on the Van Krevelen diagram (Figure 5) to characterize the fuel type in which the solid fraction is classified. The different colored signs represent the arrangement of the 120–180 °C tests for one hour, while the brown signs represent the experiments at 180 °C for different hours. It is observed that, at higher reaction times and temperatures, the residual biomass from the hydrothermal process starts to assimilate more to other types of fuels, in this case classifying the 180 °C 5 h test as a peat-type fuel, which has applications for cleaning oil spills and for water purification filters, as well as a fuel.

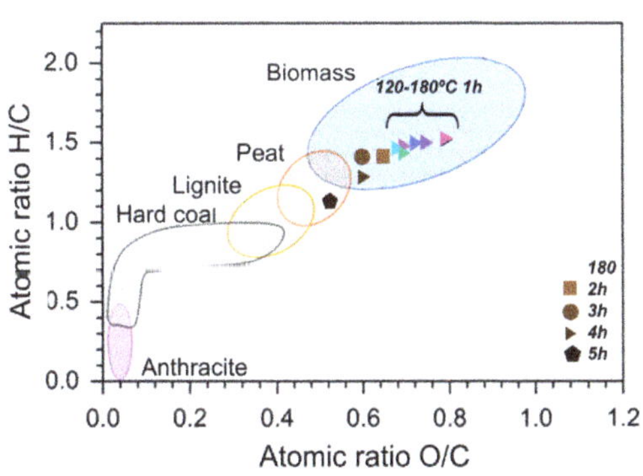

Figure 5. Van Krevelen diagram of LHW experiments.

The C/N ratio is also modified by the hydrothermal processes, reaching 31:1. This increase in the ratio, as was shown in the initial characterization of the biomass, is above the

24:1 ratio, which allows us to classify the relative decomposition process as a slow-speed one, contrasting with what we had initially and suggesting that the biochar obtained would not be easily degraded by microorganisms.

3.2.3. Quantification of Platform Chemicals

The quantification of the platform chemicals (PC) was performed using the validated method for HPLC-RI described previously, which can be seen more in detail in the Supplementary Material, and an external calibration from 0.1 to 10 g/L of each PC (xylose, glucose, formic acid, levulinic acid, HMF, and furfural).

It was observed that in those experiments corresponding to temperatures of 120, 130, and 140 °C, there were no relevant products obtained, an aspect that goes hand in hand with what was found by monitoring the pH and conductivity of the liquid fractions. Subsequently, from 150 to 180 °C, an extraction of sugars and other compounds, possibly with characteristics of cellu-oligosaccharides, is observed. Then, quantified products in the LHW range with one-hour experiments are shown in Figure 6.

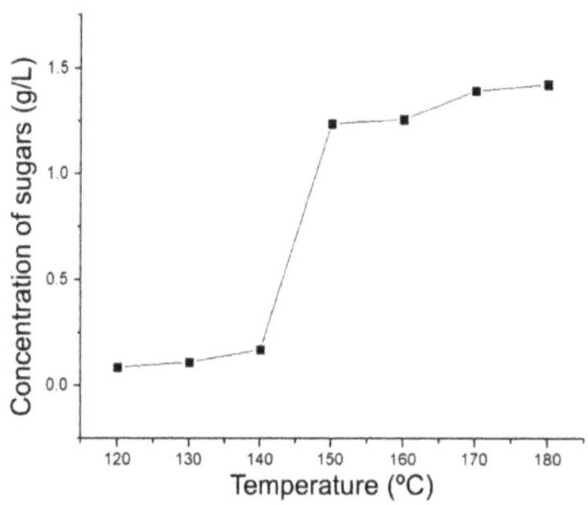

Temperature (°C)	Yield*(%)
120	0.167
130	0.213
140	0.323
150	2.347
160	2.391
170	2.654
180	2.709

Figure 6. Concentration of sugars and yield at every LHW experiment for 1 h. * Yield based on lignocellulosic structure.

It is observed that the results obtained correlate with those found in the monitoring by conductivity and pH, where there is an increase in the concentration of sugars at the temperature of 150 °C, where the process of hydrolysis of the hemicellulose/cellulose begins to take place. After this, hydrolysis continues to take place, but in a uniform manner as the temperature is raised to 180 °C. It is worth noting that no other chemical platform compounds were produced at these one-hour temperatures due to the inability to favor reaching products requiring further chemical processes such as hydrogenation, dehydrogenation, decarboxylations, and aldol condensations.

The yield of sugar production from two aspects was calculated in the same way, initially from the initially weighed biomass, in this case, 3 g, obtaining a maximum of 1.086%w sugar. The same process was also used, but taking the percentage of the biomass that corresponds to the lignocellulosic structure from which the sugars come (cellulose and hemicellulose) as the initial quantity for the process, as other components such as lignin, lipids, pigments, and others do not contribute to and could even inhibit the hydrolysis process. With this ratio, a yield of 2.709%w was obtained, which can also be maximized. It is worth mentioning that in the LHW range and with only one hour, it was not possible to obtain more PC, only the extraction of sugars. Subsequently, the same PC quantification and monitoring process was carried out for the experiments performed at constant temperature (180 °C) and modifying the reaction time, obtaining the results presented in Figure 7.

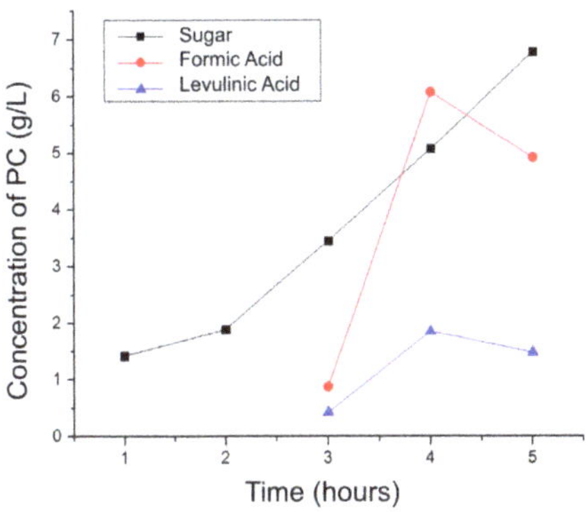

Time (Hours)	Total Yield (%)
1	2.716
2	3.209
3	7.502
4	23.56
5	21.59

Figure 7. Concentration of sugars at 180 °C for 1–5 h. Yield based on lignocellulosic structure.

It is observed that as the reaction time increases, the processes of hydrolysis, dehydration, rehydration, and condensation are favored (Figure 7), which leads to the production of PC. After 3 h, formic and levulinic acids start to be extracted, and after 4 h, HMF and furfural are obtained. After this, at 5 h, the efficiency of all products except sugars decreases due to possible degradation to smaller molecules and the instability of some of the previously quantified molecules.

On the other hand, Figure 7 shows the yield of the process based on the amount of initial biomass and its corresponding lignocellulosic structures. It is concluded that the best condition to valorize coffee without the implementation of catalysts and in the LHW range is a temperature of 180 °C for 4 h, achieving a valorization of about 25%w. As previously mentioned, continuing with this process for more hours would decrease the yield of the reaction, and, therefore, HMF and furfural would not be obtained, and the acids would start to break down into ketones and smaller acids, and would undergo decarboxylation, eliminating CO_2.

3.3. Hydrothermal Valorization via HTC

To cover another area within the hydrothermal processes (HTC), reactions were carried out between 200 and 260 °C for one hour and at 200 °C for 5 h, starting in all cases with 10 g of biomass and 90 g of water. The solid, liquid, and gas phase results were obtained, as shown in Table 7.

Table 7. Fractions obtained via HTC.

Temperature (°C)	Time (Hours)	Solid Fraction (g)	Liquid Fraction (g)	Liquid Fraction (g)
200	1	3.578	60.012	36.410
200	2	4.279	64.120	31.601
200	3	4.531	60.234	35.235
200	4	4.531	65.015	30.454
200	5	4.452	66.925	28.624
220	1	3.407	59.901	36.692
240	1	3.334	66.290	30.376
260	1	2.972	83.312	13.714

It is observed that when the temperature is modified between the 200 and 260 °C tests, there is an increase in the amount of liquid fraction and a significant decrease in the gaseous fraction. This can be attributed to the fact that the reaction is mainly going toward the production of biochar, an aspect that is denoted by the decrease in the weight of the solid fraction due to the loss of oxygen and hydrogen molecules from the initial biomass. By preferring the production of biochar, no significant extraction processes are carried out in the liquid fraction (which increases its quantity), and there are no dehydration, decarboxylation, or hydration processes that lead to the expulsion of gases into the reaction medium.

On the other hand, for the experiments carried out at constant temperatures for different times, there is an increase in the solid fraction, and the liquid fraction remains relatively constant; the opposite behavior is presented by the higher HTC valorization temperatures, which may imply that the biochar is not yet fully favored, and the organic species are maintained in the liquid phase.

3.3.1. Preliminary Follow-Ups on Liquid Fraction

The liquid fraction was monitored the same way as in the LHW process, using conductivity and pH properties to understand the behavior of dissociated, acidic, basic, and neutral species in the medium.

Figure 8 shows how the pH behavior in HTC is opposite to that found in LHW. As the temperature increases in HTC, the pH increases and becomes more basic; this implies a decrease in the acidic species in the medium. As for the conductivity, a maximum occurs at 220 °C, where the highest concentration of a species that dissociates in the medium probably occurs, and then, the conductivity decreases following the behavior of the pH, allowing us to conclude that the species of interest produced decrease. This goes hand in hand with the analysis in Table 7, where it can be assumed that higher temperatures favor the production of biochar and decrease some of the organic species of interest in the medium.

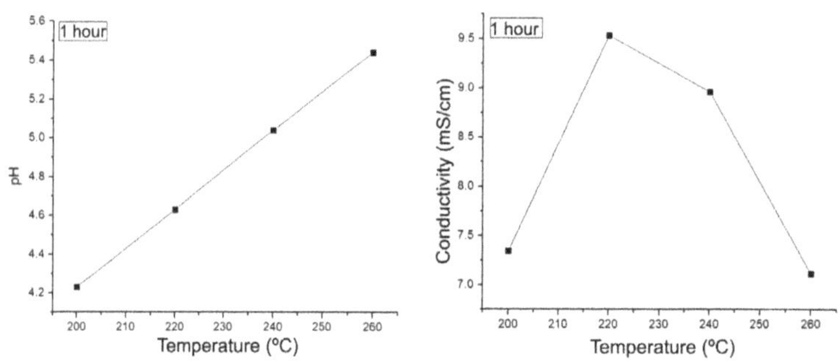

Figure 8. Preliminary follow-up of experiments from 200 to 260 °C for 1 h.

A similar behavior is presented in Figure 9, where the reaction at 200 °C is followed. There is a minimum pH in the one-hour reaction, implying that most of the acidic species are present at this time and then decrease as the reaction continues until 5 h. As for the conductivity, it keeps oscillating after 2 h of reaction, which can happen due to the equilibrium present in the system, but it does not decay as it happens from 240 °C onwards.

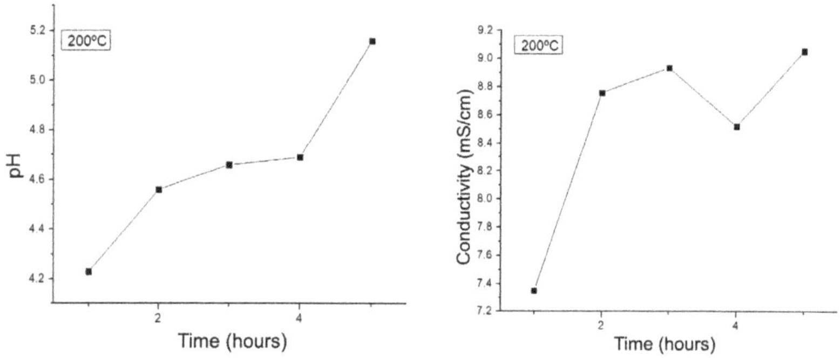

Figure 9. Preliminary follow-up of experiments from 1 to 5 h at 200 °C.

3.3.2. Characterization of Solid Fraction

The same technique as in LHW was implemented to make a preliminary follow-up to the solid phase. Figure 10 shows the spectra of the hydrothermal processes carried out between 200 and 260 °C and the signal of the initial biomass. Table 8 shows the signals that are maintained, although they change in intensity in the biochar spectra. Initially, it can be observed that there is a decrease or disappearance of the signal at 3200 cm^{-1}, characteristic of the OH vibration; there is also a significant decrease in the signals of the aromatic components and C=O (ketone and acid), as well as a decrease in the signals between 1500–1350 cm^{-1} of amorphous polysaccharides. Finally, and the most significant change, is the decrease in the signal at 1030 cm^{-1} corresponding to cellulose, hemicellulose, and lignin vibrations, showing that significant hydrolysis processes of the fundamental structures of the lignocellulosic biomass are being carried out in order to proceed to be transformed into biochar or extracted to the aqueous fraction.

The same monitoring procedure was repeated for the experiments performed at 200 °C at different times, resulting in Figure 10b. In this figure, the same changes described above can be seen with respect to the initial biomass, and unlike the LHW process, there are no major changes between the spectra at different times, which suggests that the greatest change in the solid is found between the temperatures of 180–200 °C where the lignocellulosic structures of the biomass are broken. In order to monitor and partially characterize the solid fraction, elemental analysis of the solid obtained in each of the hydrothermal valorization processes was carried out, obtaining Table 8.

Table 8. Elemental analysis of HTC experiments.

Temperature (°C)	Time (Hours)	C (%)	H (%)	N (%)	O (%)	O/C	H/C	C/N
	Initial biomass	44.65	5.57	2.61	47.16	0.79	1.50	19.98
200	1	66.99	5.54	2.42	25.05	0.28	0.99	32.26
220	1	67.21	6.03	2.36	24.40	0.27	1.08	33.17
240	1	70.42	7.06	1.59	20.94	0.22	1.20	51.56
260	1	79.57	2.20	0.73	17.50	0.16	0.33	126.52

Figure 11 shows the location of the H/C and O/C ratios in the Van Krevelen diagram, where the different fractions were characterized as lignite, hard coal, and anthracite, with the highest temperature achieving a significant biomass transformation due to the loss of oxygen and hydrogen. Anthracite can be used as a fuel, and hard coal can be used as a coating for paints and binders. On the other hand, as the temperature increases, the organic characteristics of the biomass are lost, leaving mostly inorganic components that affect the C/N ratio and make these biochars unsuitable for implementation in processes related to microorganism treatments. Comparing Table 8 with that previously obtained for LHW, it is found that there is a greater change in the percentages of carbon, hydrogen, and nitrogen, and, therefore, that processes above 200 °C lead to carbonization of the biomass and obtaining new characteristics in the biomass.

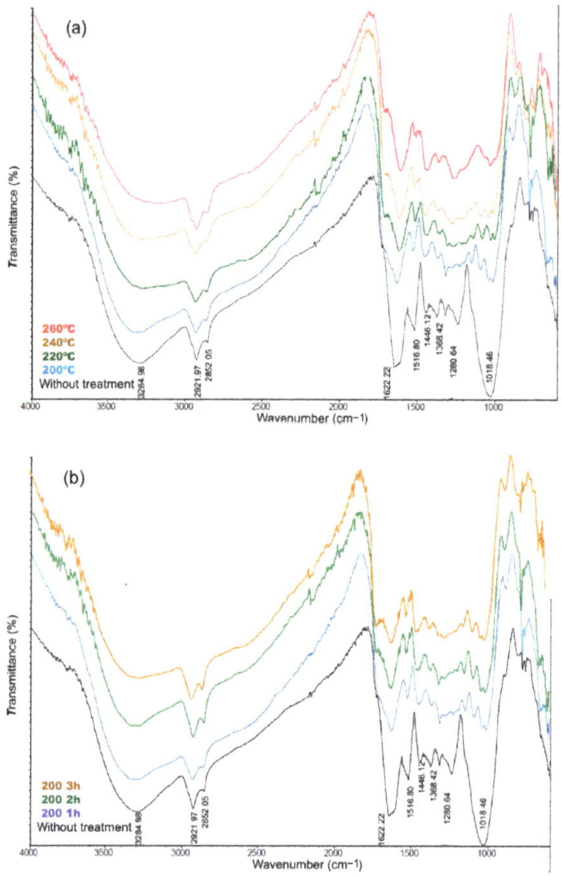

Figure 10. Infrared spectra of HTC valorization (**a**) from 200 to 260 °C for 1 h, (**b**) from 1 to 3 h at 200 °C.

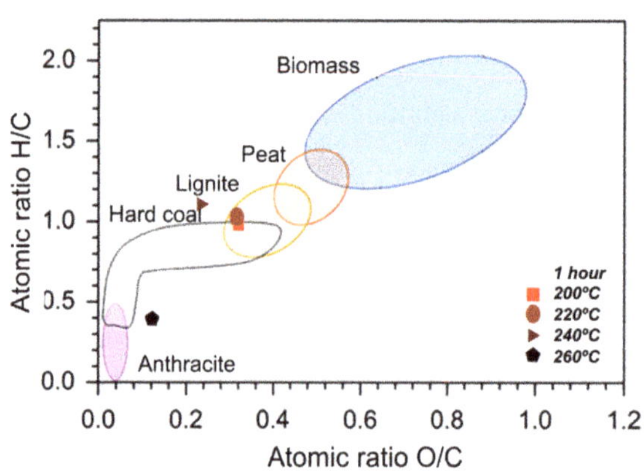

Figure 11. Van Krevelen characterization of HTC solid fractions.

In order to observe the morphology, samples of the solid fractions obtained after hydrothermal processes at 200, 220, 240, and 260 °C were examined by SEM. The macrostructure of samples and the images (Scheme 2) of the internal structure of biomass after HTC were analyzed. At 220 °C, carbonization begins to break down the cell walls; in contrast, the sample carbonized for one hour at 260 °C looks darker and has a rough, disintegrating surface. The imaging at 260 °C shows a significant development of pores that start to form at 240 °C, which makes the material more reactive for processes of combustion and gasification, an aspect that goes hand in hand with the previously performed Van Krevelen characterization. It is worth noting that a BET analysis or adsorption assays can be carried out with the obtained biochar, but the scope of the present manuscript is the obtention of platform chemicals and not biochar.

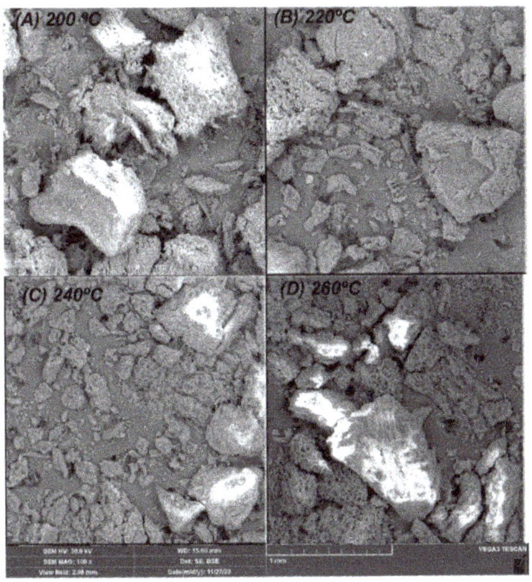

Scheme 2. SEM imaging of HTC solid fraction.

3.3.3. Quantification of Platform Chemicals

For the quantification process, undiluted liquid fractions were injected into the chromatograph using the previously validated method to qualitatively monitor the production of platform chemicals of interest via the hydrothermal HTC processes from 200 °C to 260 °C for one hour, resulting in Figure 12.

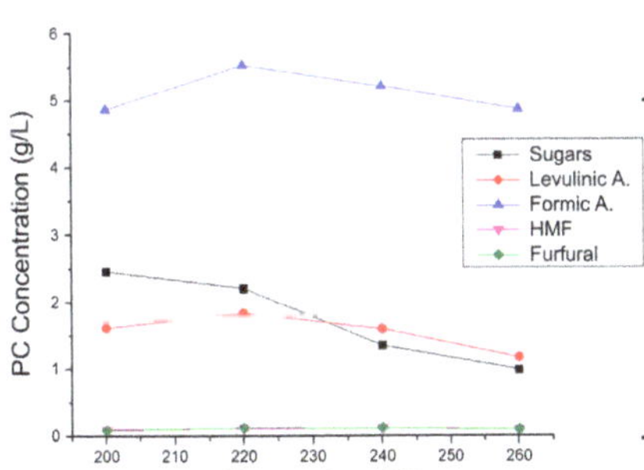

Temperature (°C)	Total Yield (%)
200	18.486
220	20.151
240	19.499
260	21.311

Figure 12. Platform chemicals production in HTC experiments. Yield based on lignocellulosic structure.

Figure 12 shows that the highest concentrations were obtained at 220 °C (for acids) and at 200 °C for sugars, while HMF and furfural remained relatively constant; this information goes hand in hand with the results previously obtained for the conductivity of these experiments, where a maximum of conductivity was presented at 220 °C caused by the increase in acidic species. From that temperature onward, the concentrations start to decrease for all compounds except HMF and furfural due to degradation of the PCs. On the other hand, when observing Figure 12, the yield (calculated based on the corresponding lignocellulosic structure) of all the processes is very close, but 260 °C is the one with the highest yield, although the concentration of PCs is low compared to the others. This is because the final volume of the reaction at 260 °C (83.312 mL) is greater than that of the other temperatures (62 mL on average), and because the way of evaluating the yield is (g of PC/g of lignocellulosic structure) × 100%, having a greater final volume, the grams in these mL are greater than those of other reactions with higher concentrations but smaller volumes. It is also worth noting that HTC processes with coffee residues lead mostly to the formation of formic acid, and the concentration of sugars decreases compared to LHW processes. Figure 13 shows the PC results for the hydrothermal processes at constant temperatures and varying times.

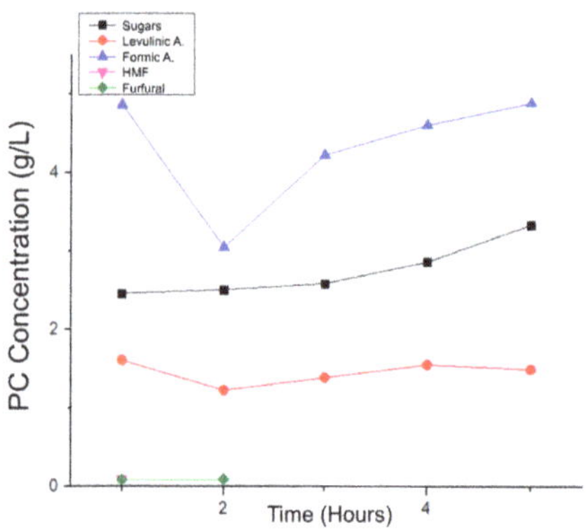

Time (hours)	Total yield (%)
1	18.483
2	14.519
3	16.165
4	19.169
5	21.049

Figure 13. Platform chemical production in HTC at different times. Based on lignocellulosic structure.

In the case of sugars and organic acids, an increase is observed as the reaction hours go by, while on the other hand, HMF concentrations decrease to the point of being below the calibration curve and not being relevant compared to the other species produced. A maximum yield of 21.049%w was achieved, a value similar to that obtained for the process at 260 °C for 1 h but obtaining a greater quantity of sugars instead of products such as HMF and furfural.

3.4. *Kinetics*

3.4.1. Kinetics of LHW

In order to analyze the behavior of hydrothermal reactions, the reason why they are carried out must be understood. Water is implemented as a reactant in these reactions due to its capacity to self-ionize, generating hydronium and hydroxyl ions that are free in the aqueous phase. As the temperature of the water increases, the ratio of hydronium and hydroxyl ions is maintained, but the concentration of these ions changes, causing both the Kw and the pH of the water to change as the temperature increases. The higher the temperature, the higher the concentration of free ions because the dissociation of water requires energy, this being an endothermic reaction, and the higher the concentration of hydronium ions, the more hydrolysis processes are favored by generating a positive charge in the oxygen of the β-1-4 glycosidic bonds and favoring the breaking of the bonds.

To analyze the kinetics of the initial hydrolysis of the lignocellulosic biomass to sugars, the concentrations obtained were taken as a function of time at 180 °C by HPLC-RI, and the respective mathematical treatments were made at the kinetics of orders 0, 1, 2, and 3 to obtain Figure 14.

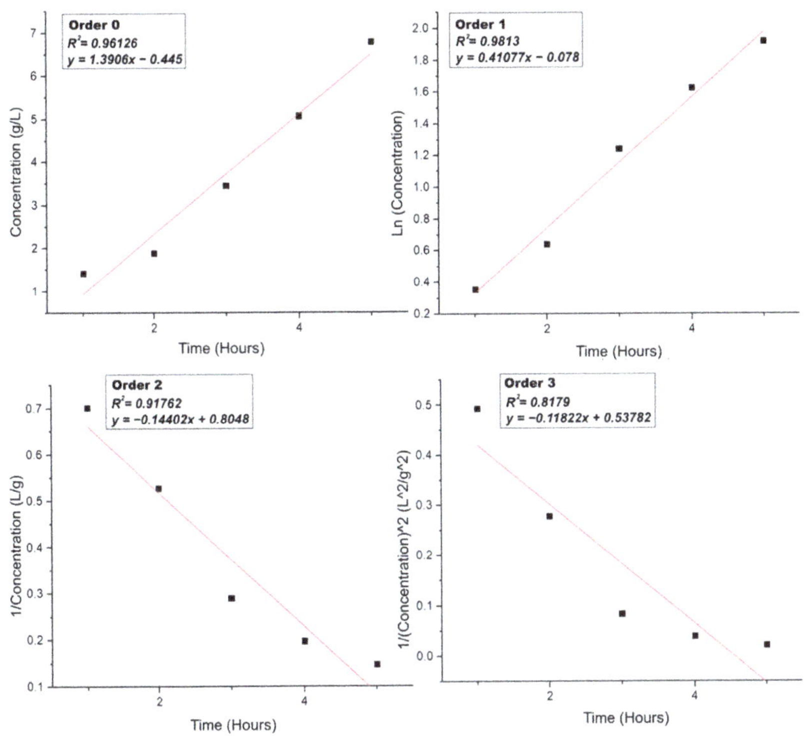

Figure 14. Mathematical treatments to evaluate the reaction order of kinetics of LHW.

When comparing the correlation with the linear regression of each order, it is observed that the best R^2 is that of order 1, which allows us to say that it is a pseudo-first-order reaction as it behaves like a first-order reaction because the rate equation of the reaction is expressed as a first-order-rate equation. The reaction is higher order because it involves more than one reactant (water and lignocellulosic biomass); because one of the reactants is present in excess, its concentration remains apparently unchanged throughout the reaction. Thus, the reaction, when dependent on the concentration of the other reactant, apparently follows a first-order-rate equation.

The fact that it is a pseudo-first-order reaction means that the rate of reaction (hydrolysis to sugars) is directly proportional to the concentration of the lignocellulosic biomass; therefore, the equation governing the process of hydrolysis of biomass to sugars in LHW is Equation (2) and the rate constant is the corresponding slope in Figure 14 order 1 ($0.41077\ s^{-1}$).

$$[A] = [A]_0 e^{-kt} \tag{2}$$

3.4.2. Kinetics of HTC

To describe the kinetics of the HTC process, the same technique previously introduced for LHW was implemented at a constant temperature of 200 °C, and at different times, the relevant mathematical treatments were carried out, and Figure 15 was obtained.

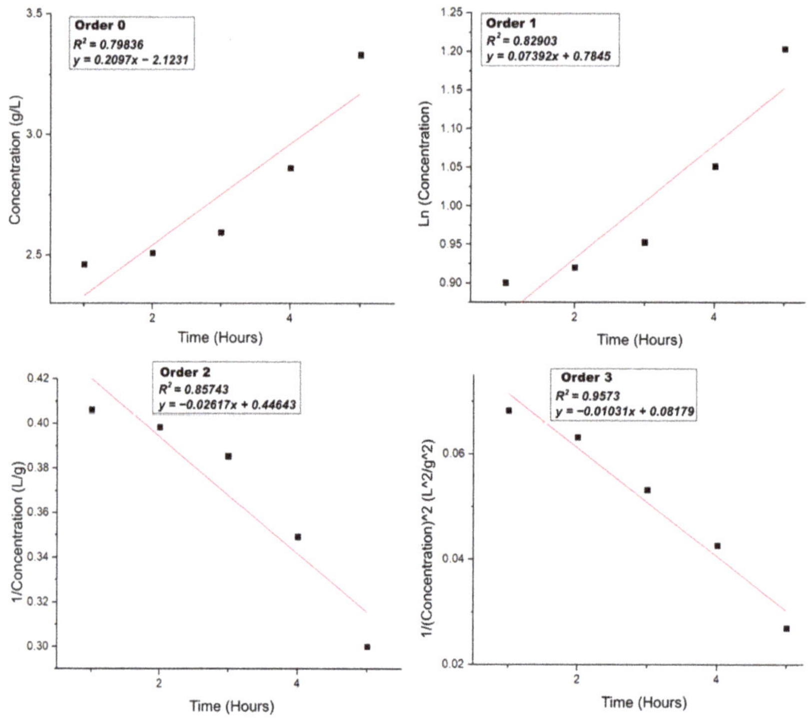

Figure 15. Mathematical treatments to evaluate the reaction order of kinetics of HTC.

In the case of HTC, it is observed that the behavior is similar to that of higher-order kinetics; in these reactions, more than two species are involved, an aspect that can be explained by the fact that the higher temperature makes hydrolysis more effective, but in the same way, it does not stop at the sugars as it did in LHW, but the reaction continues until levulinic and formic acids form and then other compounds are derived from the degradation of these. It is also important to note that at the working temperature, biochar is also produced, which is another species that affects the kinetics of the reaction. The rate constant corresponds to the slope of the higher-order graph (0.0103 L/g), and Equation (3) governing the process is presented below:

$$\frac{1}{2}\left(\frac{1}{[A]^2} - \frac{1}{[A_0]^2}\right) = 3kt \tag{3}$$

This shows that the reactions carried out in HTC are more complex than those in LHW; the hydrolysis process is accompanied by other species and does not stop at the sugars but continues the process up to the other PCs. The production of acids, furfural, HMF, gases, biochar, and other products makes it impossible to follow the kinetics of specific reactions, as the chemical processes of hydrothermal valorization are simultaneous and can be favored or inhibited by the species present in the medium, making the kinetics and their description more complex.

3.5. Optimization Based on Platform Chemicals

Considering the experiments carried out, the mechanism, and the yields based on each platform's chemical composition, the optimal conditions to favor the specific obtaining of each PC are reported.

For sugars, Figure 16 shows that the use of H_2SO_4 favors the production of sugars at 180 °C in less time and more selectively, but that similar concentrations can be reached with reactions at 180 °C, but for a longer time (5 h), because these processes allow for the hydrolysis of the lignocellulosic structures to take place, but the energy provided to the system is not enough for the subsequent reactions to take place. It is important to emphasize that the energy expenditure to carry out the reaction for 5 h is higher than that involved in a one-hour reaction. Weak basic catalysts also act favorably on the process and could be successfully implemented. On the other hand, the temperature increases above 220 °C, and the weak acid catalysts act as inhibitors of the sugar production process and favor the production of other products.

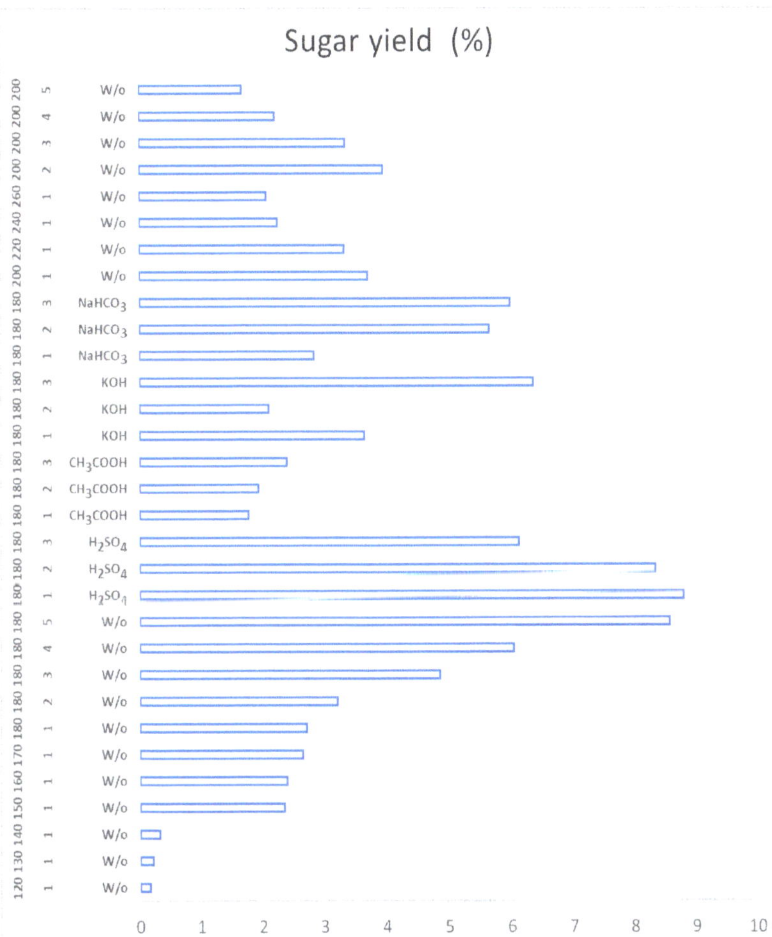

Figure 16. Optimization of reaction conditions for sugar obtention.

The production of levulinic acid is also not favored at low temperatures as there is not enough energy for the reactions characteristic of hydrothermal processes to take place. Figure 17 shows how the process of obtaining levulinic acid requires the use of a catalyst that increases the yield and displaces the hydrothermal reactions to obtain it. With the results of the present work, the use of weak acid catalysts such as acetic acid is suggested since, as previously presented, strong acid catalysts lead to the majority production of sugars and not the breakdown of these to levulinic acid.

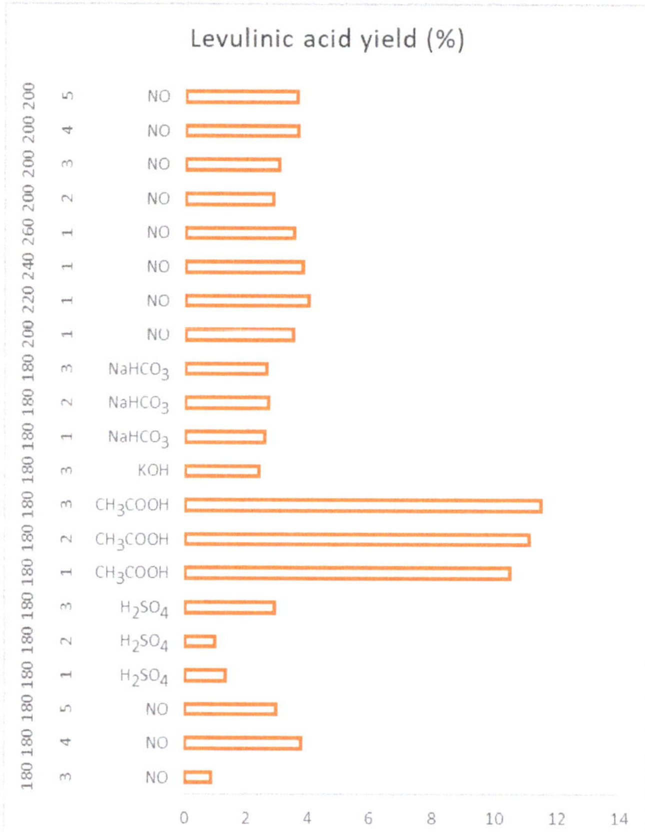

Figure 17. Optimization of reaction conditions for levulinic acid obtention.

The increase in reaction time and temperature are factors that do not drastically change the yield of levulinic acid, so a catalyst is highly recommended. As for formic acid, Figure 18 shows that the process is not favored at low temperatures and short reaction times, which is why it is not produced below 180 °C for 3 h. Although there is enough energy to hydrolyze even sugars, the condensation and hydration reactions that form formic acid are not fully favored. The best condition for obtaining this PQP is the implementation of HTC, especially at temperatures above 240 °C, although significant concentrations can also be obtained by increasing the reaction time at temperatures not as high as 180 and 200 °C. In the same way, the implementation of weak basic catalysts leads to the formation of formic acid in less time and under LHW conditions.

HMF and furfural are the minority products of the hydrothermal reactions carried out with coffee residues, achieving, as shown in Figure 19, a maximum yield of 1.480%w based on the corresponding lignocellulosic structures. In order to obtain these products, temperatures above 180 °C must be used, supported either by a longer reaction time or by strong acid catalysts. Reactions above 200 °C and for more than one hour start to cause degradation of the products and are, therefore, not suggested, and it is desired to stop the reaction on these specific products.

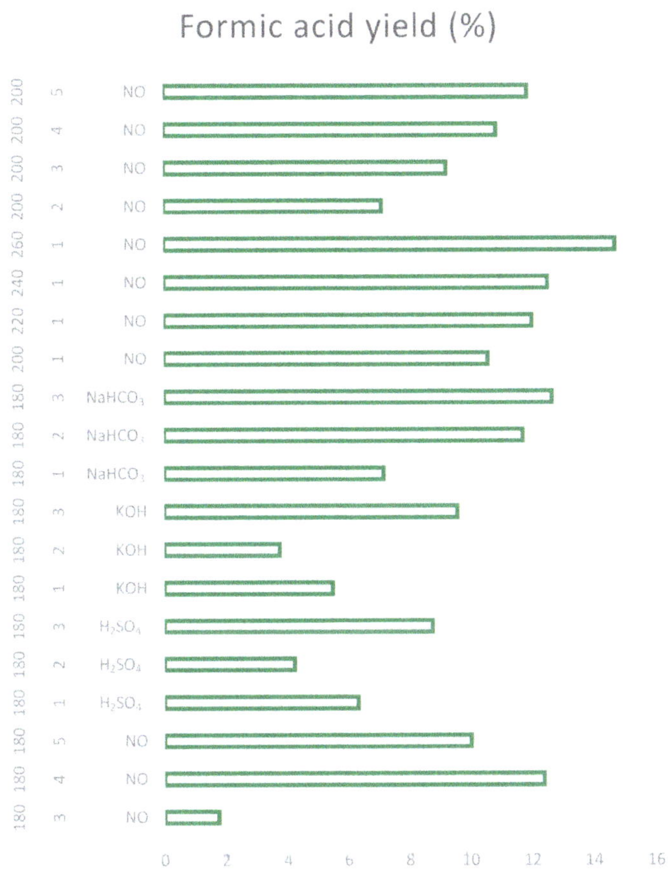

Figure 18. Optimization of reaction conditions for formic acid obtention.

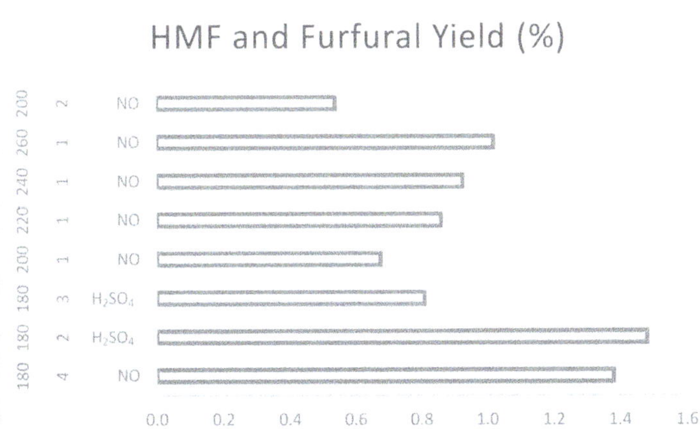

Figure 19. Optimization of reaction conditions for HMF and furfural obtention.

Catalysts Influence

Table S10 of the Supplementary Material presents in detail the total yields and concentrations of catalysts used in the optimization previously shown. It is then observed that in all the hours, the total yield of all the catalysts exceeds the yield of the reaction without a catalyst, an aspect that demonstrates the efficiency of the catalysis. On the other hand, it can be observed how the different catalysts lead to the majority production of different PCs (as seen in Figures 16–19).

In the case of basic catalysts, both potassium hydroxide and sodium bicarbonate favor the production of formic acid in addition to producing biochar with lignite characteristics and sodium in its structure, characteristics that are not favored by implementing acid catalysts. These conclusions go hand in hand with what has been reported in the literature, where acid catalysts decrease the quality of the biochar or biocrude obtained but improve the extraction of organic compounds [41]. It is also observed that the acidic medium increases the production of levulinic acid, HMF, and furfural, benefiting the dehydration reactions [42] as they form Lewis acid sites that can effectively catalyze the dehydration reactions. Lewis acids can also activate the carbonyl groups of the reactants, which is essential to initiate key reactions. Finally, Lewis acids' sites can promote isomerization reactions, which may be involved in the conversion of intermediates into final products. This may improve the selectivity and increase the yield of HMF and levulinic acid [43]. Figure 20 shows a graph with the specific behavior of the coffee residues that allows us to know the functioning of the catalysts and their selectivity in order to select the reaction conditions depending on the desired PCs.

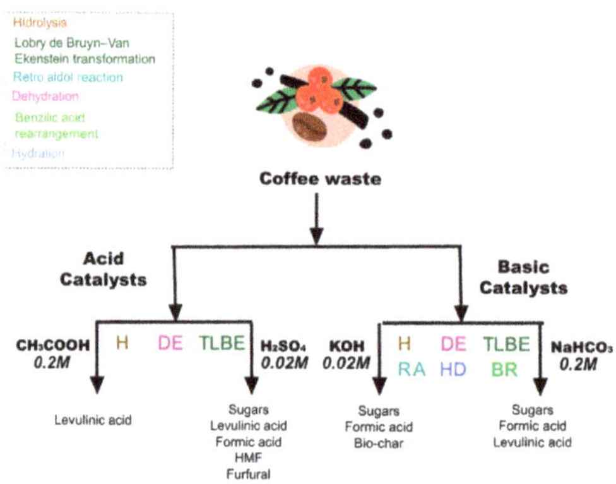

Figure 20. Catalysts influence on coffee cherry waste valorization.

In the case of sulfuric acid, the presence of a variety of platform chemicals can be observed, as well as how they decrease the concentration of sugars (by means of hydrolysis) to proceed to produce organic acids, HMF, and furfural. Acetic acid is, on the contrary, a very selective catalyst for levulinic acid, obtaining mostly levulinic acid and non-hydrolyzed sugars, and it is also the catalyst with the highest yield against levulinic acid from the first hour. The basic catalysts produce mostly formic acid and levulinic acid, but unlike the acid catalysts, the concentration of sugars does not decrease over time, and more sugars continue to be produced. This may be because the acid medium favors the hydrolysis of cellulose and hemicellulose species more efficiently and rapidly, breaking the glycosidic bonds and obtaining sugars in shorter times; on the other hand, the base does not favor these initial breaks in the biomass, so hydrolysis is slower and takes place little by little.

3.6. Purification

To perform the separation of the formic acid, the boiling point difference in the system was used, implementing a simple distillation system at 100 °C to eliminate both water and formic acid. After the formic acid was collected, it was injected into the HPLC-RI to check the purity of the separated product.

Preliminary separation tests were performed with the remaining mixture from the distillation (mixture of sugars and levulinic acid). The liquid–liquid separation with octanol to extract the levulinic acid in the organic phase and the sugars in the aqueous phase was not successful because levulinic acid had polar characteristics, which means that the partition is not completely carried out and a large amount of the sample remains in the aqueous phase. In the same way, separation was attempted by implementing column chromatography with a silica gel stationary phase and ethanol mobile phase followed by water due to the affinity of levulinic acid toward ethanol. This process was again unsatisfactory due to the similarity in polarity of the compounds to be separated.

Due to the inability to separate by polarity, ion exchange (AMBERLITE IRA 400), cationic (DOWEX-50W), and anionic:cationic (DOWEX* MARATHON* MR-3) resins were implemented. Considering the behavior of acid when it is at a pH below its pKa, as is the case of levulinic acid, it would be expected that this would be in its anionic form at the elution pH of the first buffer (4.3) and, therefore, the most suitable resin for the separation process would be AMBERLITE IRA 400, which would retain the levulinic acid in its levulinate form while leaving the sugars in the mixture to elute. This process was confirmed by obtaining zero separation of the desired compounds in the case of the DOWEX MARATHON MR-3, while in the case of the DOWEX-50W, a separation was achieved, but of some colored compounds that do not absorb in the IR (a brown colored spot was retained), but not of levulinic acid and sugars. Finally, as expected, when implementing the anion exchange column, levulinic acid was retained as levulinate when eluting the column at pH 4.3, while the sugars came out with the buffer (Figure 21). Subsequently, when the pH of the buffer was changed above the pKa of the levulinic acid, the levulinic acid started to elute without sugars, thus achieving its separation.

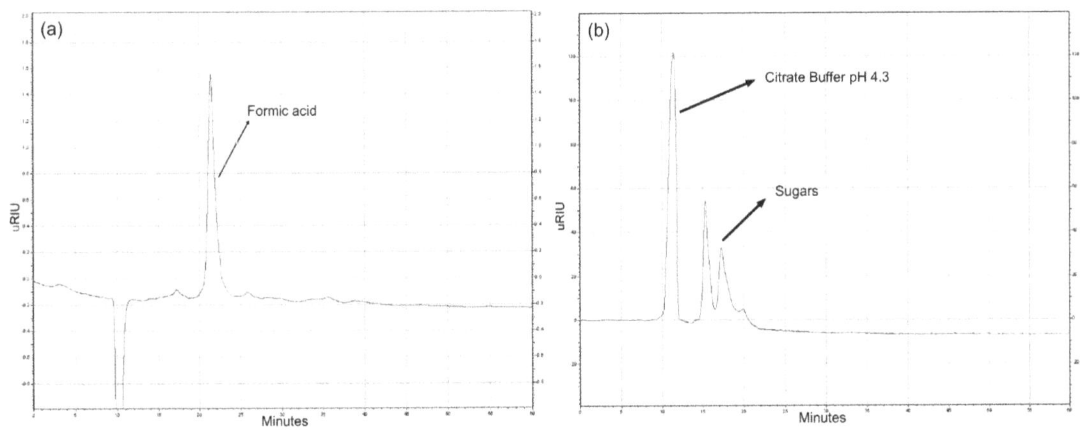

Figure 21. Purification process for (**a**) formic acid and (**b**) sugars.

It is, therefore, suggested that separation methods such as liquid–liquid partitioning and silica gel column chromatography are not suitable for the separation of the platform chemicals. On the other hand, distillation, followed by anion exchange chromatography, can be implemented to separate and purify the desired compounds. It is worth noting that in this process, neither HMF nor furfural was separated due to their low concentrations and the loss of these in the distillation process due to possible residues in the assembly.

4. Conclusions

This study characterized coffee waste biomass through ultimate, proximate, and structural tests. It was found to be highly humid (80.79%) with fuel characteristics suitable for biorefinery processes and microorganism valorization processes. The biomass contained similar percentages of hemicellulose (12.5%), lignin (13.7%), and hemicellulose (27.6%).

For the LHW range, reactions below 150 °C did not allow for the hydrolysis of lignocellulosic structures to sugars. pH and conductivity were used as properties to monitor PC production. The solid fractions were classified according to fuel type, and infrared spectroscopy was used to follow the disappearance of functional groups from lignin, cellulose, and hemicellulose between 1000 and 1500 cm^{-1} and their relationship with the production of platform chemicals. The best conditions for obtaining platform chemicals were achieved at 180 °C at 4 h, with a yield of 23.565%w. The efficiency and selectivity of homogeneous catalysts were tested, with acetic acid leading to selective production of levulinic acid, sulfuric acid leading to the highest efficiency of hydrolysis of sugars, potassium hydroxide leading to the production of biochar, formic acid, and sugars, and sodium bicarbonate producing formic acid and levulinic acid. The kinetics of the hydrolysis of lignocellulosic structures to sugars in the LHW range was evaluated, with orders of 1 with their respective rate constant. Purification of PCs was achieved using distillation techniques and anion exchange chromatography with AMBERLITE IRA 400 exchange resin.

HTC valorization trials were conducted, with the solid phase characterized as hard coal and anthracites and an overall higher formation of formic acid due to the decomposition of bigger PC and a decrease in sugar formation due to its transformation into smaller PC. Optimal conditions for obtaining sugars involved using sulfuric acid as a catalyst or 180 °C for 5 h for the production of formic acid reactions should be carried out at HTC conditions, preferably 260 °C, and levulinic acid requires weak acid catalysts like acetic acid. HMF and furfural were benefited by strong acid catalysts at 180 °C or a longer reaction time. In conclusion, coffee cherry waste as a raw material for hydrothermal biorefinery is viable and favored for obtaining glucose, xylose, levulinic acid, and furfural, with lower percentages of HMF and furfural.

The utilization of millions of tons of coffee berry waste to produce 9% glucose and xylose, 1.5% furfural and HMF, 11.5% levulinic acid, and 15% formic acid presents a significant opportunity for sustainable resource management and the development of a circular bioeconomy. The production of glucose and xylose, key monomeric sugars, offers the potential for the synthesis of biofuels, biochemicals, and biopolymers, contributing to the reduction in reliance on fossil resources. Furfural and HMF, derived from the dehydration of sugars, can be utilized as platform chemicals for the production of various value-added products, including biofuels and bioplastics. Furthermore, levulinic acid and formic acid, as versatile building blocks, can be employed in the synthesis of pharmaceuticals, agrochemicals, and polymers, thus contributing to the diversification of the chemical industry and the reduction in environmental impact. This approach aligns with the principles of a sustainable bioeconomy, where waste materials are transformed into high-value products, contributing to resource efficiency, reduced environmental impact, and economic growth. The switch from conventional waste treatment techniques, such as burning and fertilizer, to HTC for the valorization of coffee berry waste can lead to a more sustainable and circular approach to waste management, reducing environmental impact and contributing to the development of a sustainable bioeconomy.

Supplementary Materials: The following supporting information can be downloaded at https://www.mdpi.com/article/10.3390/su16072854/s1. Equations (S1)–(S5). Image S1. LHW Batch 100 mL reactor for hydrothermal processes. Image S2. HTC 500 mL reactors for hydrothermal valorization. Table S1: Retention time for standards used; Figure S1: Standards used for analytic method development. Table S2: Information of calibration curves for platform chemicals (PC); Figure S2: Calibration curves done for every PC.; Figure S3: Residuals of calibration curves; Table S3: Repeatability of the system based on the retention time; Table S4: Repeatability of the system based on the area; Table S5: Results for the repeatability of the method; Table S6: Results for intermediate precision for the method developed; Table S7: Accuracy of the results obtained for the method; Table S8: Detection and quantification limit for the method; Table S9: Results for robustness of the method; Table S10: Yield of platform chemicals produced from LHW valorization; Table S11: Yields and concentration of platform chemicals produces at 180 °C for 1–5 h; Table S12: HTC yields and concentration; Table S13: HTC yields and concentration at 200 °C; Table S14: Yields and concentrations of PC using homogeneous catalysts.

Author Contributions: Conceptualization, A.S.L.-P.; methodology, A.S.L.-P.; validation, A.S.L.-P.; formal analysis, A.S.L.-P.; investigation, A.S.L.-P.; resources, C.A.G.-F.; data curation, A.S.L.-P.; writing—original draft preparation, A.S.L.-P.; writing—review and editing, C.A.G.-F.; supervision, C.A.G.-F.; project administration, C.A.G.-F.; funding acquisition, A.S.L.-P. All authors have read and agreed to the published version of the manuscript.

Funding: This research was funded by MINCIENCIAS, Contrato de financiamiento de recuperación contingente No. 80740-101-2022.

Data Availability Statement: Data are contained within the article and supplementary materials.

Acknowledgments: We thank the Universidad Nacional de Colombia and the Departamento de Química for their support and the possibility of using the equipment and techniques that allowed for the development of this article.

Conflicts of Interest: The authors declare no conflict of interest.

References

1. Word Population Review. Coffee Consumption by Country 2023. Available online: https://worldpopulationreview.com/country-rankings/coffee-consumption-by-country (accessed on 1 December 2023).
2. International Coffee Organization. Trade Statistics—December 2021. Available online: http://www.ico.org/#:~:text=Total%20production%20for%20coffee%20year,remains%201.2%25%20below%20world%20production (accessed on 28 November 2023).
3. Apriani, D.; Marissa, F.; Gustriani, G.; Sinta, H. Determinants of Indonesia's Coffee Commodity with Trading Partner Countries. *J. Ekon. Pembang. Kaji. Masal. Ekon. Dan. Pembang.* **2022**, *23*, 98–109. [CrossRef]
4. Arruda, A.C.; Minim, V.P.R.; Ferreira, M.A.M.; Minim, L.A.; da Silva, N.M.; Soares, C.F. Justificativas e motivações do consumo e não consumo de café. *Ciência E Tecnol. De Aliment.* **2009**, *29*, 754–763. [CrossRef]
5. Folch, A.; Planas, J. Cooperation, Fair Trade, and the Development of Organic Coffee Growing in Chiapas (1980–2015). *Sustainability* **2019**, *11*, 357. [CrossRef]
6. FAO Statistics. *World Coffee Production by Country*; FAO Statistics: Rome, Italy, 2019.
7. Federación de Cafeteros. Colombian Coffee Production Closes 2021 at 12.6 Million Bags. Available online: https://federaciondecafeteros.org/wp/listado-noticias/colombian-coffee-production-closes-2021-at-12-6-million-bags/?lang=en (accessed on 1 December 2023).
8. Peshev, D.; Mitev, D.; Peeva, L.; Peev, G. Valorization of spent coffee grounds—A new approach. *Sep. Purif. Technol.* **2018**, *192*, 271–277. [CrossRef]
9. Mazzafera, P. Degradation of caffeine by microorganisms and potential use of decaffeinated coffee husk and pulp in animal feeding. *Sci. Agric.* **2002**, *59*, 815–821. [CrossRef]
10. Kasongo, R.K.; Verdoodt, A.; Kanyankagote, P.; Baert, G.; Van Ranst, E. Coffee waste as an alternative fertilizer with soil improving properties for sandy soils in humid tropical environments. *Soil Use Manag.* **2011**, *27*, 94–102. [CrossRef]
11. Abdeltaif, S.; SirElkhatim, K.; Hassan, A. Estimation of Phenolic and Flavonoid Compounds and Antioxidant Activity of Spent Coffee and Black Tea (Processing) Waste for Potential Recovery and Reuse in Sudan. *Recycling* **2018**, *3*, 27. [CrossRef]
12. Rodriguez, N. Manejo de Residuos en la Agroindustria Cafetera. Available online: https://silo.tips/download/manejo-de-residuos-en-la-agroindustria-cafetera (accessed on 1 December 2023).
13. Asociación Interamericana de Ingeniería Sanitaria y Ambiental. *Gestión Integral de Residuos Sólidos Urbanos*; Asociación Interamericana de Ingeniería Sanitaria y Ambiental: São Paulo, Brazil, 2018.
14. Gallego, D.; Montoya, J.; Valverde, J. Funcionamiento hidráulico de un filtro anaerobio de flujo ascendente. *Ing. Química* **2007**, *450*, 172–183.
15. Gomez, L.M. Manual de Lombricultura. 2008. Available online: https://www.yumpu.com/es/document/read/14512054/manual-de-lombricultura-cecytech (accessed on 21 December 2023).
16. De Bomfim, A.S.C.; de Oliveira, D.M.; Walling, E.; Babin, A.; Hersant, G.; Vaneeckhaute, C.; Dumont, M.-J.; Rodrigue, D. Spent Coffee Grounds Characterization and Reuse in Composting and Soil Amendment. *Waste* **2022**, *1*, 2–20. [CrossRef]
17. Vijayalaxmi, S.; Jayalakshmi, S.K.; Sreeramulu, K. Polyphenols from different agricultural residues: Extraction, identification and their antioxidant properties. *J. Food Sci. Technol.* **2015**, *52*, 2761–2769. [CrossRef]

18. Solomakou, N.; Tsafrakidou, P.; Goula, A.M. Valorization of SCG through Extraction of Phenolic Compounds and Synthesis of New Biosorbent. *Sustainability* **2022**, *14*, 9358. [CrossRef]
19. Silva, N.C.S.; da Fonseca, Y.A.; de Camargos, A.B.; Lima, A.L.D.; Ribeiro, M.C.; Gurgel, L.V.A.; Baêta, B.E.L. Pretreatment and enzymatic hydrolysis of coffee husk for the production of potentially fermentable sugars. *J. Chem. Technol. Biotechnol.* **2022**, *97*, 676–688. [CrossRef]
20. Kemp, K.C.; Baek, S.B.; Lee, W.-G.; Meyyappan, M.; Kim, K.S. Activated carbon derived from waste coffee grounds for stable methane storage. *Nanotechnology* **2015**, *26*, 385602. [CrossRef] [PubMed]
21. Kim, D.; Lee, K.; Bae, D.; Park, K.Y. Characterizations of biochar from hydrothermal carbonization of exhausted coffee residue. *J. Mater. Cycles Waste Manag.* **2017**, *19*, 1036–1043. [CrossRef]
22. Xiao, L.-P.; Shi, Z.-J.; Xu, F.; Sun, R.-C. Hydrothermal carbonization of lignocellulosic biomass. *Bioresour. Technol.* **2012**, *118*, 619–623. [CrossRef] [PubMed]
23. Plaza, M.; Turner, C. Pressurized hot water extraction of bioactives. *TrAC Trends Anal. Chem.* **2015**, *71*, 39–54. [CrossRef]
24. Çam, M.; Yüksel, E.; Alaşalvar, H.; Başyiğit, B.; Şen, H.; Yılmaztekin, M.; Ahhmed, A.; Sağdıç, O. Simultaneous extraction of phenolics and essential oil from peppermint by pressurized hot water extraction. *J. Food Sci. Technol.* **2019**, *56*, 200–207. [CrossRef]
25. Massaya, J.; Pickens, G.; Mills-Lamptey, B.; Chuck, C.J. Enhanced Hydrothermal Carbonization of Spent Coffee Grounds for the Efficient Production of Solid Fuel with Lower Nitrogen Content. *Energy Fuels* **2021**, *35*, 9462–9473. [CrossRef]
26. Kumar, N.; Weldon, R.; Lynam, J.G. Hydrothermal carbonization of coffee silverskins. *Biocatal. Agric. Biotechnol.* **2021**, *36*, 102145. [CrossRef]
27. Mariotti-Celis, M.; Martínez-Cifuentes, M.; Huamán-Castilla, N.; Vargas-González, M.; Pedreschi, F.; Pérez-Correa, J. The Antioxidant and Safety Properties of Spent Coffee Ground Extracts Impacted by the Combined Hot Pressurized Liquid Extraction–Resin Purification Process. *Molecules* **2017**, *23*, 21. [CrossRef]
28. Xu, J.L.; Kim, T.J.; Kim, J.-K.; Choi, Y. Simultaneous roasting and extraction of green coffee beans by pressurized liquid extraction. *Food Chem.* **2019**, *281*, 261–268. [CrossRef]
29. Murthy, P.S.; Naidu, M.M. Sustainable management of coffee industry by-products and value addition—A review. *Resour. Conserv. Recycl.* **2012**, *66*, 45–58. [CrossRef]
30. *NREL/TP-510-42621*; Determination of Total Solids in Biomass and Total Dissolved Solids in Liquid Process Samples. NREL: Golden, CO, USA, 2008.
31. *NREL/TP-510-42622*; Determination of Ash in Biomass. NREL: Golden, CO, USA, 2005.
32. *ISO 18123:2015*; Solid Biofuels—Determination of the Content of Volatile Matter. ISO: Santry, Dublin, 2015.
33. *ASTM D 3172-89*; Standard Practice for Proximate Analysis of Coal and Coke. ASTM: West Conshohocken, PA, USA, 2002.
34. Phuong, D.V.; Quoc, L.P.T.; Van Tan, P.; Duy, L.N.D. Production of bioethanol from Robusta coffee pulp (*Coffea robusta* L.) in Vietnam. *Foods Raw Mater.* **2019**, *7*, 10–17. [CrossRef]
35. Manrique, R.; Vásquez, D.; Ceballos, C.; Chejne, F.; Amell, A. Evaluation of the Energy Density for Burning Disaggregated and Pelletized Coffee Husks. *ACS Omega* **2019**, *4*, 2957–2963. [CrossRef] [PubMed]
36. Park, J.E.; Lee, G.B.; Jeong, C.J.; Kim, H.; Kim, C.G. Determination of Relationship between Higher Heating Value and Atomic Ratio of Hydrogen to Carbon in Spent Coffee Grounds by Hydrothermal Carbonization. *Energies* **2021**, *14*, 6551. [CrossRef]
37. Poudel, J.; Karki, S.; Oh, S. Valorization of Waste Wood as a Solid Fuel by Torrefaction. *Energies* **2018**, *11*, 1641. [CrossRef]
38. Frómeta, R.A.R.; Sánchez, J.L.; García, J.M.R. Evaluation of coffee pulp as substrate for polygalacturonase production in solid state fermentation. *Emir. J. Food Agric.* **2020**, *32*, 117–124. [CrossRef]
39. Colantoni, A.; Paris, E.; Bianchini, L.; Ferri, S.; Marcantonio, V.; Carnevale, M.; Palma, A.; Civitarese, V.; Gallucci, F. Spent coffee ground characterization, pelletization test and emissions assessment in the combustion process. *Sci. Rep.* **2021**, *11*, 5119. [CrossRef]
40. Gouvea, B.M.; Torres, C.; Franca, A.S.; Oliveira, L.S.; Oliveira, E.S. Feasibility of ethanol production from coffee husks. *Biotechnol. Lett.* **2009**, *31*, 1315–1319. [CrossRef]
41. Yang, W.; Li, X.; Liu, S.; Feng, L. Direct hydrothermal liquefaction of undried macroalgae Enteromorpha prolifera using acid catalysts. *Energy Convers. Manag.* **2014**, *87*, 938–945. [CrossRef]
42. Yin, S.; Tan, Z. Hydrothermal liquefaction of cellulose to bio-oil under acidic, neutral and alkaline conditions. *Appl. Energy* **2012**, *92*, 234–239. [CrossRef]
43. Liu, S.; Wang, K.; Yu, H.; Li, B.; Yu, S. Catalytic preparation of levulinic acid from cellobiose via Brønsted-Lewis acidic ionic liquids functional catalysts. *Sci. Rep.* **2019**, *9*, 1810. [CrossRef] [PubMed]

sustainable chemistry

MDPI

Article

Waste Lignocellulosic Biomass as a Source for Bioethanol Production

Klemen Rola, Sven Gruber, Darko Goričanec and Danijela Urbancl *

Faculty of Chemistry and Chemical Engineering, University of Maribor, Smetanova 17, 2000 Maribor, Slovenia; klemen.rola@um.si (K.R.); sven.gruber@um.si (S.G.); darko.goricanec@um.si (D.G.)
* Correspondence: danijela.urbancl@um.si

Abstract: Synthetically produced biofuels play a critical role in the energy transition away from fossil fuels. Biofuels could effectively lower greenhouse gas (GHG) emissions and contribute to better air quality. One of these biofuels is bioethanol, which could act as a gasoline replacement. For this purpose, a simulation of bioethanol production through lignocellulosic biomass fermentation, focused on distillation, was carried out in simulation software Aspen Plus. Since the possibility of absolute ethanol production through distillation is limited by the ethanol–water azeotrope, pressure swing distillation (PSD) was used to obtain fuel-grade ethanol (EtOH) with a fraction of 99.60 wt.%. The flowsheet was optimised with NQ analysis, which is a simple optimisation method for distillation columns. We found that the PSD has the potential to concentrate the EtOH to a desired value, while simultaneously removing other unwanted impurities whose presence is a consequence of pretreatment and fermentation processes.

Keywords: bioethanol; distillation; lignocellulosic biomass; azeotrope; Aspen Plus

1. Introduction

Lignocellulosic biomass is the most widespread form of renewable raw material on Earth. It accounts for half of the plant resources obtained through photosynthesis [1]. Due to its abundance, availability, and global renewability, it represents the most important alternative to fossil fuels, especially for the production of second generation biofuels, chemicals, and materials from biological sources, without influencing global food security. Lignocellulosic biomass includes agricultural residues and energy crops as well as woody biomass. On the other hand, transportation is responsible for 15% of global greenhouse gas emissions and 23% of total energy-related CO_2 emissions [2]. To reduce the dependence on petroleum-based fuels and mitigate climate change, biofuels are widely seen as promising alternative transportation fuels. Biofuels such as bioethanol have been proven to emit significantly less GHG than petroleum-based fuels.

Second generation bioethanol is produced from lignocellulosic biomass [3]. The production consists of four major steps. First, pretreatment is conducted to break down the lignocellulosic structure, consisting of lignin, cellulose, and hemicellulose. This ensures the accessibility of cellulose and hemicellulose for hydrolysis, which hydrolyses the two components in fermentable sugars. Both pentoses and hexoses are then ideally fermented into EtOH in the third step. Lastly, the purification of the fermentation product is required to provide fuel-grade EtOH [4].

Even though the EtOH can be purified by distillation, purity higher than 95.6 wt.% EtOH cannot be achieved with conventional distillation, due to the EtOH-H_2O azeotrope [5]. Among the challenges in second-generation bioethanol production, the main ones are financially achievable pretreatment for the deconstruction of lignocellulosic structure and the optimisation of enzymes to maximise fermentable sugar yields [3]; the dehydration step also represents a major challenge in bioethanol production plants [6], due to its high energy demand [4]. To achieve fuel-grade purity of 99.60 wt.%, numerous technologies are available, like pressure swing adsorption (PSA) [7,8], membrane separation [6], and extractive, azeotropic, and pressure swing distillation [4,5]. The latter has an advantage among the distillation techniques, since it does not depend on an entrainer to separate the EtOH-H_2O azeotrope [5]. Membrane separation has the same advantage, besides its low energy consumption, but requires a high capital investment [6].

Second-generation bioethanol production has been studied extensively, and even though it has seen rapid advancements, its production is still inefficient [3]. Due to the high energy requirements of the purification step, the present work focuses on the PSD, since according to our knowledge, not many studies have been conducted on that topic. The presented research therefore deals with the development of the process for bioethanol purification.

Few processes have been developed. Mulia-Soto and Flores-Tlacuahuac developed a model of internal heat-integrated pressure swing distillation. The process was used for the separation of an azeotropic mixture of EtOH and water (H_2O) and the production of high-purity EtOH suitable for the transportation sector. The system consisted of 30 trays; the feed was fed into the 23rd tray of the first column, while the reflux was fed to the 13th tray. The vapour from the first column was fed into the 14th tray. The feed and distillate of the first column contained 20 mol% (38.99 wt.%) and 85.9 mol% EtOH, respectively [9]. Iqbal and Ahmad used the same conditions [10]. In the flowsheet of Kiran and Jana, a 20 mol% EtOH solution was used as feed for the 21st tray of the first column. The column was used to produce EtOH at 85.65 mol%; the recycle tray was fed into the eleventh tray, and the distillate from the first column was fed into the twelfth tray of the second column. The PSD was used to produce EtOH with a purity of 99.62 mol% [11].

In our work, the stream from the fermentation unit contained other components, whereas in the literature only water and EtOH were present in the stream. The stream in our work was first concentrated to a higher, 50 wt.% of EtOH. Then, the PSD was applied for a possibility of fuel grade bioethanol production. To reduce GHG emissions and achieve sustainability of the processes, the reduction in energy consumption and optimisation play influential roles. To promote bioethanol production and its alternative to fossil fuels, the work explores many trade-offs that are present in designing the distillation columns and attempts to optimise the columns with a simple yet effective method.

2. Materials and Methods

2.1. Process Feed and Thermodynamic Properties

The simulation of bioethanol distillation was carried out using the Aspen Plus V8 computer program. In accordance with [9,10], we used the Redlich–Kwong equation of state (Wilson RK) to describe the interaction between the components. According to the source, it agrees well with the experimental data.

The data used for the simulations were obtained from Cardona and Quintero, from the outlet of the fermentation unit [12]. Apart from H_2O and EtOH, the stream included sulfuric acid (H_2SO_4), Hydroxymethylfurfural (HMF), furfural, and calcium hydroxide ($Ca(OH)_2$). The bacterial culture Z-Mobilis was also present in the stream, which Aspen Plus does not have in its component database, so we had to self-report it along with its properties. To avoid this, we added its proportion to the water content and excluded it from the simulations. The input data are listed in Table 1 and correspond to the feed to the first distillation column RC-1 (process stream F-1 in Figure 1). We used the obtained mass flow rate, temperature, and pressure of 50,188.90 kg/h, 33 °C, and 1 bar, respectively.

Table 1. Feed composition, adapted from Cardona and Quintero [12].

Component	w (wt.%)
H_2O	0.9320
EtOH	0.0540
H_2SO_4	0.0007
HMF	0.0119
Furfural	0.0010
$Ca(OH)_2$	0.0004

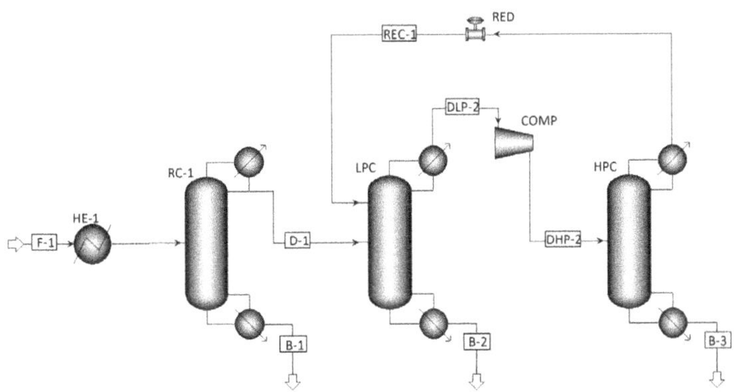

Figure 1. Process scheme used for the simulations of EtOH distillation, adapted from Karimi et al. [7].

2.2. Process Scheme

The entire process shown in Figure 1 consists of three distillation columns and is taken from Karimi et al. [7]. The first column RC-1 is intended for the concentration of EtOH up to 50% by weight. The second and third columns are used to concentrate EtOH to 99.60% by weight.

It is known that the azeotropic composition of mixtures changes as a function of pressure, but in the case of a mixture of EtOH and water this change is not so significant [7]. This effect on the azeotrope is shown in Figure 2, in the xy diagrams for EtOH-H_2O, where x_{EtOH} and y_{EtOH} represent the fractions of EtOH in equilibrium for liquid and vapour phases, respectively. By properly utilising this property, it is possible to obtain absolute EtOH by distillation without organic solvents.

The basis of PSD is that the two columns operate at different pressures. In our example, the low-pressure column (LPC) operates at 1 bar, while the high-pressure column (HPC) operates at 10 bar. The LPC is used to concentrate EtOH to a composition that is close to the azeotrope, in our case up to 94% by weight. The pressure of the distillate vapour was increased to 10 bar with the compressor COMP, which allows the distillate to reach another, previously inaccessible distillation area that does not have an upper limit regarding azeotropic composition [13]. In the HPC, therefore, EtOH at 99.60 wt.% in the bottom of the column was obtained, while the distillate was close to the azeotropic composition corresponding to the operating pressure of the HPC. In our case, the vapour with a composition of 91.50 wt.% EtOH was fed back into the LPC through the RED reduction valve. The reduction valve lowers the pressure of the vapours to 1 bar. The Radfrac model available in Aspen Plus V8 was used to simulate the distillation columns.

The HE-1 heat exchanger was used to preheat the feed mixture to boiling point. The Heater model, which is available in Aspen Plus V8, was used for the heat exchanger.

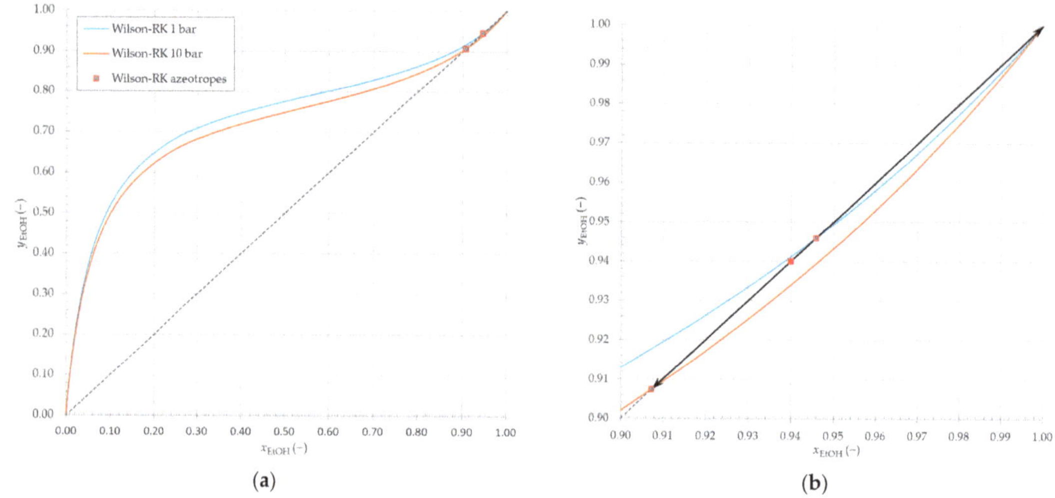

Figure 2. (**a**) EtOH-H_2O xy diagram; (**b**) EtOH-H_2O xy diagram magnified at the location of azeotrope shift (full red square shows EtOH fraction of distillate, achieved in LPC, while the black line shows the possible operation of HPC).

2.3. Simulation Procedure

2.3.1. Distillation Column RC-1

First, a simulation of column RC-1 with feed stream F-1 was carried out. The composition of the feed is shown in Section 2.1. The column was simulated with the specifications that the distillate D-1 and bottom B-1 contain 50 wt.% and 0.01 wt.% EtOH, respectively. These values were achieved automatically by varying the reflux ration (RR) and the distillate flow rate.

The RC-1 column was optimised by NQ analysis, which determines the optimum RR and tray numbers N, assuming for simplicity that they represent the operating and investment costs of a distillation column. Assuming that both costs are equally weighted, the optimal column design is the minimum of the diagram RR × N = f(N) [14]. For this purpose, Aspen Plus has the built-in function »NQ curves«, which provides the results of the heat flows, RR, and the optimum feed tray for a selected range of the total distillation column trays.

2.3.2. Pressure Swing Distillation

Based on the results obtained for column RC-1, the PSD was simulated separately. First, the columns were simulated with 50 trays, where the 40th tray was the feed tray (counted from the top of the column to the bottom). In both columns, the proportion of EtOH that the distillate must contain was determined. Values were determined near the azeotropic points with an EtOH content of 94.00 wt.% and 91.50 wt.% in the DLP-2 (distillate from LPC) and DLP-3 (distillate from HPC) streams, respectively. In addition, a specification for EtOH recovery of 99.90 wt.% was added in the LPC column. For the HPC column, the proportion of 99.60 wt.% EtOH in the B-3 was determined. The specifications were set so that they depend on the flow rate in the sump and the RR of the respective columns.

To ensure convergence, the simulation of the PSD was initially carried out as shown in Figure 3, where the streams DLP-3 and A-1 were disconnected. Later, the streams were connected as shown in Figure 1. For simplicity, both streams A-1 and D-1 were combined with the mixer MX, resulting in stream FD, as in the document of AspenTech [15]. In this way, the NQ analysis on the LPC was enabled. After the first simulation, the composition and flow rate of stream A-1 in further simulations were adjusted manually until they corresponded to stream DLP-3.

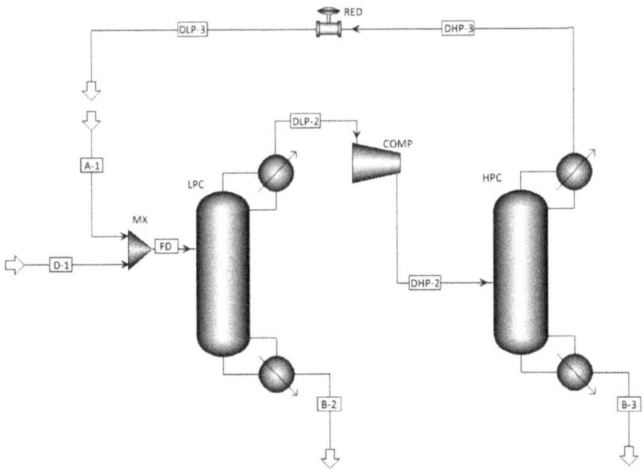

Figure 3. Initial process scheme for the simulation of PSD.

Using the process scheme in Figure 3, the NQ analysis was carried out and the optimum feed trays for the FD and DHP-2 streams were determined. After the analysis, the mixer MX and the stream FD were removed and the optimal feed trays for the streams A-1 and DLP-3 were determined manually. Since all simulations were performed with the above specifications, the compositions and flow rates of the process streams remained unchanged. During the manual search for the feed trays, the RR and heat flows of the distillation columns were automatically adjusted to meet the desired specifications. After determining the optimal feed trays, the A-1 and DLP-3 streams were combined with Wegstein convergence to form the REC-1 stream, resulting in the final process scheme shown in Figure 1.

2.4. Sensitivity Analysis

Sensitivity analyses on the LPC and HPC were performed. For the LPC, the effect of the recovery and the EtOH fraction in the distillate was analysed, while the effect of EtOH fraction in the distillate of HPC was examined. The specification to achieve 99.60 wt.% of EtOH in B-3 was constant since it was set as a desired value. The effect of the variables on the compressor work, heat demand in the evaporators, EtOH losses, RR, and the mass flow of stream REC-1 were analysed.

2.5. Restrictions and Assumptions

- The distillation columns are modelled with the Radfrac mode.
- Pressure drops are neglected.
- The compressor is isentropic with isentropic efficiency of 0.72, based on commercially available compressors from certified manufacturers.
- The simulation was carried out in a steady state.
- For optimisation with the NQ analysis, the same weight of operating and investment costs was assumed.

3. Results and Discussion

3.1. Distillation in Column RC-1

The results of the NQ analysis are shown graphically in Figure 4, for the range of column trays from 11 to 20. The optimum column from the analysis point of view is the column with 17 trays, where the third tray is the optimal feed tray.

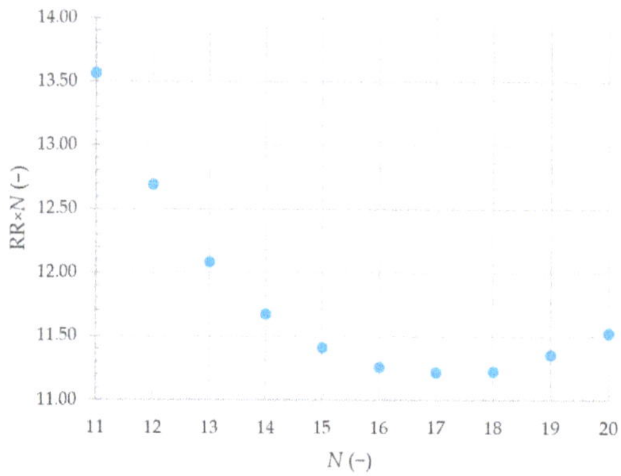

Figure 4. NQ analysis results for distillation column RC 1.

Column design corresponds to that of Morales et al. [16], where the column with 18 trays was used for the production of 45 wt.% EtOH. Feed was also introduced to the column on the third tray. Columns with a lower or higher number of trays have also been reported, as in [12], where 10 trays with an RR of 3 were used. Lassmann et al. used 32 trays with an RR of 3 [17]. Of course, all of the columns were capable of concentrating EtOH up to 50 wt.%. In the case of using a column with a lower number of trays, the required RR will be higher, whereas when the column has a higher number of trays, the RR needed will be lower. This is ultimately reflected in the investment and operating costs.

3.2. Pressure Swing Distillation

The results of the NQ analysis performed with the process scheme in Figure 3 are shown graphically in Figure 5. For the LPC, it turned out that the optimal inlet tray is the 24th tray. With this design, 9171 kW of heat needs to be supplied to the LPC evaporator, and 8 470 kW needs to be removed at the condenser. A further reduction in heating and cooling demand was possible with a manual adjustment of inlet trays for flows REC-1 and D-1. For this purpose, the process scheme was altered slightly, and the MX mixer and the FD flow were removed, which resulted in the process scheme shown in Figure 1. The number of column trays was not changed. The final designs of the LPC and HPC are shown in Table 2, where ϕ_b represents boiler heat flow, ϕ_c cooling need in the condenser, N is number of trays, and N_f is the feed tray. The negative sign for the values of ϕ_c emphasises the heat released from the condenser. The table also shows the column heat flows. The preheating demand in HE-1 and compressor work needed were excluded from the table. The values for process units amount to 3569 kW and 593 kW, respectively.

Table 2. Final design of distillation columns.

	RC-1	LPC: After NQ Analysis [1]	LPC: After Additional Feed Tray Optimisation [2]	HPC
ϕ_b (kW)	4109	9171	8377	5910
ϕ_c (kW)	−3850	−8470	−7852	−6436
RR (−)	0.61	3.70	3.43	4.68
N (−)	17	28	28	32
N_f (−)	3	24	REC-1: 18 D-1: 25	14

[1] Results refer to the scheme in Figure 3; [2] results refer to the scheme in Figure 1.

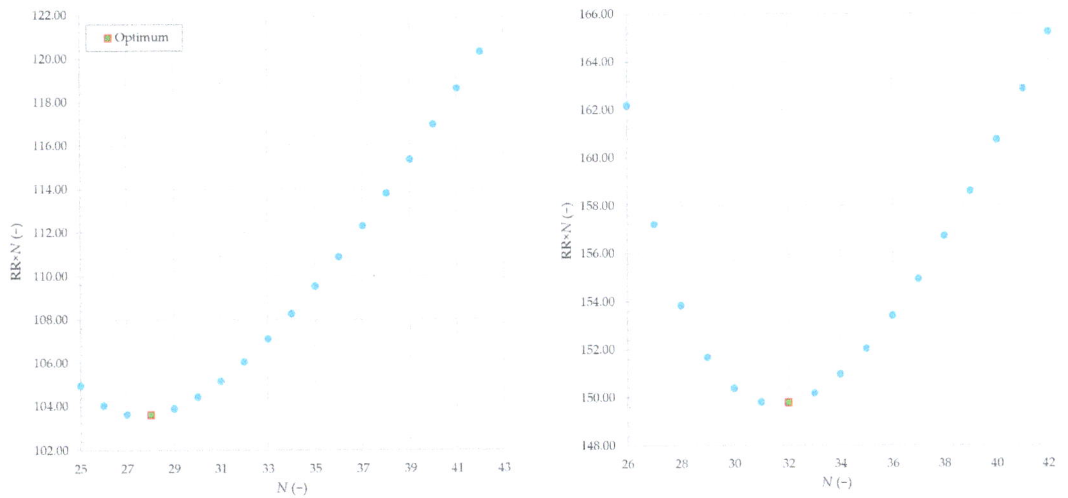

(a) (b)

Figure 5. NQ analysis results for (**a**) LPC and (**b**) HPC.

The results of all process streams are shown in Table 3, where q_m represents mass flow, T temperature, p pressure, and w mass fraction of the process flow component.

Table 3. Process flow results for the optimal process scheme.

	q_m (kg/h)	T (°C)	p (bar)	w_{H2O} (−)	w_{EtOH} (−)	w_{H2SO4} (−)	w_{HMF} (−)	$w_{Furfural}$ (−)	$w_{Ca(OH)2}$ (−)
F-1	50,188.90	33.00	1	0.9320	0.0540	0.0007	0.0119	0.0010	0.0004
F-2	50,188.90	94.47	1	0.9320	0.0540	0.0007	0.0119	0.0010	0.0004
D-1	5411.45	81.83	1	0.4928	0.5000	3.47×10^{-12}	2.24×10^{-09}	0.0072	5.47×10^{-54}
B-1	44,777.45	100.05	1	0.9851	0.0001	0.0008	0.0133	0.0002	0.0004
DLP-2	8774.91	77.99	1	0.0600	0.9400	3.64×10^{-132}	3.82×10^{-108}	6.42×10^{-19}	0
DHP-2	8774.91	217.72	10	0.0600	0.9400	3.64×10^{-132}	3.82×10^{-108}	6.42×10^{-19}	0
B-2	2703.14	99.12	1	0.9824	0.0031	6.95×10^{-12}	4.48×10^{-09}	0.0145	0
DHP-3	6066.60	149.43	10	0.0850	0.9150	0	0	0	0
REC-1	6066.60	138.46	1	0.0850	0.9150	0	0	0	0
B-3	2708.30	150.70	10	0.0040	0.9960	0	0	0	0

The desired content of 99.60 wt.% of EtOH in the final product was achieved, while the remaining 0.40 wt.% was mainly water, with a low proportion of other components.

3.3. *Sensitivity Analysis*

3.3.1. The Effect of EtOH Recovery

The results of the first sensitivity analysis are shown graphically in Figure 6. Upon varying the recovery of EtOH (R_{EtOH}) from D-1, it was observed that the increase in recovery resulted in linear decrease in EtOH losses in B-2 ($q_{m,\,EtOH,\,B\text{-}2}$). Simultaneously, the REC-1 mass flow ($q_{m,\,REC\text{-}1}$) increased. This was due to the specification for the EtOH fraction of 94 wt.% in the DLP-2 stream, which also resulted in the increase in the stream's flow rate.

As more vapour flows into the HPC, the boiler heat demand increases linearly, while the RR remains constant due to the constant composition of the DHP-2 process flow. Because the condenser of the column has the same relationship, it is not additionally illustrated. The same applies for the following sensitivity analyses.

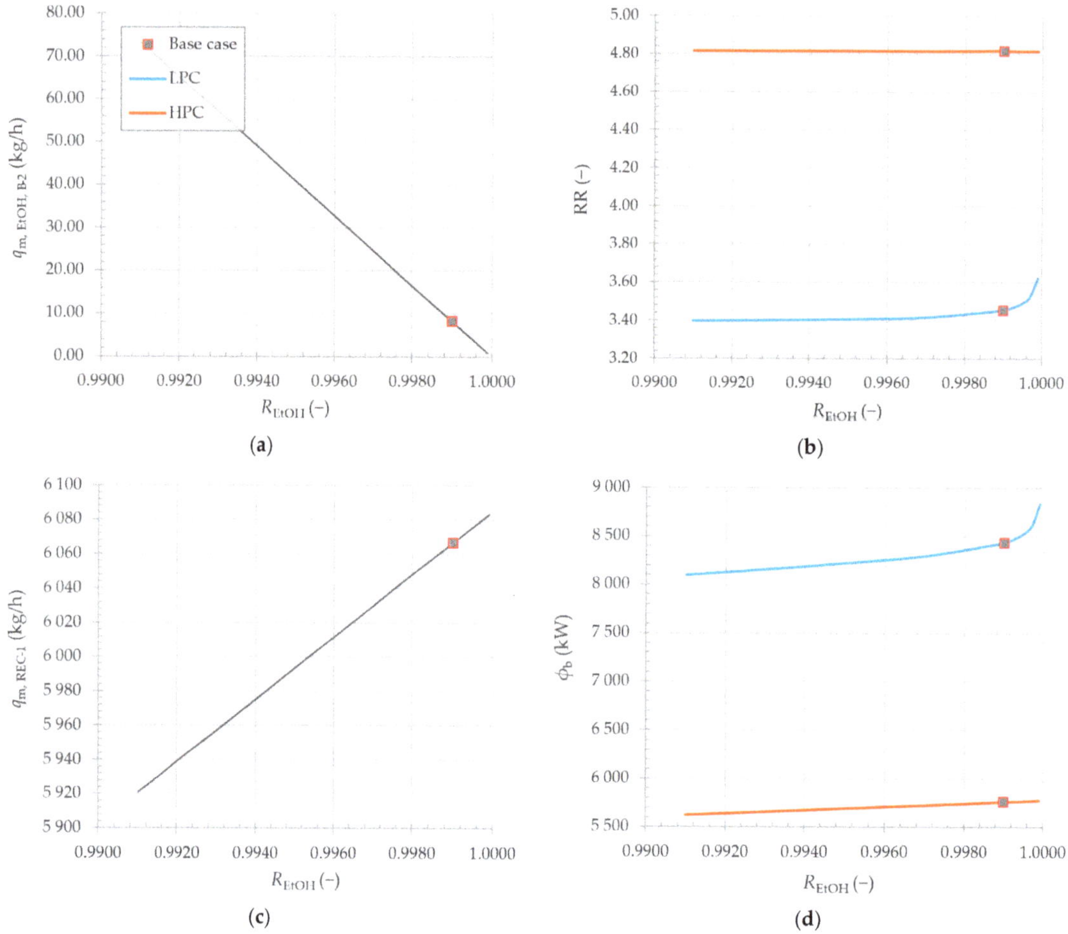

Figure 6. The effect of EtOH recovery on: (**a**) losses of EtOH in distillation bottom B-2; (**b**) RR of distillation columns; (**c**) mass flow of REC-1; and (**d**) heat demand in distillation column boilers.

Larger effects of recovery were observed in the LPC. As the recovery of EtOH increases, heat flows and RR increase exponentially. Increasing the recovery of EtOH increases the difficulty of recovery, which is reflected in higher RR. This results in the circulation of a larger mass of the mixture in the column, which increases the heat demand for the vaporisation of the mixture. It is worth noting that 100% recovery is not achievable.

3.3.2. The Effect of the EtOH Content in the Distillate of the LPC

Changing the EtOH content of the distillate has a much more complex effect on the overall system. Increasing the mass fraction results in a decrease in the flow rate of DLP-2 due to the lower flow rate of water in the distillate. Consequently, the flow rate of REC-1 recycle is reduced, as shown in Figure 7.

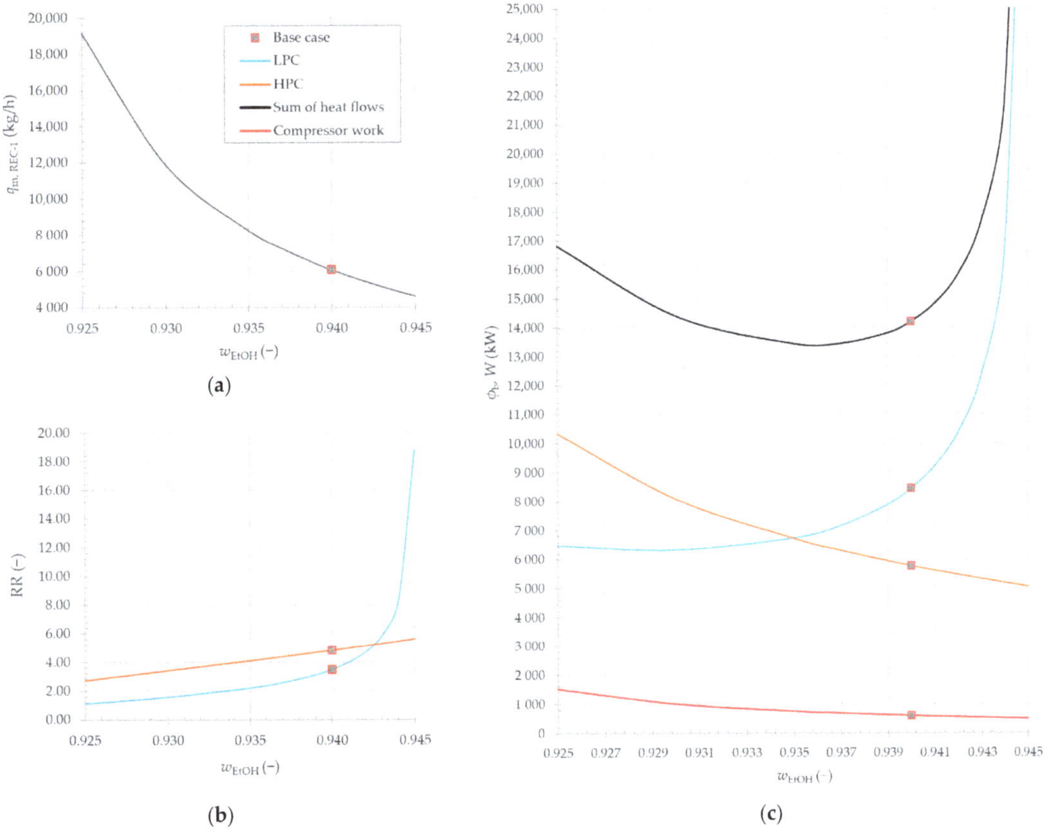

Figure 7. The effect of EtOH fraction in the distillate of LPC on (**a**) mass flow rate of recycle REC-1; (**b**) RR of distillation columns; (**c**) boiler heat demand and necessary compressor work.

Because the recovery of EtOH remains constant, the mass flow rate of EtOH losses in the bottoms decreases. The non-linearity of the relationship is mainly influenced by the second column. It is observed that as the EtOH fraction decreases, the mass flow rate of REC-1 starts to increase exponentially. This is due to the specification of the second column, which requires 91.50 wt.% of EtOH in the distillate. Although the RR decreases linearly, the closer the fraction of EtOH in the distillate is to this value, more distillate DHP-3 is sent to recycle. We conclude that the linear decrease in RR in the HPC is a result of approaching the value of 91.50 wt.%, since it is easier to reach the required value by adjusting the RR, and since the composition of the feed to the HPC column is already close to the specified value of the HPC. On the other hand, as the specification for EtOH fraction in DLP-2 increases, the RR in the LPC increases exponentially, which is caused by the approach to the azeotropic point.

The complex relationship between heat flows in the LPC column is caused by the simultaneous variation in RR and recycle flow rate. At one extreme, as we approach the azeotrope at 1 bar, the heat flow increases exponentially due to the RR in the LPC. At the other extreme, as we approach the specification 91.50 wt.% of EtOH in the HPC distillate, the increase in heat flow increases the flow rate of the recycling. The EtOH fraction of the DLP-2 distillate is therefore confined between the azeotrope at 1 bar and the specification for the EtOH fraction of the HPC distillate.

Compressor work (*W*), shown by the red line, is dependent on the mass flow rate of the LPC vapour distillate.

3.3.3. The Effect of the EtOH Content in the Distillate of the HPC

Changing the EtOH fraction in the distillate of HPC has a similar effect as the fraction of EtOH in the distillate of LPC. This can be seen in Figure 8. When the purity of the distillate is increased, an increase in recycle flow rate is observed. This is a result of approaching the specification of 94 wt.% of EtOH in LPC distillate DLP-2. As we increase the EtOH content in the distillate DHP-3, the difference between EtOH fractions becomes smaller, resulting in a larger volume of distillate being sent to the DLP-2 stream. At the same time, the RR decreases, because from the RR point of view it becomes easier to obtain the target value of 94 wt.% EtOH. The EtOH losses increase due to the increase in recycle flow rate. Because of higher DLP-2 flow rates, the compressor also requires more energy.

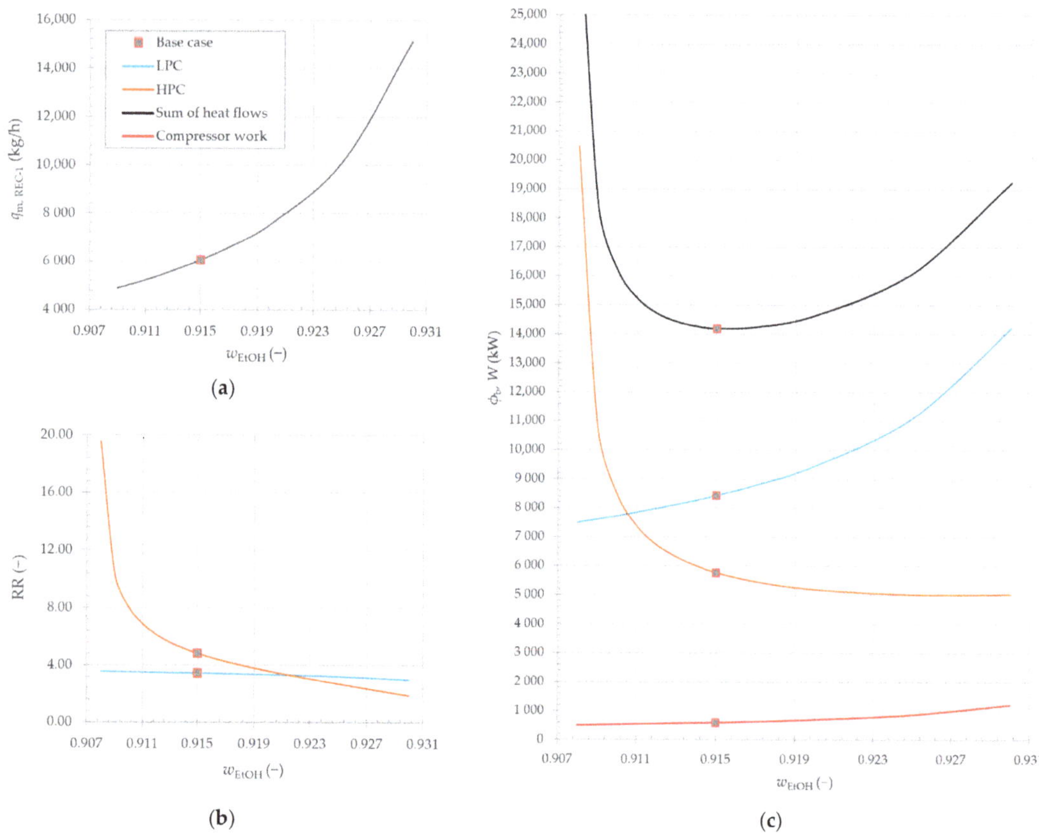

Figure 8. The effect of EtOH fraction in distillate of HPC on (**a**) mass flow rate of recycle REC-1; (**b**) RR of distillation columns; and (**c**) boiler heat demand and necessary compressor work.

In the HPC, the RR increases as the EtOH fraction in the distillate decreases. This is due to the approach of the azeotropic point at 10 bars. Consequently, at the two extremes, we can again observe an increase in heat flows. As the investigated specification approaches 94 wt.%, the heat flows increase due to the high recycle flows. As we decrease the specification, the heat flows increase due to the increase in RR in the HPC.

Comparing the heat flow plots in Figures 7 and 8, we see that the minima of the sum lie at different values. From an operational point of view, the 93.70 wt.% of EtOH in the distillate of LPC might be better, but the recycle flow rate would increase, leading to higher EtOH losses. This would also lead to higher flow rates of DLP-2, leading to an increase in compressor work. The change in the specification of LPC would require further optimisation of the HPC feed tray, while it is also not necessary that the optimum specification of the HPC distillate would be 91.50 wt.%. Due to the complexity of the interactions between the columns, the problem requires a more detailed optimisation and economic analysis, which is very difficult with a sensitivity analysis, and is beyond the scope of the present work.

The design of the columns is comparable with the literature. In Mulia-Soto and Flores-Tlacuahuac [9], Iqbal and Ahmad [10], and Kiran and Jana [11], 30 trays were used for both columns, which was very close to our results. The inlet trays differed due to minor differences between the compositions of all streams. According to Karimi et al., PSD is often used to separate azeotropic mixtures of EtOH and water [7], but we have not found any other reliable data on its use to produce anhydrous EtOH in industry. In any case, the separation of azeotropic mixtures by such distillation is possible but is more difficult when the azeotropic composition with pressure varies only slightly [7].

4. Conclusions

The present work presents the possibility of using PSD for the production of high-purity bioethanol obtained through the fermentation of biomass. The produced renewable ethanol could be used as a fuel instead of conventional gasoline.

NQ analysis, a simple method for determining the optimal design of distillation columns, was used for the process design. The results of the analysis showed a process design with good agreement with the data reported in the literature. The components present in the feed stream were successfully removed by distillation, producing high-purity EtOH with 99.60 wt.% and a water content of 0.40 wt.%. However, the process consumes a lot of heat and could be further improved by heat integration.

Author Contributions: Conceptualisation, methodology, software, validation, investigation, formal analysis, writing—original draft: K.R.; investigation, formal analysis, writing—original draft: S.G.; formal analysis, writing—review and editing, supervision: D.G.; writing—review and editing, supervision: D.U. All authors have read and agreed to the published version of the manuscript.

Funding: This research received no external funding.

Institutional Review Board Statement: Not applicable.

Informed Consent Statement: Not applicable.

Data Availability Statement: All the data generated during work is provided through tables, diagrams and figures.

Conflicts of Interest: The authors declare no conflicts of interest.

References

1. Chen, W.-H.; Lin, B.-J.; Lin, Y.-Y.; Chu, Y.-S.; Ubando, A.T.; Show, P.L.; Ong, H.C.; Chang, J.-S.; Ho, S.-H.; Culaba, A.B.; et al. Progress in Biomass Torrefaction: Principles, Applications and Challenges. *Prog. Energy Combust. Sci.* **2021**, *82*, 100887. [CrossRef]
2. Environmental Protection Agency. *Inventory of U.S. Greenhouse Gas Emissions and Sinks: 1990–2019*, U.S. Environmental Protection Agency: Washington, DC, USA, 2021.
3. Lamichhane, G.; Acharya, A.; Poudel, D.K.; Aryal, B.; Gyawali, N.; Niraula, P.; Phuyal, S.R.; Budhathoki, P.; Bk, G.; Parajuli, N. Recent Advances in Bioethanol Production from Lignocellulosic Biomass. *Int. J. Green Energy* **2021**, *18*, 731–744. [CrossRef]
4. Amornraksa, S.; Subsaipin, I.; Simasatitkul, L.; Assabumrungrat, S. Systematic Design of Separation Process for Bioethanol Production from Corn Stover. *BMC Chem. Eng.* **2020**, *2*, 10. [CrossRef]
5. Beluhan, S.; Mihajlovski, K.; Šantek, B.; Ivančić Šantek, M. The Production of Bioethanol from Lignocellulosic Biomass: Pretreatment Methods, Fermentation, and Downstream Processing. *Energies* **2023**, *16*, 7003. [CrossRef]
6. Conde-Mejía, C.; Jiménez-Gutiérrez, A. Analysis of Ethanol Dehydration Using Membrane Separation Processes. *Open Life Sci.* **2020**, *15*, 122–132. [CrossRef] [PubMed]
7. Karimi, S.; Karri, R.R.; Tavakkoli Yaraki, M.; Koduru, J.R. Processes and Separation Technologies for the Production of Fuel-Grade Bioethanol: A Review. *Environ. Chem. Lett.* **2021**, *19*, 2873–2890. [CrossRef]

8. Rumbo Morales, J.Y.; Ortiz-Torres, G.; García, R.O.D.; Cantero, C.A.T.; Rodriguez, M.C.; Sarmiento-Bustos, E.; Oceguera-Contreras, E.; Hernández, A.A.F.; Cerda, J.C.R.; Molina, Y.A.; et al. Review of the Pressure Swing Adsorption Process for the Production of Biofuels and Medical Oxygen: Separation and Purification Technology. *Adsorpt. Sci. Technol.* **2022**, *2022*, 3030519. [CrossRef]
9. Mulia-Soto, J.F.; Flores-Tlacuahuac, A. Modeling, Simulation and Control of an Internally Heat Integrated Pressure-Swing Distillation Process for Bioethanol Separation. *Comput. Chem. Eng.* **2011**, *35*, 1532–1546. [CrossRef]
10. Iqbal, A.; Ahmad, S.A. Pressure Swing Distillation of Azeotropic Mixture—A Simulation Study. *Perspect. Sci.* **2016**, *8*, 4–6. [CrossRef]
11. Kiran, B.; Jana, A.K. A Hybrid Heat Integration Scheme for Bioethanol Separation through Pressure-Swing Distillation Route. *Sep. Purif. Technol.* **2015**, *142*, 307–315. [CrossRef]
12. Quintero, J.A.; Cardona, C.A. Process Simulation of Fuel Ethanol Production from Lignocellulosics Using Aspen Plus. *Ind. Eng. Chem. Res.* **2011**, *50*, 6205–6212. [CrossRef]
13. Thomas Rodgers Plotting Approximate Residue Curves and Pressure Swing. Available online: https://www.youtube.com/watch?v=o6zOi_Pp_80 (accessed on 25 December 2023).
14. Roman, C.; García-Morales, M. Achieving a Better Understanding of Binary Azeotropic Mixtures Distillation through Aspen Plus Process Simulations. *Comp. Applic. Eng.* **2019**, *27*, 1453–1464. [CrossRef]
15. Aspen Technology Inc. *Pressure Swing Distillation with Aspen Plus®V8.0*; Aspen Technology Inc.: Bedford, MA, USA, 2012.
16. Morales, M.; Arvesen, A.; Cherubini, F. Integrated Process Simulation for Bioethanol Production: Effects of Varying Lignocellulosic Feedstocks on Technical Performance. *Bioresour. Technol.* **2021**, *328*, 124833. [CrossRef] [PubMed]
17. Lassmann, T.; Kravanja, P.; Friedl, A. Simulation of the Downstream Processing in the Ethanol Production from Lignocellulosic Biomass with ASPEN Plus®and IPSEpro. *Energy Sustain. Soc.* **2014**, *4*, 27. [CrossRef]

molecules

MDPI

Article

Catalytic Degradation of Lignin over Sulfonyl-Chloride-Modified Lignin-Based Porous Carbon-Supported Metal Phthalocyanine: Effect of Catalyst Concentrations

Fangli Du, Xuequan Xian, Peiduo Tang and Yanming Li *

High Performance Materials Research Institute, Guangxi Academy of Sciences, Nanning 530007, China; dfl@gxas.cn (F.D.)

* Correspondence: liyanming@gxas.cn

Abstract: A sulfonyl-chloride-modified lignin-based porous carbon-supported metal phthalocyanine catalyst was prepared and used to replace the traditional Fenton's reagent for lignin degradation. The catalyst underwent a detailed characterization analysis in terms of functional group distributions, surface area, morphological structure, via FT-IR, XPS, BET, and SEM. The catalyst possessed a specific surface area of 638.98 m^2/g and a pore volume of 0.291 cm^3/g. The prepared catalyst was studied for its ability of oxidative degradation of lignin under different reaction conditions. By optimizing the reaction conditions, a maximum liquid product yield of 38.94% was obtained at 135 °C with 3.5 wt% of catalyst and 15×10^{-2} mol/L H_2O_2; at the same time, a maximum phenols selectivity of 32.58% was achieved. The compositions and properties of liquid products obtained from lignin degradation using different catalyst concentrations were studied comparatively via GC-MS, FT-IR, ^{1}H-NMR, and EA. Furthermore, the structure changes of solid residues are also discussed.

Keywords: lignin; supported metal phthalocyanine; lignin degradation; liquid products

1. Introduction

Facing the problem of energy shortage and environmental pollution caused by excessive consumption of fossil energy, it is especially important to development and employ sustainable renewable resources such as renewable biomass resources. According to the predictions of the International Energy Agency (IEA), bioenergy will account for about 10% of total primary energy use by 2035, and by 2050, bioenergy will account for around 27% of the world's total transport energy [1]. Thus, it can be seen that there is considerable urgency and demand for biomass renewable energy that can replace traditional fossil energy. One of the most popular approaches is to utilize industrial waste lignocellulose [2].

Lignin is the most abundant three-dimensional phenolic polymer bearing three aromatic structures: syringyl (S), guaiacyl (G), and hydroxyphenyl (H) units. Each unit has been verified to stem from one of the monolignols: p-hydroxycinnamyl, coniferyl alcohol, and sinapyl alcohol [3,4]. The three basic monolignols are mainly linked by C-C or C-O types; however, other chemical bonds are also presented in lignin such as aryglycerol-β-ether dimer (β-O-4, accounting for 45–50%), biphenyl/dibenzodioxocin (5-5′, accounting for 18–25%), pino/resinol (β-5, accounting for 9–12%), diphenylethane (β-1, accounting for 7–10%), aryglycerol-α-ether dimer (α-O-4, 6–8%), phenylcoumaran (β-β′, 0–3%), and siaryl ether (4-O-5, 4–8%) [5]. Because of this peculiar phenylpropanol structure, and the recent emergence of biorefineries, lignin attracts global attention and was thought to be a low-value energy source and a viable means to manufacture chemicals and biofuels [6]. However, various strategies have been implemented to transform lignin into small chemicals via cleavage of C-C and C-O linkages [4].

Among these strategies, liquid phase catalytic degradation of lignin to produce small chemicals has been found to be an effective measure because of the mild reaction conditions and high product selectivity. So far, numerous studies have examined the liquid phase catalytic degradation of lignin [7,8], including acid-catalyzed [9–12] and base-catalyzed depolymerization [13,14]. Although acid- and base-catalyzed depolymerization of lignin have been studied extensively, there are still some drawbacks that need to be faced; for instance, it is difficult to isolate the homogeneous acid and base catalyst after the reaction is completed, thus resulting in environmental pollution. Furthermore, carbocation occurring during acid-catalyzed depolymerization or the quinone structure formed during base-catalyzed depolymerization favor the production char, and so reduce the selectivity of aromatic monomers.

In order to avoid the disadvantages of acid- or base-catalyzed depolymerization, researchers gradually started to use Fenton's reagent for the lignin treatments. G. Bentivenga et al. found that the application of $FeSO_4{\cdot}7H_2O/H_2O_2$ could accelerate the degradation of lignin [15]. Francesc Torrades [16] and co-workers investigated the degradation of black liquor using $[Fe(II)]/[H_2O_2]$; the best conversion of 94.8% was achieved when $[H_2O_2]$ = 44.1 mM, [Fe(II)] = 4.655 mM, at 298 K and pH = 3. Phisit Seesuriyachan et al. [17] reported that only about 10% of lignin was present after the degradation using the synergistic catalysts 7.61 mg/L Fe^0, 9.89 mg/L Fe^{2+}, 14.27 mg/L Fe^{3+}, and 376.88 mg/L H_2O_2 at pH = 3. Recently, some studies on the treatment of lignin with Fenton-like reagent have been conducted. For instance, YuYun Lin and co-workers [18] investigated the cleavage of a lignin model compound into aromatic chemicals using $CuFe_2O_4$@rGO-150 and H_2O_2; the best yield of guaiacol was 72.6% and the best yield of 2-methoxy-4-propyl phenol was 52.5%. In these systems, OH· was generated from the decomposition of H_2O_2 by Fe^{2+} (or other transition metals) catalysis at low pH (−2–3) (Figure 1a); subsequently a large number of subsequent reactions were triggered by HO· (Figure 1b) [19–22].

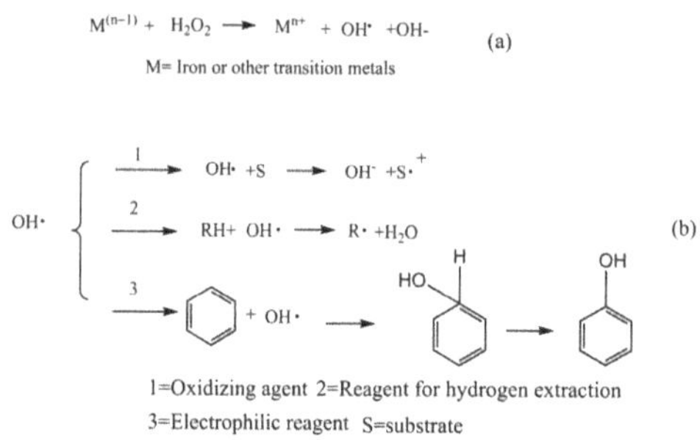

Figure 1. Reaction formulas: (**a**) formation of HO· from H_2O_2; (**b**): the reaction triggered by HO·.

It can be seen from the above that Fenton's reagent or Fenton-like reagent has feasibility in the depolymerization of lignin. In our previous literature investigation, we found that metal phthalocyanines exhibit high catalytic capacity for various types of reactions [23]. Tao X et al. found that the introduction of metal phthalocyanine can ensure the Fenton's reagent maintains high catalytic activity even under neutral or weakly alkaline conditions [24]. In addition, in the metal phthalocyanine structure, the metal atom (M) binds to the N atom to form M-N chelate; when the metal atom M is strongly anchored, this kind of structure can avoid the metal loss and agglomeration. The above advantages of metal phthalocyanine create possibilities for its further application to replace the traditional Fenton's reagent. In the presence of metal phthalocyanine or supported metal phthalocyanine, the generation of HO· abidesby the principle described in Figure 2 [25,26]:

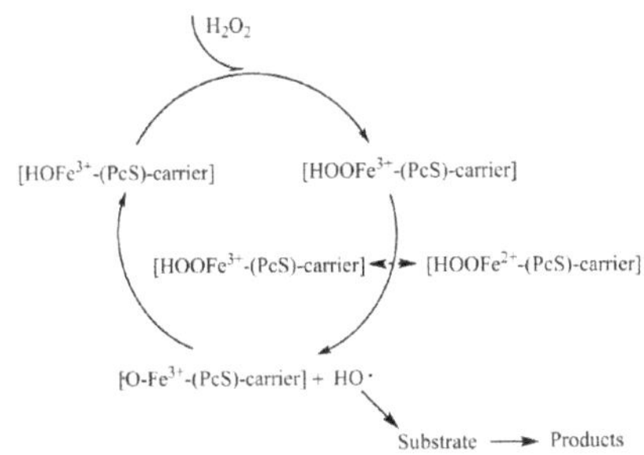

Figure 2. The route for HO· generation.

Therefore, in the present study we aimed to synthesize the sulfonyl-chloride-modified lignin-based porous carbon-supported metal phthalocyanine as a Fenton catalyst to replace the traditional ferrous ion, and develop a new Fenton process for lignin degradation, taking this supported metal phthalocyanine as the catalyst and H_2O_2 as the oxidant. The detailed experiment process is summarized in Figure 3. This experiment injects new vitality into the development of Fenton's reagent; additionally, the research theoretically provides evidence of the new technology for the resource utilization of lignin.

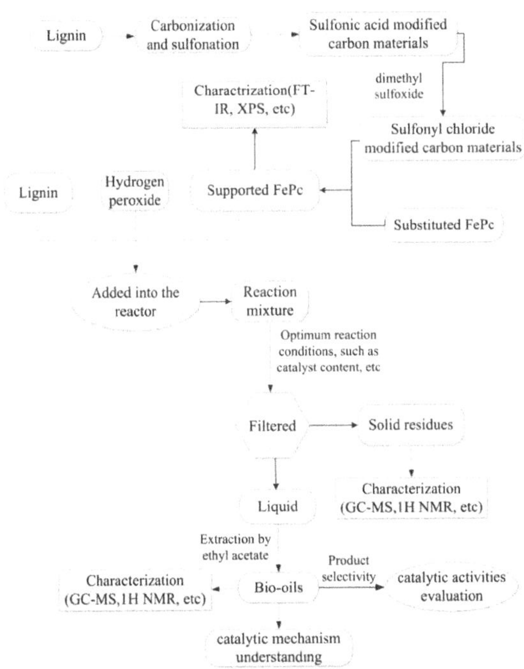

Figure 3. The detailed experiment process in this paper.

2. Results and Discussions

2.1. Characterization of Catalyst

2.1.1. FTIR Analysis

The structural changes of FePc and FePc@SO_2Cl-C_{lig} were confirmed using FT-IR analyses; the results are exhibited in Figure 4.

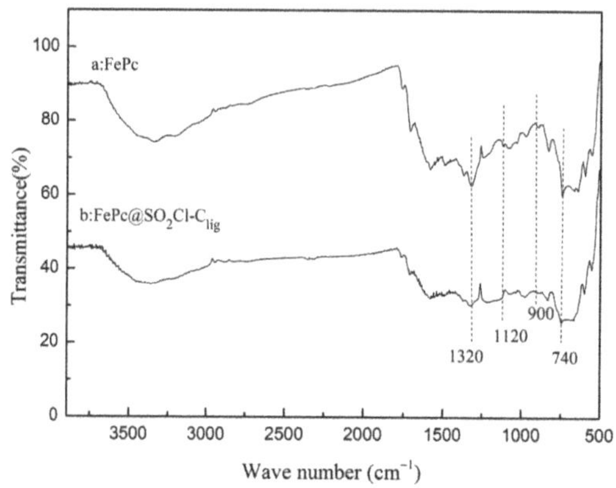

Figure 4. FT-IR spectra of FePc and FePc @ SO_2Cl-C_{lig}.

The FT-IR spectra revealed that the characteristic peaks of FePc appeared at 1320 cm^{-1}, 1120 cm^{-1}, and 740 cm^{-1}, which are assigned to the vibration of the phthalocyanine ring [27,28]. Similar characteristic bands were exhibited obviously in the spectrum of FePc@SO_2Cl-C_{lig}, indicating that FePc clung to the surface of SO_2Cl-C_{lig} without changes in the chemical feature. A new weak signal derived from S-N stretching vibration appeared around 900 cm^{-1} [29], demonstrating that the FePc successfully introduced into the surface of SO_2Cl-C_{lig} via chemical reaction.

2.1.2. XPS Analysis

The surface composition of FePc@SO_2Cl-C_{lig} was investigated using X-ray photoelectron spectroscopy (XPS), and the spectra are illustrated in Figure 5. The N1s spectrum (A1) could be split into two main peaks at 399.2 eV and 401.2 eV, which correspond to N of C-N (399.2 eV) and S-N (401.2 eV) [30,31]; for Fe2p spectrum (A2), four sets of signals could be captured in the region of 700–730 eV, which is consistent with the $2p_{3/2}$ orbital of Fe^{2+} (710.7 eV), $2p_{3/2}$ orbital of Fe^{3+} (714.6 eV), $2p_{1/2}$ orbital of Fe^{2+} (723.6 eV), and $2p_{1/2}$ orbital of Fe^{3+} (726.3 eV) [28,32,33]. Figure 5-A3 shows the spectrum of S 2p; a binding energy around 168.3 eV was observed, which is consistent with S 2p of typical sulphonamides [34,35]. It can be inferred from these data that iron phthalocyanine was successfully attached to the SO_2Cl-C_{lig}. Significantly, another sulfur signal at a binding energy around 164.2 eV corresponding to sulfoxide (–SO–) was captured, indicating that some of the sulfoxide was intact on the surface of the catalyst after acyl chlorination [36].

2.1.3. BET Analysis

To check the structure and morphology, BET analyses including specific surface areas and pore volume were performed for SO_2Cl-C_{lig} and FePc@SO_2Cl-C_{lig}.

As shown in Figure 6, a steep slope can be seen when P/P0 was below 0.2. This can confirm that the two N_2 adsorption–desorption isotherms accord with the type I isotherm adsorption and H4-type hysteresis loop; moreover, this can also imply that a large number of micro-pores are present in the samples [37,38]. The detailed surface area and pore volume of SO_2Cl-C_{lig} and PcFe@SO_2Cl-C_{lig} are listed in Table 1. The sample of FePc@ SO_2Cl-C_{lig} displayed lower surface area and pore volume, that is, the specific surface area and pore volume of two samples abide by a slight declining trend: SO_2Cl-C_{lig} > FePc@ SO_2Cl-C_{lig}. When FePc was loaded on SO_2Cl-C_{lig}, the BET surface area decreased from 643.67 m^2/g to 638.98 m^2/g, while the total pore volume decreased from 0.337 cm^3/g to 0. 291 cm^3/g. This drop in the BET surface area and the total pore volume may be because some pores of the carrier were blocked by the FePc [39]. The pore size distribution curves of SO_2Cl-C_{lig} and FePc@SO_2Cl-C_{lig} are displayed in Figure 7. For SO_2Cl-C_{lig}, one strong peak and two weak peaks can be seen in the range of 0.6–2.0 nm, but regarding FePc@SO_2Cl-C_{lig}, two strong peaks can be observed in the range of 0.85–1.42 nm. The two plots confirmed the large number of micro-pores, demonstrating that after the treatment by FePc, the pore structure did not change significantly. Clearly, a peak around 4.84 nm in plot a and another peak around 3.38 nm in plot b can be observed, which belong to meso-pores. Regarding porous catalysts, micro-pores mainly provide sufficient activity sites for catalysts. Opportune meso-pores enable efficient mass transport, and the catalytic performance can be improved only if the balance is kept between active sites and mass transport [40].

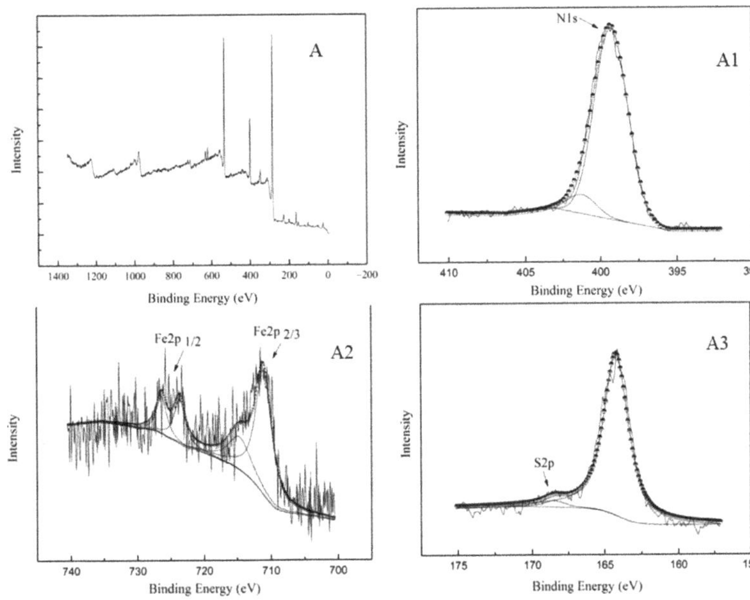

Figure 5. XPS spectrum of FePc@SO_2Cl-C_{lig} ((**A**): XPS spectrum of FePc@SO_2Cl-C_{lig}; (**A1**): N 1s photopeaks; (**A2**): Fe 2p photopeaks; (**A3**): S2p photopeaks).

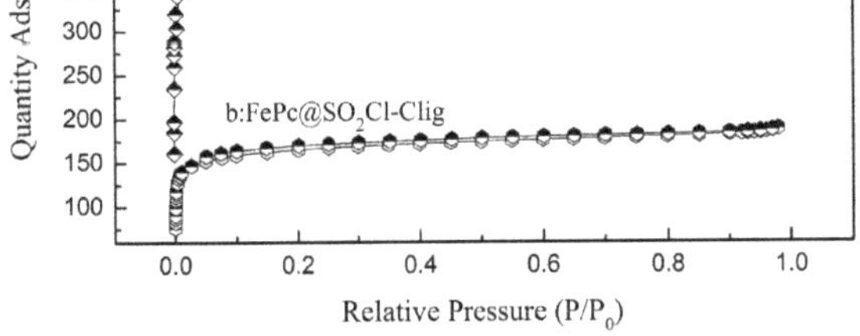

Figure 6. N_2 adsorption–desorption isotherms of SO_2Cl-C_{lig} and PcFe@SO_2Cl-C_{lig}.

Table 1. Characterization of the porous texture of $SO_2Cl\text{-}C_{lig}$ and $PcFe@SO_2Cl\text{-}C_{lig}$.

Samples	S_{BET} (m^2/g)	S_{Mic} (m^2/g)	V_{total} (cm^3/g)	V_{Mic} (cm^3/g)
$SO_2Cl\text{-}C_{lig}$	643.67	540.79	0.337	0.210
$FePc@ SO_2Cl\text{-}C_{lig}$	638.98	603.57	0.291	0.246

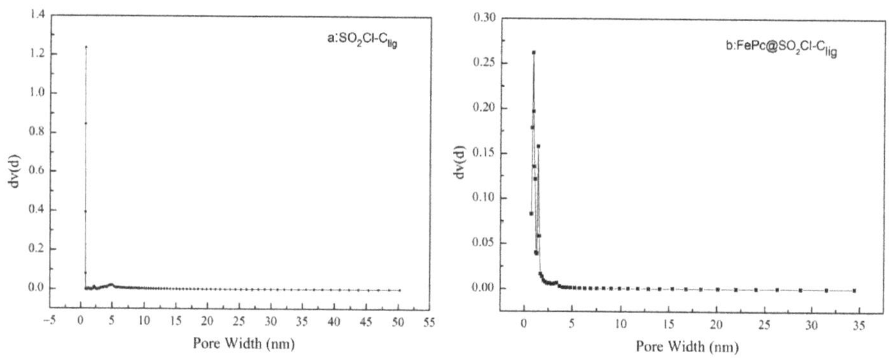

Figure 7. Pore size distributions of curves of (**a**) $SO_2Cl\text{-}C_{lig}$ and (**b**) $FePc@SO_2Cl\text{-}C_{lig}$.

2.1.4. SEM Analysis

SEM analysis was conducted to investigate the morphology changes in $SO_2Cl\text{-}C_{lig}$ and $FePc@SO_2Cl\text{-}C_{lig}$; the images are displayed in Figure 8.

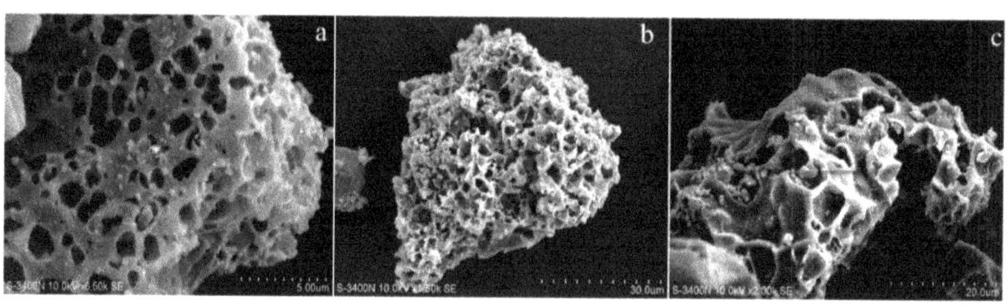

Figure 8. SEM images of $SO_2Cl\text{-}C_{lig}$ and $FePc@SO_2Cl\text{-}C_{lig}$ ((**a**). $SO_2Cl\text{-}C_{lig}$; (**b**). $FePc@SO_2Cl\text{-}C_{lig}$; (**c**). $FePc@SO_2Cl\text{-}C_{lig}$; magnified 2000 times).

As shown in Figure 8a, the pore structure was clearly visible and more uniformly dispersed on the surface of $SO_2Cl\text{-}C_{lig}$ before the loading of FePc. It can be seen obviously from Figure 8b that the sample already had abundant pores; however, when the samples were magnified 2000 times, as shown in Figure 8c, it was clearly seen that some pores were blocked by FePc, and this phenomenon confirmed the BET analysis.

2.2. *Optimizing the Reaction Conditions*

The degradation efficiency of lignin and the composition properties of liquid products were significantly affected by various reaction conditions, such as reaction temperature, reaction time, the type of catalyst and content, and the feedstock content [41]. In this study, the effects of catalyst content, reaction time, and hydrogen peroxide concentration on the degradation efficiency of lignin were investigated; the results are displayed in Figures 9–11.

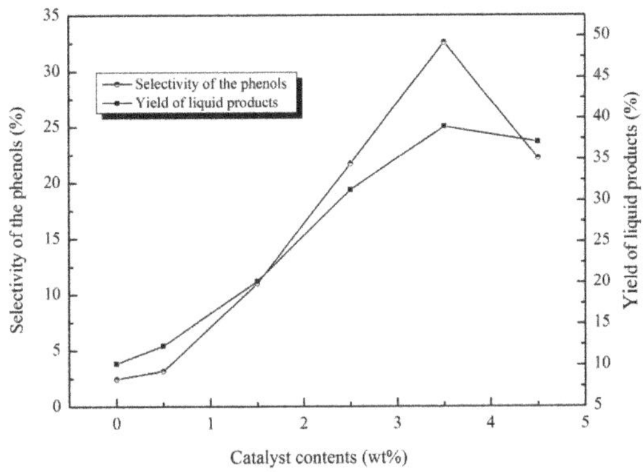

Figure 9. Effects of catalyst contents on the liquid product yields and the selectivity of phenolic compounds (catalyst contents: 0 wt%, 0.5 wt%, 1.5%, 2.5 wt%, 3.5 wt%, 4.5 wt%; H_2O_2 concentration: 15×10^{-2} mol/L; reaction time: 120 min; T = 135 °C pH = 3.0.).

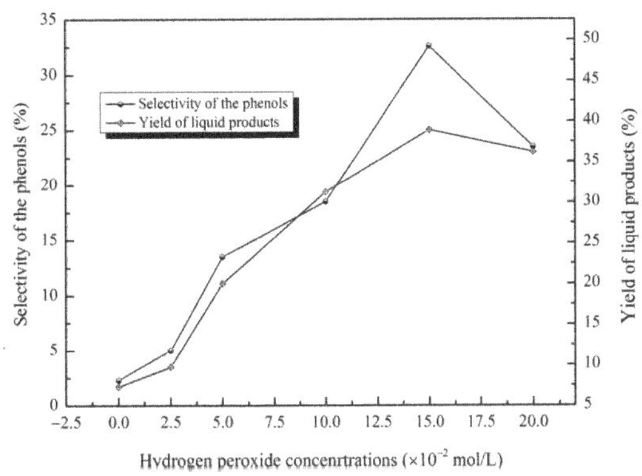

Figure 10. The effects of hydrogen peroxide concentrations on the yield of liquid products and the selectivity of phenolic compounds (catalyst contents: 3.5 wt%; H_2O_2 concentration: 0 mol/L, 2.5×10^{-2} mol/L, 5×10^{-2} mol/L, 10×10^{-2} mol/L, 15×10^{-2} mol/L, 20×10^{-2} mol/L; reaction time: 120 min; T = 135 °C; pH = 3.0).

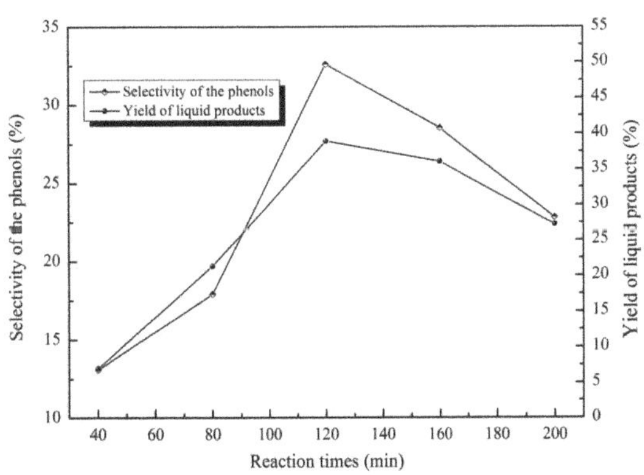

Figure 11. The effects of reaction time on the yield of liquid products and the selectivity of phenolic compounds (catalyst contents: 3.5 wt%; H_2O_2 concentration: 15×10^{-2} mol/L; reaction time: 40 min, 80 min, 120 min, 160 min, 200 min; T = 135 °C; pH = 3.0).

Figure 9 illustrates the catalyst content effects on the liquid product yields and phenol selectivity. As is known, most of the reactions were affected by the activity sites and basic properties of the catalysts. Keeping this in mind, six different catalyst concentrations were tried in this work: 0 wt%, 0.5 wt%, 1.5 wt%, 2.5 wt%, 3.5 wt%, and 4.5 wt%. It can be observed from Figure 9 that with the increase in catalyst content, the yield of liquid products and the selectivity of the phenols were both higher compared to those of the non-catalyst degradation. A maximum liquid product yield of 38.94% was observed when 3.5 wt% of catalyst was employed, while the selectivity of the phenols also reached the highest value of 32.58%. When the catalyst content increased to 4.5 wt%, the liquid product yield and the selectivity of the phenols both decreased.

These observations implied that the proper amount of the catalyst had a significant contribution to the yields of liquid products and the selectivity of the phenols. Once the amount of the catalyst increases beyond the optimum value, the yields of liquid products and the selectivity of the phenols may decrease. The catalyst not only can offer ample activity sites, but also facilitate the generation of HO· (hydroxyl) and HOO· (perhydroxyl) radicals from H_2O_2, and HO· can weaken the C-O or C-C linkages in lignin, subsequently resulting in the conversation of lignin into phenolic compounds and other small chemicals [42]. In this study, the maximum proportion of the phenolic compounds was found with 3.5 wt% of catalyst, suggesting that this amount of catalyst favored the generation of the amount of available OH·, and so enhanced the breaking of the C-O or C-C bonds in lignin. When the amount of catalyst exceeded 3.5 wt%, the production of more OH· contributed to the over-oxidation of the aromatic products and the coupling and condensation of the intermediates, thereby reducing the yield of liquid products and proportion of phenolic compounds [18].

Figure 10 illustrates the influence of the amount of hydrogen peroxide on the yield of liquid products and the selectivity of phenolic compounds. Six different levels (0 mol/L, 2.5×10^{-2} mol/L, 5×10^{-2} mol/L, 10×10^{-2} mol/L, 15×10^{-2} mol/L, 20×10^{-2} mol/L) were used with 3.5 wt% of catalyst at 135 °C and pH 3.0 for 120 min. Similar changing trends were observed for the yield of liquid products and the selectivity of phenolic compounds with the increase in the H_2O_2 concentration from 0 mol/L to 20×10^{-2} mol/L. The yield of liquid products and the selectivity of phenolic compounds increased as the hydrogen peroxide concentration increased from 0 mol/L to 15×10^{-2} mol/L. A maximum yield of 38.94% and a highest selectivity of 32.58% were obtained when 15×10^{-2} mol/L of hydrogen peroxide was used, further increasing the hydrogen peroxide concentration and resulting in a non-obvious change in liquid product yields, but the selectivity of phenolic compounds obviously decreased.

At a lower H_2O_2 concentration, the generation of HO· was not enough to induce the reaction. With the increase in the hydrogen peroxide concentrations, more HO· was generated; thus, more C-O or C-C linkages were weakened and finally cracked to form phenols or other chemicals [43,44]. However, when the hydrogen peroxide exceeded the optimum concentration, the excessive hydrogen peroxide could react with HO· to form more $HO_2^\bullet$ radicals, which had a weaker activity compared to HO·. Furthermore, the concentration of HO· was reduced by the reaction with H_2O_2, which slowed, or even stopped, the breaking of chemical bonds in lignin (the detailed reactions are described as Formulas (1) and (2)) [16,45,46]. Therefore, the yield of liquid products and the selectivity of phenolic compounds did not increase further when 20×10^{-2} mol/L hydrogen peroxide was used.

$$H_2O_2 + HO \rightarrow H_2O + HOO\cdot \quad (1)$$

$$HOO\cdot + HO \rightarrow H_2O + O_2 \quad (2)$$

The yield of liquid products and the selectivity of phenolic compounds achieved at different reaction times are illustrated in Figure 11. A gradual increase was observed in the yield of liquid products and the selectivity of phenolic compounds with the increase in the reaction time to 2 h. When the reaction time was 2 h, the two values reached their maximum; with the further increase in the reaction time to 160 min and 200 min, the yield of liquid products and the selectivity of phenolic compounds declined significantly. When the reaction time was 200 min, the yield of liquid products and the selectivity of phenolic compounds exhibited their lowest values of 27.21% and 22.83%, respectively. This indicates that when the reaction time exceeded the optimal time of 2 h, it was not conducive to breaking the C-O or C-C linkage in lignin [47]. These variation trends suggest that as the reaction time increased, more HO· was generated, and the over-oxidation of the phenolic compounds, as well as the coupling and condensation of intermediates, became more pronounced [48,49].

2.3. *Analysis of Liquid Products*

2.3.1. GC-MS Analysis

The reaction process and the properties of liquid products were mainly examined under five different catalyst concentrations (0.5 wt%, 1.5 wt%, 2.5 wt%, 3.5 wt%, 4.5 wt%). The liquid products obtained were extracted with ethyl acetate, and the ethyl acetate-soluble aqueous phases are shown in Figure 12. The product distributions were detected via GC-MS, and the results are displayed in Figure 13.

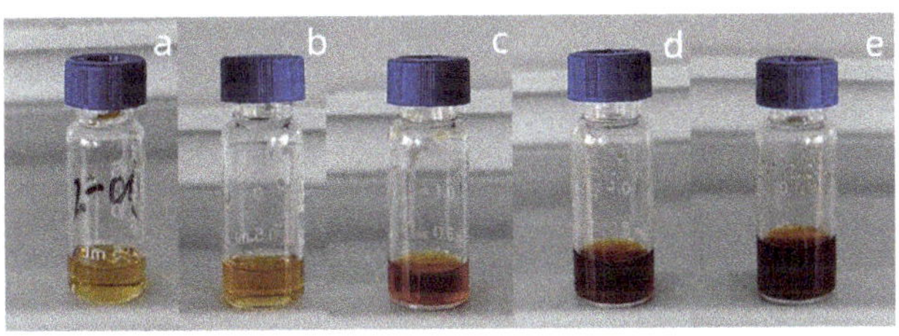

Figure 12. Ethyl acetate-soluble phases obtained from different catalyst contents ((**a**): 0.5 wt%, (**b**): 1.5 wt%, (**c**): 2.5 wt%, (**d**): 3.5 wt%, (**e**): 4.5 wt%. H_2O_2 concentration: 15×10^{-2} mol/L; reaction time: 120 min; T = 135 °C; pH = 3.0).

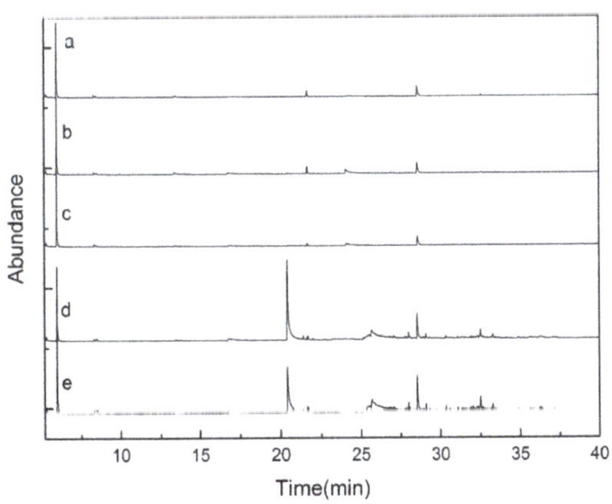

Figure 13. GC-MS of the depolymerization products from different catalyst concentrations ((**a**): 0.5 wt%, (**b**): 1.5 wt% (**c**): 2.5 wt%, (**d**): 3.5 wt%, (**e**): 4.5 wt%. H_2O_2 concentration: 15×10^{-2} mol/L; reaction time: 120 min; T = 135 °C; pH = 3.0).

All the samples exhibited different colors in Figure 12, and when the content of catalyst increased, the colors turned from light yellow to dark brown. From Figure 13 we can see that the degradation products obtained when using diverse catalyst concentrations had obvious differences. The products detected mainly included phenolic compounds, esters, and aldehydes, etc. The detailed information for the degradation products is summarized in Table 2.

Table 2. GC-MS analysis results for bio-oils using different catalyst contents.

Retention Time/s	Compound Name	Catalyst Concentrations (wt%) Quantitative Content (Area %)				
		0.5	1.5	2.5	3.5	4.5
5.712	Propanoic acid, 3-mercapto-, methyl ester	-	0.14	-	-	0.35
5.958	Propanoic acid, ethyl ester	5.54	17.42	65.18	51.59	59.06
8.344	Acetic acid, butyl ester	0.33	0.92	3.28	2.65	2.92
8.489	Acetosyringone	0.82	0.84	1.49	1.38	1.88
10.753	Cyclohexanone	-	0.17	0.11	0.08	-
12.449	2-Anthracenamine	0.08	-	-	-	-
13.431	Phenol	0.16	0.24	3.11	4.22	3.70
13.734	1,2,4-trimethoxybenzene	-	0.14	0.46	0.41	0.49
15.611	Eicosyl nonyl ether	-	0.09	-	-	-
16.818	Phenol, 2-methoxy-	-	0.61	0.72	1.56	0.61
16.923	Mequinol	0.13	-	-	-	-
18.42	formic acid	-	-	0.16	0.10	0.14
19.095	2-Isopropylpyrazine	-	-	0.08	-	-
19.765	Dodecane	-	0.13	-	-	-
20.154	Formic Acid, 4-methoxyphenyl ester	-	0.08	-	-	-
20.408	Benzofuran, 2,3-dihydro-	13.72	48.16	-	-	-
20.44	Benzaldehyde, 3,4-dimethyl-	-	-	0.74	1.23	0.24
21.202	Benzonitrile, 2-(4-methylphenyl)-	-	-	-	-	0.11
21.375	Benzene, 1,3-bis(1,1-dimethylethyl)-	0.16	0.94	0.27	0.16	0.40
21.655	Naphthalene, 2,6-bis(1,1-dimethylethyl)-	0.27	1.05	2.56	5.78	5.41
21.985	(S)-(−)-2-Amino-3-phenyl-1 -propanol	-	-	-	0.09	-
21.991	Dodecane, 1-iodo-	-	0.40	-	-	0.11
22.004	1,4-Benzenediamine, *N*,*N*′-bis(1-methylethyl)-	0.18	-	-	-	-
22.363	Octane, 2,3,6,7-tetramethyl-	-	0.07	-	-	-
22.666	Ethanone, 2-(2-benzothiazolylthio)-1-(3,5-dimethylpyrazolyl)-	-	-	-	-	0.09
22.737	Acetaldehyde, bis(1-methylethyl)hydrazone	0.11	0.27	-	-	-
22.809	Benzaldehyde, 2,4,5-trimethyl-	-	-	-	-	0.07
23.198	Linoleic acid, ethyl ester	-	-	0.18	0.17	-
23.204	Undecanoic acid		-	-	-	0.18
23.21	Hexadecane, 1-iodo-	-		-	--	-
24.119	Phenol, 2,6-dimethoxy-	-	-	2.15	10.78	0.90
25.186	Tetradecane	0.12	0.68	-	-	-
25.556	[1,2,4]Oxadiazole, 5-(4-tert-butylphenoxymethyl)-3-(thiophen-2-yl)-	-	-	-	-	0.224
25.632	Benzaldehyde, 4-hydroxy-	0.29	-	-	-	-
25.798	Benzaldehyde, 3-hydroxy-4-methoxy-	0.47	5.38	-	-	-
26.872	desaspinidol	-	-	0.52	0.84	0.69
27.049	3-Ethyl-2,6,10-trimethylundecane	-	0.33	-	-	-
27.072	Dimethyl phthalate	-	-	0.34	0.20	0.12
27.472	2-Heptanone, 3-propylidene-	-	-	-	0.12	-
27.753	Tetradecane, 2,6,10-trimethyl-	-	0.25	-	-	-
27.982	Octacosane, 1-iodo-	0.362	1.62	-	-	-
27.993	Hexacosane	-	-	-	-	0.18
27.999	2,3-Dimethoxyphenol	-	-	-	0.08	-
28.056	Phenol, 3,4-dimethoxy-	-	-	-	0.18	-
28.062	Pentadecane	-	0.69	-	-	-
28.073	Nonahexacontanoic acid	-	-	-	-	0.20
28.216	Vanillin	-	-	-	-	-
28.545	2,4-Di-tert-butylphenol	3.01	10.12	15.73	14.84	17.08

Table 2. *Cont.*

Retention Time/s	Compound Name	Catalyst Concentrations (wt%) Quantitative Content (Area %)				
		0.5	1.5	2.5	3.5	4.5
29.075	Eicosane	-	1.11	-	-	0.07
29.083	Decane, 3,8-dimethyl-	0.29	-	-	-	-
29.311	Vanillin	0.11	0.42	-	-	-
30.351	Hexadecane	0.21	0.54	-	-	-
30.362	Pyrrolidine, 1-(1-cyclohexen-1-yl)-	--	-	-	-	0.11
30.366	Benzaldehyde, 4-(methylthio)-	-	0.20	-	0.07	-
31.089	Eicosane	-	0.42	-	-	-
31.118	3-Methoxy-4-[3-oxo-3-(pyrrolidin-1-yl)propoxy]benzaldehyde	-	-	-	-	0.28
31.93	Dotriacontane	-	0.51	-	-	-
31.966	Tetracosane, 1-iodo-	0.19	-	-	-	-
32.073	Octadecane	-	0.55	-	-	-
32.104	Docosane, 1-iodo-	0.14	-	-	-	-
32.514	Heneicosane	0.78	3.09	-	-	-
32.554	Hexanoic acid, 4-tridecyl ester	-	-	-	0.83	1.04

Table 2 shows that some chemicals accounted for more of the total integral area, such as propanoic acid ethyl ester, acetic acid butyl ester, acetosyringone, phenol, 2,6-bis(1,1-dimethylethyl)- naphthalene, and 2,4-di-tert-butylphenol. Those chemicals were watched in the products obtained when different catalyst concentrations were chosen. When the catalyst concentration was 0.5 wt%, the main compound was 2,3-dihydro-benzofuran, which shared 13.72% of the total integral area. The second compound was propanoic acid ethyl ester, which accounted for 5.54% of the total integral area. Next was 2,4-di-tert-butylphenol, which made up 3.01% of the total integral area. Regarding the catalyst concentration of 1.5 wt%, the compound 2,3-dihydro-benzofuran made up most of the total integral area, that is 48.16%; this was followed by propanoic acid ethyl ester and 2,4-di-tert-butylphenol, which accounted for the total integral area of 17.42% and 10.12%, respectively. The compounds 3-hydroxy-4-methoxy-benzaldehyde and heneicosane shared 5.38% and 3.09% of the total integral area. Regarding the catalyst concentration of 2.5, 3.5, and 4.5 wt%, the compound propanoic acid ethyl ester all shared most of the total integral area, that is 65.18%, 51.59%, and 59.06%, respectively; next was 2,4-di-tert-butylphenol, which accounted for the total integral area of 15.73%, 14.84%, and 17.08%, respectively.

It is worth noting that as the catalyst concentration increased to 3.5 wt%, the total proportion of the phenolic compounds reached a maximum. The kind of phenolic compound also changed, and the obtained phenolic compounds included phenol, 2-methoxy-phenol, 2,4-di-tert-butylphenol, 2,3-dimethoxyphenol, 3,4-dimethoxy-phenol, and 2,6-dimethoxy-phenol. 2,4-di-tert-butylphenol accounted for the maximum proportion of total integral area, that is, 14.84%; 2,6-dimethoxy-phenol accounted for 10.78% of the total integral area. The compounds 2,3-dimethoxyphenol and 3,4-dimethoxy-phenol accounted for relatively less of the total integral area, and could only be identified in the case of 3.5 wt% catalyst treatment. Regarding 2,4-di-tert-butylphenol, with the increase in the catalyst content from 0.5 wt% to 4.5%, its proportion increased from 3.01% to 17.08%, indicating that this catalyst encouraged the production of 2,4-di-tert-butylphenol. Regarding phenol, with the increase in the catalyst content from 0.5 wt% to 4.5%, its proportion increased from 0.16% to 4.22% first and then declined to 3.7% when 4.5 wt% of catalyst was used. When 0.5 wt% and 1.5 wt% of catalyst was used, there was no 2,6-dimethoxy-phenol, and when the amount of catalyst increased from 2.5 wt% and 3.5 wt%, the proportion of 2,6-dimethoxy-phenol increased to 2.15% and 10.78%, respectively. However, when the catalyst concentration further increased to 4.5 wt%, the proportion of 2,6-dimethoxy-phenol declined to 0.9%. The proportions of 0.61%, 0.72%, and 1.56% for 2-methoxy-phenol could be observed when the catalyst was 1.5 wt%, 2.5 wt%, and 3.5 wt%, respectively. When the catalyst concentration was further increased to 4.5%, its proportion decreased to 0.61%. As discussed above, the greater catalyst content can accelerate more HO· generated from H_2O_2, and the excessive HO· can result in the over-oxidation of the phenols (Figure 14a, taking phenol for instance) [49], thereby increasing the proportion of the phenols reduced when the catalyst concentration further increased. The identified aromatic monomers are summarized in Figure 14b.

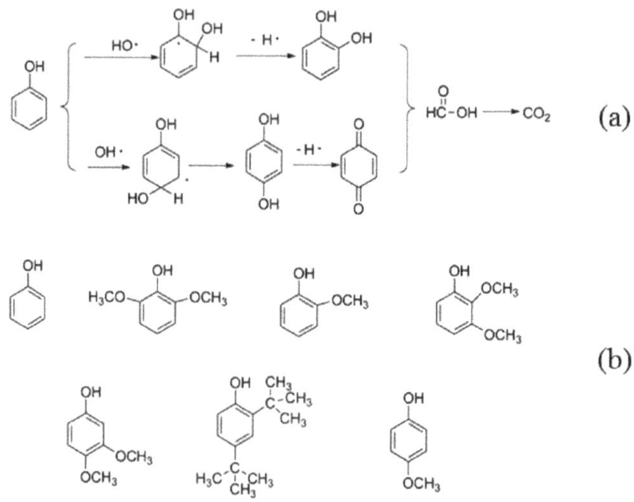

Figure 14. The over-oxidation of the phenols (**a**) and summary of the phenolic compounds (**b**).

2.3.2. FT-IR and ^{1}H-NMR Analysis

In order to further confirm the components of the degradation products, FT-IR analysis and ^{1}H-NMR analysis were conducted; the spectra are displayed in Figures 15 and 16, respectively.

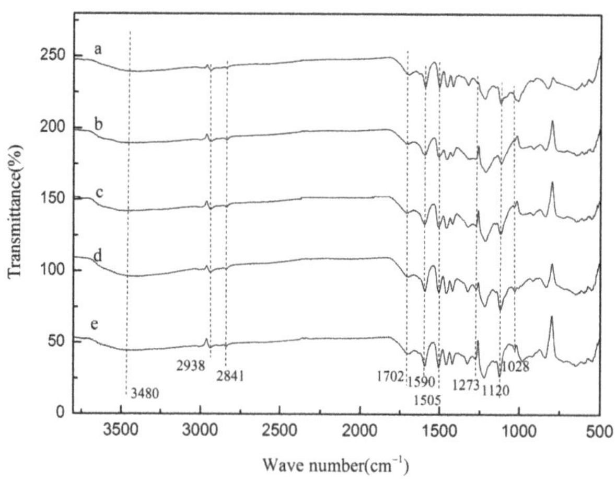

Figure 15. FTIR spectra of products obtained from different catalyst contents (catalyst contents: (**a**): 0.5 wt%, (**b**): 1.5 wt% (**c**): 2.5 wt%, (**d**): 3.5 wt%, (**e**): 4.5 wt%. H_2O_2 concentration: 15×10^{-2} mol/L; reaction time: 120 min; T = 135 °C; pH = 3.0).

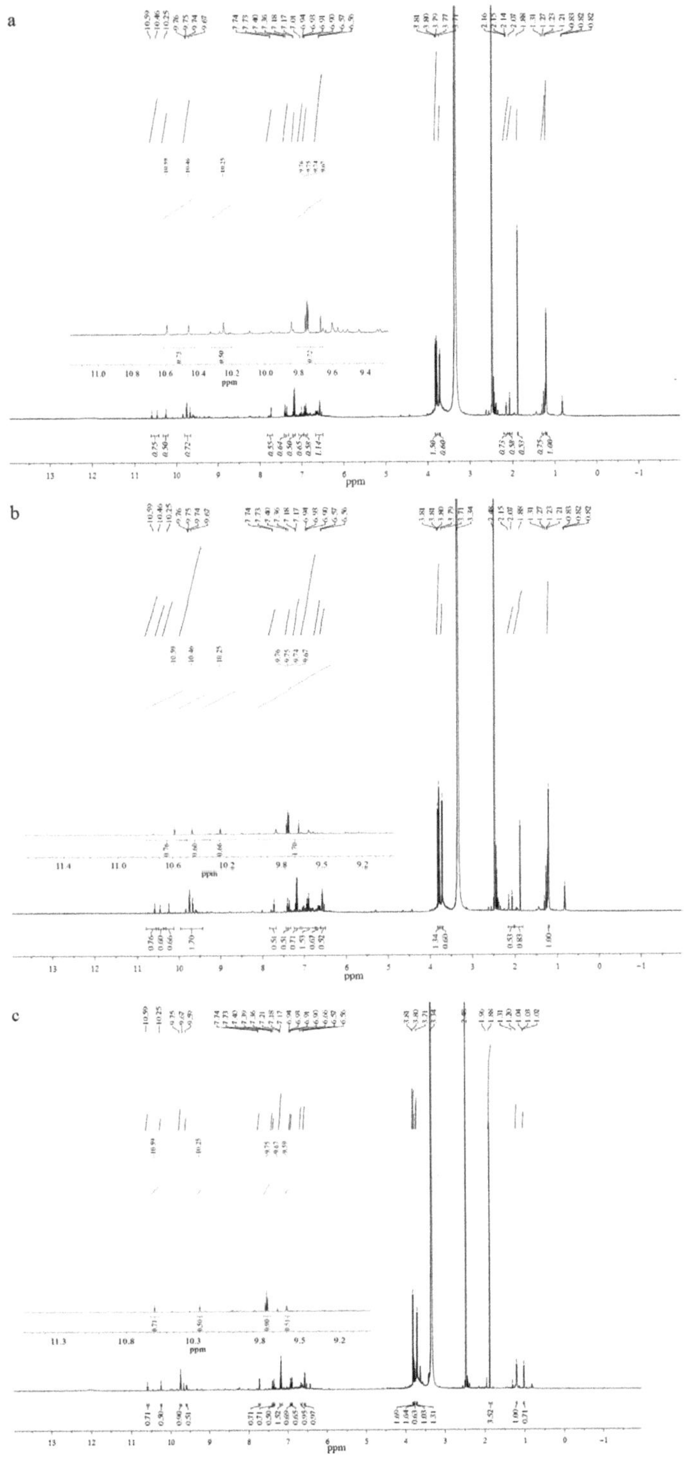

Figure 16. *Cont.*

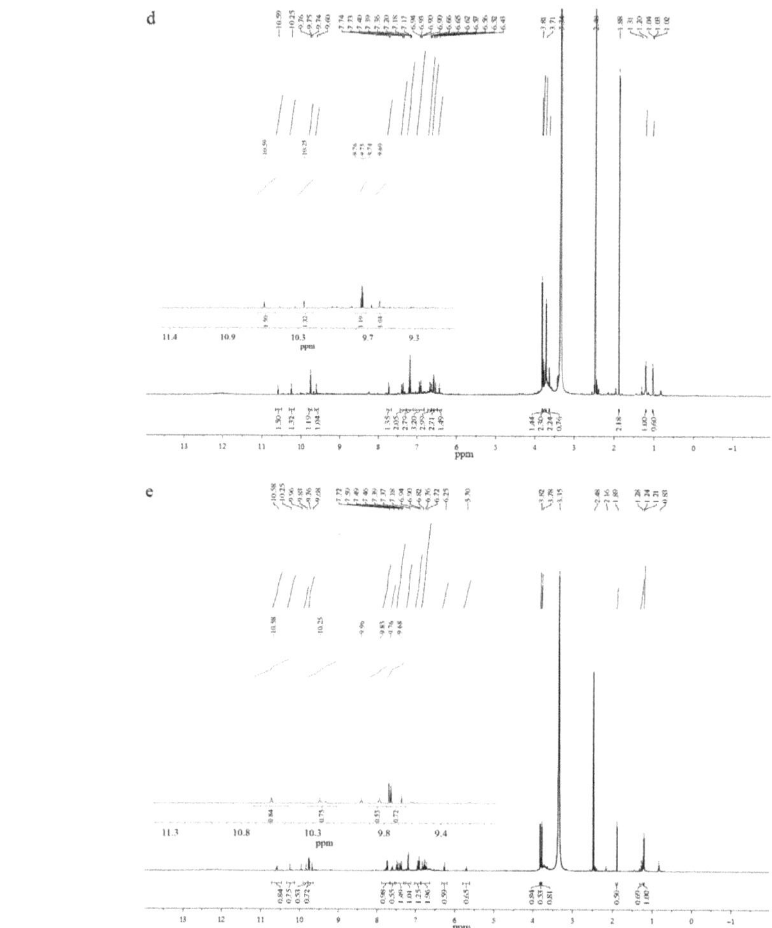

Figure 16. ^{1}H-NMR spectra of ethyl acetate-soluble products (catalyst contents: (**a**): 0.5 wt%, (**b**): 1.5 wt% (**c**): 2.5 wt%, (**d**): 3.5 wt%, (**e**): 4.5 wt%. reaction time: 120 min; H_2O_2 concentration: 15×10^{-2} mol/L; T = 135 °C; pH = 3.0).

As shown in Figure 15, the products had relatively similar chemical groups, and the position of each vibrational peak was basically the same. These results imply that the main composition categories of the products were alike. The wide band at 3480 cm^{-1} was assigned to the combination and overlap of the stretching vibrations of O-H, =C-H, and N–H in aromatic or aliphatic groups, implying the presence of phenols or benzene in the products. At 2938 and 2841 cm^{-1}, the weak vibration peaks of aliphatic C-H were generated by the asymmetric stretching vibration of C-H and symmetric stretching vibration of methyl and methylene, respectively, indicating the formation of saturated aliphatic hydrocarbons from lignin depolymerization [50]. The small and broad bands around 1702 cm^{-1} were due to the stretching vibration of C=O from ketone or carboxylic acid derivatives. The peaks around 1590 and 1505 cm^{-1} were assigned to the vibrations of C=C bonds in the aromatic skeleton. These results showed that the aromatic skeleton structure of lignin was well retained in the products [51], and these aromatic skeletons mainly existed in the form of phenols or phenol derivatives. The signals appearing at 1273 and 1120 cm^{-1} were regarded as Ar-O linkages of methoxy groups containing the phenols. Ester groups were confirmed by the peak at 1028 cm^{-1} [52].

However, some differences could be found in products obtained with different quantities of catalysts: (1) The intensity of the peak at 3480 cm^{-1} was relatively higher for the products obtained with 3.5 wt% catalysts, suggesting that, under this condition, more β-O-4 linkages were cracked to yield phenolic compounds. (2) The intensity of the peaks at 1273, 1120, and 1028 cm^{-1} was also higher in products with 3.5 wt% catalyst compared to that in the products with 0.5 wt% to 2.5 wt% catalysts, verifying the above analysis results. (3) The intensity of the peaks at 1273, 1120, and 1028 cm^{-1} in products with 3.5 wt% catalysts was somewhat similar to that in the products with 4.5 wt% catalysts, according to GC-MS analysis. With the increase in catalyst content from 3.5 wt% to 4.5 wt%, the proportion of the phenolic compounds only dropped slightly from 32.58% to 22.29%. As a result, a similar FT-IR spectrum was gained. (3) The intensity of the peaks at 1028 cm^{-1} enhanced gradually in the products using 0.5 wt% to 2.5 wt% catalyst, and was finally similar to that of the products with 3.5 wt% and 4.5 wt% catalyst. According to the GC-MS results, when 0.5 wt% and 1.5 wt% catalyst were used, the proportion of major ester compounds was 5.54% and 17.64%, respectively; when the amount of catalyst increased to 2.5 wt%, 3.5 wt%, or 4.5 wt%, the proportion of major ester compounds increased to about three or four times. This well explained the FT-IR changes for the intensity of the peaks at 1028 cm^{-1}.

Clearly, the proton signals were mainly distributed in four regions: δ0.8–1.9 ppm, δ3.3–4.0 ppm, δ6.0–8.0 ppm, and δ9.5–11 ppm. The peaks occurring at δ0.8–1.9 ppm were assigned to methyl and methylene mainly from the aliphatic chain. The next strong signal, appearing at δ3.3–4.0 ppm, was attributed to the protons from methoxy [47,53]. It is obvious that a relatively high proportion of methoxy protons could be captured with 3.5 wt% catalyst. On the one hand, this phenomenon confirms the presence of a high area of methoxy chemicals such as 2,6-dimethoxy-phenol, 2-methoxy- phenol, and acetosyringone. On the other hand, the presence of the compounds 2,6-dimethoxy-phenol and 2-methoxy-phenol can concurrently prove the breaking of guaiacol and syringol sub-aromatic units of lignin under the action of the catalyst. The presence of aromatic protons was verified by the signals between the regions of 6.0 and 8.0 ppm [54], which exhibited the presence of the aromatic compounds such as phenol, 2-methoxy-phenol, acetosyringone, 2,6-dimethoxy-phenol, and 2,4-di-tert-butylphenol. The relatively high proportion of aromatic protons was observed in the case of 3.5 wt% catalyst, which is in agreement with the GC-MS results. The weak peaks observed at δ9.5–11 ppm were attributed to the protons form aldehyde or acidic groups, which validated the GC-MS analysis of the emergence of formic acid, 3,4-dimethyl-benzaldehyde, etc.

2.3.3. Elemental Analysis

The elemental compositions in the liquid products were investigated and the results are shown in Table 3.

Table 3. The elemental compositions of liquid products obtained from different catalyst contents.

Samples	Elemental Composition (wt%)				HHV
	C	H	O	N	(MJ/kg)
Lignin	61.3	6.4	29.1	1.7	20.5
0.5 wt% catalyst	63.2	6.9	26.7	0.9	23.3
1.5 wt% catalyst	65.5	6.8	25.2	0.7	24.9
2.5 wt% catalyst	66.9	6.8	24.7	0.6	27.1
3.5 wt% catalyst	69.2	7.0	21.8	0.7	30.2
4.5 wt% catalyst	68.6	7.2	22.4	0.6	29.0

The elemental analysis results showed that the content of carbon increased in liquid products compared to that in lignin. The highest carbon content was obtained when 3.5 wt% catalyst was used, and the hydrogen content increased slightly compared to that in lignin. The oxygen content decreased compared to that in lignin, and the lowest oxygen content of 21.8% was found in the products with 3.5 wt% catalyst. This result implies that the deoxygenation or decarbonylation were improved by the catalyst [54]. Additionally, the 'higher heating value' (HHV) was enhanced in products, with lignin showing HHV of 20.5 MJ/kg. The value of HHV increased to 23.3–30.2 MJ/kg in the products and the highest HHV of 30.2 MJ/kg was found in the products with 3.5 wt% catalyst. As is known, the higher content of carbon can result in a higher HHV value [55].

2.4. *Characterization of the Solid Residue Fractions*

2.4.1. FT-IR Analysis

To investigate the changes in the chemical structures of lignin before and after the FePc@SO_2Cl-C_{lig} treatment, FT-IR analyses of lignin and residues were performed and the results are depicted in Figure 17.

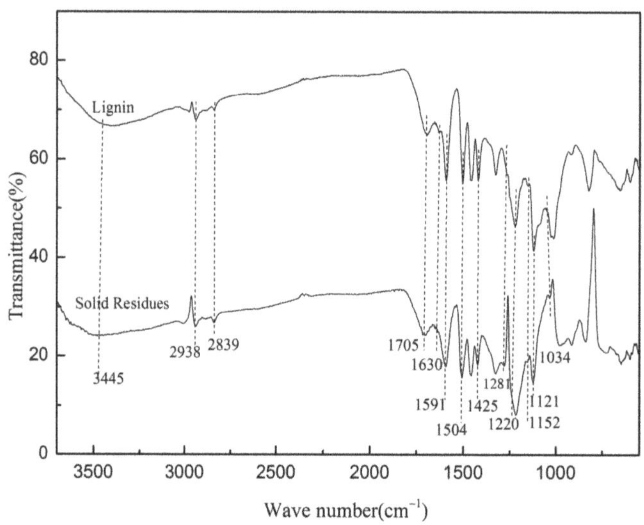

Figure 17. FT-IR spectrum of lignin and solid residues obtained when 3.5 wt% catalyst was used (reaction time:120 min; H_2O_2 concentration = 15 × 10^{-2} mol/L; T = 135 °C; pH = 3.0).

The peaks occurring around 3400–3500 cm^{-1} were assigned to O-H stretching vibration. The weak peaks at 2938 cm^{-1} and 2839 cm^{-1} issued from C-H stretching vibrations in methyl and methylene groups [56]; the signals around 1705 cm^{-1} were attributed to the ester groups; and the weak peak at 1630 cm^{-1} corresponded to C=O stretching of carbonyl. The bands at 1591, 1504, and 1425 cm^{-1} were attributed to the skeletal vibrations of the benzene ring, and the peaks appearing at 1121 cm^{-1} and 1152 cm^{-1} corresponded to the C-O-C stretching vibration [57];

As is well known, the solid residues were mainly generated from the decomposition of the lignin and polycondensation of the intermediates. In comparison, some dramatic changes can be discovered for the solid residues: (1): The intensity in the region corresponding to the skeletal vibrations of the benzene ring slightly declined. The presence of this region pointed to the fact that the solid residues retained the core structure of lignin; furthermore, the weaker intensity of these peaks indicated the degradation of lignin after FePc@SO_2Cl-C_{lig} treatment. (2) The weak peak observed in lignin at 1630 cm^{-1}, corresponding to C=O stretching of carbonyl, disappeared in solid residues, suggesting that the de-carbonylation reaction occurred during the degradation of lignin. (3) The typical signals for C-O-C stretching vibrations at 1121 cm^{-1} and 1152 cm^{-1} became weaker, respectively, indicating that the β-O-4 ether bond cleaved during the degradation process under the action of FePc@SO_2Cl-C_{lig}. (4) The band at 1220 cm^{-1} in lignin shifted to a lower wavenumber of 1216 cm^{-1} in the solid residues, indicating some new structures generated during the degradation process. (5) Some new peaks were observed in solid residues at 1281 cm^{-1} and 1034 cm^{-1}, which also indicated that the new structures were formed; this was because, during degradation, the small intermediate molecules reacted with the chains of lignin and generated new chain structures [58].

2.4.2. ^{1}H-NMR Analysis

In order to gain further information about the structure changes of lignin and solid residues, ^{1}H-NMR analysis was conducted, and the results are displayed in Figure 18.

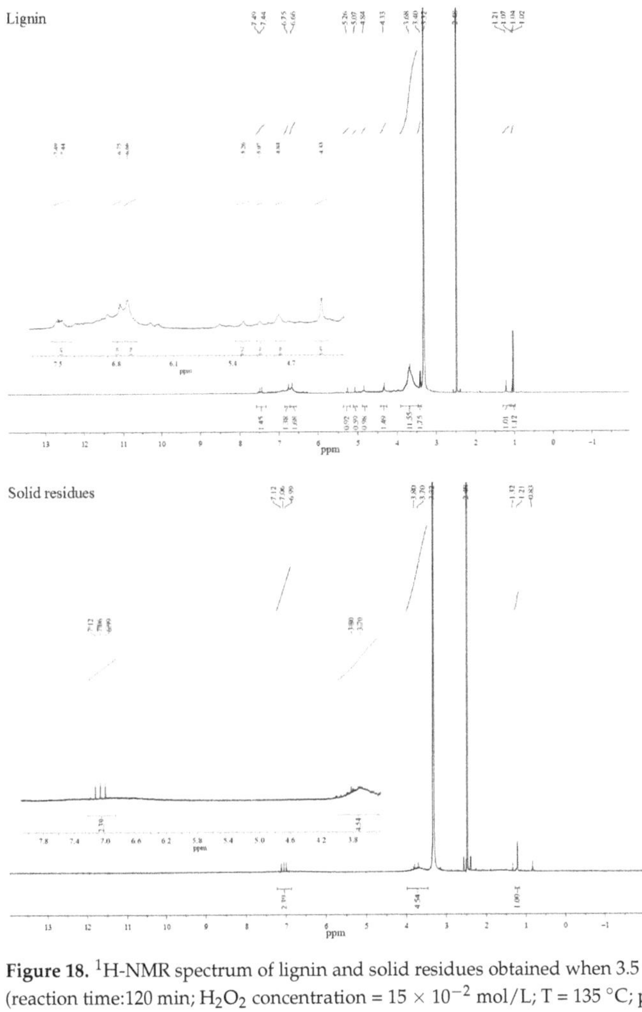

Figure 18. ^{1}H-NMR spectrum of lignin and solid residues obtained when 3.5 wt% catalyst was used (reaction time:120 min; H_2O_2 concentration = 15×10^{-2} mol/L; T = 135 °C; pH = 3.0).

As can be seen from Figure 18, lignin exhibited strong peaks as compared to the solid residues. For lignin, the strong signals mainly occurred at four regions: the signals corresponding to methylene groups of the aliphatic chain were observed at the region from 1 to 1.3 ppm; the methoxyl group signals occurred around 3.7 ppm; the signals appearing at 4–5.3 ppm were assigned to H-α and H-β in β-O-4 substructures; and the typical signals corresponding to aromatic protons were observed in their regions of 6.5–7.5 ppm [59,60]. However, for solid residues, the signals for methoxyl groups around 3.7 ppm and the typical signals corresponding to aromatic protons between 6.5 and 7.5 ppm diminished significantly. Furthermore, the signals assigned to β-O-4 substructures almost disappeared compared with lignin. The findings further confirmed the FT-IR results, also indicating that the breaking of β-O-4 linkages was improved by FePc@SO_2Cl-C_{lig}.

2.4.3. GPC Analysis

To investigate if the degradation of lignin increased under the treatment of FePc@SO_2Cl-C_{lig}, the molecular weights of lignin and solid residues obtained at different reaction times (40 min, 80 min, 120 min, 150 min) were measured with the assistance of GPC analysis. The reported data including weight-average (Mw) and number-average (Mn) molecular weights and polydispersityare (Mw/Mn) of lignin and solid residues, which are listed in Table 4.

Table 4. Weight-average (Mw) and number-average (Mn) molecular weights, and polydispersity (Mw/Mn) of lignin and solid residues (catalyst: 3.5 wt%; reaction time:40 min, 80 min, 120 min, 150 min; H_2O_2 concentration = 15×10^{-2} mol/L; T = 135 °C; pH = 3.0).

Samples	Mw (g/mol)	Mn (g/mol)	Mw/Mn
Lignin	66,671	36,726	1.81
Residues (40 min)	62,256	36,208	1.72
Residues (80 min)	41,063	27,202	1.51
Residues (120 min)	31,421	25,633	1.22
Residues (150 min)	33,468	14,597	2.29

During the degradation process of lignin, small molecule chemicals, superpolymers, or char could all be formed by a range of reactions. In this process, the degradation of lignin and the polycondensation of intermediates take place simultaneously and compete with each other. The cleavage of β-O-4 results in the reduction in the molecular weight; conversely, the formation of superpolymers or char may result in an increase in the molecular size [61]. As can be seen from Table 4, lignin had an Mw of 66,671 g/mol and the polydispersityare value was 1.81. In comparison, a distinct reduction in the molecular weight for solid residues was observed; the value of Mw fell by almost half for the solid residues obtained at 80 min and 120 min. However, the molecular weight showed a slight increase to 33,468 g/mol for the solid residues obtained with 150 min reaction time. This can be explained because the polycondensation gradually became the major reaction at this stage. At the same time, the polydispersity index of solid residues declined to 1.51 and 1.22 when the reaction time was 80 min and 120 min, demonstrating that the molecular weight distributions of solid residues were relatively narrow at this two stage. The index then increased to 2.29 with the extension in reaction time to 150 min. This phenomenon also confirmed that the polycondensation reaction was dominant at this stage. In addition, the low polydispersity index values (<3) manifested that lignin molecularly belongs to a kind of uniform-sized polymer [62]. All these changes validate that the lignin was subjected to the cleavage of the ether bonds when using FePc@SO_2Cl-C_{lig} as a catalyst.

2.4.4. SEM Analysis

SEM analysis was performed to investigate the native microstructure alteration of lignin during the degradation process at different reaction times (40 min, 80 min, 120 min, 150 min); the results are depicted in Figure 19.

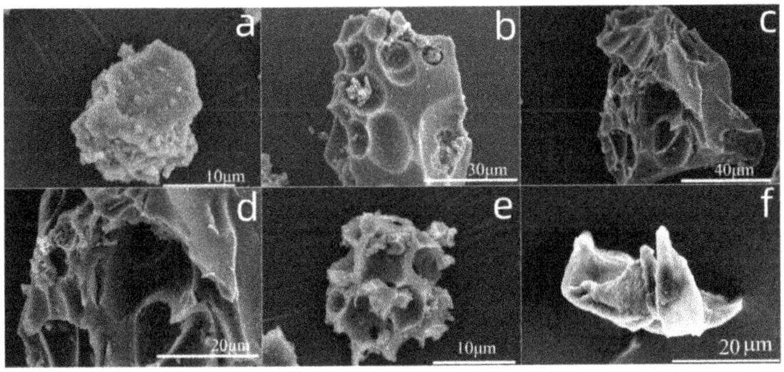

Figure 19. SEM images of solid residues at different reaction times (conditions: 3.5 wt% of catalyst, T = 135 °C, concentration of H_2O_2: 15 × 10^{-2} mol/L; pH = 3.0. (**a**): lignin, (**b**): residues at 40 min, (**c**): residues at 80 min, (**d**): enlargement of residues at 80 min, (**e**): residues at 120 min, (**f**): residues at 150 min).

As can be seen from Figure 19a, the original lignin had a relatively smooth and tight surface structure. When the reaction time was increased to 40 min (Figure 19b), the smooth surface was slightly destroyed; some shallow pits were observed and a small amount of dross was deposited on the surface. This observation indicated that the degradation of lignin begins on the surface. At the same time, this GPC analysis result made it clear that when the reaction time was 40 min, the Mw value only fell slightly to 62,256 g/mol; with a further increase in reaction time to 80 min (Figure 19c), the surface damage intensified and some deeper pits arose. Magnification of this figure (Figure 19d) shows that more fragments attached on the surface or in the pits. This phenomenon demonstrated that at this stage the main reaction was depolymerization. When the reaction time continued to increase to 120 min (Figure 19e), slight caking could be captured. This implies that the coupling and polycondensation reaction gradually began to become obvious with the increase in the reaction time. When the reaction time was further increased to 150 min (Figure 19f), conglomerations could be observed. The longer time promoted the coupling and polycondensation to a much greater extent. Combined with the GPC analysis, this also explained why the weight average of solid residues increased after the reaction time of 2 h.

3. Experimental

3.1. Materials

The key experiment reagents are listed as follows:

Lignin (prepared in the lab); iron chloride hexahydrate ($FeCl_3{\cdot}6H_2O$), hydrogen peroxide (H_2O_2), thionyl chloride ($SOCl_2$), 4-nitrophthalic acid ($C_8H_5NO_6$), urea ($CO(NH_2)_2$), sodium sulfide ($NaS_2{\cdot}9H_2O$), and *N,N*-dimethylformamide (C_3H_7NO) were provided by Aladdin. Potassium hydroxide (KOH, purity of ≥98%), sodium hydroxide (NaOH, purity of ≥97%), sulfuric acid (H_2SO_4, concentration of 98%), and ethanol (C_2H_5OH, 99.9%) were provided by JiKun Chemicals Company (Nanning, China).

3.2. Experimental Procedure

3.2.1. Catalyst Preparation

Manufacture of Sulphonyl Chloride Derivative of Lignin-Base Carbon

Lignin used in this work was refined from black pulping liquor with H_2SO_4 (concentration: 65%) and then dried at a constant temperature of 105 °C for 12 h. The sulphonyl chloride derivative of lignin-base carbon was synthesized via three steps: carbonization, sulfonation, and acyl chlorination. First, carbonization of lignin was conducted in a tube furnace under the protection of N_2. Lignin and KOH were placed in a nickel disc in proportion and then placed in a tube furnace; the tube furnace was heated to 550 °C with the ramp rate of 15 °C/min and finally maintained for 90 min. After the reaction, the obtained black sample (named C_{lig}) was washed and filtered with hot water, then dried at 105 °C for 12 h. Second, sulfonation of C_{lig} was conducted in a three-necked round-bottom flask under the protection of N_2. C_{lig} and H_2SO_4 were mixed in the round-bottom flask in proportion, and the reaction was kept for 6 h at 150 °C. The final black sample (named SO_3H-C_{lig}) was washed and filtered with water, and finally dried at 105 °C for 12 h. Third, the sulphonyl chloride derivative was synthesized using the method in a previous report [63]; SO_3H-C_{lig} and an overabundance of $SOCl_2$ were added to the mixed solvents of *N,N′*-dimethylformamide (DMF) and water. The reaction was performed at 90 °C for 24 h; then the residual $SOCl_2$ and solvent were removed by distillation. The solid products (named SO_2Cl-C_{lig}) were washed and dried for the next application. The synthesis method of SO_2Cl-C_{lig} is described in Figure 20.

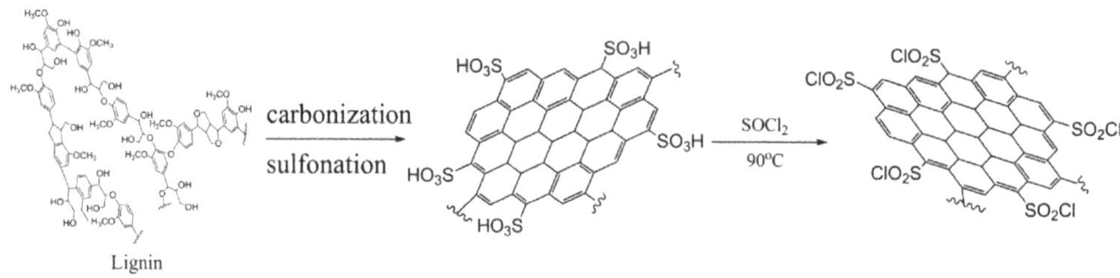

Figure 20. The synthesis route of SO_2Cl-C_{lig}.

The Synthesis of Iron Amino Phthalocyanine (FePc-NH_2)

Iron amino phthalocyanine was synthesized through two steps: first, iron tetranitro phthalocyanine was prepared according to a previous method [64]; second, the prepared iron tetranitro phthalocyanine was ammoniated by the reaction with sodium sulfide non-ahydrate (NaS·$9H_2O$) at 65 °C for 4 h to finally yield the objective product [65]. In this reaction, the molar ratio of iron tetranitro phthalocyanine and NaS·$9H_2O$ was 1:15. The final reaction mixture was treated with 0.5 M HCl and 1.0 M NaOH. The obtained blue-green product (named FePc-NH_2) was washed with water and ethanol, and dried at 65 °C for 12 h. The detailed synthesis route of FePc-NH_2 is described in Figure 21.

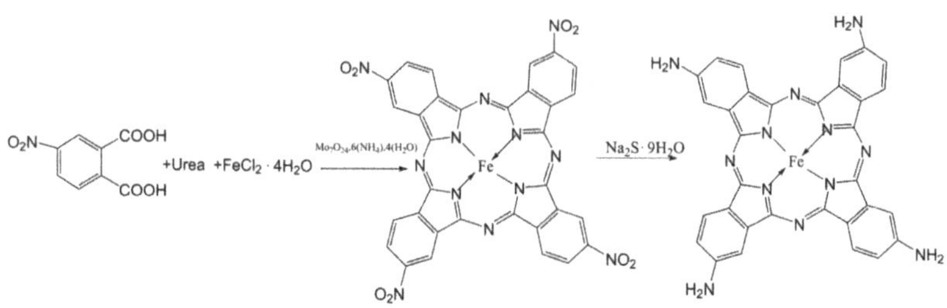

Figure 21. The synthesis route of iron amino phthalocyanine.

Synthesis of Sulfonyl-Chloride-Modified Lignin-Based Porous Carbon-Supported Iron Phthalocyanine

The above prepared SO_2Cl-C_{lig} and FePc-NH_2 were added to a mixture of DMF and H_2O and sonicated for 2 h; then, the mixture was stirred for 72 h at 120 °C [30]. Finally, the reaction mixture was filtered off, and the solid residues were washed with water and acetone, and dried at 65 °C to yield a black-blue powder (named FePc@SO_2Cl-C_{lig}). The synthesis route of FePc@SO_2Cl-C_{lig} is displayed in Figure 22.

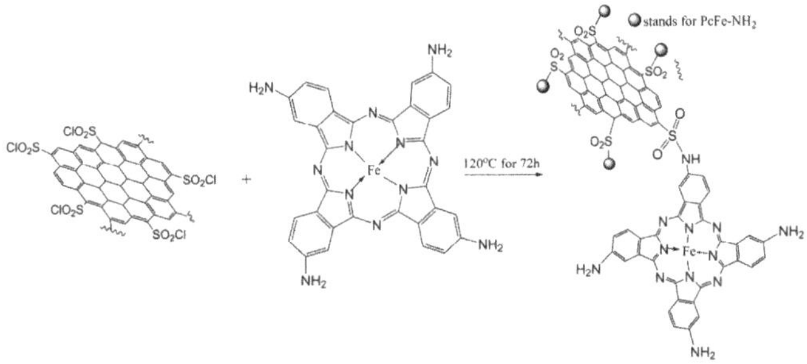

Figure 22. The synthesis route of FePc@SO_2Cl-C_{lig}.

3.2.2. Treatment of Lignin with FePc@ SO_2Cl-C_{lig}

The final experiment was performed in a 150 mL reactor. First, the appropriate amount of FePc@SO_2Cl-Clig catalyst (catalyst-to-lignin ratio from 0 wt% to 4.5% wt%) was dispersed in a combined solvent of water and acetonitrile; second, the appropriate amount of lignin (1 g) was added to the above mixture, and the pH of the mixture was adjusted to the designed value (pH = 3.0 in this work) with sulfuric acid. Subsequently, the mixture was vibrated for 20–30 min to fully disperse the FePc@SO_2Cl-C_{lig} and lignin. Finally, the mixture was transferred into the reactor; simultaneously, H_2O_2 (from 0 mol/L to 20×10^{-2} mol/L) was added to the reaction mixture and the mixture was stirred at 135 °C for 40–200 min. After the reaction finished, the solid residues and liquid products were gained by centrifugal separation; the liquid products were extracted by ethyl acetate and then used for analysis.

3.3. *Characterizations*

The type of chemical functional group of the catalyst was recorded using Fourier transform infrared spectroscopy (FTIR) on Nicolet IS10 (Waltham, MA, USA). Element compositions and speciation were detected by X-ray photoelectron spectroscopy (XPS) on a thermo EscaLab 250 Xi (Norristown, PA, USA). The BET surface area was tested using N_2 adsorption/desorption isotherms (Quantachrome autosorb IQ-C analyzer, Boynton Beach, FL, USA). The surface structure of catalysts was recorded using a Hitachi s-3400 Scanning Electron Microscope (Hitachi, Tokyo, Japan).

Chemical distributions of degradation products were investigated via GC/MS chromatograms using an Agilent (Santa Clara, CA, USA) 6890A/5973N series instrument fitted with a HP-5MS (30 m × 0.25 mm × 0.25 um) column. A quantity of 2 μL of sample was injected into the system at 40 °C. The column was maintained at 40 °C for 2 min; the first ramp was for 5 °C/min to 150 °C and maintained for 2 min, followed by a ramp for 8 °C/min to 280 °C, which was held for 5 min. Chemical structures and distributions of degradation products were characterized by ^{1}H-NMR determinations on an Agilent DD2 800 MHz NMR spectrometer (Santa Clara, CA, USA)with deuterated dimethyl sulfoxide (Dimethyl sulfoxide-d6) as the standard solvent, and FTIR on a Nicolet IS10 (Waltham, MA, USA). Element analysis (EA) was used to analyze the element compositions of the products.

The solid residues were characterized via FTIR on a Nicolet IS10 (Waltham, MA, USA). ^{1}H-NMR determinations were carried out on an Agilent DD2 800 MHz NMR spectrometer (Santa Clara, CA, USA) with deuterated dimethyl sulfoxide (Dimethyl sulfoxide-d6) as the standard solvent. GPC analysis was conducted using a 1260 Infinity II GPC/MDS instrument (Santa Clara, CA, USA) with DMF as mobile phase, and SEM analysis was conducted on a Hitachi s-3400 Scanning Electron Microscope (Hitachi, Tokyo, Japan).

The liquid product yield was calculated as follows:

$$\text{liquid product (wt\%)} = W_{liquid}/W_{lignin} \times 100\% \quad (3)$$

where W_{lignin} and W_{liquid} represent the weight of lignin and the weight of liquid products, respectively.

The selectivity of the phenols was presented as the ratio of the peak area of the corresponding products to the total area of the liquid products.

4. Conclusions

In this study, sulfonyl-chloride-modified lignin-based porous carbon-supported metal phthalocyanine catalyst with a high surface area of 638.98 m^2/g and a pore volume of 0.291 cm^3/g was manufactured and employed for lignin degradation via Fenton reaction. The catalyst exhibited good ester and phenol selectivity according to GC-MS analysis. The highest liquid product yield of 38.94% and the highest phenol selectivity of 32.58% were observed at 135 °C with 3.5 wt% of catalyst and 15×10^{-2} mol/L H_2O_2 for 120 min. The main phenols were phenol, 2-methoxy-phenol, 2,4-di-tert-butylphenol, 2,3-dimethoxyphenol, 3,4-dimethoxy-phenol, and 2,6-dimethoxy-phenol. The analysis results for liquid products showed that (1) the composition of liquid products was different under various catalyst concentrations; and (2) the highest calorific value of 30.2 MJ/kg was acquired with 3.5 wt% catalyst. The analysis results for solid residues showed that (1) the M_W of residues was almost half that of the lignin; and (2) the degradation of the lignin started on the surface. These findings offer a significant theoretical reference for lignin degradation using a lignin-based porous carbon-supported metal phthalocyanine catalyst via Fenton reaction.

Author Contributions: F.D.: Writing—original draft, Funding acquisition. X.X.: Data curation. P.T.: Visualization, Investigation. Y.L.: supervision, Funding acquisition. All authors have read and agreed to the published version of the manuscript.

Funding: This work was supported by Natural Science Foundation of Guangxi Zhuang Autonomous Region (2021GXNSFBA075018); National Natural Science Foundation of China (Regional Fund) (22368012).

Institutional Review Board Statement: Not applicable.

Informed Consent Statement: Not applicable.

Data Availability Statement: The data presented in this study are available in article.

Conflicts of Interest: The authors declare no conflicts of interest.

References

1. Liu, X.; Duan, X.; Wei, W.; Wang, S.; Ni, B. Photocatalytic conversion of lignocellulosic biomass to valuable products. *Green. Chem.* **2019**, *21*, 4266–4289. [CrossRef]
2. Hilde, V.H.; Tanja, B. Quantitative NMR analysis of the aqueous phase from hydrothermal liquefaction of lignin. *J. Anal. Appl. Pyrolysis* **2020**, *151*, 104919–104927.
3. Zhang, Y.; Maryam, N. Lignin: A review on structure, properties, and applications as a light-colored UV absorber. *ACS Sustain. Chem. Eng.* **2021**, *9*, 1427–1442. [CrossRef]
4. Mahdi, M.A.; Katalin, B.; Gregg, T.B.; Jeremy, S.L.; John, R.; Roberto, R.; Yuriy, R.L.; Samec, J.S.M.; Sels, B.F.; Wang, F. Guidelines for performing lignin-first biorefining. *Energy Environ. Sci.* **2021**, *14*, 262–292.
5. Yao, L.; Lu, Y.C.; Hu, H.Q.; Xie, F.J.; Wei, X.Y.; Fan, X. Structural characterization of lignin and its degradation products with spectroscopic methods. *J. Spectrosc.* **2017**, *2017*, 8951658.
6. Patryk, J.; Maurice, N.C.; Teofl, J.; Łukasz, K. The role of lignin and lignin-based materials in sustainable construction—A comprehensive review. *Int. J. Biol. Macromol.* **2021**, *187*, 624–650.
7. Yokoyama, T. Revisiting the Mechanism of β-O-4 bond cleavage during acidolysis of lignin. Part 6: A review. *J. Wood Chem. Technol.* **2014**, *35*, 27–42. [CrossRef]
8. Constant, S.; Basset, C.; Dumas, C.e; Di Renzo, F.; Robitzer, M.; Barakat, A.; Quignard, F. Reactive organosolv lignin extraction from wheat straw: Influence of lewis acid catalysts on structural and chemical properties of lignins. *Ind. Crop. Prod.* **2015**, *65*, 180–189. [CrossRef]
9. Deepa, A.K.; Dhepe, P.L. Lignin depolymerization into aromatic monomers over solid acid catalysts. *ACS Catal.* **2015**, *5*, 365–379. [CrossRef]
10. Deuss, P.J.; Scott, M.; Tran, F.; Westwood, N.J.; de Vries, J.G.; Barta, K. Aromatic monomers by in situ conversion of reactive intermediates in the acid-catalyzed depolymerization of lignin. *J. Am. Chem. Soc.* **2015**, *137*, 7456–7467. [CrossRef]

11. Yang, X.J.; Feng, M.Q.; Choi, J.S.; Meyer III, H.M.; Yang, B. Depolymerization of corn stover lignin with bulk molybdenum carbide catalysts. *Fuel* **2019**, *244*, 528–535. [CrossRef]
12. Chen, L.; Dou, J.; Ma, Q.; Li, N.; Wu, R.; Bian, H.; Yelle, D.J.; Vuorinen, T.; Fu, S.; Pan, X.; et al. Rapid and near-complete dissolution of wood lignin at </=80 degrees C by a recyclable acid hydrotrope. *Sci. Adv.* **2017**, *3*, e1701735.
13. Miller, J.E.; Evans, L.; Littlewolf, A.; Trudell, D.E. Batch microreactor studies of lignin and lignin model compound depolymerization by bases in alcohol solvents. *Fuel* **1999**, *78*, 1363–1366. [CrossRef]
14. Alfonso, C.; Fernando, B.; Rui, M.; Karina, H.; Íñigo, G.Y.; Maitane, M.; António, P.; Gandía, L.M.; Víctor, M.M. Production of aromatic compounds by catalytic depolymerization of technical and downstream biorefinery lignins. *Biomolecules* **2020**, *10*, 1338–1398.
15. Bentivenga, G.; Bonini, C.; ĎAuria, M.; Bona, A.D. Degradation of steam-exploded lignin from beech by using Fenton's reagent. *Biomass Bioenerg.* **2003**, *24*, 233–238. [CrossRef]
16. Torrades, F.; Saiz, S.; García-Hortal, J.A. Using central composite experimental design to optimize the degradation of black liquor by Fenton reagent. *Desalination* **2011**, *268*, 97–102. [CrossRef]
17. Phisit, S.; Ampin, K.; Arthitaya, K.; Charin, T.; Thanongsak, C.; Noppol, L. Improvement in efficiency of lignin degradation by Fenton reaction using synergistic catalytic action. *Ecol. Eng.* **2015**, *85*, 283–287.
18. Lin, Y.Y.; Lu, S.Y. Selective and efficient cleavage of lignin model compound into value-added aromatic chemicals with $CuFe_2O_4$ nanoparticles decorated on partially reduced graphene oxides via sunlight-assisted heterogeneous Fenton processes. *J. Taiwan Inst. Chem. Eng.* **2019**, *97*, 264–271. [CrossRef]
19. Pecci, L.; Montefoschi, G.; Cavallini, D. Some new details of the copperhydrogen peroxide interaction. *Biochem. Biophys. Res. Commun.* **1997**, *235*, 264–267. [CrossRef]
20. Valdeir, A.; Adriane, M.F.M. The effect of a catecholate chelator as a redox agent in Fenton-based reactions on degradation of lignin-model substrates and on COD removal from effluent of an ECF kraft pulp mill. *J. Hazard. Mater.* **2007**, *141*, 273–279.
21. Kim, H.E.; Lee, J.; Lee, H.; Lee, C. Synergistticc effects of TiO_2 photo catalysis in combination with Fenton-like reaction oxideation of organic compounds at circumneutral pH. *Appl. Catal. B Environ.* **2012**, *115–116*, 219–224. [CrossRef]
22. Martín, R.; Navalon, S.; Alvaro, M.; Hermenegildo, G. Optimized water treatment by combining catalytic Fenton reaction using diamond supported gold and biological degradation. *Appl. Catal. B Environ.* **2011**, *103*, 246–252. [CrossRef]
23. Colomban, C.; Kudrik, E.V.; Afanasiev, P.; Alexander, B.S. Degradation of chlorinated phenols in water in the presence of H_2O_2 and water-soluble μ-nitrido diiron phthalocyanine. *Catal. Today* **2014**, *235*, 14–19. [CrossRef]
24. Tao, X.; Ma, W.; Zhang, T.; Zhao, J.C. A novel approach for the oxidative degradation of organic pollutants in aqueous solutions mediated by iron tetrasulfophthalocyanine under visible light radiation. *Chem. Eur. J.* **2002**, *8*, 1321–1326. [CrossRef] [PubMed]
25. Tang, X.; Liu, Y. Heterogeneous photo-Fenton degradation of methylene blue under visible irradiation by iron tetrasulphophthalocyanine immobilized layered double hydroxide at circumneutral pH. *Dyes Pigments* **2016**, *134*, 397–408. [CrossRef]
26. Huang, L.; Denis, T.G.S.; Xuan, Y.; Huang, Y.Y.; Masamitsu, T.; Andrzej, Z.; Tadeusz, S.; Michael, R. HamblinParadoxical potentiation of methylene blue-mediated antimicrobial photodynamic inactivation by sodium azide: Role of ambient oxygen and azide radicals. *Free Radic. Biol. Med.* **2012**, *53*, 2062–2071. [CrossRef]
27. Shi, M.; Chen, Z.M.; Guo, L.X.; Liang, X.H.; Zhang, J.L.; He, C.Y.; Wang, B.; Wu, Y.Q. A multiwalled carbon nanotube/tetra bisoheptyloxyphthalocyanine cobalt(II) composite with high dispersibility for electrochemical detection of ascorbic acid. *J. Mater. Chem. B* **2014**, *2*, 4876–4882. [CrossRef]
28. Yan, X.M.; Xu, X.; Zhong, Z.; Liu, J.J.; Tian, X.M.; Kang, L.T.; Yao, J.N. The effect of oxygen content of carbon nanotubes on the catalytic activity of carbon-based iron phthalocyanine for oxygen reduction reaction. *Electrochim. Acta* **2018**, *281*, 562–570. [CrossRef]
29. Topacli, A.; Kesimli, B. Investigation on sulfanilamide and its interation eith some metals and lincomycin by infrared spectroscopy. *Spectrosc. Lett.* **2001**, *34*, 513–526. [CrossRef]
30. AmalAl, A.; Sajjad, K. Cross-linked chitosan aerogel modifed with Pd(II)/phthalocyanine: Synthesis, characterization, and catalytic application. *Sci. Rep.* **2019**, *9*, 13849.
31. Watkins, J.D.; Lawrence, R.; Taylor, J.E.; Bull, S.D.; Nelson, G.W.; Foord, J.S.; Wolverson, D.; Rassaei, L.; Evans, N.D.; Gascond, M.S.A.; et al. Carbon nanoparticle surface functionalisation: Converting negatively charged sulfonate to positively charged sulfonamide. *Phys. Chem. Chem. Phys.* **2010**, *12*, 4872–4878. [CrossRef] [PubMed]
32. Kumar, A.; Yasin, G.; Korai, R.M.; Slimani, U.Y.; Muhammad, F.A.; Mohammad, T.; Nazir, M.T.; Nguyen, T.A. Boosting oxygen reduction reaction activity by incorporating the iron phthalocyanine nanoparticles on carbon nanotubes network. *Inorg. Chem. Commun.* **2020**, *120*, 108160. [CrossRef]
33. Zhang, Z.P.; Dou, M.L.; Ji, J.; Wang, F. Phthalocyanine tethered iron phthalocyanine on graphitized carbon black as superior electrocatalyst for oxygen reduction reaction. *Nano Energy* **2017**, *34*, 338–343. [CrossRef]
34. Schick, G.A.; Sun, Z.Q. Spectroscopic characterization of sulfonyl chloride immobilization on silica. *Langmuir* **1994**, *10*, 3105–3110. [CrossRef]
35. Exner, O.; Juška, T. X-Ray Photoelectron Spectroscopic Study of Sulphonamides: Charge distrbution and tautomerism. *Collect. Czechoslov. Chern. Commun.* **1984**, *49*, 51–57. [CrossRef]
36. Catherine, D.C.; Haskouri, S.; Cachet, H. Study by XPS of the chlorination of proteins aggregated onto tin dioxide during electrochemical production of hypochlorous acid. *Appl. Surf. Sci.* **2007**, *253*, 5506–5510.

37. Hou, L.J.; Zhang, J.L.; Pu, Y.F.; Li, W. Effects of nitrogen-dopants on Ru-supported catalysts for acetylene hydrochlorination. *RSC Adv.* **2016**, *6*, 18026–18032. [CrossRef]
38. Wang, H.Q.; Hana, J.; Zhao, B.; Qin, L.B.; Wang, Y.; Yu, F. Non-thermal plasma enhanced dry reforming of CH_4 with CO_2 over activated carbon supported Ni catalysts. *Mol. Catal.* **2019**, *475*, 110486–110495. [CrossRef]
39. Liu, Y.; Fan, Y.S.; Liu, Z.M. Pyrolysis of iron phthalocyanine on activated carbon as highly efficient nonnoble metal oxygen reduction catalyst in microbial fuel cells. *Chem. Eng. J.* **2019**, *361*, 416–427. [CrossRef]
40. Yang, T.; Li, K.; Pu, L.; Liu, Z.; Ge, B.; Pan, Y.; Liu, Y. Hollow-spherical Co/N-C nanoparticle as an efficient electrocatalyst used in air cathode microbial fuel cell. *Biosens. Bioelectron.* **2016**, *86*, 129–134. [CrossRef]
41. Lappalainen, J.; Baudouin, D.; Hornung, U.; Schuler, J.; Melin, K.; Bjelíc, S.; Vogel, F.; Konttinen, J.; Joronen, T. Sub- and supercritical water liquefaction of kraft lignin and black liquor derived lignin. *Energies* **2020**, *13*, 3309. [CrossRef]
42. Rangarajan, G.; Koh, J.M.; Farnood, R. Touch wood: Conversion of lignin to dicarboxylic acids using biochar-based solid-acid photo-Fenton catalysts. *J. Clean. Prod.* **2023**, *415*, 137816–137830. [CrossRef]
43. Peng, F.; Peng, P.; Xu, F.; Sun, R.C. Fractional purification and bioconversion of hemicelluloses. *Biotechnol Adv.* **2012**, *30*, 879–903. [CrossRef] [PubMed]
44. Yu, H.T.; Chen, B.Y.; Li, B.Y.; Tseng, M.C.; Han, C.C.; Shyu, S.G. Efficient pretreatment of lignocellulosic biomass with high recovery of solid lignin and fermentable sugars using Fenton reaction in a mixed solvent. *Biotechnol. Biofuels* **2018**, *11*, 287–298. [CrossRef] [PubMed]
45. Schrank, S.G.; Santos, J.N.R.; Souza, D.S.; Souza, E.E.S. Decolourisation effects of Vat Green 01 textile dye and textile wastewater using H_2O_2/UV process. *J. Photochem. Photobiol. Chem.* **2007**, *186*, 125–129. [CrossRef]
46. Xu, F.; Lu, Q.W.; Li, K.P.; Zhu, T.T.; Wang, W.K.; Hu, Z.H. Green synthesis of magnetic mesoporous carbon from waste-lignin and its application as an efficient heterogeneous Fenton catalyst. *J. Clean. Prod.* **2021**, *285*, 125363–125372. [CrossRef]
47. Wang, Y.S.; Tang, Z.Y.; Chen, M.Q.; Zhang, J.H.; Shi, J.J.; Wang, C.S.; Yang, Z.L.; Wang, J. Effect of Mo content in Mo/Sepiolite catalyst on catalytic depolymerization of Kraft lignin under supercritical ethanol. *Energy Convers. Manag.* **2020**, *222*, 113227–113240. [CrossRef]
48. Leybo, D.V.; Konopatsky, A.S.; Fang, X.S.; Shtansky, D.V. Photocatalytic phenol oxidation over ball milled hexagonal boron nitride. *J. Water Process. Eng.* **2023**, *51*, 103367–103377. [CrossRef]
49. Huang, X.M.; Korányi, T.I.; Boot, M.D.; Hensen, E.J.M. Ethanol as capping agent and formaldehyde scavenger for efficient depolymerization of lignin to aromatics. *Green Chem.* **2015**, *17*, 4941–4950. [CrossRef]
50. Yan, L.; Wang, Y.; Li, J.; Zhang, Y.; Ma, L.; Fu, F.; Chen, B.; Liu, H.J. Hydrothermal liquefaction of Ulva prolifera macroalgae and the influence of base catalysts on products. *Bioresour. Technol.* **2019**, *292*, 121286. [CrossRef]
51. Song, Y.; Motagamwale, A.H.; Karlen, S.D.; Dumesic, J.A.; Ralph, J.; Mobley, J.K.; Crocker, M. A comparative study of secondary depolymerization methods on oxidized lignin. *Green Chem.* **2019**, *21*, 3904–3947. [CrossRef]
52. Zhu, Y.F.; Song, W.Q.; Yao, R.J.; Zhao, Y.Z.; Xu, G.Y. In-situ catalytic hydropyrolysis of lignin for the production of aromatic rich bio-oil. *J. Energy Inst.* **2022**, *101*, 187–193. [CrossRef]
53. Biswas, B.; Kumar, A.A.; Bisht, Y.; Singh, R.; Kumar, J.; Bhaskar, T. Effects of temperature and solvents on hydrothermal liquefaction of *Sargassum tenerrimum* algae. *Bioresour. Technol.* **2017**, *242*, 344–350. [CrossRef] [PubMed]
54. Biswasa, B.; Kumara, A.; Fernandesc, A.C.; Sainia, K.; Negia, S.; Muraleedharanc, U.D.; Bhaskar, T. Solid base catalytic hydrothermal liquefaction of macroalgae: Effects of process parameter on product yield and characterization. *Bioresour. Technol.* **2020**, *307*, 123232–123239. [CrossRef] [PubMed]
55. Raikova, S.; Allen, M.J.; Chuck, C.J. Hydrothermal liquefaction of macroalgae for the production of renewable biofuels. *Biofuels Bioprod. Biof.* **2019**, *13*, 1483–1504. [CrossRef]
56. Yang, Z.X.; Lei, H.W.; Zhang, Y.Y.; Qian, K.Z.; Villot, E.; Qian, M.; Yadavalli, G.; Sun, H. Production of renewable alkyl-phenols from catalytic pyrolysis of Douglas fr sawdust over biomass-derived activated carbons. *Appl Energy* **2018**, *220*, 426–436. [CrossRef]
57. Chen, H.T.; He, Z.X.; Zhang, B.; Feng, H.; Kandasamy, S.; Wang, B. Efects of the aqueous phase recycling on bio-oil yield in hydrothermal liquefaction of Spirulina Platensis, a-cellulose, and lignin. *Energy* **2019**, *179*, 1103–1113. [CrossRef]
58. Liu, H.M.; Li, M.F.; Yang, S.; Sun, R.C. Understanding the mechanism of cypress liquefaction in hot-compressed water through characterization of solid residues. *Energies* **2013**, *6*, 1590–1603. [CrossRef]
59. Li, Z.; Zhang, J.; Qin, L.; Ge, Y. Enhancing antioxidant performance of lignin by enzymatic treatment with laccase. *ACS Sustain. Chem. Eng.* **2018**, *6*, 2591–2595. [CrossRef]
60. Ramires, E.C.; Megiatto, J.D.; Gardrat, J.C.; Castellan, A.; Frollini, E. Valorization of an industrial organosolv-sugarcane bagasse lignin: Characterization and use as a matrix in biobased composites reinforced with sisal fibers. *Biotechnol. Bioeng.* **2010**, *107*, 612–621. [CrossRef]
61. Wang, W.Y.; Qin, Z.; Liu, H.M.; Wang, X.D.; Gao, J.H.; Qin, G.Y. Structural changes in milled wood lignin (MWL) of chinese quince (*Chaenomeles sinensis*) fruit subjected to subcritical water treatment. *Molecules* **2021**, *26*, 398. [CrossRef] [PubMed]
62. Glasser, W.G.; Kaar, W.E.; Jain, R.K.; Sealey, J.E. Isolation options for non-cellulosic heteropolysaccharides (HetPS). *Cellulose* **2000**, *7*, 299–317. [CrossRef]
63. Zhu, J.H.; Li, Y.X.; Chen, Y.; Wang, J.; Zhang, B.; Zhang, J.J.; Werner, J.B. Graphene oxide covalently functionalized with zinc phthalocyanine for broadband optical limiting. *Carbon* **2011**, *49*, 1900–1905. [CrossRef]

64. Nemykin, V.N.; Polshyna, A.E.; Borisenkova, S.A.; Strelko, V.V. Preparation, characterization, and catalytic activity of synthetic carbon-supported (phthalocyaninato)cobalt-containing complexes in dodecane-1-thiol oxidation reaction. *J. Mol. Catal A Chem.* **2007**, *264*, 103–109. [CrossRef]
65. Achar, B.N.; Lokesh, K.S. Studies on tetra-amine phthalocyanines. *J. Organomet. Chem.* **2004**, *689*, 3357–3361. [CrossRef]

sustainability

MDPI

Article

Energy and Business Synergy: Leveraging Biogenic Resources from Agriculture, Waste, and Wastewater in German Rural Areas

Moritz Pollack [1,*], Andrea Lück [2], Mario Wolf [2], Eckhard Kraft [2] and Conrad Völker [1]

[1] Department of Building Physics, Bauhaus-University Weimar, 99423 Weimar, Germany; conrad.voelker@uni-weimar.de

[2] Department of Biotechnology in Resource Management, Bauhaus-University Weimar, 99423 Weimar, Germany; andrea.lueck@uni-weimar.de (A.L.); ulrich.mario.wolf@uni-weimar.de (M.W.); eckhard.kraft@uni-weimar.de (E.K.)

* Correspondence: moritz.pollack@uni-weimar.de; Tel.: +49-3643-584847

Abstract: The imperative to transform current energy provisions is widely acknowledged. However, scant attention has hitherto been directed toward rural municipalities and their innate resources, notably biogenic resources. In this paper, a methodological framework is developed to interconnect resources from waste, wastewater, and agricultural domains for energy utilization. This entails cataloging existing resources, delineating their potential via quantitative assessments utilizing diverse technologies, and encapsulating them in a conceptual model. The formulated models underwent iterative evaluation with engagement from diverse stakeholders. Consequently, 3 main concepts, complemented by 72 sub-concepts, were delineated, all fostering positive contributions to climate protection and providing heat supply in the rural study area. The outcomes' replicability is underscored by the study area's generic structure and the employed methodology. Through these inquiries, a framework for the requisite energy transition, with a pronounced emphasis on the coupling of waste, wastewater, and agriculture sectors in rural environments, is robustly analyzed.

Keywords: energy transition; energy concepts; energy systems; rural area; biogenic residues; biogenic waste materials; agriculture; waste management; wastewater management; regional development

1. Introduction

To ensure the future livelihood of humankind, the international community adopted the 2030 Agenda in 2015. Consisting of 17 Global Sustainable Development Goals (SDGs) for socially, economically, and ecologically sustainable development, this agenda is a roadmap for the future that will enable people around the world to live in dignity while preserving the natural foundations of life in the long term. As part of the transition toward a more sustainable future for humankind, goal 7 seeks to strive away from the use of fossil fuels, such as natural gas, coal, and oil, by increasing the share of renewable energy in the global energy mix to significantly enhance energy efficiency [1]. Researchers have intensified their work to deliver solutions to meet the defined goals of the 2030 Agenda in terms of heat and electricity supply. New technologies are being developed, such as hydrogen-based combined heat and power units (CHP) [2], and the efficiency of existing technologies, such as solar panels, windmills, and gas turbines, has been continuously increased [3–7]. However, achieving the goals set forth by the 2030 Agenda appears to be challenging. In the past few decades, the world has experienced a rapid increase in primary energy consumption; from 1950 until the year 2023, it had more than doubled [8]. For the future, studies (e.g., [8]) have predicted an increase of 28% in global primary energy consumption from 13.8 billion tons of oil equivalent (bn toe) in 2016 to 17.6 bn toe in 2040. Of that predicted total increase, 40% will be covered by renewable energies. Therefore, the share of primary energy consumption accounted for by conventional energies will only decrease by 6% from 86% in 2016 to 80% in 2040. While the increase in the electricity sector in recent years is notable, the heating sector lacks an accelerating momentum [9].

The necessary shift toward renewable energies and higher energy efficiency does not only pose a technological challenge. It also requires a complex process of altering our current energy consumption and production methods at national, continental, and global scales [10]. The process of planning and implementing this transition is confronted by a range of challenges, including conflicting objectives and values, and it requires a navigation of the complexities and uncertainties associated with political and administrative considerations. Additionally, it involves the coordination of multiple stakeholders throughout the transition process [11].

Municipalities and local authorities play a pivotal role in driving energy planning and energy transition efforts [12,13]. In the past, the primary focus was on larger municipalities that have well-equipped staff with the necessary capabilities [14]. Recently, a gradual shift toward rural areas was observed [15,16], leading to the current implementation of a national law requiring each municipality in Germany to design a Communal Energy Plan (CEP). In light of the pursuit for renewable energies and heat waste as sole sources of usage for heating purposes, the aim of a CEP is to determine a climate-friendly and progressive heat supply at the municipal level as a long-term prospect and to identify measures of its implementation. The transformation toward a more renewable energy system must be ecological, economical, and socially acceptable [17]. Thus, each CEP should incorporate the local conditions of its municipality, such as the technical infrastructure systems, environmental and economic aspects, as well as human resources [18]. This shift is not surprising, given that approximately 90% of the European Union's territory is rural or predominantly rural, and slightly over half of the European population resides in such areas [19]. Rural areas offer the essential resources and suitable locations for renewable energy production [20–22].

In rural areas, a significant accumulation of agricultural residues and animal manure is evident [23,24]. Moreover, organic and green waste is generated, with some of it either remaining uncollected or being transported over substantial distances for treatment. Furthermore, additional biogenic resources can be derived from the wastewater sector, which frequently lacks modern technical infrastructure in Germany's rural regions [25,26]. Despite the inherent benefits, these three sources of biogenic resources are seldom integrated into a cohesive management system. Their utilization and coordinated management face a multitude of non-technical challenges, encompassing social, economic, and legislative dimensions [27,28]. These challenges include resource competition, conflicts over land use, environmental considerations, regulatory and policy obstacles, community acceptance, economic viability, and issues related to skills and training.

Managing entities like municipalities and municipal utilities is confronted with constrained resources in the forms of limited time, personnel, and expertise to proactively steer the development and execution of renewable energy concepts. This dilemma emerges from the juxtaposition of external challenges and the internal need for competence, all within the constraints of finite resources [29]. Consequently, there is a pressing need for guidance to assist local stakeholders in structuring the implementation of rural energy concepts that harmonize the wastewater, waste, and agricultural sectors, while being tailored to their unique circumstances.

In this context, the aim of this article is to demonstrate a systematic stakeholder-oriented approach to define viable energy concepts based on the integrated management of wastewater, biogenic waste, and agricultural residues. Based on a rural area in Germany, the suggested approach consists of several steps, including an in-depth analysis of the given situation in the case study area as well as an assessment of the economic and ecological consequences of executing these diverse concepts, considering the distinct material flows within the study area. Thus, the motivation of this study lies in offering rural stakeholders a guideline for investigating the individual local energetic potentials of utilizing residues from wastewater, waste, and agricultural sectors. In this way, this study adds a conceptual energy component that should be considered in the design process of CEPs for rural areas.

This article is structured into five sections. Following the introduction, a description of the study area, the methodological steps, and the analytical framework is provided in Section 2. In Section 3, details of how the methodological approach is applied and the identified energy concepts are presented. A discussion of the results is presented in Section 4. Finally, the findings are summarized in Section 5.

2. Materials and Methods

2.1. Study Area Description

The case study area (Figure 1) is the administrative union Am Ettersberg, consisting of 19 villages and three smaller towns in Germany. Additionally, three neighboring villages are part of the study area, as they are also members of the local wastewater association responsible for the administrative union Am Ettersberg. Located within a typical rural area in Thuringia, the study area holds approximately 9000 inhabitants and spans an area of nearly 100 km^2. Since 1990, the demographic change led to a significant decrease in population, whereas for the state of Thuringia, a loss of 8.7% has been predicted for the period from 2021 to 2042; this trend can be also observed in the study area [30].

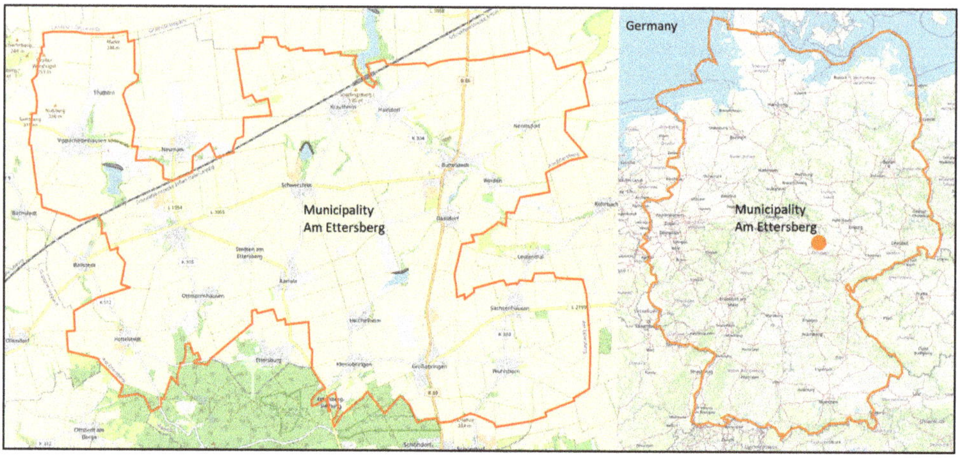

Figure 1. Map of the study area, the municipality Am Ettersberg, and its location in Germany [31].

With the purpose of providing reliable tasks in the public sector despite the demographic challenges, the administrative union Am Ettersberg was established in the year 2019 [32]. Prior to its establishment, an administrative union consisting of a smaller number of villages governed the area. From a communal perspective, today's administrative union is responsible for the energy transition and design of a CEP.

Aiming to overcome the dependence on fossil fuels, local authorities must define the future energy solutions for all areas under their jurisdiction (e.g., decentralized or grid-bound). This includes options for how renewable energies and industrial heat can be used in generating and distributing energy locally [33]. Another public task lies in the management of biogenic waste. This includes public loads, such as green cuttings, and waste in public areas, e.g., parks and roadsides. However, this public green waste is not collected properly and is disposed of in a non-compliant way to current state of the art. The same applies to private green biowaste.

Wastewater management in the municipality Am Ettersberg is widely decentralized, which is typical for the rural areas of Thuringia [34]. A wastewater association is responsible for providing adequate wastewater treatment in the study area. However, due to limited financial resources, the required infrastructure projects are considered long-term endeavors [35]. Currently, only a minority of residents within the operational zone have access to biological wastewater treatment. Fecal sludge disposal is costly, though it could potentially be harnessed for on-site energy generation. The remaining inhabitants lack access to state-of-the-art wastewater treatment, necessitating urgent action. Simultaneously, there exists an opportunity to implement more resource-efficient systems that facilitate synergy across various sectors.

Economically, agriculture is a driving force for the development of livable rural areas in Thuringia [36]. In addition to basic food production, the range of services includes the development of alternative sources of income. Specifically, the need for an increasing link with the energy industry is given. In 2016, Thuringian farmers operated 260 biogas plants, collectively supplying electricity to nearly 340,000 households [37]. As of 2023, the number of biogas plants has barely changed [38]. These dynamics are similarly observed in the farms situated within Am Ettersberg. Given the absence of other major corporations in the region, farms serve as the primary employers. The economic emphasis among them varies; some exclusively prioritize arable land products, while others engage in intensive livestock farming, frequently complemented by the operation of biogas plants.

2.2. Methodology

To identify suitable energy concepts for the study area, a systematic methodological approach was applied by the researchers (adopted from [39,40]). This approach is primarily based on the recommended steps for municipal energy planning [41,42], which involves conducting a thorough inventory analysis, assessing potential, and identifying actionable options. While previous models and methodologies for renewable energy supply in urban areas and communities have considered different kinds of biomass (e.g., [43–45]), there has been a notable absence of a combined examination of material flows despite its significant potential, particularly in rural areas [25].

Therefore, this study integrates the conventional approach of life cycle assessments (Goal and Scope–Life Cycle Inventory (Cradle to Grave)–Impact Assessment–Interpretation), which is commonly utilized for evaluating waste disposal pathways (see [46,47]) and resource planning [48]. The intention is to bridge the gap by applying this established methodology to the assessment of biomass in the context of renewable energy supply for municipalities. This novel approach not only extends the existing framework but also addresses a crucial aspect of rural energy planning by considering material flows and their potential impact on sustainable energy solutions.

This approach consists of the four following steps, which will be depicted in more detail below (Figure 2):

- Inventory analysis to assess the current technical infrastructures and biogenic material flows in the analyzed sectors in the study area;
- Potential analysis to identify suitable technology options to transfer biogenic residues into energetic potential;
- Conceptualization to define the energy concepts and their sub-concepts;
- Concept evaluation to identify the (dis)advantages of each sub-concept in reference to the current system.

Inventory analysis: The inventory analysis involves a comprehensive assessment of infrastructures and material flows within the waste management, wastewater management, and agricultural sectors, as well as the energy demand in the study area.

In the waste management sector, particular focus is placed on biowaste and private or municipal green waste, along with the respective collection systems and disposal routes, including the distances travelled. Additionally, any existing treatment facilities, especially those with energy output, such as biogas plants, are taken into account.

Within the wastewater management sector, the quantities and compositions of sewage sludge are of primary importance. This includes a consideration of pre-treatments and disposal routes, along with corresponding distances. Furthermore, sewage sludge treatment plants (including dewatering, drying, and digestion) within the area under consideration are documented. Future developments like the construction of municipal sewage treatment plants or the decommissioning of small-scale sewage treatment plants and related changes in sewage sludge generation are also considered.

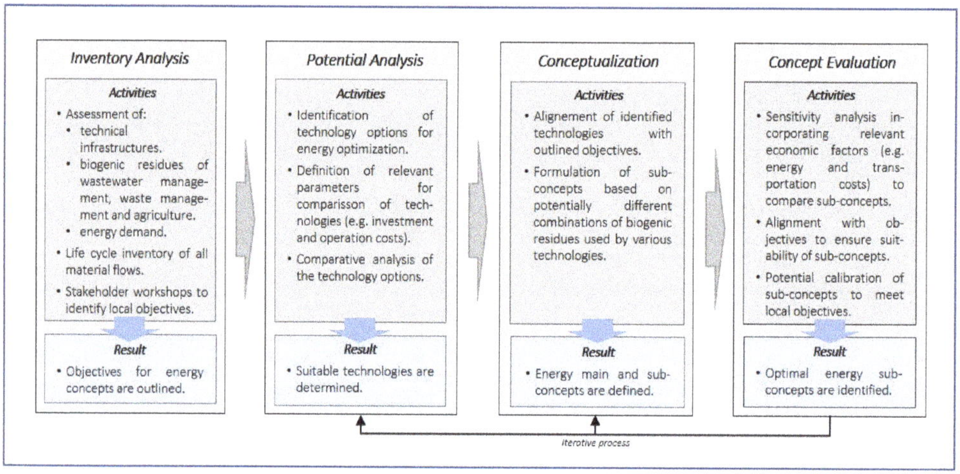

Figure 2. Conceptual framework of the methodology used.

In agriculture, the focus lies on previously unutilized material flows. This encompasses manure and compost, as well as residual straw. Additionally, existing agricultural biogas plants and their respective input materials are recorded. The energy demand within the area of interest plays a crucial role in determining the utilization of energies generated through new energy concepts. Local consumable energies such as heat and fuel are particularly relevant. The determination of heat demand should be spatially resolved and can be simplified through the use of statistical indicators (e.g., [49]) or can be derived from existing municipal heat planning. If there are no aggregated data available for calculating the heat demand of private households in rural municipalities due to a decentralized supply structure, it can be derived at the building level using the online tool TRAIL (see [44]). TRAIL provides the heat demand of private households in kWh/(m^2 of living space per year) in a 100 × 100 m grid for all Thuringian municipalities based on publicly accessible data (statistical data on building age [50] and typical heat demand by building age and type [50], with 3D buildings on the level of detail 2 (LoD2 city model) [51]). From the LoD2 city model, the living area for each residential building was determined following the TRAIL calculation algorithm [49]. Then, the heat demand per building was calculated as follows:

$$Q_{building} = A \cdot 0.8 \cdot n \cdot C_{heat} \quad (1)$$

where $Q_{building}$ is the annual heat demand of a specific building, A is the building floor area, n is the number of floors, and C_{heat} is the specific heat demand in kWh/(m^2·a) of the corresponding tile in the 100 × 100 m grid [49]. The heat demand of non-residential buildings, for which no consumption data are available, as well as commercially used buildings, can be estimated based on the gross floor area of these buildings and their characteristic values depending on the type of use [52].

In addition to capturing the infrastructure, material flows, and energy demands of public entities and private households, these aspects should also be taken into account for relevant commercial and industrial enterprises. An overview of the captured data for the inventory analysis is given in Table 1.

Table 1. Captured data for inventory analysis.

	Energy Sector	Wastewater Sector	Waste Sector	Agriculture Sector
Material flows	Energy demand in households (heat, fuel)	Sewage sludge	Biowaste from households	Manure
			Private green waste	Compost
	Energy demand in industry (heat, fuel)		Municipal green waste	Residual Straw
			Landscaping materials	Further input material
Infrastructure		Pre-treatments Disposal routes Treatment facilities	Collection system(s) Disposal routes Treatment facilities	Biogas plant(s)

In the next step, a life cycle inventory of the current utilization pathways for all considered material flows in terms of costs, revenues, greenhouse gas emissions, carbon credits, and energy consumption are compiled using the available literature or operating data. Then, the abovementioned parameters are summed up across the pathways to determine the costs, revenues, emissions, credits, and cumulative energy expenditures for the existing system. The balance framework of the life cycle inventory is schematically shown in Figure 3.

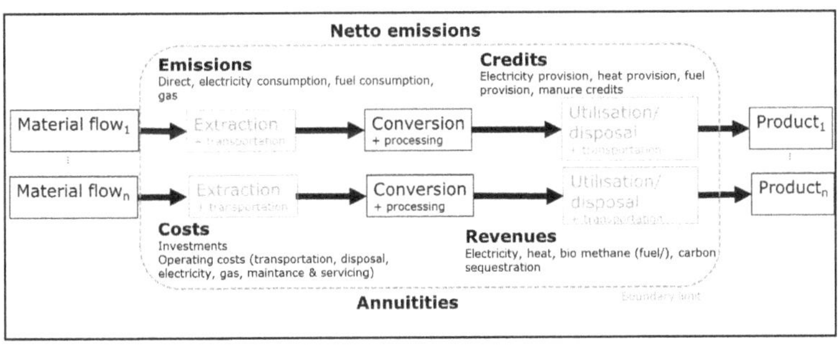

Figure 3. Balance framework of the life cycle inventory.

The energy content of the material flows should be determined to identify effective utilization routes at a later stage.

Utilizing this flow balance enables the identification of weaknesses in the current system, pinpointing areas such as outdated treatment technologies or unutilized material flows. Simultaneously, it is crucial to assess the objectives of local stakeholders in the study area concerning resource utilization, heat planning, and future developments, particularly in the context of energy resources.

Potential analysis: Potential analysis aims to identify various technology options (conversion technologies) for energy-optimized treatment and the effective utilization of material flows (adopted from [53]). This includes a special focus on technologies that allow for the combined treatment of multiple material flows from different sectors. Additionally, it involves identifying any potentially necessary technologies for the pre-treatment (provision technologies) of material flows. Furthermore, the analysis seeks to identify potential utilization technologies to harness the products generated through the application of conversion technologies.

In this context, relevant parameters for all identified technology options (provision, conversion, and utilization) should be determined. These parameters include investment costs, operating costs, revenues, lifespan, greenhouse gas emissions, carbon credits, and energy consumption. These factors are evaluated in relation to local specifics, such as the quantities of material flows to be treated and transportation distances.

Conceptualization: During this phase, the main concepts are formulated by aligning the identified technology options with the objectives set by regional stakeholders. Any material flows not addressed in the central conversion technology will have their existing utilization pathways incorporated into the main concepts to ensure that a complete life cycle inventory is prepared for each concept.

Additionally, sub-concepts are developed through the exploration of different combinations of treated material flows using the conversion technologies and potential utilization technologies to obtain the resulting products.

Concept evaluation: It is important to select a time frame (TF) for assessing energy concepts. This period is typically determined by the longest lifespan among all considered technology options. However, deviations may occur in cases of different planning scenarios at the local or higher levels.

Next, greenhouse gas emissions, carbon credits, and cumulative energy expenditures are calculated on an annual basis for each utilization pathway. The calculation of necessary investment costs, discounted expenses, and revenues are aggregated to a present value (PV) and then converted into equivalent annual cost (EAC) according to current guidelines for dynamic cost comparisons for water management facilities [54] using the Formulas (2) and (3):

$$PV = \frac{I_0}{(1+p)^i} + \Sigma \frac{\text{operating costs}_i * (1+g)^i - \text{revenues}_i}{(1+p)^i} \quad (2)$$

where I_0 is the investment cost, p is the interest rate, g is the annual price increase rate, and i is the year.

$$EAC = \frac{PV}{TF} \quad (3)$$

Following this, the parameters mentioned above are aggregated across all material flows to determine the net emissions, EAC, and cumulative energy expenditures for each sub-concept. In the subsequent step, the greenhouse gas abatement potential (AP) and the greenhouse gas abatement costs (ACs) for all sub-concepts are computed as indicators for their climate change mitigation effects as well as economical performances. This is carried out by comparing the results of the calculations and energy savings obtained in the previous step with the results for the existing system using Formula (4):

$$AC = \frac{\Delta\ EAC}{\Delta\ \text{emissions}} = \frac{EAC_j - EAC_0}{\text{emissions}_j - \text{emissions}_0} \quad (4)$$

where j indicates the values for the corresponding sub-concept, and 0 indicates the values for the existing system.

Finally, a sensitivity analysis is carried out, considering economically relevant parameters such as energy costs, transportation costs, price increases, interest rates, and the chosen time frame. This analysis helps identify the optimal concepts based on criteria like maximum greenhouse gas reduction potential, lowest greenhouse gas reduction costs, the best ratio of greenhouse gas savings to investment costs, and the best ratio of greenhouse gas reduction costs to investment costs. These findings are then aligned with local objectives and goals to identify the optimal sub-concept.

3. Results

Applying the previously described methodology to the study area requires a structured process. Therefore, the results will be presented below according to each of the four analytical steps.

3.1. Results of Inventory Analysis

Waste disposal is managed at the district level. The municipal waste management company is the entity responsible for waste disposal. Due to the rural structure of the study area and the resulting lack of economic viability for the separate collection of organic waste, the municipal waste management company set the goal of widespread on-site composting. In the study area, biogenic household waste was not yet separately collected. Therefore, it was collected together with residual waste in a pick-up system and subsequently loaded onto trucks at the district's depot located at a distance of approximately 30 km. From there, it was thermally treated in an incineration plant located about 140 km away. Since precise data on waste quantities in the municipality are not recorded, they must be estimated based on the quantities and compositions of waste for the entire district provided by the district works, as well as population figures. Information regarding the compostable fractions of household waste is available from the household waste analysis for 2016/2017 [55]. Green waste was collected through three containers located in the municipality, using a drop-off system. These containers contained private and municipal green waste.

Focusing on wastewater management, approximately 54% of the population was connected to a central municipal sewage treatment plant, 39% had access to a mechanical small-scale sewage treatment plant, and 7% had access to a fully biological small-scale sewage treatment plant. Consequently, the collected sludge primarily consisted of sludge from small-scale sewage treatment plants with a very high water content. The data on sludge production and composition were provided by the wastewater disposal entity of the study area. The sludge in the study area was not collected or treated centrally but rather disposed of at a large sewage treatment plant located approximately 30 km away.

In the study area, there are two large agricultural enterprises. In addition to plant production, their activities include the breeding of sows and dairy farming, along with the operation of a power-driven biogas plant. The heat generated by this biogas plant is utilized for heating the stables and buildings, although a significant portion remains unused.

The heat supply in the study area is characterized by a decentralized structure due to its rural nature, and there is no centralized heating network. The majority of heat generation is fossil-based, with a significant portion coming from heating oil, as only a few villages are connected to the gas network. Since there are no aggregated data available for the heat demand of private households in the rural municipality due to this decentralized supply structure, these were derived at the building level from the online tool TRAIL (see Formula (1)).

The data on heat demands were then consolidated using a heat cadaster. This was gradually expanded to include non-residential buildings for which consumption data are available (buildings owned by the municipality and industrial buildings). In addition, other municipal buildings for which no consumption data are available, as well as commercially used buildings, were added by estimating the heat demand based on the gross floor area of these buildings and their characteristic values depending on the type of use.

The electricity demand for private households in the study area was adopted from TRAIL. This was based on statistical data on average electricity demand per person. The municipality provided the electricity demand for the community (street lighting, sewage treatment plants, municipally owned buildings, etc.).

The energy demand for road transportation was extrapolated from the energy demand for Thuringia [56], using the population figures for the study area.

Based on the data collected as described above, a balance was created for the quantities and energy contents of material flows, as well as the costs, revenues, greenhouse gas emissions, carbon credits, and cumulative energy expenditures for the material flows generated in the study area, which include sludge, organic waste, green waste, cattle manure, swine manure, and residual straw. This is illustrated in Figure 4. The costs were discounted within a 30-year time frame using Formula (2) for comparison with the later energy concepts.

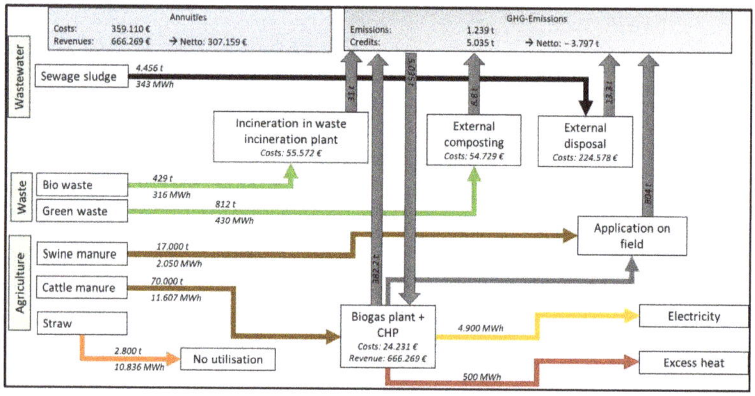

Figure 4. Schematic representation of the existing system and its life cycle inventory.

Notable negative points in the balance include particularly high emissions from the direct application of swine manure, elevated costs due to the external disposal of untreated sludge, the suboptimal disposal of high-calorific organic waste and green waste, and the underutilization of excess heat from the biogas plant.

In several workshops and discussions with stakeholders from the wastewater management sector (Managing Director of the wastewater disposal entity), waste management sector (Member of the Environmental Committee of the district), and agriculture sector (Managing Director/Department Head of the agricultural enterprises), as well as members of the municipal administration (Mayor, Energy Manager of the municipality, and Head of the Building Department), the objectives of each sector were discussed. The objectives were outlined as follows:

- Wastewater management: the aim is to reduce the high costs of external sludge disposal by implementing in-house treatment solutions.
- Waste management: this sector is open to cross-sector collaboration and partnerships.
- Agriculture: This sector's focus is on the continued operation of the existing biogas plant that is economically viable beyond the expiration of the state Renewable Energy Law (Erneuerbare-Energien-Gesetz (EEG)) subsidies. Also, this sector aims to explore opportunities for utilizing swine manure, especially for meeting their own energy needs.
- Municipality: the aim is to ensure a sustainable heat supply within the municipal area, with a preference for leveraging regional resources.

3.2. Results of Potential Analysis

To investigate how to transform the biogenic resources into energy, different technologies were catalogued, and their impacts were quantified. Based on a literature review (see [57–59]), biogas plants with power generation in combined heat and power units (CHP), fermentation with the upgrading to biomethane, combustion, gasification, hydrothermal carbonization (HTC), and pyrolysis were identified as suitable conversion technologies for the partial joint treatment of the relevant material flows in the study area. Their specific characteristics were documented in technology profiles.

In some cases, (pre-) processing steps are required for certain material flows. Likewise, only some of the technologies are suitable for certain material flows. Different products can be provided using the various conversion technologies identified from the literature review. These findings are summarized in the table below (Table 2).

Table 2. Investigated technology options, with (pre-) processing steps and outputs.

	Biogas Plant and CHP	Biomethane Plant	Combustion	Gasification	Hydrothermal Carbonization	Pyrolysis
Input						
Organic Waste	Shredding, separation of impurities	Shredding, separation of impurities	No	Drying	Yes	Drying
Green Waste	Shredding, separation of impurities	Shredding, separation of impurities	Drying	Drying	Yes	Drying
Sewage sludge	Yes	Yes	Drying	Drying	Yes	Drying
Cattle manure	Yes	Yes	No	No	Yes	No
Swine manure	Yes	Yes	No	No	Yes	No
Straw	Shredding	Shredding	Yes	Yes	Yes	Yes
Output						
Heat	Yes	No	Yes	Yes	No	No
Electricity	Yes	No	No	Yes	No	No
Fuel	No	Yes	No	No	Yes	Yes
Fertilizer/soil conditioner	Yes	Yes	No	No	Yes	Yes

To effectively harness the respective material flows and the resultant products, a range of extraction and utilization technologies is imperative. Regarding extraction technologies, consideration encompasses a separate collection of biowaste and straw extraction. As for utilization technologies, integration involves a heating network, mobile heat storage, a biomethane filling station, and biomethane injection into the gas grid.

3.3. *Results of Conceptualization*

Derived from the conversion technologies and aligned with the goals of local stakeholders, three primary concepts were discerned. By modifying the pertinent material flows and corresponding technology options for each concept, a total of 72 sub-concepts were delineated in accordance with the three main concepts.

3.3.1. Concept 1: Maximum Heat Provision

Based on the high heat demand identified in the inventory analysis and considering the structural suitability of a selected village in the study area, Concept 1 aims to provide the maximum amount of heat in the area through the establishment of a district heating network. Possible heat sources include the utilization of excess heat from the existing biogas plant, an expansion of this biogas plant for the utilization of treatment-requiring material flows, and the construction of a new heating plant. By varying the heating plant (combustion/gasification), the materials used (option 1: straw; option 2: green waste; option 3: straw + green waste), the configuration of the existing biogas plant (option 1: existing fermenter; option 2: new fermenter; option 3: without excess heat), and the biogas substrates used (option 1: cattle manure; option 2: cattle manure + sludge; option 3: cattle manure + organic waste; option 4: cattle manure + sludge + organic waste), a total of 48 sub-concepts were considered for Concept 1 (see Figure 5).

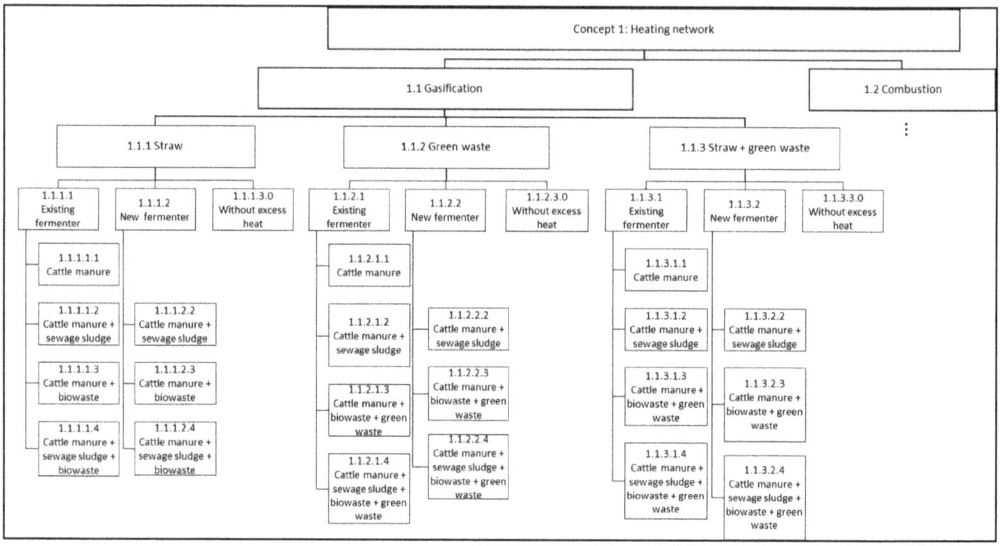

Figure 5. Schematic of sub-concepts for Concept 1.

3.3.2. Concept 2: Emission Minimization through Swine Manure Utilization

Due to the identified high greenhouse gas emissions from untreated swine manure in the study area, the digestion of this material presents significant potential for minimizing greenhouse gas emissions. Through the co-digestion of other treatment-requiring materials, synergistic effects can be harnessed, leading to significant cost savings and enhanced emission reductions. Moreover, this approach presents the prospect of producing climate-neutral energy products, including upgraded biogas, as well as electricity and heat. By varying the utilization technologies for biogas (option 1: CHP + mobile heat storage; option 2: upgrading to fuel; option 3: upgrading + injection into gas grid) and the biogas substrates used (option 1: swine manure; option 2: swine manure + sludge; option 3: swine manure + organic waste; option 4: swine manure + sludge + organic waste), 12 sub-concepts were identified for Concept 2 (see Figure 6).

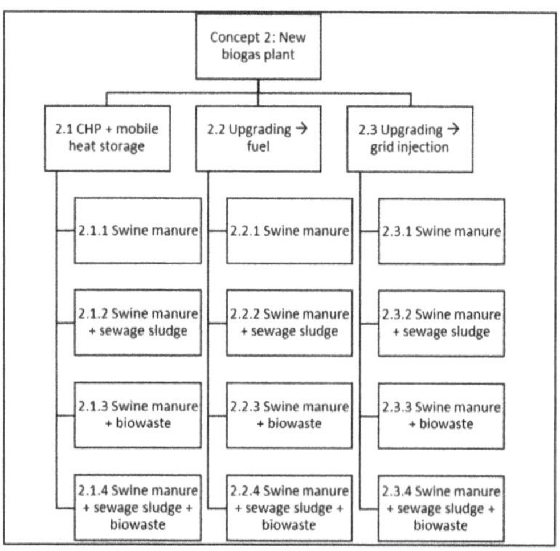

Figure 6. Schematic of sub-concepts for Concept 2.

3.3.3. Concept 3: Optimized Sludge Management

External sludge disposal constitutes a significant cost factor for the considered sectors. Achieving a more efficient local treatment of sludge is feasible by tapping into existing waste heat sources, such as the available biogas plant, or previously unexplored biomass potentials. Moreover, this approach opens up the possibility for producing high-quality biochar. Concept 3 introduces 12 sub-concepts by varying the sludge treatment technology (option 1: dewatering + drying; option 2: hydrothermal carbonization; option 3: pyrolysis), the heat provision (option 1: new biogas plant + CHP; option 2: existing biogas plant + CHP), the utilization of biochar from hydrothermal carbonization or pyrolysis (option 1: sludge disposal; option 2: soil conditioner; option 3: fuel), and the materials involved in hydrothermal carbonization (option 1: sludge; option 2: sludge + organic waste; option 3: sludge + organic waste + green waste) (see Figure 7).

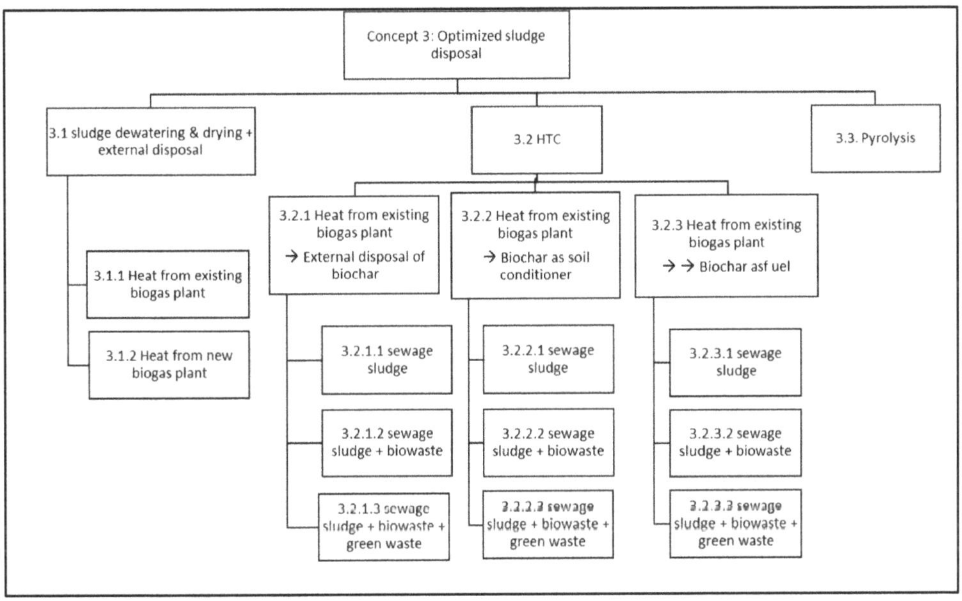

Figure 7. Schematic of sub-concepts for Concept 3.

3.4. *Results of Concept Evaluation*

Based on an Excel tool created for this case study, the necessary investment costs, discounted costs, and revenues for each utilization pathway, as well as greenhouse gas emissions and cumulative energy expenditures, were calculated on an annual basis for all the sub-concepts. These values were then summed for comparison within each main concept. By comparing them with the existing system, the required investment costs, greenhouse gas emission abatement potentials, greenhouse gas abatement costs, and energy savings were derived. The results are presented in a diagram (Figure 8). The diagram depicts the annual GHG abatement potential on the x-axis and the specific GHG abatement costs on the y-axis. The size of the circles corresponds to the investment costs required for the implementation of each sub-concept, and the color indicates their alignment with the main concepts. It is important to note that almost all the sub-concepts have negative GHG abatement costs, indicating that these sub-concepts are associated with financial savings or revenues.

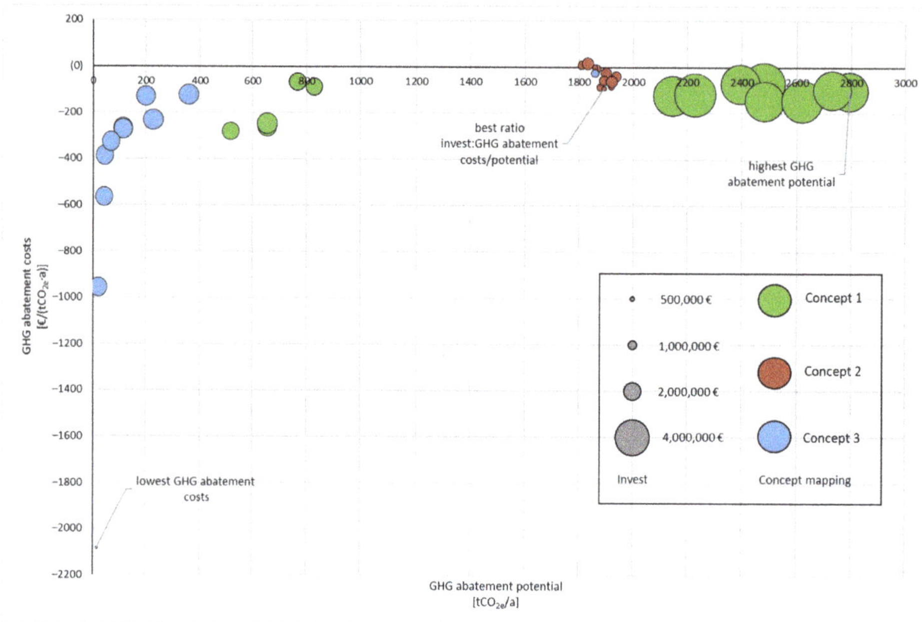

Figure 8. GHG abatement costs and potentials and investment costs of the evaluated concepts.

Depending on the prioritized objective, different concepts represent the optimum:

1. Establishing a district-heating network with the incineration of straw and green waste, along with the treatment of cattle manure, organic waste, and green waste, enables the highest greenhouse gas emission reductions. However, this option entails very high investments. As was the case before, sewage sludge would continue to be externally disposed without prior treatment, and swine manure would not be utilized in this sub-concept (see Figure 9).
2. The dewatering and drying of sewage sludge, using the excess heat from the existing biogas plant, offers the lowest GHG abatement costs, resulting in significant financial savings with manageable efforts. For this purpose, sewage sludge dewatering and drying facilities would be installed, which would receive the required heat from the existing biogas plant. Subsequently, the dried sewage sludge would continue to be disposed externally. The remaining material flows would be treated or disposed of in the same manner as in the existing system (see Figure 10).
3. The co-digestion of swine manure and sewage sludge presents the best ratio of investment costs to mitigation potential and mitigation costs. The generated biogas would be upgraded and subsequently distributed as fuel. The disposal routes for the remaining material flows remain the same in this sub-concept as in the existing system (see Figure 11).

A one-dimensional sensitivity analysis was conducted for the parameters including transportation costs, energy costs, interest rate, evaluation period, and price increase, as these parameters have a significant influence on the economic performance of the concepts aside from the investment costs. The remaining parameters that form the basis of the calculations were retained, since they are precisely known due to local conditions (e.g., distances and material flows) or can be considered secure based on thorough research (e.g., emissions), and no significant alterations are to be expected. The GHG abatement costs were used as the case study variable. In light of the significant uncertainties prevailing in recent years, particularly concerning costs, especially in the energy sector, the estimation of realistic parameter ranges proved to be a challenging task. Consequently, wide parameter ranges were chosen for the following parameters: interest rate, price increase, transportation costs, and energy costs. The volatile nature of these variables underscores the complexity of forecasting, making it difficult to establish well-founded parameter ranges.

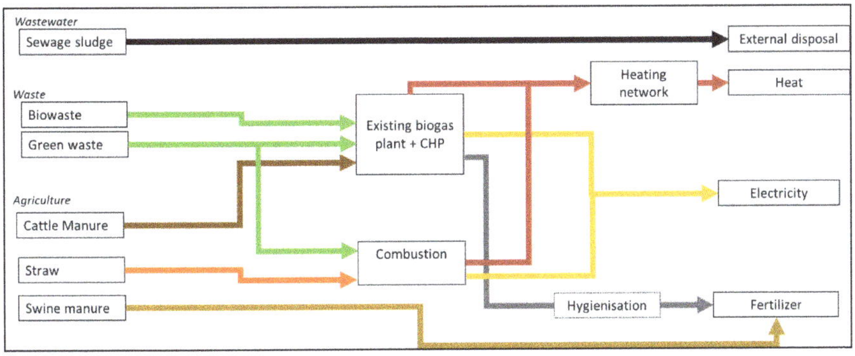

Figure 9. Schematic representation of the concept with the highest GHG reduction.

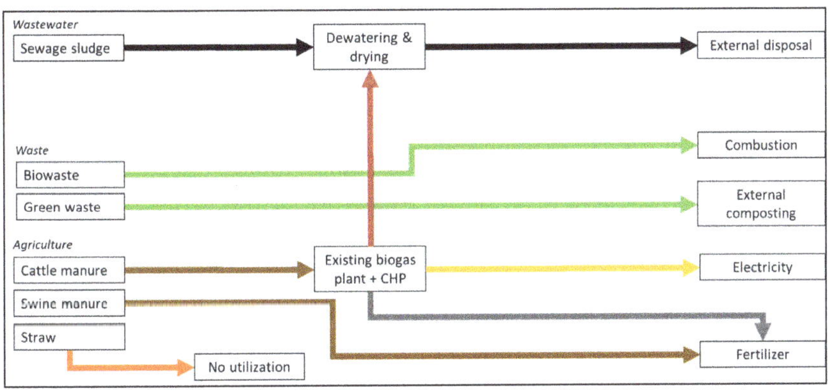

Figure 10. Schematic representation of the concept with the lowest GHG abatement costs.

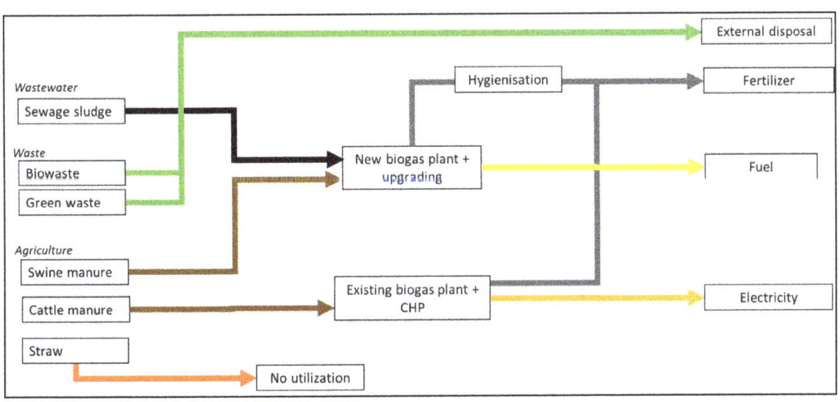

Figure 11. Schematic representation of the concept with the best ratio of investment costs to GHG abatement costs and potential.

The selection of the evaluation period typically aligns with the longest depreciation time of the system components under consideration. However, due to the extended operational lifetimes observed in certain technologies, such as heat networks, the evaluation period was expanded to up to 50 years. This expansion was motivated by the recognition that some technologies exhibit longer lifespans in practical applications than initially anticipated. Conversely, given the anticipated significant shift in the role of biomass within the future energy system, it is necessary to consider substantially shorter evaluation periods in this context. The ranges of the parameters used are depicted in Table 3.

Table 3. Parameters of sensitivity analysis.

Parameter	Default Value	Range
Interest rate	3%	0–10%
Evaluation period	30 yr	10–50 yr
Price increase	0%	0–10%
Transportation costs	EUR 0.58/(t·km)	EUR 0.20–1.50/(t·km)
Energy costs	100%	30–300% energy price limit

A rising interest rate, especially in concepts where depreciation costs constitute a significant portion of the annuities (concepts with HTC), leads to significantly higher abatement costs. The magnitude of the influence depends on the proportion of depreciation to annual costs, which, in turn, is determined by the ratio of investment costs to ongoing expenses. Overall, the interest rate exerts no significant influence on the ranking of sub-concepts, resulting in low sensitivity overall (see Appendix A, Figure A1).

The evaluation period also affects the results of some concepts. It is evident that, with a duration of 20 years and especially in concepts involving HTC, significant inflection points occur due to necessary reinvestments after the depreciation period. The distribution of concepts remains largely unchanged, indicating that the considered evaluation period does not have a decisive influence on the choice of concepts. With the exception of the aforementioned cases, the evaluation period does not lead to noteworthy changes in the ranking of concepts and is therefore considered not highly sensitive (see Appendix A, Figure A2).

Price increases primarily impact external sludge disposal, making it considerably more expensive. This leads to increased savings compared to the existing system in concepts where sludge is utilized. A price increase has a substantial impact on the ranking of the main concepts but only a minor effect on the ranking of sub-concepts, making it conditionally sensitive (see Appendix A, Figure A3).

Regarding energy costs, it is observed that an increase in these expenses makes concepts with high electricity production or heat generation more economical. However, energy costs do not have a significant impact on the overall results, thus also being conditionally sensitive (see Appendix A, Figure A4).

Transportation costs play a crucial role in concepts where sludge is utilized. In cases of external disposal, transportation constitutes a significant portion of the total costs. Therefore, a reduction in transportation expenses leads to greater savings compared to the existing system. The influence of transportation costs on the overall results is particularly pronounced. Transportation costs have substantial effects on Concept 3 and negligible impacts on other concepts, with no influence on the ranking of sub-concepts, making such costs conditionally sensitive as well (see Appendix A, Figure A5).

Overall, sensitivity analysis reveals that significant changes in the ranking of the main concepts only occur at the edges of the parameter ranges, and these mostly affect the sub-concepts within a given concept. Therefore, the examined parameters are classified as slightly sensitive for comparing the concepts with each other.

4. Discussion

This manuscript delineates a systematic approach for the comprehensive integration of biogenic resources across diverse sectors. Functioning as a nexus, this approach establishes a collaborative framework involving both private and public stakeholders within waste management, wastewater management, and agricultural domains, which are recognized as potential contributors to the biogenic resources essential for energy generation. This paper provides an exhaustive and foundational elucidation of the procedural steps a rural municipality should undertake to harness biogenic resources for energy production in rural environments.

Employing an iterative strategy, the methodology's development underwent recurrent cycles of inventory analysis, potential analysis, conceptualization, and concept evaluation. This iterative design unfolded in tandem with the active engagement of local stakeholders representing the pertinent sectors. Key figures, including the mayor, local energy manager, and representatives from industrial and civil sectors, participated in a collaborative fashion. This approach facilitated the development of a pragmatic methodology.

Crucially, the conceptualization phase remained an ongoing consultative process with local stakeholders to ensure that the continuous results aligned with real-world implementability. This proactive engagement sought to validate the viability and feasibility of the proposed methodology within the temporal constraints of practical application.

By focusing on the interconnected dynamics of waste management, wastewater management, and agriculture, this study selectively delves into a particular facet and a designated option space pertinent to energy transition. As a consequence, no deliberate juxtaposition or assessment is made vis à vis with alternative sustainable energy supply concepts. Nevertheless, it is imperative to recognize that the outlined methodology possesses the versatility to be incorporated into a more comprehensive and holistic perspective. Notably, industrial biomass (e.g., from breweries) or forestry biomass is not considered as in other approaches (see, e.g., [53,60]) and must be analyzed on a case-by-case basis. On the other hand, the energy potential of wastewater and its partial material flows are rarely integrated in biomass investigations combining more than two sectors (see, e.g., [61]). This study, therefore, provides a new type of added value for research into the energetic utilization of biogenic resources.

The scope of this study demarcates the spatial confines within which the analyses and conceptual frameworks were tailored. Consequently, the exploration overlooked potentially synergistic biogenic resources from neighboring areas that could amplify the efficacy of the investigated concepts. Importantly, the formulated concepts are elaborated for three distinct target settings within the study domains. Any extension of these investigations to alternative study areas necessitates a bespoke adaptation, as there exists no universal "recipe" applicable to all locales. Rather, the configuration of technology options (and their combinations) must be individually modeled to ascertain their efficacy in achieving specific objectives.

However, the developed methodology remains highly replicable in rural settings, with the exception of noted limitations and location-specific considerations. The used calculations and values were derived from the recognized technical rules of the sectors involved in this study and were verified with the specific stakeholders in the study area. Furthermore, methodological and technical focal points were discussed and defined jointly in workshops.

Implementing the developed methodology yielded reliable results within the study area. The three core concepts, along with their sub-concepts, which were collaboratively crafted with local stakeholders, contribute positively to both climate protection and local heat supply. This underscores the effectiveness of interconnecting waste management, wastewater management, and agriculture to generate pertinent energy solutions.

Notably, the obtained outcomes prompted immediate efforts from local stakeholders to realize Concept 3. However, discussions with local stakeholders and experts affirmed the desirability of a combined approach involving all three core concepts. The study area's generic structure for rural municipalities enhances the potential for widespread applicability. As the study area consists of 19 municipal districts, the findings also hold significance for inter-municipal cooperation in diverse local settings.

To advance the scientific evolution of the research findings, dynamic exploration is warranted. This entails various analyses, e.g., analyzing the impact of changing population figures on the achieved results. Additionally, there is a need for addressing unresolved issues surrounding legal implementation and financial management for local stakeholders. Given the ongoing transformation of the institutional framework for energy concepts, the role of biomass in local energy generation requires continuous scrutiny in future studies.

Applying the presented approach should be supported by adjusting policies at the local and national levels. At the local level, the paradigm of the cross-sectoral management of biogenic residues should be incorporated into municipal statutes and administrative guidelines to initialize a shift from predominant sectoral management solutions toward energy-optimized approaches based on combinations of biogenic residues of different sectors. Furthermore, to be able to implement each identified sub-concept, adjustments of national policies are needed. The current institutional framework is not aligned with the aim of the systematic usage of local residues for energy production and the implementation of corresponding concepts. For example, from an operator's perspective, it is almost impossible to secure the (long-term) procurement of local residues due to competitive tendering requirements under the law, which makes investing in approaches to utilize local residues considerably more difficult.

5. Conclusions

The transition toward sustainable energy systems requires a more efficient use of energy and the systematic coupling of different energy sources from various sectors. From a socio-economic perspective, a transition and future energy supply at the municipal level needs to be affordable, consumer-friendly, efficient, and environmentally friendly. In order to pursue energy transition, urban and rural municipalities are challenged with designing Communal Energy Plans describing their long-term perspective on heat supply. In comparison to renewable energy sources such as solar power, wind energy, and geothermal heat, the systematic coupling of biogenic residues from waste, agriculture, and wastewater sectors for energy production has barely been considered by community decision makers until now. Aiming to support the design of suitable energy systems for rural areas, this paper outlines a systematic approach consisting of four steps, starting from an in-depth analysis of local technical and biogenic aspects to the conceptualization and evaluation of potential energy concepts. The presented approach supports decision makers to systematically investigate the local energy potential of biogenic residues from waste, agriculture, and wastewater management and define the energy concepts suitable for the local conditions. While it can be assumed that most municipalities rely on other sources of energy to meet their energy demand, biogenic residues can be considered a relevant component of the transition toward locally based, diversified, and climate-friendly energy systems.

Author Contributions: Conceptualization, A.L., M.W. and M.P.; methodology, M.P.; formal analysis, M.P.; investigation, M.W., A.L. and M.P.; data curation, M.P.; writing—original draft preparation, M.P., A.L. and M.W.; writing—review and editing, E.K. and C.V.; visualization, M.P. and A.L.; supervision, E.K. and C.V.; project administration, E.K.; funding acquisition, A.L. and C.V. All authors have read and agreed to the published version of the manuscript.

Funding: The research for this paper was part of the project OLE (Organisation ländlicher Energiekonzepte/organizing rural energy concepts), which is financially supported by the German Federal Ministry of Education and Research (BMBF) within the measure "Kommunen Innovativ" (grant number: 033L229).

Institutional Review Board Statement: Not applicable.

Informed Consent Statement: Not applicable.

Data Availability Statement: No new data were created or analyzed in this study. Data sharing is not applicable to this article.

Conflicts of Interest: The authors declare no conflict of interest. The funders had no role in the design of the study; in the collection, analyses, or interpretation of data; in the writing of the manuscript; or in the decision to publish the results.

Appendix A

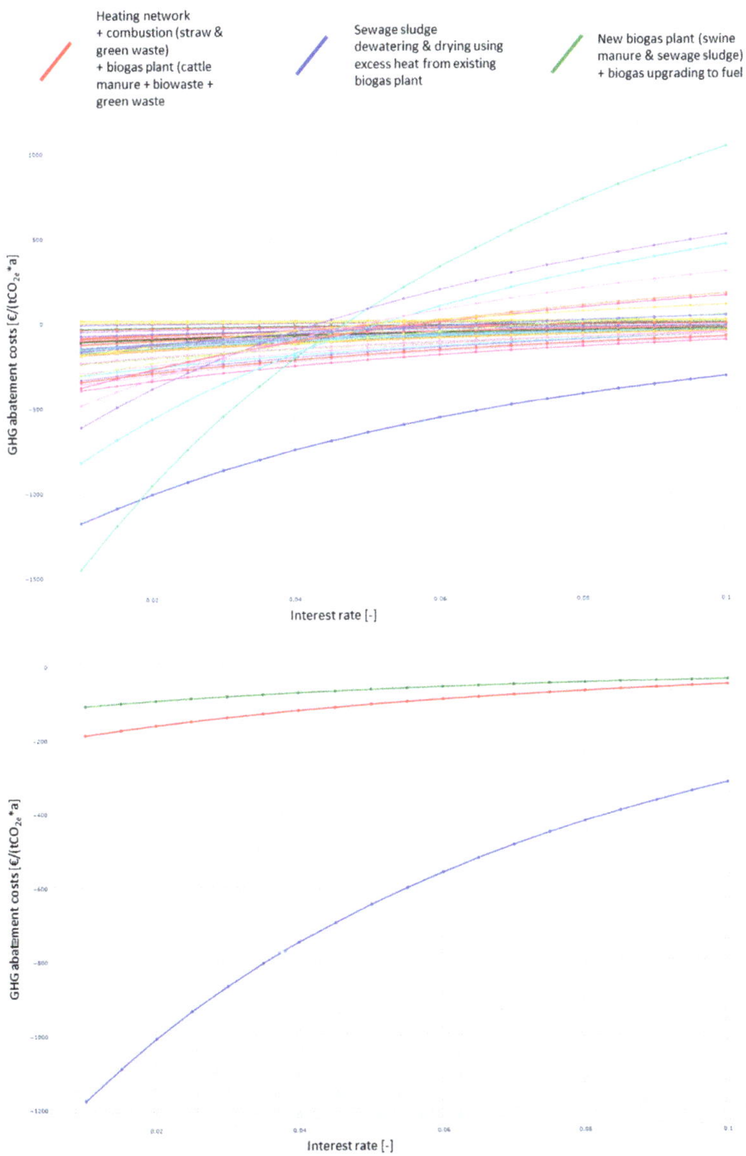

Figure A1. Interest rate in the sensitivity analysis.

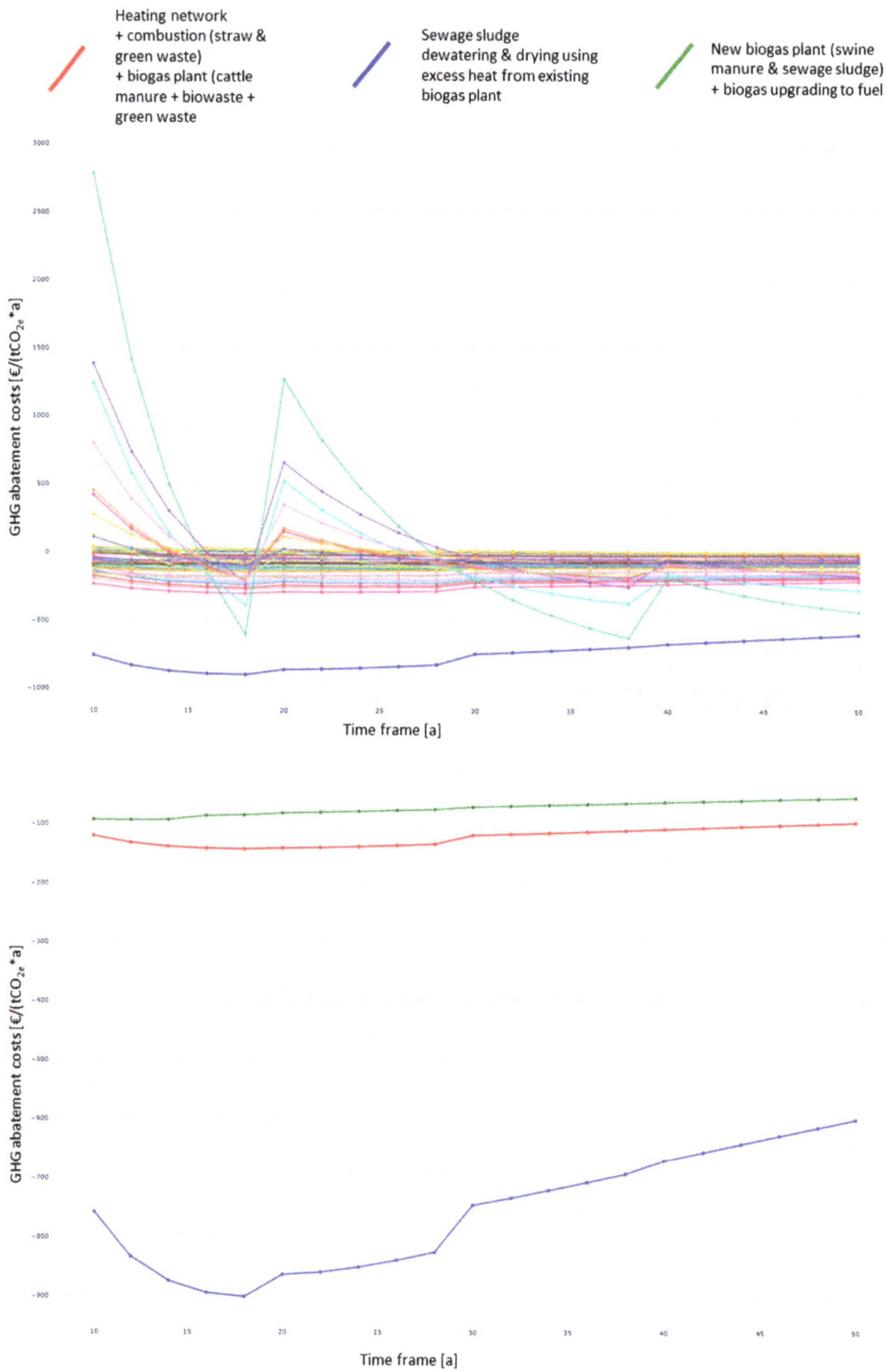

Figure A2. Time frame of the sensitivity analysis.

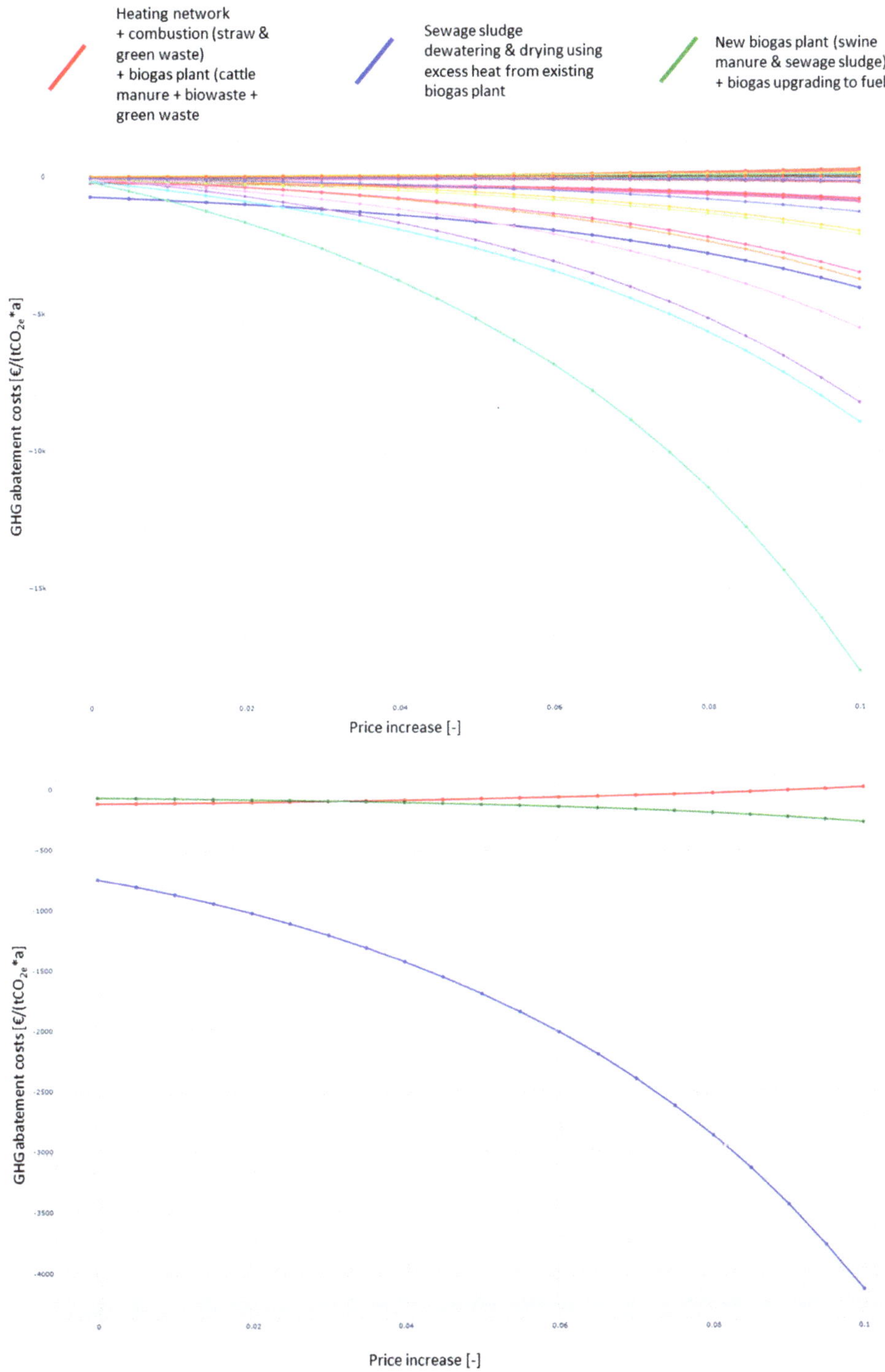

Figure A3. Price increase in the sensitivity analysis.

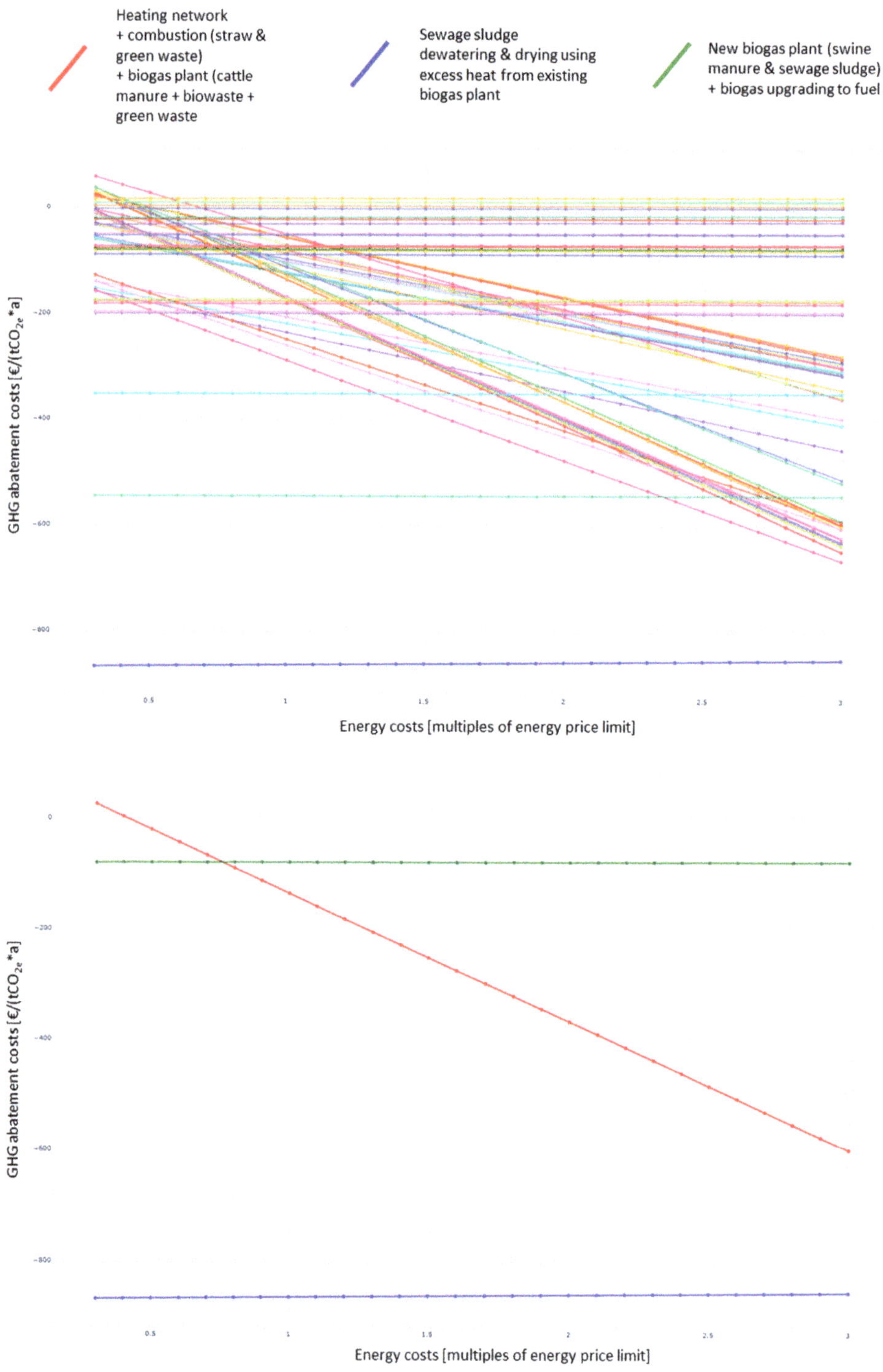

Figure A4. Energy costs in the sensitivity analysis.

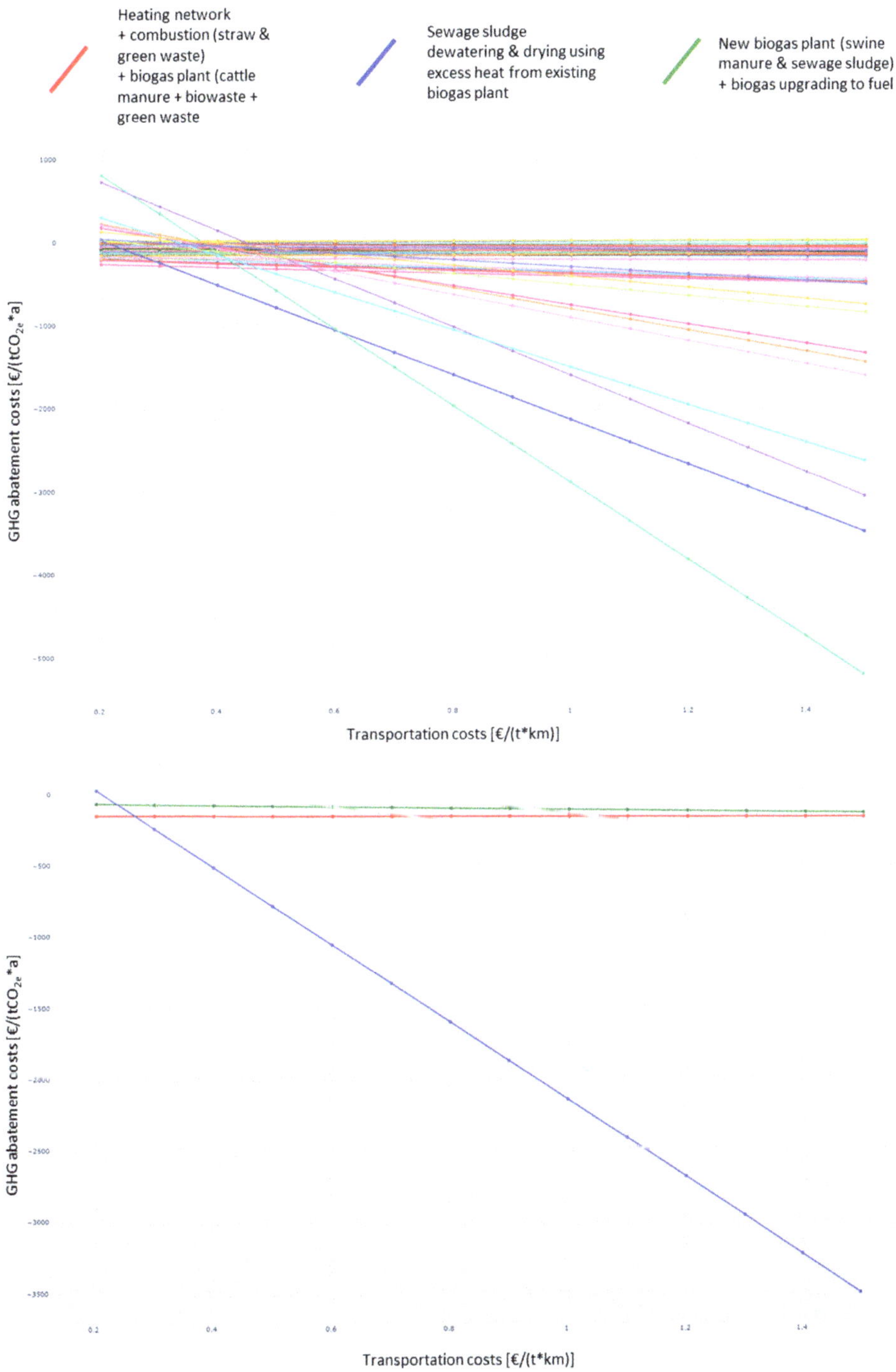

Figure A5. Transportation costs in the sensitivity analysis.

References

1. United Nations. *Transforming Our World: The 2030 Agenda for Sustainable Development*; United Nations: New York, NY, USA, 2015.
2. Bundesministerium für Bildung und Forschung-BMBF Innovation & Strukturwandel, Strom und Wärme Durch Wasserstoff. 2021. Available online: https://www.innovation-strukturwandel.de/strukturwandel/de/report/_documents/artikel/r-t/strom-und-waerme-durch-wasserstoff.html (accessed on 27 November 2023).
3. Kurz, R.; Brun, K. Gas Turbine Performance. In *Asia Turbomachinery & Pump Symposium*; Turbomachinery Laboratories, Texas A&M Engineering Experiment Station: College Station, TX, USA, 2016. [CrossRef]
4. Shourangiz-Haghighi, A.; Diazd, M.; Zhang, Y.; Li, J.; Yuan, Y.; Faraji, R.; Ding, L.; Guerrero, J.M. Developing More Efficient Wind Turbines: A Survey of Control Challenges and Opportunities. *IEEE Ind. Electron. Mag.* **2020**, *14*, 53–64. [CrossRef]
5. Florescu, A.; Barabas, S.; Dobrescu, T. Research on Increasing the Performance of Wind Power Plants for Sustainable Development. *Sustainability* **2019**, *11*, 1266. [CrossRef]
6. Parthiban, R.; Ponnambalam, P. An Enhancement of the Solar Panel Efficiency: A Comprehensive Review. *Front. Energy Res.* **2022**, *10*, 937155. [CrossRef]
7. Dhilipan, J.; Vijayalakshmi, N.; Shanmugam, D.B.; Jai Ganesh, R.; Kodeeswaran, S.; Muralidharan, S. Performance and efficiency of different types of solar cell material—A review. *Mater. Today Proc.* **2022**, *66*, 1295–1302. [CrossRef]
8. Energy Institute. *Statistical Review of World Energy*; Energy Institute: London, UK, 2023.
9. United Nations. *The Sustainable Development Goals Report 2023*; United Nations: New York, NY, USA, 2023.
10. Lachal, B. *Energy Transition*; ISTE Ltd.: London, UK; John Wiley and Sons Inc.: London, UK, 2019.
11. Cajot, S.; Peter, M.; Bahu, J.-M.; Guignet, F.; Koch, A.; Maréchal, F. Obstacles in energy planning at the urban scale. *Sustain. Cities Soc.* **2017**, *30*, 223–236. [CrossRef]
12. Sperling, K.; Hvelplund, F.; Mathiesen, B.V. Centralisation and decentralisation in strategic municipal energy planning in Denmark. *Energy Policy* **2011**, *39*, 1338–1351. [CrossRef]
13. Brugmann, J. Planning for sustainability at the local government level. *Environ. Impact Assess. Rev.* **1996**, *16*, 363–379. [CrossRef]
14. Hoppe, T.; Miedema, M. A Governance Approach to Regional Energy Transition: Meaning, Conceptualization and Practice. *Sustainability* **2020**, *12*, 915. [CrossRef]
15. Poggi, F.; Firmino, A.; Amado, M. Planning renewable energy in rural areas: Impacts on occupation and land use. *Energy* **2018**, *155*, 630–640. [CrossRef]
16. Marsden, T. Exploring the Rural Eco-Economy: Beyond Neoliberalism. *Sociol. Rural.* **2016**, *56*, 597–615. [CrossRef]
17. Wagner, O.; Thomas, S. Wärmewende darf nicht zu Lasten der Armen gehen. Die Wärmewende durch kommunale Wärmeplanung sozial gestalten. *Kommunale* **2023**, *20*, 12–13.
18. Energieeffizienzverband für Wärme, Kälte und KWK e.V.; Deutscher Verein des Gas-und Wasserfaches e.V. Praxisleitfaden: Kommunale Wärmeplanung, Frankfurt/Bonn, Germany. 2023. Available online: https://www.dvgw.de/medien/dvgw/leistungen/publikationen/leitfaden-kommunale-waermeplanung-dvgw-agfw.pdf (accessed on 5 November 2023).
19. Eurostat. *Eurostat Regional Yearbook 2019*; Publications Office of the European Union: Luxembourg, 2019.
20. Naumann, M.; Rudolph, D. Conceptualizing rural energy transitions: Energizing rural studies, ruralizing energy research. *J. Rural. Stud.* **2020**, *73*, 97–104. [CrossRef]
21. Woods, M. Rural geography III. *Prog. Hum. Geogr.* **2012**, *36*, 125–134. [CrossRef]
22. Markantoni, M.; Woolvin, M. The role of rural communities in the transition to a low-carbon Scotland: A review. *Local Environ.* **2015**, *20*, 202–219. [CrossRef]
23. Gregg, J.S.; Smith, S.J. Global and regional potential for bioenergy from agricultural and forestry residue biomass. *Mitig. Adapt. Strateg. Glob. Change* **2010**, *15*, 241–262. [CrossRef]
24. Singh, A.K.; Pal, P.; Rathore, S.S.; Sahoo, U.K.; Sarangi, P.K.; Prus, P.; Dziekański, P. Sustainable Utilization of Biowaste Resources for Biogas Production to Meet Rural Bioenergy Requirements. *Energies* **2023**, *16*, 5409. [CrossRef]
25. Lueck, A.; Haupt, T.; Kraft, E.; Londong, J. Energetische und Stoffliche Potentiale der Synergetischen Verwertung von Abwasser und Abfall in Thüringen. In Proceedings of the DWA Landesverbandstagung Sachsen/Thüringen, Leipzig, Germany, 9 June 2019.
26. Kalcher, J.; Naegeli de Torres, F.; Gareis, E.; Cyffka, K.-F.; Brosowski, A. *Dashboard Biogene Rohstoffe in Deutschland*; Open Agrar Repositorium: Greifswald, Germany, 2022.
27. Ramirez Camargo, L.; Stoeglehner, G. Spatiotemporal modelling for integrated spatial and energy planning. *Energy Sustain. Soc.* **2018**, *8*, 32. [CrossRef]
28. Müller, M.; Kubli, M.; Ulli-Beer, S. Interdisciplinary Modeling of Energy Transition in Rural and Urban Systems. In Proceedings of the International Conference CISBAT 2015, Lausanne, Switzerland, 9–11 September 2015.
29. Naumann, M.; Reichert-Schick, A. Ländliche Infrastrukturen–Risiken, Anpassungserfordernisse und Handlungsoptionen. *Raumforsch. Raumordn.* **2015**, *73*, 1–3. [CrossRef]
30. Thüringer Ministerium für Infrastruktur und Landwirtschaft. *Demografiebericht_2023: Teil 1: Bevölkerungsentwicklung des Freistaats Thüringen und Seiner Regionen, Erfurt, Germany*. 2023. Available online: https://www.serviceagentur-demografie.de/service/publikationen/demografieberichte/ (accessed on 10 November 2023).
31. © OpenStreetMap Contributors. 2023. Available online: www.openstreetmap.org/copyright (accessed on 29 November 2023).

32. Gemeinde Am Ettersberg. Herzlich Willkommen: Leben Am Ettersberg, das Bedeutet Echte Lebensqualität. Available online: https://www.am-ettersberg.de/ (accessed on 10 November 2023).
33. Bundesministerium für Wohnen, Stadtentwicklung und Bauwesen. Kommunale Wärmeplanung: Für eine Deutschlandweit Zukunftsfeste und Bezahlbare Wärmeversorgung. Available online: https://www.bmwsb.bund.de/Webs/BMWSB/DE/themen/stadt-wohnen/WPG/WPG-node.html (accessed on 10 November 2023).
34. Thüringer Ministerium für Umwelt, Energie und Naturschutz. *Beseitigen von Kommunalem Abwasser in Thüringen: Berichtszeitraum 2021/2022, Erfurt, Germany*. 2023. Available online: https://umwelt.thueringen.de/fileadmin/001_TMUEN/Unsere_Themen/Boden_Wasser_Luft_Laerm/Abwasser/Lageberichte_Abwasserentsorgung/Thueringer_Lagebericht_Kommunalawasser_2023_01.pdf (accessed on 5 November 2023).
35. Wolf, M.; Londong, J. Transformation der Siedlungswasserwirtschaft-Steuerungsmechanismen im Diskurs Ressourcenorientierter Systemansätze am Beispiel von Thüringen. *Raumforsch. Raumordn.* **2020**, *78*, 397–411. [CrossRef]
36. Thüringer Landesamt für Landwirtschaft und Ländlichen Raum. *Jahresbericht 2021/22, Jena, Germany*. 2023. Available online: https://tlllr.thueringen.de/fileadmin/TLLLR/Service/Publikationen/Broschueren/2023_JB21.pdf (accessed on 10 November 2023).
37. Thüringer Landesanstalt für Landwirtschaft. *Integration der Biogaserzeugung in die Landwirtschaft in Thüringen: Abschlussbericht*; Thüringer Landesanstalt für Landwirtschaft: Jena, Germany, 2017.
38. Thüringer Landesamt für Landwirtschaft und Ländlichen Raum. Biogas. Available online: https://tlllr.thueringen.de/landwirtschaft/pflanzenproduktion/nawaro/biogas (accessed on 8 November 2023).
39. Schluck, T.; Hangartner, D.; Facchinetti, E.; Sulzer-Worlitschek, S.; Mennel, S.; Sulzer, M. The Uettligen Case—A Step Towards a Role Model in Energy Concept Development. *Energ. Proc.* **2017**, *122*, 1069–1074. [CrossRef]
40. Schluck, T.; Sulzer, M. Towards a Holistic Methodology: A Practical Approach to Local Energy Planning. In Proceedings of the World Sustainable Built Environment Conference 2017, Hong Kong, China, 5–7 June 2017.
41. Berger, K.; Hauer, S.; Peters-Anders, J.; Schmidt, R.-R.; Anna Shadrina, A.; Widl, E.E. *Digitale Energieplanung & Optimierung urbaner Regionen-Herausforderungen, Best-Practice-Beispiele & Handlungsempfehlungen*; Austrian Institute of Technology: Vienna, Austria, 2020.
42. Antoni, O.; Kluge, C.; Knies, J.; Balling, V.; Benz, S.; Schneller, A.; Wegner, N. *Handlungsempfehlungen für ein Planungsmodell der kommunalen Wärmeplanung auf Grundlage Kommunaler Erfahrungswerte und Dessen Rechtlicher Implementierung*; Hochschule Bremen: Bremen, Germany, 2022.
43. Drebes, C.; Meinberg, T.; Schulze, J.; Sieber, S.; Ousmane, S.; Hegger, M.; Dettmar, J. *UrbanReNet II-Weiterentwicklung des Bilanzmodells der Phase I zu einem Quatiersbezogenen Vernetzungsmodell*; Technische-Universität Darmstadt: Darmstadt, Germany, 2015.
44. Caemmerer-Seibel, U.; Kiesel, G.; Voelker, C.; Cebulla, D. Ein online-tool zur Erprobung bezahlbarer Handlungsoptionen für Energieeffizienz im ländlichen Raum. In Proceedings of the BauSIM 2020-8th Conference of IBPSA Germany and Austria, Graz, Austria, 23–25 September 2020.
45. Blesl, M.; Stehle, M.; Brodecki, L.; Groß, P.; Grassl, G.; Reiser, S.; Karajan, J.; Christ, A.; Müller, E. *Systemanalyse für die Städtische Energieplanung mit einem Modularen Planungsinstrument-Methodische Grundlagen und Fallbeispiele*; Endbericht; Universität Stuttgart: Stuttgart, Germany, 2019.
46. Zhang, J.; Qin, Q.; Li, G.; Tseng, C.-H. Sustainable municipal waste management strategies through life cycle assessment method: A review. *J. Environ. Manag.* **2021**, *287*, 112238. [CrossRef]
47. Sarigiannis, D.A.; Handakas, E.J.; Karakitsios, S.P.; Gotti, A. Life cycle assessment of municipal waste management options. *Environ. Res.* **2021**, *193*, 110307. [CrossRef]
48. Hörnschemeyer, B.; Söfker-Rieniets, A.; Niesten, J.; Arendt, R.; Kleckers, J.; Stretz, C.; Klemm, C.; Budde, J.; Wagner, R.; Vonhoegen, L.; et al. *Leitfaden RessourcenPlan–Teil 1: Konzeption RessourcenPlan. Ergebnisse des Projekts R2Q RessourcenPlan im Quartier*; Fachhochschule Münster: Münster, Germany, 2023.
49. Caemmerer-Seibel, U.; Lück, A.; Osman, A.; Kiesel, G.; Völker, C.; Alfen, H.W.; Cebulla, D. Energy Transition in Rural Areas–Supporting Local Energy Planning by the Development of an Online-Tool for Identification and Promotion of Energy-Efficiency and the Use of Renewables. In Proceedings of the Building Simulation 2019-16th Conference of IBPSA, Rome, Italy, 2–4 September 2019.
50. Statistische Ämter des Bundes und der Länder. ZENSUS 2011. Available online: https://www.zensus2011.de (accessed on 10 November 2023).
51. Thüringer Landesamt für Vermessung. Thüringer ALKIS®-Objektartenkatalog: Auf Basis der GeoInfoDok Version 6.0.1. Available online: https://www.thueringen.de/mam/th9/tlvermgeo/landesamt/thuralkis-ok.pdf (accessed on 7 January 2022).
52. Gesellschaft für Energieplanung und Systemplanung mbH. *Verbrauchskennwerte 2015*; Gesellschaft für Energieplanung und Systemplanung mbH: Münster, Germany, 2015.
53. Yang, L.; Wang, X.-C.; Dai, M.; Chen, B.; Qiao, Y.; Deng, H.; Zhang, D.; Zhang, Y.; Villas Bôas de Almeida, C.M.; Chiu, A.S.; et al. Shifting from fossil-based economy to bio-based economy: Status quo, challenges, and prospects. *Energy* **2021**, *228*, 120533. [CrossRef]
54. German Association for Water Management, Sewage and Waste e. v. *Leitlinien zur Durchführung Dynamischer Kostenvergleichsrechnungen (KVR-Leitlinien)*; German Association for Water Management, Sewage and Waste e.v.: Hennef, Germany, 2012.

55. Thüringer Landesamt für Umwelt und Geologie. *Ergebnisse der Hausmüllanalysen der Öffentlich-Rechtlichen Entsorgungsträger im Zeitraum 2011–2017: Kompostierbare Abfälle*; Thüringer Landesamt für Umwelt und Geologie: Jena, Germany, 2018.
56. Thüringer Landesamt für Statistik. *Energiebilanz Thüringen 2020*; Thüringer Landesamt für Statistik: Erfurt, Germany, 2022.
57. Kaltschmitt, M.; Hartmann, H.; Hofbauer, H. *Energie aus Biomasse*, 3rd ed.; Springer: Berlin/Heidelberg, Germany, 2016.
58. Bulach, W.; Dehoust, G.; Möck, A.; Oetjen-Dehne, R.; Kaiser, F.; Radermacher, J.; Lichtl, M. *Ermittlung von Kriterien für Hochwertige Anderweitige Verwertungsmöglichkeiten von Bioabfällen*; Umweltbundesamt: Dessau-Rosslau, Germany, 2020.
59. Kern, M.; Raussen, T.; Funda, K.; Lootsma, A.; Hofmann, H. *Aufwand und Nutzen einer Optimierten Bioabfallverwertung Hinsichtlich Energieeffizienz, Klima-und Ressourcenschutz*; Umweltbundesamt: Dessau-Roßlau, Germany, 2010.
60. Long, H.; Li, X.; Wang, H.; Jia, J. Biomass resources and their bioenergy potential estimation: A review. *Renew. Sustain. Energy Rev.* **2013**, *26*, 344–352. [CrossRef]
61. Iqbal, A.; Ekama, G.A.; Zan, F.; Liu, F.X.; Chui, H.-K.; Chen, G.-H. Potential for co-disposal and treatment of food waste with sewage: A plant-wide steady-state model evaluation. *Water Res.* **2020**, *184*, 116175. [CrossRef]

sustainability

MDPI

Article

Comparison of the Techno-Economic and Environmental Assessment of Hydrodynamic Cavitation and Mechanical Stirring Reactors for the Production of Sustainable *Hevea brasiliensis* Ethyl Ester

Olusegun David Samuel [1,2], Peter A. Aigba [1], Thien Khanh Tran [3,4,*], H. Fayaz [5], Carlo Pastore [6], Oguzhan Der [7], Ali Erçetin [8], Christopher C. Enweremadu [2] and Ahmad Mustafa [9,10]

1 Department of Mechanical Engineering, Federal University of Petroleum Resources, P.M.B 1221, Effurun 330102, Delta State, Nigeria; samuel.david@fupre.edu.ng (O.D.S.); paaigba@yahoo.com (P.A.A.)
2 Department of Mechanical Engineering, University of South Africa, Science Campus, Private Bag X6, Florida 1709, South Africa; enwercc@unisa.ac.za
3 Advanced Applied Sciences Research Group, Dong Nai Technology University, Bien Hoa City 76100, Vietnam
4 Faculty of Technology, Dong Nai Technology University, Bien Hoa City 76100, Vietnam
5 Power Generation Unit, Institute of Power Engineering (IPE), Universiti Tenaga Nasional, Kajang 43000, Malaysia; katperfayaz@ieee.org
6 Water Research Institute (IRSA), National Research Council (CNR), via F. de Blasio 5, 70132 Bari, Italy; carlo.pastore@ba.irsa.cnr.it
7 Department of Maritime Vehicles Management Engineering, Maritime Faculty, Bandırma Onyedi Eylul University, 10200 Bandırma, Turkey; oder@bandirma.edu.tr
8 Department of Naval Architecture and Marine Engineering, Maritime Faculty, Bandırma Onyedi Eylul University, 10200 Bandırma, Turkey; aercetin@bandirma.edu.tr
9 Faculty of Engineering, October University for Modern Sciences and Arts (MSA), Giza 12451, Egypt; ammhamed@msa.edu.eg
10 Center of Excellence, October University for Modern Sciences and Arts (MSA), Giza 12451, Egypt
* Correspondence: tranthienkhanh@dntu.edu.vn

Abstract: Even though the hydrodynamic cavitation reactor (HCR) performs better than the mechanical stirring reactor (MSR) at producing biodiesel, and the ethylic process of biodiesel production is entirely bio-based and environmentally friendly, non-homogeneous ethanol with the triglyceride of underutilized oil, despite the many technical advantages, has discouraged the biodiesel industry and stakeholders from producing ethylic biodiesel in HCRs. This study examines the generation of biodiesel from rubber seed oil (RSO) by comparing the ethyl-based HCR and MSR. Despite ethyl's technical advantages and environmental friendliness, a lack of scalable protocols for various feedstocks hinders its global adoption. The research employs Aspen HYSYS simulations to investigate the ethanolysis process for RSO in both HCRs and MSRs. The HCR proves more productive, converting 99.01% of RSO compared to the MSR's 94.85%. The HCR's exergetic efficiency is 89.56% vs. the MSR's 54.92%, with significantly lower energy usage. Removing catalytic and glycerin purification stages impacts both processes, with HC showing lower exergy destruction. Economic analysis reveals the HCR's lower investment cost and higher net present value (USD 57.2 million) and return on investment (176%) compared to the MSR's. The HCR also has a much smaller carbon footprint, emitting 7.2 t CO_2 eq./year, while the MSR emits 172 t CO_2 eq./year. This study provides database information for quickly scaling up the production of ethanolic biodiesel from non-edible and third-generation feedstocks in the HCR and MSR.

Keywords: biodiesel; exergy; transesterification; Aspen HYSYS; reactor technology; ethyl ester; sustainability; mechanical stirring; hydrodynamic cavitation; techno-economics

1. Introduction

Increasing the production and use of biofuels, such as biodiesel, in the transportation industry is now thought to be a workable way to cut back on the use of fossil fuels and the pollution that they cause. The deployment of biofuels as a sustainable substitute for fossil fuels has gained greater recognition during the last few decades. Biodiesel (BD), one of the most renowned biofuels, has tremendous potential to replace the scarce and non-renewable resources of fossil fuels [1]. Research on biodiesel is being driven by worldwide policies that fit with the implementation of Sustainable Development Goal 7 (SDG 7) [2]. These policies highlight the mitigation of climate change, renewable energy, and environmental sustainability as priorities. Improving farming practices, advocating for clean energy, and reducing carbon emissions are the main goals. Biodiesel technology research and development are being actively supported by international agreements and efforts. Sustainable production practices, waste reduction, and the utilization of a variety of feedstocks are all emphasized by these initiatives. Important international accords, such as the United Nations Framework Convention on Climate Change (UNFCCC) Paris Agreement, in conjunction with regional partnerships, are actively promoting sustainable energy alternatives [3]. Under SDG 2030, regulations encouraging biodiesel research are being implemented more quickly due to the urgent need to mitigate climate change.

The demand for BD adoption and its use in diesel engines could stem from its reduced emission profiles, non-toxicity, carbon neutrality, and renewability [1]. Despite these benefits, biodiesel's higher charge than diesel fuel has persisted. A higher percentage of biodiesel production costs emanate from edible oils [4]. Hence, the deployment of inedible and non-competitive generational oils (NCGOs) for BD production has been extensively investigated [5]. Yang et al. [6] hinted that the most common inedible and NCGO oily feedstock are jatropha, jojoba, mahua, moringa, Tung oil, camelina, castor oil, croton oil, milk bush, algae, neem, tobacco, RSO, etc. RSO is a preferred choice for BD over the other stated oils due to its similar oil qualities to regular diesel. Commercial BDs are synthesized using methylic and ethylic methods (ME and ER, respectively), as reported by Samuel et al. [7] and Yusuff et al. [8]. Samuel et al. [9] and Altamirano et al. [10] stated that ER will surpass ME due to its eco-friendliness, reduced emission profiles, and renewability. Mandari and Devarai [11] attributed the immiscibility of inedible and NCGOs during ER to a notable transesterification and a reduction in mass transfer rate. To enhance mass transfer and increase the contact surface area between reactants during BD manufacturing, MSRs have been used on an industrial scale [12,13]. However, the MSR cannot adequately mix, and long reaction times, high energy consumption, excess molar ratios, and excessive catalyst doses constrain its effectiveness [14]. Most industrial biodiesel facilities use conventional batch reactors, but they have significant capital, operating, and reaction time costs and unneeded operational and production expenses [15,16].

Various technologies are employed in the production of biodiesel. Conventional reactors require longer reaction times, and using an extra-large reactor adds unnecessary costs to production and operation [16]. Several technologies have been developed to address the difficulty outlined above. The most prevalent types are hydrodynamic, acoustic, optical, and particle cavitation [17]. HCRs have recently emerged as a promising, extremely inventive technique for steadying the production of biodiesel [18]. The peculiar superiority of the HCR is due to its capacity to be better set up and effective in mixing non-miscible fluids compared to prolonged, unadventurous mechanical procedures [7] The technique, known as hydro cavitation, occurs when a fluid passes through an opening and a pressure drop causes cavities to form appropriately. This method yields higher yields and shorter reaction times. The most cutting-edge technology for creating a cleaner biodiesel synthesis process is expected to be the HCR, ensuring a cleaner fuel production. Regarding conversion, alcohol-to-oil molar ratios, catalyst dose, reaction duration, and oily feedstock consumption, Maddikeri et al. [19], Chuah et al. [20], and Laosuttiwong et al. [21] proposed that HCRs might be more effective than MSRs. However, the techno-economic feasibility and environmental impact of HCR technology are not well understood by biofuel and processing factories that employ it to produce ethyl ester biodiesel. The long-term success of biodiesel depends heavily on these variables.

Recent studies on the production of methyl biodiesel from feedstock in most developing nations utilizing HCRs and MSRs indicate that the technology has a promising future [22]. This result is consistent with previous studies on the production of biodiesel from RSO. A review of the literature indicates that the earlier studies on this area of study did not include basic process engineering studies, such as process simulation, scale-up studies, process design, and techno-economic studies of HCR and MSR-based ethylic biodiesel, which can increase the viability of RSO biodiesel production and commercialization. The technical and financial performance of a chemical process or product design can be assessed using the techno-economic analysis (TEA) technique. Process simulation and economic models use TEA to assess the feasibility and sustainability of new or emerging technologies [23]. It also considers a technology's worldwide cost–profit value, which helps prospective investors to make important financial decisions [24]. TEAs are useful not only for evaluating the overall economic viability before deployment but also for scaling up small-scale BD facilities.

The techno-economic elements of producing biodiesel from different oils have been analyzed and upgraded through the use of commercial process simulators such as Aspen Plus®, HYSYS®, and Superpro [24,25]. By incorporating these simulators with the Aspen capital cost estimator, researchers and stakeholders may define material, mass, and energy balances and build an extensive cost database. HCRs are preferred over conventional reactors and other intensification techniques, including tube, microwave, ultrasonic, spinning disk, and supercritical conditions. HCRs are desired due to their superior mixing capabilities, reduced reaction duration, and ability to enhance fuel properties [7]. The TEA of biodiesels from oily feedstocks, particularly in HCRs, has a research gap that is highlighted in Appendix Table A1. The TEA of methylic BD catalyzed by lipase and a sonicator has received less attention than methylic-based HCRs in previous investigations. Notably, there is a lack of TEA studies on second-generation oily feedstocks like rubber seed oil and ethanol for transesterification. Furthermore, no study has been conducted on the TEA, environmental assessment, or comparison with conventional reactors of ethylic BD produced by an HCR. To assure 100% bio-based scaling up and economically viable biofuel industries, the following measures were implemented in response to the lack of such studies in the literature: (i) modelling a cavitation system and a cleaner rubber seed oil ethyl ester based on a conventional reactor, (ii) developing an in-depth technical, financial, and environmental analysis contrasting the well-established MSR, and (iii) comparing the long-established MSR [13] with the suggested HCR [26] in depth on technological, financial, and environmental fronts.

2. Materials and Methods

Process Description and Assumption

The present study utilized simulation models to conduct a detailed comparative analysis of the biodiesel synthesis processes of MSRs and HCRs. The simulation process generated a large amount of design data, including mass balance and energy, in addition to other thermodynamic operating data. The transesterification reaction between RSO and ethanol was carried out with potassium hydroxide acting as a catalyst. RSO is represented by triolein while the biodiesel product is represented by ethyl oleate ($C_{19}H_{36}O_2$). The downstream processes comprises catalyst neutralization and unreacted ethanol removal (only for MS process). As glycerin and biodiesel are hydrophilic and hydrophobic, water washing was used as a separation technique. After applying phosphoric acid to the acquired potassium hydroxide phase to neutralize its pH, the sodium phosphate that was left behind was isolated. The principal reaction outputs were thought to be glycerin and biodiesel. Vacuum distillation was used to remove the unreacted ethanol (only in MS-based processes) and return it to the mixture along with fresh ethanol as fuel.

Conversely, the HC process scenario is a case study inquiry that employs HCR, utilizing the same quantitative features and process parameters as the MS process. For the first time, the HC system was modelled in this study as a pump, throttling valve, and plug flow reactor (PFR) [26]. The hydrodynamic cavitation phenomenon is caused by mechanical constriction brought on by a throttling valve (VLV-100) that creates a pressure differential large enough to result in the creation of vapour bubbles or cavities. High energy is released when the pressure drops from 4 bar to 0.2 bar below the reaction mixture's vapour pressure (0.3 bar). This causes vapour cavities to collapse. By overcoming the immiscibility of the oil and ethanol phases and improving the physical and chemical transformation mixture, this energy increases mass transfer, which in turn increases reaction rate and decreases residence time. With the exception of the tower requirement, the remaining steps of the process are essentially similar to those of the MS process. Figure 1 depicts the entire reaction process.

The following are the assumptions on the thermodynamic characteristics of ethyl ester biodiesel:

- The atmospheric temperature and pressure are fixed at 25 °C and 1 bar, respectively.
- Pressure drops in pipelines and heat exchangers are disregarded.
- The RSO is modelled using triolein, with an assumed conversion rate of 95% [27].
- Every process is assumed to be in a steady-state, adiabatic condition.
- Any changes to the fluid's kinetic and potential energy are regarded as negligible.
- Pumps and compressors are supposed to have an adiabatic efficiency of 85%, the distillation tray tower to have an efficiency of 65%, and the electrical generator (motor) to have an efficiency of 96%.
- The transesterification process operates continuously in a steady-state mode.
- The MS process consists of a transesterification unit, glycerine purification unit, and biodiesel distillation unit. The glycerine purification unit involves washing the biodiesel with water to remove glycerine, potassium hydroxide removal through phosphoric acid addition, and the subsequent removal of potassium phosphate.
- A glycerine purification unit and a transesterification unit make up the HC process.
- Ethanol recovery is considered only in the MS process, while it is disregarded in the HC process due to the high conversion rate of over 99%.

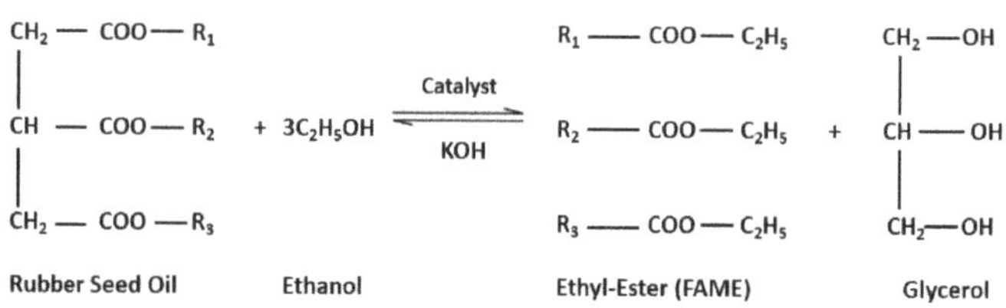

Figure 1. Transesterification reaction process [28].

3. Modelling and Analysis

Aspen HYSYS, v10, which is widely recognized for its broad applicability in biofuels and other similar emerging feedstock research, was used to model the transesterification processes for biodiesel generation using the MSR and HCR techniques [29]. This simulation was approached through certain fundamental steps such as transesterification, washing with water or neutralization or catalytic removal, ethanol recovery, and biodiesel purification. A nonrandom two-liquid (NRTL) model was employed in Aspen HYSYS as the thermodynamic fluid package since the reactants, ethanol and glycerol, are polar substances. The alcohol-to-oil molar ratio is adjusted to 3:1 [30]. In order to prevent the production of ethanol vapour, RSO is injected into the preheated continuously stirred tank reactor (CSTR) at a temperature of 60 °C and a pressure of 4 bar as triolein (feedstock) in the MS process.

The conversion reactor, represented by the CSTR in this model, is necessary for handling thermodynamic and kinetic data, including the activation energy (E) and Arrhenius constant (A), which are produced by RSO during research. RSO ethyl esters (biodiesel) and glycerol are produced at this stage when the RSO reacts with ethanol in the presence of a homogeneous catalyst, specifically KOH. This stream enters a separation column (Splitter-1) where the content is cleaned with water to help to separate the alcohol and ethyl oleate mixture at the top of the separator into a distillation tower, where the salt is processed into the catalytic reactor at the bottom and the ethanol is recycled back into the process to produce biodiesel. However, the salt is further processed in a catalytic reaction vessel (Cata-RXR) where H_3PO_4 is encountered to neutralize the content and release water vapour at the top and transfer the recovered mixture from the bottom into the Splitter-2 separator, where glycerol and K_3PO_4 are recovered at the top and bottom, respectively. Figure 2 depicts the distinct process streams.

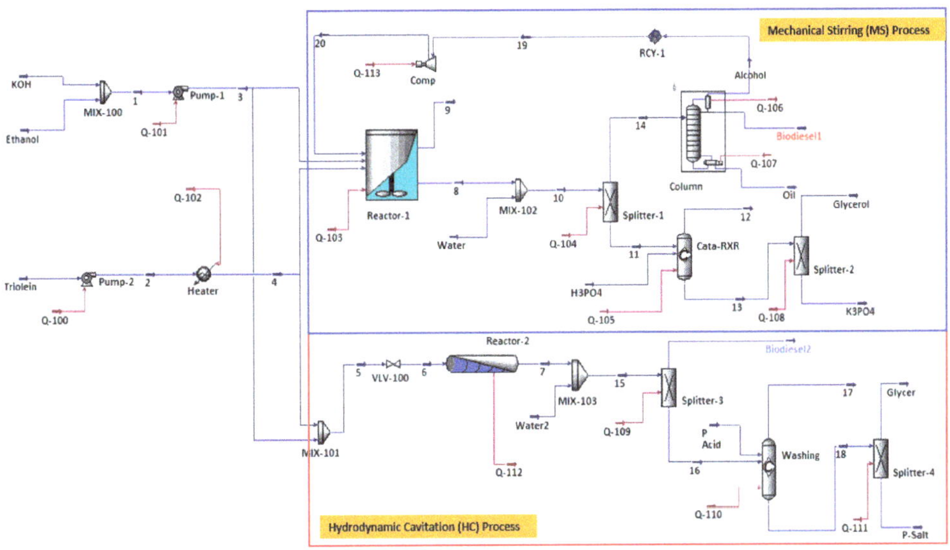

Figure 2. Combined MS and HC same-source processes feed streams.

3.1. Energy Analysis

The Arrhenius equation predicts that a small increase in reaction temperature will result in a significant increase in the reaction rate constant magnitude, k, defined by Equation (1). This prediction is based on the kinetic data of the transesterification reaction obtained from the literature [31].

$$k = A * exp\left(\frac{-E_a}{RT}\right) \quad (1)$$

where A is the pre-exponential factor, E_a is the activation energy, R is the ideal gas constant 8.314 (J/Kmol), and T is the absolute temperature. The reaction rate constant and the activation energy are considered as 0.063 min^{-1} and 29,800 kJ/kg. mol, respectively [31]. As discussed, when a pressure differential is established across the upstream valve, which is utilized to enhance the cavitation phenomena in the HC process, care must be taken to prevent any jet erosion occurrence. Equation (2) is utilized to determine the cavitation number, σ, which is crucial for forecasting cavitation and its possible impacts, or its effectiveness, in the HC reaction process [32,33] from Equation (2)

$$\sigma = \frac{P_f - P_v}{\frac{1}{2}\rho V^2} \quad (2)$$

where P_f *and* P_v (pa) are the outlet and vapour pressure of the reaction mixture, the density of the reaction mixture is ρ, (kg/m^3), and V (m/s) is the flow velocity through the constriction section, which can be evaluated by knowing the upstream flow rate and diameter of the constriction hole. A higher conversion yield is produced when the cavitation number is decreased because the reactive mixture stays in the cavitation zone for a longer period of time. With the exception of preheating the feed before introducing it to the HCR, the energy created is added to the reaction mixture. This highlights the HC process's intriguing potential for energy savings over the MS process. Furthermore, Equation (3) can be used to evaluate the energy released, E, in the HC reactor [34].

$$\dot{E} = Q_F * \Delta P \quad (3)$$

where Q_F is the volumetric flow of the liquid reaction mixture (m^3/s) and ΔP is the pressure change (kpa) through the constriction.

3.2. Exergy Analysis

The mass, energy, and energy balance general expressions of the rubber seed oil biodiesel plant at a control volume (CV) are estimated in this section. In other words, mass balance (continuity) is expressed by Equation (4), and any energy changes over the boundary are considered to be negligible.

$$\sum \dot{m}_{in} = \sum \dot{m}_{Out} \quad (4)$$

To evaluate the energy balance of the streams, Aspen Plus was used. Energy is induced in the system by the mass flow rate, work output, and energy released into the surrounding environment. Equations (5)–(7) can therefore be utilized to give energy balance in control volume at the steady state.

Equation (8) expresses the entropy balance produced by the same process, while Equation (7) is obtained by applying the first law of thermodynamics at a control volume (CV).

$$\dot{E}_{in} - \dot{E}_{Out} = \frac{dE_{system}}{dt} = 0 \Longrightarrow steady \quad (5)$$

$$\sum \dot{E}_{in} = \sum \dot{E}_{Out} \quad (6)$$

$$\sum Q - \sum \dot{W} = \sum \dot{E}_{Out} - \sum \dot{E}_{in} \quad (7)$$

$$\sum \dot{m}_{in} s_{in} + \sum \frac{Q_{CV}}{T} + \dot{S}_{gen} = \sum \dot{m}_{Out} s_{Out} \quad (8)$$

Energy losses, energy destruction, component improvement potentials, and energy efficiencies are the main ideas examined from an exergy fundamental point of view. Equation (9) is used to compute the exergy balance for each component under steady-state conditions [35].

Equation (9), in contrast to energy, states that exergy is not irreversibly destructible within a control container. In order to create a thermodynamic process flowsheet that will aid in the various destroyed exergy calculations of the system's components, the biodiesel synthesis process is simulated in this work. To define exergy destruction, Equation (9) is modified as shown in Equation (10).

Equations (12) and (13) describe, respectively, the exergetic work performed and the heat flow across the control volume of each component.

Equation (14) can be employed for expressing flow exergy. Equation (14) transforms into Equation (15) due to the negligible kinetic and potential components of flow exergy. Additionally, Equation (16) represents the physical exergy component of flow exergy.

$$Ex_{in} - Ex_{Out} - Ex_D = \frac{dE_{system}}{dt} = 0 \Longrightarrow steady \quad (9)$$

$$Ex_D = Ex_{in} - Ex_{Out} \quad (10)$$

Or, more broadly, as

$$Ex_D = \sum \dot{m}_{in} ex_{in} + \sum Ex_{Q,\, in} + \sum Ex_{W,\, in} - \left(\sum \dot{m}_{out} ex_{out} + \sum Ex_{Q,\, out} + \sum Ex_{W,\, out}\right) \quad (11)$$

where $\dot{m}$, $\dot{E}$, Ex, Ex_W, and Ex_Q represent the inlet and outlet flow of mass, energy, exergy, work, and power transfer rates, respectively.

$$Ex_W = W \quad (12)$$

$$Ex_Q = \left(1 - \frac{T_o}{T}\right) Q \quad (13)$$

$$e_x = Ex_{KE} + Ex_{PE} + Ex_{CHM} + Ex_{PHY} \quad (14)$$

$$ex_{flow} = ex_{CHM} + ex_{PHY} \quad (15)$$

$$ex_{PHY} = (h - h_o) - T_o(s - s_o) \quad (16)$$

$$Ex_{CHM} = \sum_i y_i \overline{ex}_i^{CHM} + \overline{R} T_o \sum_i y_i ln(y_i) \quad (17)$$

where y_i is the component i molar fraction in the gaseous mixture gas, $\overline{ex}_i^{CHM}$ is the standard exergy of the constituent, h and s are the enthalpy and entropy generated, and R is the universal gas constant.

Equations (18) and (19) present the ratio of the chemical exergy of fuel to the LHV. The chemical exergy of fuel can also be expressed in terms of the LHV of solid or liquid fuel [36,37].

The chemical exergy of a stream is computed using Equation (20), and it can also be computed as flow, as illustrated in Equation (21).

Equation (22) calculates the ratio of energy intake to energy destruction in a component, which is known as energy fuel depletion. Additionally, the rate of irreversibility is shown as a ratio in Equation (23). Equation (24), however, defines the components' potential for improvement.

The system's overall exergy efficiency and its component exergy efficiencies can be determined using Equations (25) and (26).

For operational component assessment, Tsatsaronis and Lazzaretto [38] proposed and established a fuel and product exergy model. As shown in Table 1, the transesterification and neutralization processes were assessed in this study using the strategy by Boyano et al. [39].

$$\varphi = \frac{ex_F^{CH}}{\text{LHV}} \quad (18)$$

$$\varphi = 1.0401 + 0.1728\left(\frac{\text{h}}{\text{c}}\right) + 0.0432\left(\frac{\text{o}}{\text{c}}\right) + 0.2169\left(\frac{\text{s}}{\text{c}}\right) * \left(1 - 2.0628\frac{\text{h}}{\text{c}}\right) \quad (19)$$

where c, h, s, o, and φ are the mass fractions of carbon, hydrogen, sulfur, and oxygen, respectively [37, 40,41].

$$Ex_{CHM} = \dot{m}_i ex^{CHM} \quad (20)$$

$$Ex_{flow}^{CH} = \Delta G_f + \sum_i ex_i N_i \quad (21)$$

where ΔG_f is the standard Gibbs free energy of the formation of the reactants, ex_i is the ith pure elemental component chemical exergy of the substance, and N_i is the ith pure elemental molar fraction of the substance.

Table 1. Biodiesel process exergy balance equations.

Processes	Techniques	Exergy Balance Equation
Transesterification	MS	$Ex_D = (Ex_3 + Ex_4 + Ex_{20} + Ex_Q) - Ex_8 + Ex_9$
	HC	$Ex_D = Ex_6 - (Ex_{Q-112P} + Ex_7)$
Neutralization (catalyst removal)	MS	$Ex_D = (Ex_{11} + Ex_{H_3PO_4} + Ex_Q) - Ex_{12} + Ex_{13}$
	HC	$Ex_D = Ex_{16} + Ex_{PACID} + Ex_{Q-110P} - (Ex_{17} + Ex_{18})$
Ethanol recovery	MS	$Ex_D = (Ex_{14} - Ex_{19}) + Ex_{Q-107p} - ((Ex_{oil} - Ex_{14}) + Ex_{Q-106p} + Ex_{biodiel1})$
Glycerol purification	MS	$Ex_D = Ex_{13} + Ex_{Q-108p} - Ex_{K3PO4} + Ex_{GLYCEROL}$
	HC	$Ex_D = Ex_{18} + Ex_{Q-111p} - Ex_{PSALT} + Ex_{GLYCER}$
SPLITTER-1	MS	$Ex_D = Ex_{10} + Ex_{Q-104p} - Ex_{12} + Ex_{13}$
MIX-100	MS/HC	$Ex_D = (Ex_{KOH} + Ex_{ETHANOL}) - Ex_1$
MIX-102	MS	$Ex_D = (Ex_8 + Ex_{WATER}) - Ex_{10}$
PUMP-1	MS/HC	$Ex_D = Ex_1 + Ex_{Q-101P} - Ex_3$
COMPRESSOR	MS	$Ex_D = Ex_{19} + Ex_{Q-113P} - Ex_{20}$
VLV-100	HC	$Ex_D = Ex_5 - Ex_6$
MIX-101	HC	$Ex_D = (Ex_3 + Ex_4) - Ex_5$
MIX-103	HC	$Ex_D = (Ex_7 + Ex_{WATER-2}) - Ex_{15}$
HEATER	MS/HC	$Ex_D = (Ex_2 + Ex_{Q-102P}) - Ex_4$
PUMP-2	MC/HC	$Ex_D = Ex_{TRIOLEIN} + Ex_{Q-100P} - Ex_2$

Equations (22)–(24) provide the relative ratios of the component's exergetic destruction to its total exergy destruction and component improvement in the plant [35]. Equations (25) and (26) provide the energy and energy efficiency, respectively.

$$y_D = \frac{Ex_D}{Ex_{in,tot}} \tag{22}$$

$$y^*_{D,i} = \frac{Ex_{D,i}}{Ex_{D,tot}} \tag{23}$$

$$IP_i = \left(1 - \frac{\eta_{ex}}{100}\right) Ex_{D,i} \tag{24}$$

$$\text{Energy efficiency, } \eta_{\dot{E}} = \frac{Energy\ Output}{Energy\ Input} \tag{25}$$

$$\text{Exergy efficiency, } \psi_{Ex} = \frac{product\ Exergy}{Fuel\ exergy} = 1 - \frac{Ex_D}{Fuel\ exergy} \tag{26}$$

4. Economic Model Analysis

In order to evaluate the project's viability, the sources of income, project expenses, and cash flow analysis are all determined using the economic models in this study. By providing a variable summary that may be utilized as a selection tool for the most suitable and economical (MS vs. HC) process for investment objectives, this techno-economic comparison study helps the biodiesel community. Due to a capacity factor of 90.4% for the biodiesel plant, the economic assessment projected 7920 operating hours annually.

4.1. Summary of Total Capital Investment

The importance of accurately computing the total capital investment for this fictitious project by adding the necessary fixed capital investment, working capital, and initial expenditure is examined [42]. Therefore, it is crucial to consider the long-term total spending investment in light of how it affects the plant's planning [43]. The necessity of avoiding unrealistic budgets highlights the importance of precise costing, which includes engineering, inside and outside battery limits (ISBLs and OSBLs), and contingency expenses for various production-related accessories and equipment. To calculate the ISBL cost, a variety of estimating techniques can be applied, including those by Bridgewater, Taylor, Gore, Stallworthy, Klumpar, Brown, and Fromme [44]. Bridgewater's approach was used for the MS and HC plants. Accurate results from Bridgewater's methodologies depend on knowing the reactor's conversion rate, plant capacity, and number of main units. For the HC and MS plants, the productivity of the plants and the conversion rates were found to be 224 and 207 tons, respectively, based on Aspen estimates. Lastly, Bridgewater's method was used to compute the ISBL cost via Equation (27).

The OSBL cost encompasses off-site developments for plant operation and normally ranges from 10% to 100% [24]. However, a 30% OSBL proportion was applied in this work. Since the company's research team already had the required package, engineering fees were not included. To offset unforeseen costs, a contingency sum of 10% was set out for MS technology and 15% for HC technology. A total of 15% of the direct capital cost was set aside as working capital [43]. To sum up, 10% of the combined costs of OSBL and ISBL were allocated to start-up costs. Table 2 presents other economic hypotheses used in the present study.

$$C = 280,000N\left(\frac{Q}{s}\right)^{0.3} \tag{27}$$

where C represents the ISBL capital cost (USD), N is the number of main units, Q is the plant capacity (t/year), and s is the reactor conversion rate.

$$\text{Total Project Cost} = \text{Direct Costs} + \text{Indirect Costs} \tag{28}$$

Table 2. Economic assumptions and evaluation parameters [24,43].

Economic Assumptions	Parameters
Cost of equity	25%
Cost of debt	5%
Cost of capital	15%
Debt ratio	0.5
Discount rate	11%
Tax rate	22.50%
Depreciation method	Straight-line
Depreciation period	10 years
Depreciation rate	10%
1st year direct fixed capital (DFC)	30%
2nd year direct fixed capital (DFC)	70%
Project life	15 years

4.2. Operating Cost Expenses

Costs associated with both fixed and variable production are included in operating expenses. Whatever the project's efficiency, there are always fixed costs: labour, overhead, maintenance, insurance, taxes, rent, and environmental fees [24]. On the contrary, variable costs are contingent upon production rate and output. These expenses include raw materials (such as ethanol, RSO, and potassium hydroxide), utilities (such as energy, water for heating and cooling, and transportation), packaging, and disposal of waste (such potassium phosphate). Overall variable costs can be decreased by making effective use of resources, such as reducing raw material losses and energy consumption [35]. Because they are so expensive, raw materials make up a large amount of the cost of production. The labour cost estimate for the factory was USD 64,512, taking into account six (6) operators working eight-hour shifts each day for 48 weeks at a rate of USD 20 per hour, with three operators per operator and one supervisor (at a labour cost of 25% overall).

4.3. Economic Viability Indicators

Sales of the investment's products, including both the primary output and any byproducts, generate income. In this instance, the created glycerin was judged to be worth credits. Gross margin, which is determined by deducting the cost of raw materials from product sales revenues, is another crucial metric for assessing economic viability. Beyond production costs, the retained revenues are revealed by the gross margin. The cash cost of production (CCOP) is subtracted from biodiesel revenues to determine profit. This profit is known as the gross profit, from which the net profit is calculated by subtracting the corporate tax, which was calculated in this study to be 22.5%. Equation (29), when applied, yields the tax amount.

As indicated by Equation (29), the simple payback is computed by dividing the fixed investment by the average annual cash flow, which should only comprise revenue-generating years 1 through 15.

As long as it is included in the range of revenue-generating years, working capital is not included in the average cash flow calculation.

$$\text{Tax Paid} = \text{Taxable Income} \times \text{Tax Rate} \tag{29}$$

where taxable income is Total Income—Deductions—Exemptions

$$Pay - back\ time = \frac{Total\ Investment}{Average\ annual\ cash\ flow} \tag{30}$$

Depreciation charges are computed using the decreasing balance depreciation method in situations where cash flow is predominant. In this research, a depreciation rate of 10% was considered throughout a project lifetime of 15 years, with a 5-year project recovery phase. The return on investment (ROI) is determined by deducting expenditures from income, and the resultant amount is known as the net profit. Both the HC and MS procedures have a 15-year recovery period, which reflects the project's economic viability [24]. However, an 11% discount rate was proposed in this study. The economic statistic known as the net current value (NPV) estimates the difference between the current values of cash inflows and outflows [45]. The annualization of an interest rate is used to account for the time worth of money. Equation (31) is used to compute the NPV.

$$NPV = \sum_{n} \frac{C_t}{(1+r)^n} - C_o \tag{31}$$

where C_t and C_o are the net cash inflow and initial investment costs, respectively, n = projects lifespan, and r = discount rate

In other words, a high-percentage ROI indicates a comparable investment's returns relative to its costs. The return on investment, or ROI, is computed as the ratio between net income and investment. Thus, it is a metric for evaluating the effectiveness of investments and may be applied to compare the relative performance efficiencies of many expenditures. Equation (32) provides an additional definition for ROI.

$$ROI = \frac{C_{NET}}{n + C_o} \tag{32}$$

where C_{NET} and C_o are the cumulative net profit and initial investment costs, respectively, and n is the plant life.

4.4. Environmental Analysis

Environmental effects, especially those related to greenhouse gas emissions and carbon footprint, must be taken into account for the new procedure under consideration. The carbon footprint, commonly expressed as $kgCO_2$ eq, is a measure of the amount of greenhouse gases released per unit of product. During the production of biodiesel, energy consumption contributes significantly to greenhouse gas emissions. The MS method employs energy for preheating, ethanol recovery, and biodiesel distillation, including the utilization of a high-duty steam boiler, whereas the HC process mainly relies on energy for raw material preheating. This study examined the production of steam for heating and the consumption of electricity as the two primary sources of CO_2 emissions. Equation (33) illustrates the specific equation that was used to determine the CO_2 emissions from energy usage.

$$Q_E CO_2 = CW_{el} * SW_{el} \tag{33}$$

CW_{el} represents electricity consumption and SW_{el} denotes the electricity supplier identified emission factor (given as 438.64 g CO_2 eq/kWh [45].

The emission resulting from the creation of steam can be calculated using Equation (34), assuming that the boiler (steam generator) is powered by natural gas.

$$T_E CO_2 = Q_F * \dot{E}_S * F_{CO_2} \tag{34}$$

where Q_F is the fuel quantity, $\dot{E}_S$ is energy per unit mass associated with steam production (0.0471 GJ/kg), and F_{CO_2} is the carbon equivalent per unit energy emitted (0.05582 t CO_2/GJ [46].

5. Results and Discussion

5.1. Production Process of *Hevea brasiliensis* Ethyl Ester

Figure 2 depicts an integrated production plant for producing MS and HC based biodiesel, demonstrating how the reactants are treated equally in both processes with respect to pressure, temperature, composition, flow rates, and molar ratio before reaching the transesterification reactor step. Reaction stoichiometry is used to maintain the alcohol to oil molar ratio at 3:1.

Hevea brasiliensis ethyl ester from rubber trees produces excellent biodiesel. It avoids food competition, is perennial, and produces a great amount of latex. In addition, it is climate-adaptable, has a lower carbon intensity, and contributes to emission reduction. The HCR and the MSR systems produce biodiesel at a mass flow rate of 9325 kg/h and 8625 kg/h, respectively, with the former reaching a purity that is about 5% higher than the latter. This is because the HCR delivers higher product purity and conversion than the MSR. Similar findings were reported by Gholami et al. [22], where the HCR produced biodiesel with a 9.6% greater purity than the MSR.

Biodiesel is primarily combined with residues and traces from unconverted RSO in the Splitter-1 overhead of the MS process. This mixed stream is created in a vacuum distillation column to yield pure biodiesel (>99%). The HC technique eliminates the need for a column, saving energy and money. This 10-stage column has a reflux ratio of 2, with vacuum pressures of 0.2 and 0.1 bar in the reboiler and condenser, respectively. To separate the hydrophobic and hydrophilic phases that occur during the emulsion of glycerol and biodiesel, water washing is utilized. Potassium-hydroxide-containing waste streams
(11 and 16 for MS and HC, respectively) are neutralized and transformed into potassium phosphate. The HC reactor uses cavitation processes to produce high-intensity cavitation, whereas the CSTR uses a mechanical agitator that consumes a large amount of energy (88.24 kW). The HC reactor uses cavitation processes to produce a high-intensity cavitation of 0.313 [22], whereas the CSTR requires a mechanical agitator that consumes a large amount of energy (88.24 kW). Consequently, the contact surface area increased, resulting in greater productivity and conversion as compared to the stirred tank reactor. This phenomenal increase in contact surface area results in a reduction in the mass transfer resistance and almost full conversion in a brief residence time. Thus, the HC process outperformed the MS stirred tank reactor process in terms of conversion and productivity.

5.2. Energy and Exergy Discussion

Mass, energy, and energy flow analysis provide a way to evaluate operations and are invaluable tools for determining material losses, waste, energy loss, and irreversibility. Assessing a system's energy and energy efficiency helps to pinpoint areas that need to be improved. Exergy is a useful metric for gauging material reactions and energy quality, as well as for identifying renewable energy sources and processes. Mass, energy, and exergy are evaluated for a variety of chemical components, including mixes and utility systems, as part of the analysis. The study includes the overall process exergy, which is calculated by summing the input/output chemical and physical exergy of the constituents in the process streams (see Table 3). Standard values are used for other substances such as phosphoric acid, potassium phosphate, potassium hydroxide, and ethanol. The Gibbs free energy of the formation and chemical exergy of the glycerides, fatty acids, and biodiesel are evaluated by adopting an assumed average molecular weight for the acids and ethyl carbon chains in the reaction process. Appendix Table 2 gives the findings of the thermodynamic study for the streams used in the production processes of biodiesel in the MSR and HCR. Tables 3 and 4 present the exergetic destruction, exergetic fuel depletion, exergetic efficiency, and exergy improvement potential of each component of the two processes, respectively. These can be seen as percentage irreversibility ratios, y * D, and exergetic destruction.

The ease of installation, easy of scaling up, and simple configuration of the HC process (see Appendix Figures 1 and 2) provide a clear advantage over the MS process in the comparison of the two process scenarios. There was a 99.01% conversion rate set for both processes in the model configuration and conditions. The HC system, however, offers higher conversion rates than the MS method, as confirmed by the sensitivity analysis, which is a component of the simulation findings. The emulsion effect created in the HC reactor, which improves the contact between the oil and alcohol phases, may be responsible for the HC process's higher conversion rate. As shown in Figure 2, the MS design has 94.85% recovery and 88.24 kW energy consumption at the same reactor volume of 15 m^3 for both processes, whereas the HC reactor achieves a conversion rate of 99.01% and uses only 2.274 kW.

Table 3. Exergetic data of streams for the HC process.

Component	Fuel Exergy Ex_{FUEL} [MW]	Product Exergy $Ex_{PROD.}$ MW]	Destroyed Exergy $Ex_{DESTROYED}$ [MW]	Exergy Destruction [%]	Exergetic Fuel Depletion Ratio, y_D	Irreversibility Ratio, y*D [%]	Exergy Efficiency [%]	Improvement Potential [MW]
Pump-1	0.00	0.00	0.00	99.56	0.00	0.08	0.44	0.00
Pump-2	0.00	0.00	0.00	99.99	0.00	0.38	0.01	0.00
Heater	97.10	97.10	0.00	0.00	0.00	0.05	100.00	0.00
Splitter-3	108.41	108.41	0.00	0.00	0.00	0.02	100.00	0.00
Splitter-4	4.76	4.53	0.23	5.09	0.21	92.48	95.15	0.01
Reactor-2	108.42	108.40	0.02	0.01	0.01	6.19	99.99	0.00
Washing-RXR	4.53	4.53	0.00	0.03	0.00	0.50	99.97	0.00
Valve	108.42	108.42	0.00	0.00	0.00	0.30	100.00	0.00
Total	108.42	97.10	0.25	0.23	0.23	100.00	89.56	0.03

Table 4. Exergetic data of streams for the MS process.

Component	Fuel Exergy Ex_{FUEL} [MW]	Product Exergy $Ex_{PROD.}$ [MW]	Destroyed Exergy $Ex_{DESTROYED}$ [MW]	Exergy Destruction [%]	Exergetic Fuel Depletion Ratio, y_D	Irreversibility Ratio, y*D [%]	Exergy Efficiency [%]	Improvement Potential [MW]
Pump-1	0.00	0.00	0.00	99.56	0.00	0.01	0.44	0.00
Pump-2	0.00	0.00	0.00	99.99	0.00	0.05	0.01	0.00
Comp	0.01	0.01	0.00	7.60	0.00	0.06	92.40	0.00
Heater	97.10	97.10	0.00	0.00	0.00	0.01	100.00	0.00
Splitter-1	109.69	109.69	0.00	0.00	0.00	0.00	100.00	0.00
Splitter-2	5.44	4.76	0.68	12.42	0.31	38.05	87.58	0.08
Reactor-1	109.76	109.68	0.08	0.08	0.04	4.72	99.92	0.00
Catalyst-RXR	5.44	4.53	0.91	16.66	0.42	51.05	83.34	0.15
Column	106.86	106.76	0.11	0.10	0.05	6.04	99.90	0.00
Total	216.64	118.98	1.77	0.82	0.82	100.00	54.92	0.80

Figure 3 portrays that the volumetric deviation drops when the conversion rates of the MS and HC processes increase and decrease, respectively. In other words, the PFR of the HC process benefits from a lower volumetric reactor than the CSTR of the MS process. Furthermore, Chuah et al. [47] used an orifice plate with 21 holes of 1 mm diameter to produce a cavitation number of 0.3 in contrast to this investigation. In comparison to mechanical stirring for the methyl ester synthesis process, their results demonstrated an eight-fold increase in yield efficiency and a six-fold reduction in reaction time. The designed HC table system in an ethyl ester production system was validated in our investigation with a cavitation number of 0.313 achieved with a throttling valve and a PFR. Using the same reactants, Ahmad et al. [48] obtained a 98% conversion rate; however, the HC reactor in our investigation achieved a high conversion rate of 99.01%, negating the requirement for a distillation column to separate the generated oil and ethanol from the biodiesel. Lowering downstream units lowers production and capital expenses. Similar to this, Gholami et al. [22] used a constructed cavitation chamber with a cavitation value range of 0.23 to 0.64 through several cavitation zones to achieve a conversion rate of 99.9%.

On the other hand, as seen in Tables 4 and 5, the exergy balances were carried out to evaluate the material and energy-saving potential of the HC and MS processes and compare their performance. Exergetic efficiency and exergy destruction are quantified in Tables 4 and 5, respectively, by analyzing the exergy flow for the HC and MS processes in order to identify waste streams. Oil accounts for the bulk of input exergy (82.92%) in the MS process, whereas biodiesel yields the largest exergy output (80.8%). Waste streams total 12.6% and include unreacted oil (6.6%), ethanol–water–biodiesel mixture, and ethanol–water mixture. Similar to the MS, the HC process shows oil as the primary exergy input (82.92%), but no significant waste streams are observed compared to the MS process. Overall, the processes delivered a mass flow rate of 9325 kg/h and 8629 kg/h of biodiesel from the HC and MS systems, respectively. After all equipment irreversibility rates were assessed, the overall exergy efficiency for the MS process was 54.92%, while the HC process's was 89.56%.

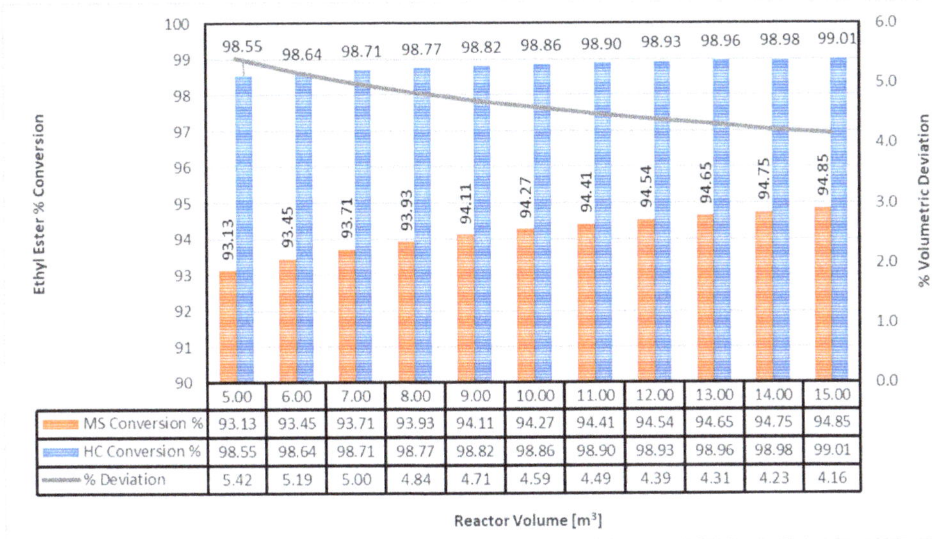

Figure 3. Conversion rate comparison (sensitivity assessment curve).

Table 5. Summary of total capital investment and total production cost [24].

Cost Parameter	Cost of MS Plant (USD)	Cost of HC Plant (USD)
I. Fixed Capital Investment (DC + IC)	12,295,681.44	8,642,910.43
A. Direct costs (DC)	10,691,896.90	7,202,425.36
1. Onsite (ISBL) cost	8,224,536.08	5,540,327.20
2. Offsite (OSBL) cost (30% of ISBL)	2,467,360.82	1,662,098.16
B. Indirect costs (IC)	1,603,784.54	1,440,485.07
1. Engineering and supervision (5% of DC)	534,594.85	360,121.27
2. Contingencies (10% & 15% of DC)	1,069,189.69	1,080,363.80
II. Other Outlays (OO)		
A. Startup costs (10% of DC)	1,069,189.69	720,242.54
B. Working capital (15% of DC)	2,004,919.90	1,458,491.14
Total capital investment	14,380,601.31	10,101,401.57
Raw materials Cost		
Rubber seed oil (86.5 ton)	196,787.50	196,787.50
Ethanol (13.5)	6750.00	6750.00
Potassium hydroxide (0.53 ton)	307.40	307.40
Phosphoric acid (0.2 ton)	167.45	167.45
Utilities		
Steam (0.0227/MJ)	9368.56	393.19
Variable production cost/day	213,380.91	204,405.54
A. Direct production costs	64,512.00	64,512.00
1. Labour (%)	30,720.00	30,720.00
a. Number of Labour (8)	15,360.00	15,360.00
b. Supervision (25% of operating labour).	18,432.00	18,432.00
c. Direct salary overhead (40% of (a+b))	4267.62	4088.11
2. Repair and maintenance (2%)	4267.62	4088.11
3. Packing (2%)	2133.81	2044.06
4. Waste stream disposal (1%)		
B. Fixed charges		
1. Depreciation (10%)	21,338.09	20,440.55
Annual Production Summary		
Total Fuel produced/day	207.00	224.00
Total production cost/day	288,561.96	279,137.81
Plant uptime for 7920 Hrs (90.4%)		
Total production cost/yr at Uptime	95,213,903.67	92,104,312.95
Gross Profit/d	320,850.00	347,200.00
Gross Profit/yr @ uptime	105,867,666.00	114,562,112.00
Net Profit Per year	10,653,762.33	22,457,799.05

Market price of biodiesel is 1550 USD/t, RBO is 2.275 USD/kg, ethanol is 0.5 USD/kg, KOH is 0.58 USD/kg, and H_3PO_4 is 0.85 USD/kg.

5.2.1. Exergy Destruction

By comparing the exergy destruction in each component to the total exergy destruction of the process, the ratio is $y^{*}_{D,i}$ and the exergy destruction ratio is $Ex_{D,\ Tot}$. The exergy destruction and efficiencies of each part of the two processes are shown in Tables 4 and 5. The values of exergy destruction for the HC and MS processes were found to be 0.25 MW and 1.77 MW, respectively. This suggests that, in comparison to the MS process, the HC process has about a sixth of the energy destruction. Analysis of the MS process's components revealed that, as shown in Figure 4, the phases of glycerin purification and catalytic removal contributed the most exergy destruction, at roughly 51% and 38%, respectively. In other words, both components contributed about 89% to the irreversibility ratio of the MS system. Additionally, it is shown that both components have improvement potentials of 0.08 MW and 0.15 MW, respectively, suggesting that by giving their designs the attention they need, these components might be made better and more effective. Conversely, the glycerin recovery (Splitter-4) unit, like the MS process, only contributed 93% exergetic degradation to the irreversibility of the system. In general, the exergetic destruction of the HC process is shown in Figure 5. Only a careful design is needed to maximize the component's exergetic improvement ratio of 0.01 MW.

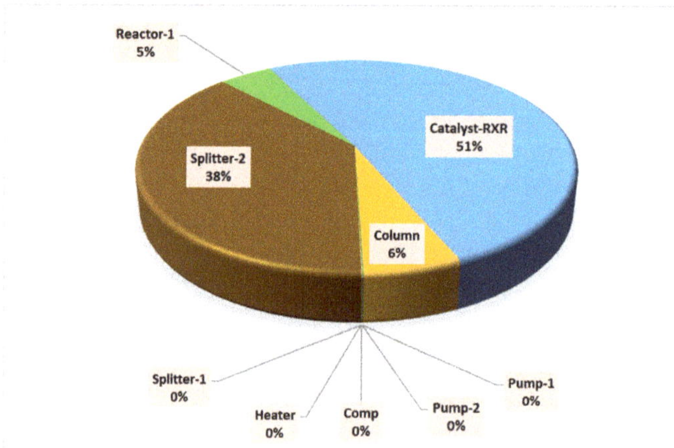

Figure 4. Components' exergetic destruction in the MS conversion process.

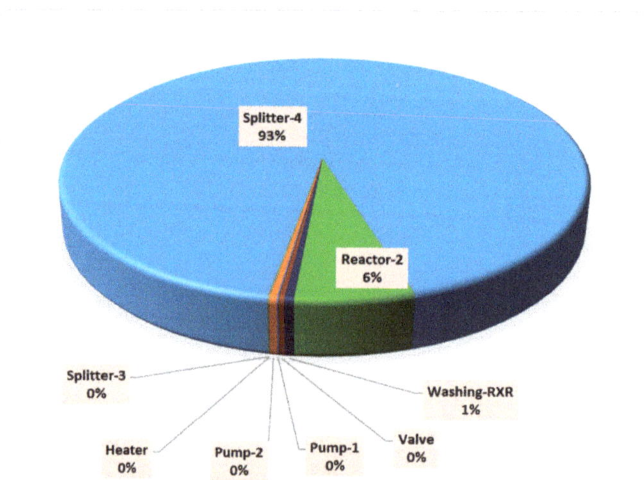

Figure 5. Components' exergetic destruction in the HC conversion process.

Reducing energy destruction in this stage has been achieved by using the solvent washing and decanting process to separate the phases of oil and biodiesel. However, the HC method has a significant advantage over the MS process in that it does away with the requirement for the biodiesel purification step because of its high conversion efficiency. Exergy destruction is largely attributed to the transesterification reactor, which comes after the biodiesel purification stage. The HC process exhibits around 8% less exergy destruction than the MS process. As compared to the MS process, the HC method uses only 3% of the electrical power in the reactor, demonstrating lower energy use. This reduction can be justified by the shorter reaction time and optimum mixture that cavitation produced at the microscopic level. The transesterification stage of ethanol recovery adds to the MS process's exergy destruction, but the HC process has less exergy destruction because less alcohol needs to be separated and it is not recycled.

Since the HC process requires a mixer to be present before the reactor, the MS method's mixing stage has a significantly lower energy destruction. The variation in reaction conditions, specifically in temperature and pressure, also affects the rates of energy destruction. As the HC process operates at room temperature and does not require preheating, there is no exergy destruction during this stage.

5.2.2. Exergy Efficiency

The energy and exergy efficiency of the systems' separate parts as well as the overall operations are presented in Figures 6 and 7. Tables 4 and 5 show that the HC process had the highest exergy efficiency, outperforming the MS process by over 35%. This was because the HC process had a greater conversion rate in the reactor and reduced waste emissions and exergy destruction. The exergy efficiencies of the major model components were generated and are displayed in Figures 6 and 7 for the MS and HC systems, respectively. The total input exergy for both processes was 108.64 MW and 216.64 MW. Despite being the stage that contributes the most to energy destruction, the catalytic removal stage demonstrated a notably high exergy efficiency because of the significant fuel exergy input and conservation in the MS process's product stream. The HC process, on the other hand, contributed the second most to energy destruction. On the other hand, as shown in Figures 6 and 7, the biodiesel pumps had the lowest exergetic efficiency because of the large pressure differential that the pump creates and the mass flow rates that result in increased pump shaft power.

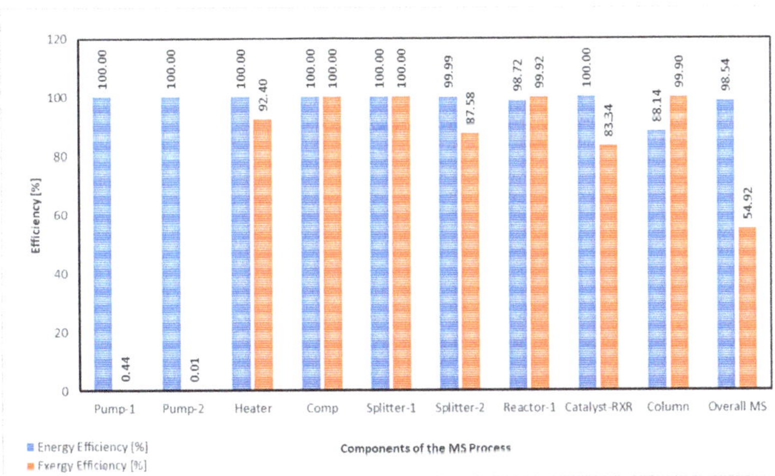

Figure 6. Components' exergetic efficiency in the MS conversion process.

For obvious reasons, the HC process does not require the column; however, this model has noted that the distillation column for the MS process requires a significant amount of steam. It has been shown that steam injected at the column accounts for approximately 49% of the fuel exergy input in the MS process. Despite the huge fuel contribution, the exergetic efficiency ranked amongst the highest in the system. This aligns reasonably well with a previous study by Blanco-Marigorta et al. [49].

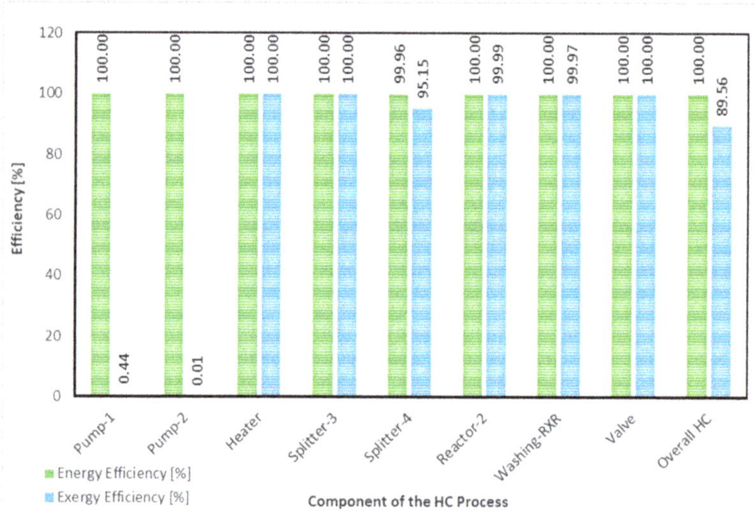

Figure 7. Components exergetic efficiency in the HC conversion process.

A notable discrepancy was noted in the transesterification phase, when the exergy efficiency of the reactor exceeded 99% in the current study, in contrast to 69% in Blanco-Marigorta et al.'s study [49] and 98% in Gholami et al.'s study [22].

This discrepancy might be ascribed to the differences in chemical exergy between rubber seed oil, canola oil, and jatropha oil employed in the several studies listed for the production of biodiesel. Even when converted 100%, the biodiesel produced from jatropha and canola oil appears to have a larger chemical exergy, which leads to a lower exergy efficiency. However, as can be seen in Figures 6 and 7 for reactors 1 and 2, respectively, RSO and its biodiesel show comparable chemical exergies that allow for a higher exergy efficiency of up to 99.9% at complete conversion. This demonstrates that, in terms of exergy efficiency, RSO is a more advantageous fuel for the manufacture of biodiesel. An essential factor in determining whether biodiesel production can be sustained in place of fossil fuels is how renewable it is in terms of energy.

The generation of ethanol and biodiesel from different feedstocks was studied in a study by Velasquez et al. [50], and their renewability performance was measured using an exergy-based indicator. The results showed that biodiesel production exhibited a renewable process with an efficiency of 74.7% and a high exergy content per unit of dried biomass. In light of this, the research indicates that producing biodiesel has the potential to be a renewable process. It also shows that the transesterification step can achieve exergy efficiencies exceeding 99%, mainly because the products have a high exergy content and low exergy destruction. Utilizing hydrodynamic cavitation processes improves production efficiency in biodiesel facilities by reducing waste emissions and exergy destruction as compared to mechanical stirring. It is interesting, however, that of all the steps of biodiesel production, the transesterification stage has one of the lowest exergy destructions. Thus, new advancements like plant species identification should be taken into consideration in order to further limit energy destruction in the biodiesel life cycle.

6. Economic Assessment

The manufacturers base their investment strategy primarily on the total cost of capital investment, with special attention to the ISBL plant cost as the most important consideration at first. Table 5 shows that the ISBL cost for the MS process was USD 8,224,536.08, whereas the HC process produced an ISBL cost of USD 5,540,327.20 using the Bridgewater's technique. The plant cost of the HC plant is approximately 1.4 times less than the plant cost of the MS plant. The MS plant's costly distillation equipment, which is necessary to separate biodiesel from unconverted rubber seed oil, is the main cause of this cost difference. The substantial divergence in investment costs indicates a promising outlook for industrial HC plants in biodiesel production. Similarly, Innocenzi and Prisciandaro [51] came to a comparable conclusion, pointing out that an HC plant's ISBL cost is 1.1 times more than that of a typical biodiesel production facility.

Regarding energy consumption, the HC method is also more appealing because it uses just 17,321 MJ/d as opposed to the MS process's 412,712 MJ/d, as shown in Table 5. As a result, the MS process's increased energy consumption is mostly caused by the addition of the energy-intensive distillation unit. This significant variation in energy use supports the HC method's superior sustainability and environmental friendliness over the MS process, in line with SDG No. 12 [52] on responsible consumption and production [53]. This SDG gives attention to maximizing efficiency and output while minimizing resources. Under this framework, the proposed HC process exhibits lower energy usage and reduced losses. When evaluating a process's viability, the total production cost is just as important as the ISBL cost. For instance, Table 5 shows that the combined cost of production for both the MS and HC procedures for biodiesel is USD 1550 per ton.

However, roughly 224 tons and 207 tons of biodiesel were converted in the HC and MS procedures, respectively, for the same quantity of materials used in the biodiesel production methods. This suggests that the HC process's production costs are approximately 9.2% greater than the MS process's. Gholami et al. [22] found a similar result and noted that the MS procedure resulted in a 10% greater cost of producing biodiesel. Notably, energy expenses make up 4.4% of the variable production cost in the MS process but only 0.2% in the HC process. The analysis includes initial cash outflows for engineering costs, equipment procurement, and plant construction. The plant's cash inflows come from product sales as soon as it is operating. The estimated cumulative net cash flow over the project's 15-year lifespan is shown in a cash flow diagram, like the one in Figure 8.

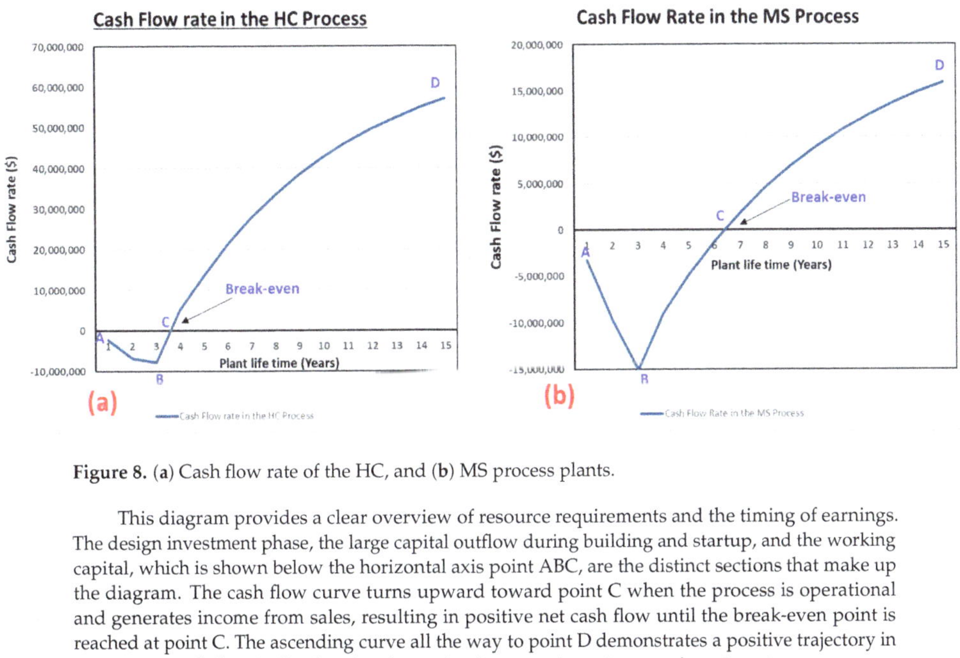

Figure 8. (**a**) Cash flow rate of the HC, and (**b**) MS process plants.

This diagram provides a clear overview of resource requirements and the timing of earnings. The design investment phase, the large capital outflow during building and startup, and the working capital, which is shown below the horizontal axis point ABC, are the distinct sections that make up the diagram. The cash flow curve turns upward toward point C when the process is operational and generates income from sales, resulting in positive net cash flow until the break-even point is reached at point C. The ascending curve all the way to point D demonstrates a positive trajectory in cumulative cash flow, indicating that the project is generating returns on investment.

The net present value (NPV) represents the difference between the present value of cash inflows and cash outflows over a specific period (in this study, 15 years). Establishing a process's economic feasibility requires a positive net present value. Appendix Tables 3 and 4 provide detailed economic analysis, whereas Table 6 summarizes the NPV for the MS and HC procedures, respectively. Year 1 of this study is the start of cash flow and is regarded as the design phase. The project receives its full fixed capital investment during the construction and installation phases of Year 2. Depreciation charges are subtracted from the gross profit once the unit reaches full capacity in Year 3. Positive NPV and return on investment (ROI) are shown for the MS and HC processes in Table 6. The MS process has an NPV of USD 15.9 million and an ROI of 55.51%, demonstrating that the investment is both feasible and profitable. On the contrary, the HC process has an ROI of 176.84% and an NPV of USD 57.2 million. Given that expenses have already exceeded revenues, this suggests that both procedures are commercially feasible. However, the higher ROI and lower NPV of the HC procedure make it clear that it is more profitable.

Table 6. Economic analysis summary.

Revenues and Production Costs	HC Process	MS Process
Product revenue [USD]	114,562,112.00	105,867,666.00
Variable cost of production (VCOP) [USD]	69,469,020.42	72,519,380.83
Fixed cost of Production (FCOP) [USD]	22,635,292.66	22,694,522.96
Economic analysis		
Average cash flow/year [USD]	17,805,196.54	8,545,148.81
Simple pay-back period [years]	0.57	1.68
Return on investment (15 yrs)	176.84%	55.51%
NPV (@ 15 years) [USD]	57,165,105.59	15,881,434.94
NPV to year [yrs]	4	7
IRR [%]	88.99	33.03

7. Environmental Assessment

Environmental effects of RSO biodiesel production include greenhouse gas (GHG) emissions, primarily CO_2. Despite these emissions, biodiesel from rubber seeds is seen as a renewable fuel with lower net CO_2 emissions than fossil fuels. This is because it is assumed that the CO_2 that rubber trees absorb during growth will balance the CO_2 released during combustion. However, the energy used in the processes involved in the production of biodiesel has a substantial impact on the environment. Thus, it is assumed that the use of renewable energy sources can notably reduce the environmental impact associated with energy use in biodiesel production. The carbon footprint of the HC process was estimated and contrasted with that of the MS process in order to determine whether the projected biodiesel production procedure could be implemented on a large scale, as shown in Table 7.

Table 7. Carbon footprint of biodiesel production using HC and MS approaches.

CO_2 Emissions (t CO_2 eq./year)	MS Process	HC Process
Steam	172.7	7.24

It is worth noting that this study did not consider emissions from plant construction materials such as steel, stainless steel, and raw materials, or other life cycle assessments (LCAs). This study focused mainly on electricity and thermal energy emissions relating to biodiesel production. Specifically, the emissions resulting from steam consumption were found to be 172 t CO_2 eq./year for the MS process and only 7.2 t CO_2 eq./year for the HC process. This significant difference indicates that the HC process generates approximately 24 times fewer CO_2 emissions than the MS process, affirming its position as a cleaner and more environmentally sustainable production pathway. It may also be relevant to exergy waste emission in relation to the environment in the production of biodiesel. In this study, the MS process resulted in a substantial waste of 5.52 MW of exergy, with a significant proportion of approximately 82% stemming from the unreacted oil present in the bottom stream of the biodiesel purification tower. Conversely, the HC process exhibits a significantly lower exergy waste emission of 1.12 MW, reflecting a reduction of 78%. Notably, this is due to the efficient conversion in the reactor and lower consumption of alcohol and oil, providing a notable improvement compared to the MS process.

In this study, the comparison of HC and MS reactors for RSO biodiesel production has been expansively discussed from varying points of view. HC proves more viable due to specific factors. HC relies on fluid flow for cavitation energy, requiring minimal additional energy, while MS needs continuous energy input from motors, making HC more energy-efficient, especially in large-scale applications. HC has fewer parts, leading to lower maintenance and less wear. It enables process intensification, speeding up reactions. Additionally, HC is environmentally friendly, reducing the need for chemicals and aligning with green chemistry practices, whereas MS emits more carbons.

MS reactors are versatile and widely adaptable to different reaction conditions in various industries relative to HC reactors. They are also easily scalable and hence suitable for both small-scale and large-scale biodiesel production. Finally, MS reactors provide uniform mixing for all reactants to ensure consistent product quality in many biodiesel production facilities.

8. Conclusions

This passage discusses the increasing interest in advanced transesterification reactors for biodiesel production, driven by the need for cost efficiency, enhanced biodiesel quality, faster reactions, and environmental benefits. However, challenges like high construction costs and energy management hinder large-scale biodiesel production from RBO. The study emphasizes the necessity for affordable and versatile reactors applicable to various feedstocks, ensuring their economic and environmental viability. The research compares tube-like plug-flow reactors (PFRs) and continuous stirred tank reactors (CSTRs) for converting RSO and fats into biodiesel, analyzing their limitations and impacts. The study also explores key parameters affecting transesterification and conducts a comparison between traditional mechanical stirring (MS) and innovative hydrodynamic cavitation (HC) methods, utilizing Aspen HYSYS version 10 software for biodiesel production analysis. An exergetic analysis technique has been adopted, it being a valuable tool for designing and evaluating energy systems and assessing the efficiencies of energy components. This methodology proves its practicality in improving energy and exergy efficiency in the production process of biodiesel from ethyl ester. The main conclusions are as follows:

- Biodiesel production processes showed energy efficiencies of 98.54% for MS and 100% for HC, with exergy efficiencies of 54.92% and 89.56%, respectively.
- Exergy analysis is vital for understanding energy use in biodiesel production. MS used 88.24 kW energy for a 94.85% conversion rate, whereas HC used 2.274 kW for 99.01% conversion at the same volumetric reactor rate.
- The study quantifies inefficiencies, helping to assess component performance and develop sustainable biodiesel production. MS's exergetic destruction is six times higher than HC's.
- Chemical exergy from RBO and ethanol is the major input, while glycerol and water cause significant exergy losses in the biodiesel production process.

Suffice to say, energy efficiency is crucial in biodiesel production, and employing exergetic techniques can significantly reduce energy consumption. This approach helps industries and stakeholders to focus on energy savings when selecting reactors. By addressing components with high exergetic destruction through proper design, performance efficiency is assured, minimizing energy costs and reducing irreversibilities of the components of the systems.

Economic and environmental assessments compared MS and HC biodiesel production. The MS plant investment cost was 1.5 times higher than HC, making HC more profitable (NPV: USD 57.2 million, ROI: 176.84% vs. USD 15.9 million, ROI: 55.51% for MS). Additionally, HC emitted significantly less CO_2 (7.2 t CO_2 eq./year) compared to MS (172 tCO_2 eq./year) due to MS's higher energy requirements. Implementing HC not only enhances efficiency and profitability but also reduces energy consumption, material usage, and waste, promising cleaner biodiesel production and sustainable energy development.

The optimization of biodiesel production from RSO often generates unrecoverable waste. A circular economy approach is crucial for enhancing the entire production chain, maximizing resource efficiency and ensuring sustainability. Stakeholders and industry merchants play a vital role in this process, particularly through sustainable rubber plantation farming practices that maintain a steady supply of seeds without depleting natural resources. To minimize waste, exploring options like utilizing by-products such as seed husks for composting or biomass energy generation is essential. This area lacks substantial research and presents a promising future study. Engaging stakeholders actively can facilitate the biodiesel industry's shift toward a circular economy, reducing environmental impact and fostering economic growth through innovation, job creation, and resource optimization.

Future Research Areas

Future research in biodiesel production via the transesterification process should focus on

1. Developing efficient and eco-friendly catalysts, exploring options like enzyme catalysts and nanocatalysts.
2. Addressing the issue of waste utilization is crucial. Research should investigate methods to utilize waste materials, such as finding uses for glycerol by-products and developing more efficient purification techniques to reduce overall waste.
3. Lastly, comprehensive life cycle assessments (LCAs) can evaluate the environmental impact of biodiesel production, guiding decision making toward more sustainable practices.

Author Contributions: Conceptualization and writing—review and editing, O.D.S. and P.A.A.; methodology, T.K.T.; software and formal analysis, H.F.; data curation and validation, C.P. and O.D.; formal analysis and writing—review and editing; A.E. and C.C.E.; writing—original draft preparation, A.M. All authors have read and agreed to the published version of the manuscript.

Funding: This research received no external funding.

Institutional Review Board Statement: Not applicable.

Informed Consent Statement: Not applicable.

Data Availability Statement: Data are contained within the article.

Conflicts of Interest: The authors declare no conflict of interest.

Nomenclature

BD	Biodiesel
CC	Contingency cost
CCPP	Combined cycle power plant
CV	Control volume
CSTR	Continuously stirred tank reactor
EC	Engineering cost
ER	Ethylic routes
FAME	Fatty acid methyl ester
FCI	Fixed capital investment
GLC	Glycerol
GWP	Global warming potential
HC(R)	Hydrodynamic cavitation reactor
HEP	Hydroelectric power
HHV	Higher heating value (MJ/kg)
IR	Interest rate
IRR	Internal rate of return
ISBL	Inside battery limit
MACRS	Modified accelerated cost recovery system
ME	Methylic route
MS(R)	Mechanical stirring reactor
NCGOs	Non-competitive generational oils
NPV	Net present value
NRTL	Nonrandom two liquid
OPEX	Operating expenditure
OSBL	Outside battery limit
PCE	Purchase cost of equipment
PFR	Plug flow reactor
PPC	Physical plant cost
ROI	Return on investment
RSO	Rubber seed oil
SDG	Sustainable development goal
TC	Total production cost
TCI	Total capital investment
TEA	Techno-economic analysis
TFC	Total fixed cost
TG	Triglyceride
TPC	Total production cost
TVC	Total variable cost
UNFCC	United Nations Framework Convention on Climate Change
WCI	Working capital investment

Greek letters	
φ	coefficient from the liquid fuel expression
$\eta_{\dot{E}}$	energy efficiency
η_{ex}	exergy efficiency
η_{Pump}	pump efficiency
η_{th}	thermal efficiency
Abbrevations	
$\dot{E}$	energy rate [kW]
Ex	exergy rate [kW]
ex	specific exergy rate of material streams (kJ/kmol)
Ex_D	exergy destruction rate
Ex_L	exergy loss rate
W_{Net}	net power (kW)
W_P	pump power (kW)
y_D	exergy destruction rate ratio
h_i	specific enthalpy at initial state (kJ/kmol)
h_o	specific enthalpy at reference state (kJ/kmol)
KE	kinetic energy
LHV	lower heating value (MJ/kg)
$\dot{m}$	mass flow rate [kg/sec]
$\dot{m}_{Fuel}$	mass flow rate of Fuel [kg/sec]
P	power output [kW]
P_o	pressure at reference state (atm)
PE	potential energy
Q	heat flow rate
S_i	specific entropy at initial state (kJ/kmol)
S_o	specific enthalpy at reference state (kJ/kmol)
T_o	temperature of reference state (K)
Subscripts	
CHM	Chemical
D	Destruction
F	Fuel
In	Inlet streams
K	kth component of system
O	Reference state
O	Ambient
Out	Outlet stream
P	Product
PHY	Physical
Th	Thermal
Tot	Total

Appendix A

Table A1. An overview of techno-economic scrutiny of biodiesels derived from oily feedstocks.

Reactor Technology	Source	Types of Alcohol Methanol	Ethanol	E_a	TE_a	Technological Challenges	Remarks	Refs.
LSC	Rice bran oil (RBAO)	✗	✓	✓	✓	The cost of lipase catalyst influenced the recovery and re-usage	The adoption of TEAs into rice bran oil refinery led to profit and a reduction in toxic chemicals and energy	Usaku et al. [54]
Sonication	Acidic oil	✓	✗	✓	✓	NS	The production cost (0.776 USD/liter), total CO_2 emissions (0.08 kg of CO_2/kg of BD), and return of investment (~330%) obtained for the medium-sized production unit	Naeem et al. [16]
Meso-OBR	Refined and low-grade vegetable oils (RLVOs)	✓	✗	✓	✓	NR	TEAs for a single step for BD derived from RLVO established.	Al-Saadi et al. [55]
LSR	Algal biomass (AB)	✓	✗	✓	✓	The economic viability of the production of the recycle of the process	Feasibility of AB biodiesel on a large scale guaranteed	Mustapha et al. [42]
HCR		✓	✗	✓	✓	NR	Feasibility of techno-economics via Aspen version 10 software established	Oke et al. [56]
LSR	Marine macroalgae Codium tomentosum	✓	✗	✓	✓	NR	The payback period (8.59 yrs.) and +ve NPV (1.38 M USD/yr.) from biodiesel production (20 MT/batch) process) recorded	Gengiah et al. [57]
LSR	Palm and Jatropha biomass (PJB)	✓	✗	✓	✓	Attractions of TEA of hybrid of PJB for bio-refinery established	Technological bottleneck in obtaining valuable products from PJB	Niño-Villalobos et al. [58]
CSTR	Calophyllum inophyllum oil (CaO)	✓	✗	✗	✓	Low feedstock cost and high biodiesel conversion	Value of TEA, annual biodiesel revenue, and payback period for CaO documented	Naveenkumar and Baskar [59]
Alkali-cat, C-SCM and L-SCM	Palm oil	✓	✗	✓	✓	High amount of methanol during C-SCM process makes financial profitability realistic in the recycling loop	Financial aspect of L-SCM process ranked the best, followed by the Alkali-cat process, then the C-SCM	Sakdasri et al. [60]
Supercritical process	Jatropha curcas oil	✓	✗	✓	✓	NS	The cost of capital investment (9.41 million USD/yr.), manufacturing (25.39 million USD/yr), and total production (31.20 million USD/yr) reported	Yusuf, and Kamarudin [8]
Continuous stirred tank reactor (CSTR)	Canola oil	✓	✗	✓	✗	Inadequate mixing, elongated reaction period, high energy utilization	Economic sustainability of biodiesel production = f Plant capacity and prices of feedstock oils	Zhang et al. [61]
Batch Reactor	Microalgal biomass	✓	✗	✗	✓	Unsatisfactory mixing, long reaction period, high energy expended, low capability, and inflexibility	Cost-effectiveness was not checked	Lee et al. [62]

SCM = supercritical methanol; L-SCM = low methanol: oil molar ratio; Alkali-cat = alkali-catalyzed process; C-SCM = conventional SCM; NR = not reported; LSC = lab-scale reactor; TEAs = techno-economic analyses.

Table 2. A. Thermodynamic material streams data for the integrated (MS and HC) biodiesel process plant.

Name	Vapour Fraction	Temperature [C]	Pressure [bar]	Mass Flow [kg/s]	Molar Flow [kg-mole/s]	Mass Enthalpy [kJ/kg]	Mass Entropy [kJ/kg-C]	Heat Flow [MW]	Specific Exergy [kJ/kg]	Physical Exergy [kW]	Chemical Exergy [kW]	Exergy Total [MW]
Ethanol	0.00	25.00	1.00	0.38	0.01	6023.85	7.52	2.31	0.00	0.00	11,316.92	11.32
	0.00	25.00	1.00	0.12	0.00	8085.31	2.02	0.99	0.00	0.00	235.46	0.24
Triolein	0.00	25.00	1.00	2.46	0.00	2330.70	6.26	5.73	0.00	0.00	97,090.33	97.09
Water2	0.00	50.00	1.00	0.01	0.00	15,760.05	8.72	0.08	4.57	0.02	11.39	0.01
H3PO4	0.00	60.00	1.00	0.00	0.00	2104.98	31.37	0.00	39.66	0.09	1.99	0.00
1	0.00	25.00	1.00	0.51	0.01	6523.37	6.10	3.31	0.00	0.00	11,316.92	11.32
2	0.00	25.22	4.00	2.46	0.00	2330.31	6.26	5.73	0.00	0.00	97,090.33	97.09
3	0.00	25.19	4.00	0.51	0.01	6522.95	6.10	3.31	0.00	0.00	11,316.92	11.32
4	0.00	70.00	4.00	2.46	0.00	2248.96	6.00	5.53	5.68	13.98	97,090.33	97.10
5	0.00	61.06	4.00	2.97	0.01	2979.03	6.00	8.84	3.96	11.73	108,407.25	108.42
6	0.09	52.83	0.20	2.97	0.01	2979.03	6.00	8.84	3.71	10.99	108,407.25	108.42
7	0.00	70.69	0.20	2.97	0.01	2979.78	5.91	8.84	6.00	17.80	108,385.26	108.40
8	0.00	60.00	2.00	3.01	0.01	3032.75	5.98	9.12	3.59	10.81	109,668.60	109.68
9	1.00	60.00	2.00	0.00	0.00	5047.86	4.51	0.00	76.76	0.00	0.00	0.00
10	0.00	59.95	1.00	3.01	0.01	3045.67	5.98	9.17	3.59	10.80	109,675.56	109.69
11	0.00	60.00	1.00	0.38	0.01	7526.60	4.88	2.86	3.29	1.25	4531.92	4.53
12	1.00	60.00	1.00	0.00	0.00	13,318.17	11.48	0.00	3227.44	0.00	0.00	0.00
13	0.00	60.00	1.00	0.38	0.01	7494.28	5.04	2.86	3.50	1.34	5437.91	5.44
14	0.00	60.00	1.00	2.63	0.01	2398.73	6.17	6.31	3.64	9.58	105,143.64	105.15
15	0.00	70.57	0.20	2.97	0.01	3001.31	5.91	8.92	5.99	17.78	108,396.65	108.41
16	0.00	60.00	0.20	0.38	0.01	7568.48	4.89	2.88	3.31	1.26	4524.51	4.53
17	1.00	60.00	0.20	0.00	0.00	9866.16	0.00	0.00	374.54	0.00	0.00	0.00
18	0.00	60.00	0.20	0.38	0.01	7583.30	4.89	2.91	3.16	1.21	4527.25	4.53
19	1.00	166.29	0.10	0.04	0.00	4401.74	4.00	0.18	73.80	3.04	1283.49	1.29
20	1.00	326.19	4.00	0.04	0.00	4056.65	3.92	0.17	392.66	16.19	1283.49	1.30
Biodiesel2	0.00	60.00	0.20	2.59	0.01	2352.54	6.16	6.09	3.61	9.34	103,872.13	103.88
P-Salt	0.00	60.00	0.09	0.12	0.00	8025.29	1.95	0.99	0.29	0.04	230.68	0.23
Glycer	0.00	60.00	0.09	0.26	0.00	7370.51	6.40	1.91	4.53	1.18	4527.25	4.53
Water	0.00	50.00	1.00	0.00	0.00	15,760.05	8.72	0.05	4.57	0.01	6.96	0.01
P-Acid	0.00	60.00	1.00	0.00	0.00	2104.98	31.37	0.00	39.66	0.09	2.08	0.00
Alcohol	1.00	166.31	0.10	0.04	0.00	4401.36	4.00	0.18	73.83	3.07	1292.65	1.30
Oil	0.00	369.58	0.20	0.02	0.00	1530.01	4.52	0.03	286.51	5.46	752.80	0.76
Biodiesel1	0.00	166.31	0.10	2.57	0.01	2125.64	5.58	5.46	54.85	140.99	103,098.18	103.24
Glycerol	0.00	60.00	0.09	0.26	0.00	7270.11	6.39	1.87	4.47	1.15	4531.92	4.53
K3PO4	0.00	60.00	0.09	0.13	0.00	7952.68	2.47	0.99	1.51	0.19	230.70	0.23

Table 3. A. Summary of NPV for MS process.

Project Year	Cap Ex	Revenue	CCOP	Gr. Profit	Deprcn	Taxbl Inc	Tax Paid	Cash Flow	PV of CF	NPV
1	3,688,704.43	-	-	-	-	-	-	(3,688,704.43)	(3,207,569.07)	(3,207,569.07)
2	8,606,977.01	-	-	-	-	-	-	(8,606,977.01)	(6,508,111.16)	(9,715,680.23)
3	2,084,919.90	52,933,833.00	58,954,213.37	(6,020,380.37)	208,491.99	(6,228,872.36)	-	(8,105,300.27)	(5,329,366.50)	(15,045,046.73)
4	-	105,867,666.00	95,213,903.79	10,653,762.21	208,491.99	10,445,270.22	-	10,653,762.21	6,091,323.12	(8,953,723.60)
5	-	105,867,666.00	95,213,903.79	10,653,762.21	208,491.99	10,445,270.22	2,350,185.80	8,303,576.41	4,128,345.01	(4,825,378.59)
6	-	105,867,666.00	95,213,903.79	10,653,762.21	208,491.99	10,445,270.22	2,350,185.80	8,303,576.41	3,589,865.23	(1,235,513.36)
7	-	105,867,666.00	95,213,903.79	10,653,762.21	208,491.99	10,445,270.22	2,350,185.80	8,303,576.41	3,121,621.94	1,886,108.57
8	-	105,867,666.00	95,213,903.79	10,653,762.21	208,491.99	10,445,270.22	2,350,185.80	8,303,576.41	2,714,453.86	4,600,562.43
9	-	105,867,666.00	95,213,903.79	10,653,762.21	208,491.99	10,445,270.22	2,350,185.80	8,303,576.41	2,360,394.66	6,960,957.09
10	-	105,867,666.00	95,213,903.79	10,653,762.21	208,491.99	10,445,270.22	2,350,185.80	8,303,576.41	2,052,517.10	9,013,474.19

Table 3. *Cont.*

Project Year	Cap Ex	Revenue	CCOP	Gr. Profit	Deprcn	Taxbl Inc	Tax Paid	Cash Flow	PV of CF	NPV
11	-	105,867,666.00	95,213,903.79	10,653,762.21	208,491.99	10,445,270.22	2,350,185.80	8,303,576.41	1,784,797.47	10,798,271.66
12	-	105,867,666.00	95,213,903.79	10,653,762.21	208,491.99	10,445,270.22	2,350,185.80	8,303,576.41	1,551,997.80	12,350,269.46
13	-	105,867,666.00	95,213,903.79	10,653,762.21	-	10,653,762.21	2,350,185.80	8,303,576.41	1,349,563.31	13,699,832.77
14	-	105,867,666.00	95,213,903.79	10,653,762.21	-	10,653,762.21	2,397,096.50	8,256,665.71	1,166,903.48	14,866,736.26
15	-	105,867,666.00	95,213,903.79	10,653,762.21	-	10,653,762.21	2,397,096.50	8,256,665.71	1,014,698.68	15,881,434.94
16	-	105,867,666.00	95,213,903.79	10,653,762.21	-	10,653,762.21	2,397,096.50	8,256,665.71	882,346.68	16,763,781.62
17	-	105,867,666.00	95,213,903.79	10,653,762.21	-	10,653,762.21	2,397,096.50	8,256,665.71	767,257.98	17,531,039.60
18	-	105,867,666.00	95,213,903.79	10,653,762.21	-	10,653,762.21	2,397,096.50	8,256,665.71	667,180.85	18,198,220.46
19	-	105,867,666.00	95,213,903.79	10,653,762.21	-	10,653,762.21	2,397,096.50	8,256,665.71	580,157.26	18,778,377.72
20	(2,084,919.90)	105,867,666.00	95,213,903.79	10,653,762.21	-	10,653,762.21	2,397,096.50	10,341,585.61	631,873.77	19,410,251.49

Table 4. A. Summary of NPV for HC process.

Project Year	Cap Ex	Revenue	CCOP	Gr. Profit	Deprcn	Taxbl Inc	Tax Paid	Cash Flow	PV of CF	NPV
1	2,592,873.13	-	-	-	-	-	-	(2,592,873.13)	(2,254,672.29)	(2,254,672.29)
2	6,050,037.30	-	-	-	-	-	-	(6,050,037.30)	(4,574,697.39)	(6,829,369.68)
3	1,458,491.14	57,281,056.00	57,369,802.87	(88,746.87)	145,849.11	(234,595.98)	-	(1,547,238.00)	(1,017,334.10)	(7,846,703.78)
4	-	114,562,112.00	92,104,313.07	22,457,798.93	145,849.11	22,311,949.81	-	22,457,798.93	12,840,319.42	4,993,615.64
5	-	114,562,112.00	92,104,313.07	22,457,798.93	145,849.11	22,311,949.81	5,020,188.71	17,437,610.22	8,669,574.12	13,663,189.76
6	-	114,562,112.00	92,104,313.07	22,457,798.93	145,849.11	22,311,949.81	5,020,188.71	17,437,610.22	7,538,760.10	21,201,949.87
7	-	114,562,112.00	92,104,313.07	22,457,798.93	145,849.11	22,311,949.81	5,020,188.71	17,437,610.22	6,555,443.57	27,757,393.44
8	-	114,562,112.00	92,104,313.07	22,457,798.93	145,849.11	22,311,949.81	5,020,188.71	17,437,610.22	5,700,385.71	33,457,779.15
9	-	114,562,112.00	92,104,313.07	22,457,798.93	145,849.11	22,311,949.81	5,020,188.71	17,437,610.22	4,956,857.14	38,414,636.29
10	-	114,562,112.00	92,104,313.07	22,457,798.93	145,849.11	22,311,949.81	5,020,188.71	17,437,610.22	4,310,310.56	42,724,946.85
11	-	114,562,112.00	92,104,313.07	22,457,798.93	145,849.11	22,311,949.81	5,020,188.71	17,437,610.22	3,748,096.14	46,473,042.98
12	-	114,562,112.00	92,104,313.07	22,457,798.93	145,849.11	22,311,949.81	5,020,188.71	17,437,610.22	3,259,214.03	49,732,257.02
13	-	114,562,112.00	92,104,313.07	22,457,798.93	-	22,457,798.93	5,020,188.71	17,437,610.22	2,834,099.16	52,566,356.17
14	-	114,562,112.00	92,104,313.07	22,457,798.93	-	22,457,798.93	5,053,004.76	17,404,794.17	2,459,796.20	55,026,152.38
15	-	114,562,112.00	92,104,313.07	22,457,798.93	-	22,457,798.93	5,053,004.76	17,404,794.17	2,138,953.22	57,165,105.59
16	-	114,562,112.00	92,104,313.07	22,457,798.93	-	22,457,798.93	5,053,004.76	17,404,794.17	1,859,959.32	59,025,064.92
17	-	114,562,112.00	92,104,313.07	22,457,798.93	-	22,457,798.93	5,053,004.76	17,404,794.17	1,617,355.93	60,642,420.85
18	-	114,562,112.00	92,104,313.07	22,457,798.93	-	22,457,798.93	5,053,004.76	17,404,794.17	1,406,396.46	62,048,817.31
19	-	114,562,112.00	92,104,313.07	22,457,798.93	-	22,457,798.93	5,053,004.76	17,404,794.17	1,222,953.45	63,271,770.75
20	(1,458,491.14)	114,562,112.00	92,104,313.07	22,457,798.93	-	22,457,798.93	5,053,004.76	18,863,285.30	1,152,551.99	64,424,322.75

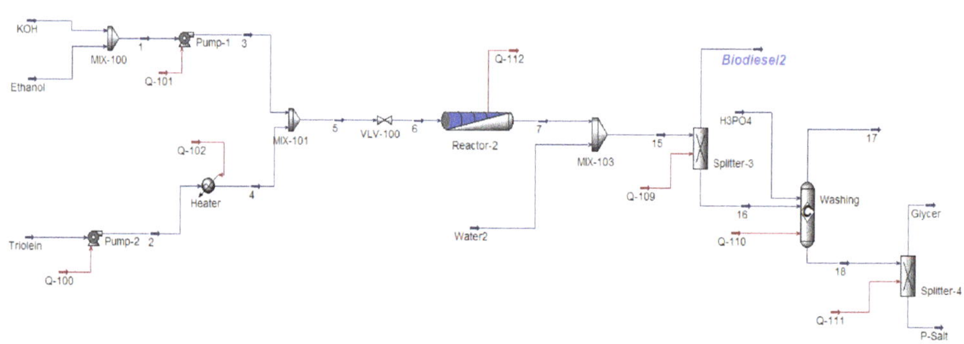

Figure 1. A. HC process feed streams simulation network.

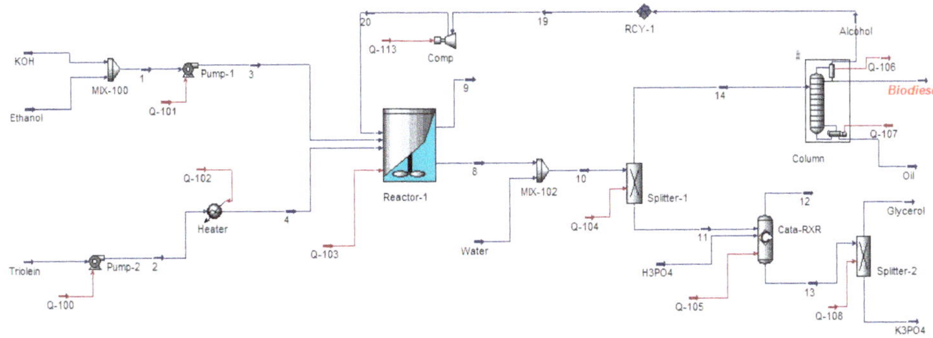

Figure 2. A. MS process feed streams simulation network.

References

1. Sajid, Z.; Khan, F.; Zhang, Y. Process simulation and life cycle analysis of biodiesel production. *Renew. Energy* **2016**, *85*, 945–952. [CrossRef]
2. IEA; IRENA; UNSD; WB; WHO. *Tracking SDG 7: The Energy Progress Report 2021*; United Nations: Washington, DC, USA, 2021.
3. Nhamo, G.; Nhemachena, C.; Nhamo, S.; Mjimba, V.; Savić, I. *SDG7–Ensure Access to Affordable, Reliable, Sustainable and Modern Energy*; Emerald Publishing: Bingley, UK, 2020; pp. 233–262.
4. Sun, X.; Liu, S.; Manickam, S.; Tao, Y.; Yoon, J.Y.; Xuan, X. Intensification of biodiesel production by hydrodynamic cavitation: A critical review. *Renew. Sustain. Energy Rev.* **2023**, *179*, 113277. [CrossRef]
5. Brahma, S.; Nath, B.; Basumatary, B.; Das, B.; Saikia, P.; Patir, K.; Basumatary, S. Biodiesel production from mixed oils: A sustainable approach towards industrial biofuel production. *Chem. Eng. J. Adv.* **2022**, *10*, 100284. [CrossRef]
6. Yang, L.; Takase, M.; Zhang, M.; Zhao, T.; Wu, X. Potential non-edible oil feedstock for biodiesel production in Africa: A survey. *Renew. Sustain. Energy Rev.* **2014**, *38*, 461–477. [CrossRef]
7. Samuel, O.D.; Okwu, M.O.; Amosun, S.T.; Verma, T.N.; Afolalu, S.A. Production of fatty acid ethyl esters from rubber seed oil in hydrodynamic cavitation reactor: Study of reaction parameters and some fuel properties. *Ind. Crops Prod.* **2019**, *141*, 111658. [CrossRef]
8. Yusuff, A.S.; Dada, T.; Olateju, I.I.; Azeez, T.M.; Azeez, S.O. Experimental investigation of influence of methyl, ethyl and methyl-ethyl ester blends of used cooking oil on engine performances and emissions. *Energy Convers. Manag. X* **2023**, *17*, 100346. [CrossRef]
9. Altamirano, C.A.A.; Yokoyama, L.; de Medeiros, J.L.; Araújo, O.D.Q.F. Ethylic or methylic route to soybean biodiesel? Tracking environmental answers through life cycle assessment. *Appl. Energy* **2016**, *184*, 1246–1263. [CrossRef]
10. Mandari, V.; Devarai, S. Biodiesel production using homogeneous, heterogeneous, and enzyme catalysts via transesterification and esterification reactions: A critical review. *BioEnergy Res.* **2022**, *15*, 935–961. [CrossRef]
11. Likozar, B.; Levec, J. Effect of process conditions on equilibrium, reaction kinetics and mass transfer for triglyceride transesterification to biodiesel: Experimental and modeling based on fatty acid composition. *Fuel Process. Technol.* **2014**, *122*, 30–41. [CrossRef]
12. Choedkiatsakul, I.; Ngaosuwan, K.; Cravotto, G.; Assabumrungrat, S. Biodiesel production from palm oil using combined mechanical stirred and ultrasonic reactor. *Ultrason. Sonochem.* **2014**, *21*, 1585–1591. [CrossRef]
13. Gholami, A.; Pourfayaz, F.; Maleki, A. Techno-economic assessment of biodiesel production from canola oil through ultrasonic cavitation. *Energy Rep.* **2021**, *7*, 266–277. [CrossRef]
14. Tabatabaei, M.; Aghbashlo, M.; Dehhaghi, M.; Panahi, H.K.S.; Mollahosseini, A.; Hosseini, M.; Soufiyan, M.M. Reactor technologies for biodiesel production and processing: A review. *Prog. Energy Combust. Sci.* **2019**, *74*, 239–303. [CrossRef]
15. Naeem, M.; Al-Sakkari, E.; Boffito, D.; Rene, E.; Gadalla, M.; Ashour, F. Single-stage waste oil conversion into biodiesel via sonication over bio based bifunctional catalyst: Optimization, preliminary techno-economic and environmental analysis. *Fuel* **2023**, *341*, 127587. [CrossRef]
16. Zheng, H.; Zheng, Y.; Zhu, J. Recent developments in hydrodynamic cavitation reactors: Cavitation mechanism, reactor design, and applications. *Engineering* **2022**, *19*, 180–198. [CrossRef]
17. Farvardin, M.; Samani, B.H.; Rostami, S.; Abbaszadeh-Mayvan, A.; Najafi, G.; Fayyazi, E. Enhancement of biodiesel production from waste cooking oil: Ultrasonic- hydrodynamic combined cavitation system. *Energy Sources Part A Recover. Util. Environ. Eff.* **2019**, *44*, 5065–5079. [CrossRef]
18. Maddikeri, G.L.; Gogate, P.R.; Pandit, A.B. Intensified synthesis of biodiesel using hydrodynamic cavitation reactors based on the interesterification of waste cooking oil. *Fuel* **2014**, *137*, 285–292. [CrossRef]

19. Chuah, L.F.; Yusup, S.; Aziz, A.R.A.; Bokhari, A.; Klemeš, J.J.; Abdullah, M.Z. Intensification of biodiesel synthesis from waste cooking oil (Palm Olein) in a Hydrodynamic Cavitation Reactor: Effect of operating parameters on methyl ester conversion. *Chem. Eng. Process. Process. Intensif.* **2015**, *95*, 235–240. [CrossRef]
20. Laosuttiwong, T.; Ngaosuwan, K.; Kiatkittipong, W.; Wongsawaeng, D.; Kim-Lohsoontorn, P.; Assabumrungrat, S. Performance comparison of different cavitation reactors for biodiesel production via transesterification of palm oil. *J. Clean. Prod.* **2018**, *205*, 1094–1101. [CrossRef]
21. Gholami, A.; Pourfayaz, F.; Saifoddin, A. Techno-economic assessment and sensitivity analysis of biodiesel production intensified through hydrodynamic cavitation. *Energy Sci. Eng.* **2021**, *9*, 1997–2018. [CrossRef]
22. Mahmud, R.; Moni, S.M.; High, K.; Carbajales-Dale, M. Integration of techno-economic analysis and life cycle assessment for sustainable process design—A review. *J. Clean. Prod.* **2021**, *317*, 128247. [CrossRef]
23. Thoppil, Y.; Zein, S.H. Techno-economic analysis and feasibility of industrial-scale biodiesel production from spent coffee grounds. *J. Clean. Prod.* **2021**, *307*, 127113. [CrossRef]
24. Lee, J.-C.; Lee, B.; Heo, J.; Kim, H.-W.; Lim, H. Techno-economic assessment of conventional and direct-transesterification processes for microalgal biomass to biodiesel conversion. *Bioresour. Technol.* **2019**, *294*, 122173. [CrossRef]
25. Gordon, R.; Gorodnitsky, I.; Grichko, V. Process for Producing Biodiesel through Lower Molecular Weight Alcohol-Targeted Cavitation. U.S. Patent 8,981,135, 17 March 2015.
26. Ihoeghian, N.A.; Usman, M.A. Exergetic evaluation of biodiesel production from rice bran oil using heterogeneous catalyst. *J. King Saud Univ. Eng. Sci.* **2018**, *32*, 101–107. [CrossRef]
27. Al-Saadi, L.S.; Eze, V.C.; Harvey, A.P. Techno-economic analysis of processes for biodiesel production with integrated co-production of higher added value products from glycerol. *Biofuels* **2020**, *13*, 489–496. [CrossRef]
28. Santori, G.; Di Nicola, G.; Moglie, M.; Polonara, F. A review analyzing the industrial biodiesel production practice starting from vegetable oil refining. *Appl. Energy* **2012**, *92*, 109–132. [CrossRef]
29. Chuah, L.F.; Klemeš, J.J.; Yusup, S.; Bokhari, A.; Akbar, M.M.; Chong, Z.K. Kinetic studies on waste cooking oil into biodiesel via hydrodynamic cavitation. *J. Clean. Prod.* **2017**, *146*, 47–56. [CrossRef]
30. Trejo-Zárraga, F.; de Jesús Hernández-Loyo, F.; Chavarría-Hernández, J.C.; Sotelo-Boyás, R. Kinetics of Transesterification Processes for Biodiesel Production. In *Biofuels—State of Development*; IntechOpen: London, UK, 2018; pp. 149–179.
31. Brennen, C.E. *Cavitation and Bubble Dynamics*; Oxford University Press: Oxford, UK, 1995.
32. Omelyanyuk, M.; Ukolov, A.; Pakhlyan, I.; Bukharin, N.; El Hassan, M. Experimental and Numerical Study of Cavitation Number Limitations for Hydrodynamic Cavitation Inception Prediction. *Fluids* **2022**, *7*, 198. [CrossRef]
33. Winterbone, D.E.; Turan, A. Chapter 15—Combustion and Flames. In *Advanced Thermodynamics for Engineers*, 2nd ed.; Elsevier: Amsterdam, The Netherlands, 2015. [CrossRef]
34. Aigba, P.A.; Emovon, I.; Samuel, O.D.; Enweremadu, C.C.; Abdeljawad, T.; Al-Mdallal, Q.M.; Afzal, A. Exergetic Assessment of Waste Gas to Energy in a Novel Integrated NGL Recovery and Power Generation Plant. *Front. Energy Res.* **2022**, *9*, 798896. [CrossRef]
35. Szargut, J. Chemical exergies of the elements. *Appl. Energy* **1989**, *32*, 269–286. [CrossRef]
36. Szargut, J.; Styrylska, T. Approximate Evaluation of the Exergy of Fuels. *Brennst. Waerme Kraft* **1964**, *16*, 589–596.
37. Tsatsaronis, G.; Lazzaretto, A. SPECO: A systematic and general methodology for calculating efficiencies and costs in thermal systems. *Energy* **2006**, *31*, 1257–1289.
38. Boyano, A.; Blanco-Marigorta, A.; Morosuk, T.; Tsatsaronis, G. Exergoenvironmental analysis of a steam methane reforming process for hydrogen production. *Energy* **2011**, *36*, 2202–2214. [CrossRef]
39. Li, Q.; Lin, Y. Exergy analysis of the LFC process. *Energy Convers. Manag.* **2016**, *108*, 348–354. [CrossRef]
40. Michalakakis, C.; Fouillou, J.; Lupton, R.C.; Gonzalez Hernandez, A.; Cullen, J.M. Calculating the chemical exergy of materials. *J. Ind. Ecol.* **2021**, *25*, 274–287. [CrossRef]
41. Mustapha, S.I.; Bux, F.; Isa, Y.M. Techno-economic analysis of biodiesel production over lipid extracted algae derived catalyst. *Biofuels* **2021**, *13*, 663–674. [CrossRef]
42. Sinnott, R.; Towler, G. *Chemical Engineering Design: Principles, Practice and Economics of Plant and Process Design*; Butterworth-Heinemann: Boston, MA, USA, 2019.
43. Tsagkari, M.; Couturier, J.; Kokossis, A.; Dubois, J. Early-Stage Capital Cost Estimation of Biorefinery Processes: A Comparative Study of Heuristic Techniques. *ChemSusChem* **2016**, *9*, 2284–2297. [CrossRef]
44. Mata, T.M.; Pinto, F.; Caetano, N.; Martins, A.A. Economic and environmental analysis of animal fats acidity reduction by enzymatic esterification. *J. Clean. Prod.* **2018**, *184*, 481–489. [CrossRef]
45. Houghton, J.T.; Meira Filho, L.G.; Lim, B.; Treanton, K.; Mamaty, I.; Bonduki, Y. *Revised 1996 IPCC Guidelines for National Greenhouse Gas Inventories: Workbook*; IPCC: Geneva, Swilzerland, 1996.
46. Chuah, L.F.; Yusup, S.; Aziz, A.R.A.; Bokhari, A.; Abdullah, M.Z. Cleaner production of methyl ester using waste cooking oil derived from palm olein using a hydrodynamic cavitation reactor. *J. Clean. Prod.* **2016**, *112*, 4505–4514. [CrossRef]
47. Ahmad, J.; Yusup, S.; Bokhari, A.; Kamil, R.N.M. Study of fuel properties of rubber seed oil based biodiesel. *Energy Convers. Manag.* **2014**, *78*, 266–275. [CrossRef]
48. Blanco-Marigorta, A.; Suárez-Medina, J.; Vera-Castellano, A. Exergetic analysis of a biodiesel production process from Jatropha curcas. *Appl. Energy* **2013**, *101*, 218–225. [CrossRef]

49. Velásquez, H.; De Oliveira, S.; Benjumea, P.; Pellegrini, L. Exergo-environmental evaluation of liquid biofuel production processes. *Energy* **2013**, *54*, 97–103. [CrossRef]
50. Innocenzi, V.; Prisciandaro, M. Technical feasibility of biodiesel production from virgin oil and waste cooking oil: Comparison between traditional and innovative process based on hydrodynamic cavitation. *Waste Manag.* **2021**, *122*, 15–25. [CrossRef]
51. Mahler, D.G.; Dominikus, B. Responsible use of natural resources: Essential for sustainable growth. In *Atlas of Sustainable Development Goals 2023*; World Bank: Washington, DC, USA, 2023.
52. Sachs, D. From millennium development goals to sustainable development goals. *Lancet* **2012**, *379*, 2206–2211. [CrossRef]
53. Usaku, C.; Yahya, A.B.; Daisuk, P.; Shotipruk, A. Enzymatic esterification/transesterification of rice bran acid oil for subsequent γ-oryzanol recovery. *Biofuel Res. J.* **2023**, *10*, 1830–1843. [CrossRef]
54. Al-Saadi, L.S.; Eze, V.C.; Harvey, A.P. A techno-economic analysis based upon a parametric study of alkali-catalysed biodiesel production from feedstocks with high free fatty acid and water contents. *Biofuels* **2022**, *13*, 401–413. [CrossRef]
55. Oke, E.O.; Okolo, B.I.; Adeyi, O.; Adeyi, J.A.; Ude, C.J.; Osoh, K.; Otolorin, J.; Nzeribe, I.; Darlinton, N.; Oladunni, S. Process Design, Techno-Economic Modelling, and Uncertainty Analysis of Biodiesel Production from Palm Kernel Oil. *BioEnergy Res.* **2021**, *15*, 1355–1369. [CrossRef]
56. Gengiah, K.; Gurunathan, B.; Rajendran, N.; Han, J. Process evaluation and techno-economic analysis of biodiesel production from marine macroalgae *Codium tomentosum*. *Bioresour. Technol.* **2022**, *351*, 126969. [CrossRef]
57. Niño-Villalobos, A.; Puello-Yarce, J.; González-Delgado, Á.D.; Ojeda, K.A.; Sánchez-Tuirán, E. Biodiesel and hydrogen production in a combined palm and jatropha biomass biorefinery: Simulation, techno-economic, and environmental evaluation. *ACS Omega* **2020**, *5*, 7074–7084. [CrossRef]
58. Naveenkumar, R.; Baskar, G. Optimization and techno-economic analysis of biodiesel production from *Calophyllum inophyllum* oil using heterogeneous nanocatalyst. *Bioresour. Technol.* **2020**, *315*, 123852. [CrossRef]
59. Sakdasri, W.; Sawangkeaw, R.; Ngamprasertsith, S. Techno-economic analysis of biodiesel production from palm oil with supercritical methanol at a low molar ratio. *Energy* **2018**, *152*, 144–153. [CrossRef]
60. Yusuf, N.; Kamarudin, S. Techno-economic analysis of biodiesel production from Jatropha curcas via a supercritical methanol process. *Energy Convers. Manag.* **2013**, *75*, 710–717. [CrossRef]
61. Zhang, Y.; Dube, M.; McLean, D.; Kates, M. Biodiesel production from waste cooking oil: 1. Process design and technological assessment. *Bioresour. Technol.* **2003**, *89*, 1–16. [CrossRef]
62. Lee, S.; Posarac, D.; Ellis, N. Process simulation and economic analysis of biodiesel production processes using fresh and waste vegetable oil and supercritical methanol. *Chem. Eng. Res. Des.* **2011**, *89*, 2626–2642. [CrossRef]

catalysts

MDPI

Article

Solketal Production Using Eco-Friendly Reduced Graphene Oxide as the Catalyst

Vinicius Rossa [1,2], Sancler da Costa Vasconcelos [1], Gisel Chenard Díaz [2], Josué de Almeida Resende [2], João Pedro Reys Mattos [1], Vinicius Gomes da Costa Madriaga [1], Fernanda Franco Massante [3], Yordanka Reyes Cruz [2], Juan Lucas Nachez [4], Yutao Xing [5], Eduardo Ariel Ponzio [3] and Thiago de Melo Lima [1,*]

1 Group of Catalysis and Biomass Valorization, Institute of Chemistry, Universidade Federal Fluminense, Niterói 24020-141, RJ, Brazil; drviniciusrossa@gmail.com (V.R.); sanclervasconcelos@id.uff.br (S.d.C.V.); jpreysmattos@id.uff.br (J.P.R.M.); vinicius_madriaga@id.uff.br (V.G.d.C.M.)

2 GreenTec Laboratory, School of Chemistry, Universidade Federal do Rio de Janeiro-EQ-UFRJ, Rio de Janeiro 21941-909, RJ, Brazil; gisemarina@yahoo.es (G.C.D.); yordanka@eq.ufrj.br (Y.R.C.)

3 Grupo de Eletroquímica e Eletroanalítica, Laboratório de Materiais, Institute of Chemistry, Universidade Federal Fluminense, Niterói 24020-141, RJ, Brazil; fernandafm@id.uff.br (F.F.M.); eduardoariel@id.uff.br (E.A.P.)

4 INFES, Institute of Physics, Universidade Federal Fluminense-IQ-UFF, Niterói 28470-000, RJ, Brazil; jlnachez@id.uff.br

5 Laboratório de Microscopia Eletrônica de Alta Resolução, Institute of Physics, Universidade Federal Fluminense-IQ-UFF, Niterói 24210-346, RJ, Brazil; xy@id.uff.br

* Correspondence: tmlima@id.uff.br

Abstract: In this study, two materials based on reduced graphene oxide (rGO_H or rGO_E) were synthesized through the Hummers methodology and a more sustainable electrochemical method. These materials were extensively characterized and tested as catalysts in solketal production. Both rGO_H and rGO_E demonstrated significant catalytic activity, achieving 66.18% and 63.97% conversion rates, respectively. The catalytic activity of the synthesized materials was 30 times more efficient than the homogeneous catalyst *p*-Toluenesulfonic acid. Pseudo-homogeneous and heterogeneous kinetic models were employed to gain further insights into the glycerol ketalization reaction with acetone. The pseudo-homogeneous model suggested that the direct rate constant was lower than the reverse rate constant. In this sense, a reversible bimolecular reaction was proposed. The heterogeneous kinetic models revealed that in the Langmuir-Hinshelwood-Hougen-Watson mechanism, the controlling step of the reaction was the glycerol-acetone surface reaction on the catalyst. In contrast, in the Eley-Rideal mechanism, the reaction was controlled by the adsorbed glycerol on the reaction surface reacting with the available acetone in the bulk fluid. In the reusability tests, the rGO_E catalyst demonstrated superior performance over five consecutive cycles, maintaining the highest activity without needing post-reaction washing or treatment.

Keywords: glycerol; solketal; reduced graphene oxide; kinetic evaluation

1. Introduction

Biofuel research has attracted attention recently as an energy transition is occurring, as seen in global discussions. Biodiesel is a notable example in this context, as its production is commonly achieved through the transesterification of triglycerides. Furthermore, biodiesel is already established in the fuel market. However, as its production increases, other issues related to the process arise, particularly concerning glycerol production [1,2]. The transesterification reaction generates glycerol as a byproduct, and it is estimated that for every liter of biodiesel produced, 100 mL of glycerol is generated. However, this compound has industrial demand, which can result in various problems related to its accumulation and transportation [3].

Glycerol chemistry has emerged as a distinct branch to enhance the value of glycerol through chemical transformations involving various reactions. These reactions include carbonation, dehydration, hydrogenolysis, and oxidation [4,5]. All of these resulting products find distinct applications across multiple industry sectors, serving as solvents, additives, plasticizers, disinfectants, and more [6]. However, the ketalization reaction stands out as it produces one of the most versatile compounds in the industry: solketal. This compound finds numerous applications in green solvents and fuel additives and is also relevant to the pharmaceutical and food industries. The production of solketal involves the ketalization of glycerol (as depicted in Scheme 1), resulting in the byproduct dioxan. In this sense, it is essential to note that an acid catalyst is necessary to facilitate this reaction [5,7].

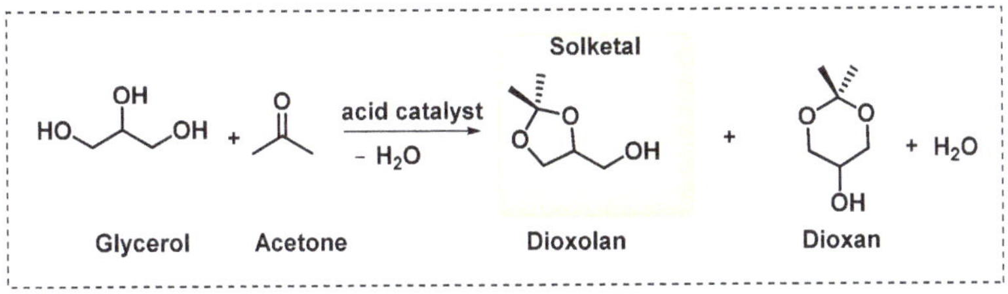

Scheme 1. Glycerol ketalization reaction with acetone producing solketal and dioxan.

As previously mentioned, selecting a catalyst is critical in successfully producing solketal. Moreover, from a green perspective, catalysis can expedite this process, reducing reaction times and temperatures in certain instances. Recently, heterogeneous catalysts have gained preference due to their favorable environmental and economic attributes. Their reusability and scalability are comparatively more straightforward, making them more practical than other types of catalysts. Additionally, these materials can be easily separated from the reaction medium, simplifying downstream processes and reducing waste generation [8,9].

Numerous heterogeneous catalysts have been proposed for glycerol ketalization, including zeolites [10,11], mesoporous silicas [11], and others. Some of these materials have achieved up to 98% conversion in certain instances, making them desirable due to their high surface area and acidity. Among these catalysts, carbon-based materials have been proposed for solketal production primarily because of their unique surface properties. For example, Rodrigues and coworkers [12] studied activated carbons as catalysts for solvent-free ketalization, achieving 52% conversion and 97% selectivity for solketal. However, despite the potential of carbon materials, only a few studies have explored these catalysts.

From this point of view, graphene and its derivatives exhibit significant potential for application as catalysts. This is attributed not only to their surface properties and thermochemical stability but also to the fact that their acidity can be adjusted through established procedures [13]. For example, graphene oxide (GO) already possesses intrinsic acidity due to phenolic and carboxylic groups on its surface. Additionally, reduced graphene oxide (rGO) is an exciting derivative since it exhibits acidity, primarily due to oxygenated functional groups at the layer edges and the presence of a graphitic aromatic structure [14]. However, the production of these materials is not environmentally friendly. As the Hummers method exemplifies, classical methods for obtaining graphene and its derivatives are frequently hazardous and result in significant waste generation. From a green chemistry perspective, adopting more sustainable approaches to minimize the environmental impact is imperative [15].

The Hummers method involves high concentrations of acidic media and manganese-based reagents [15]. However, as previously mentioned, it is imperative to discover more environmentally friendly methods for producing rGO to minimize waste generation. Electrochemical exfoliation involves the production of rGO_E through electrolysis, utilizing a graphite electrode as the anode. This process leads to the oxidation of graphite into rGO, followed by its subsequent deposition [16]. In this study, both reduced graphene oxides obtained through distinct synthesis pathways were employed and compared in relation to solketal production. Additionally, their chemical and surface properties were investigated. In this context, this paper's contribution to the specialized literature is the electrochemical exfoliation of graphite for obtaining rGO_E without hazardous chemicals, using an eco-friendly pathway for the efficient production of solketal. In addition, this paper advances catalyst reuse, in which the material can maintain its activity even after three reuse reactions without washing or purification steps. In this sense, we demonstrate here that an eco-friendly rGO can be produced with high acidity and that it is a promising material to be used as a catalyst for other acid-catalyzed reactions.

2. Results

2.1. Crystalline and Spectroscopic Characterization of Carbon-Derived Materials

Figure 1 shows the diffractograms obtained for the GFT_H, rGO_H, GFT_E, and rGO_E materials. The X-ray diffractogram of GFT_H (graphite flakes), in Figure 1a, shows a peak at $2\theta = 26.32°$ with high intensity, indicating high crystallinity in the material [17]. On the other hand, the diffractogram of rGO_H (reduced graphene oxide Hummers method), in Figure 1b, exhibits a significant reduction in this peak intensity compared to graphite, suggesting a considerable loss of crystallinity in the material, highlighting the efficiency of the graphite exfoliation process in forming rGO_H. The lower peak intensity at $2\theta = 13.29°$ in rGO_H indicates low efficiency in the samples' oxidative processes, suggesting minimal material oxidation and direct formation of reduced graphene oxide [18].

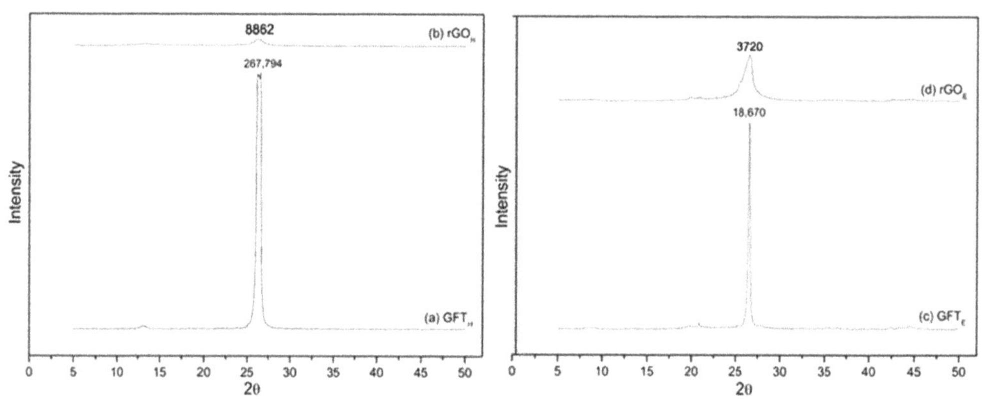

Figure 1. X-ray diffractograms of graphite flakes (**a**) GRF_H, (**b**) rGO_H, (**c**) GFT_E, and (**d**) rGO_E.

The graphite from the carpenter's pencil (GFT_E), in Figure 1c, exhibits a characteristic graphite peak diffracted at $2\theta = 26.51°$ in the (001) plane with high intensity, indicating high crystallinity for this material. After the oxidative exfoliation process, in Figure 1d (rGO_E), the peak shifted slightly to $2\theta = 26.29°$, and there was a significant loss of crystallinity and structural disorganization when compared to its starting material (GFT_E), indicating the efficiency of the exfoliation process [19].

Through XRD analysis, important information was obtained, which enabled the determination of distances between layers using Bragg's Law, average crystallite sizes through Scherrer's Equation, and the number of stacked layers in the materials GFT_H, rGO_H, GFT_E, and rGO_E, as shown in Table S1 [20–24].

Table S1 shows a slight increase in the distance difference between the layers of the carbon-based materials, indicating the low degree of graphite oxidation. However, there was a high exfoliation of 44 layers in the GFT_H precursor material and 17 layers in the rGO_H, respectively. The exfoliation via the electrochemical method proved to be more efficient than the Hummers method, with 91 layers in the $GRFT_E$ precursor and 15 layers in rGO_E.

There was also a significant decrease in the crystallite size of the precursor materials GFT_H and GFT_E (15 and 31 nm, respectively) when compared with the size of the crystallites of the rGO_H and rGO_E (5 nm).

Next, Raman spectroscopy was utilized to enhance comprehension of the oxidation processes. Generally, in the Raman spectra of graphite and reduced graphene oxide, two characteristic bands can be observed: The D Band (defects in the structure), with A_{1g} symmetry, located between 1350 and 1360 cm^{-1} [25–27], attributed to oxygenated groups in the basal plane of the graphene derivatives' structure [26] during sp^3 hybridization of carbon in the layers [28]. The G Band is present in ordered graphene networks (E_{2g} symmetry), with high intensity located between 1580 and 1610 cm^{-1} [25–27], attributed to sp^2 hybridization [27]. Figure 2 presents the Raman spectra for the GFT_H, rGO_H, GFT_E, and rGO_E materials.

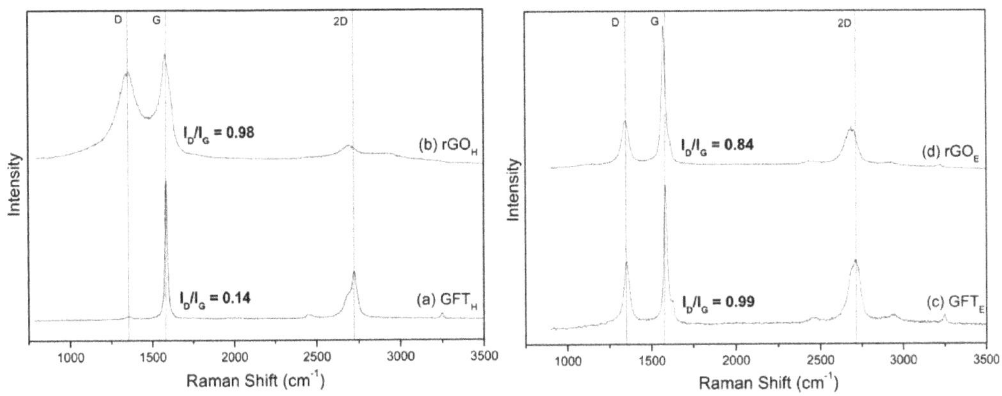

Figure 2. Raman spectra of graphite flakes (**a**) GRF_H, (**b**) rGO_H, (**c**) GFT_E, and (**d**) rGO_E.

Figure 2a presents the D band of GFT_H, corresponding to the vibrational mode of sp^3 hybridized carbon atoms with a minimum intensity at 1341 cm^{-1}, which occurs due to the high organization of carbon atoms with sp^2 hybridization [27,29,30]. Conversely, the D band of rGO_H, in Figure 2b, shows an increase in intensity at 1347 cm^{-1}, attributed to structural disorganization after the exfoliation–oxidation process. However, this increase in the intensity of the D band indicates the rupture of C=C bonds during the oxidation of GFT_H to rGO_H and an increase in sp^3 hybridized carbon atoms [27,29,30]. This is due to imperfections created via the insertion of oxygenated groups in the basal plane of the carbon layers, leading to partial structural disorder [27,29,30]. The G band decreased with the oxidation of GFT_H, seen in Figure 2a, located at 1583 cm^{-1}, while for rGO_H, seen in Figure 2b, it was observed at 1580 cm^{-1}. The rGO_H structure resulted in a higher I_D/I_G intensity ratio due to defects in GFT_H (I_D/I_G = 0.14), and after the exfoliation–oxidation process due to rGO_H (I_D/I_G = 0.98) [27,29,30]. The Raman spectrum of GFT_E presented a D band of graphite with low intensity at 1354 cm^{-1} (sp^3 hybridized carbon atoms) due to the high organization of carbon atoms with sp^2 hybridization.

On the other hand, this band in rGO_E displayed an increase in intensity due to structural disorganization after the exfoliation of GFT_E via the electrochemical method. This increase is attributed to imperfections created via the insertion of oxygenated groups in the basal plane of the carbon layers, leading to partial structural disorder. The G band, corresponding to the vibrational mode of sp^2 hybridized carbon atoms, decreased in the GFT_E material after its exfoliation to produce rGO_E [30]. The analyses of the I_D/I_G results for the rGO_H and rGO_E materials and a comparison of them with their precursor materials showed a structural rearrangement in the formation of reduced graphene oxides [30], corroborating the XRD results.

The 2D or 'G' band, ranging between 2600 cm^{-1} and 2800 cm^{-1}, is a broad hump associated with the interplanar stacking order, the number of layers, and even the in-plane order of the crystallite size for graphene materials. According to Malard and coworkers [31], the presence of the 2D (G′) band is attributed to second-order two-phonon processes of the sp^2 carbon structure. A broader 2D band implies an increased number of graphene layers.

The broader 2D region for rGO_H shows a lower intensity when compared with rGO_E. This indicates that the exfoliation of rGO_H results in more pronounced overlapping GO layers when compared with rGO_E. On the other hand, the higher intensity of the G peak compared to the D peak for rGO_E when compared with rGO_H suggests that there is more graphene oxide than graphene sheets or graphite.

To determine the oxygenated groups introduced to the graphites after the exfoliation–oxidation process, infrared spectroscopy was utilized. Figure S2 presents the infrared spectra obtained for the GFT_H, rGO_H, GFT_E, and rGO_E materials. The spectrum of GFT_H graphite, seen in Figure S2a, exhibits bands related to the vibrations of C sp^2 bonds in the graphene layers that constitute the graphite. A low-intensity band at 3488 cm^{-1} is attributed to –OH group bonds, and a band at 1409 cm^{-1} indicates the presence of –COOH bonds from the carboxyl group, suggesting that GFT_H has a low degree of oxidation. In the spectrum of rGO_H, in Figure S1b, a band appeared at 3338 cm^{-1}, corresponding to –OH group bonds. This indicates the insertion of –OH groups into the GFT_H structure during graphite's exfoliation/oxidation process. Additionally, in Figure S2b, the insertion of oxygenated groups into the graphite structure is observed, including carbonyls (1578 cm^{-1}); carboxyls (1409 cm^{-1}); alcohols and phenols (1388 cm^{-1}, 1247 cm^{-1}); and epoxide groups (1076 cm^{-1}), as reported in the literature [20,25,28,32–37].

The spectrum of GFT_E graphite, in Figure S2c, exhibits bands related to the vibrations of C sp^2 bonds in the graphene layers in the graphite structure. A low-intensity band at 3369 cm^{-1} is attributed to –OH group bonds, and a band at 1158 cm^{-1} indicates the presence of C–O bonds of alcohols and phenols, suggesting that GFT_E presents some degree of oxidation. In the spectrum of rGO_E, in Figure S2d, an intense peak appears at 3360 cm^{-1}, corresponding to –OH group bonds, indicating the insertion of –OH groups into the GFT_E structure during the exfoliation process of graphite. Figure S2d shows that the oxygenated groups inserted into the graphite structure after the exfoliation/oxidation process via the electrochemical method include carbonyls (1538 cm^{-1}); carboxyls (1444 cm^{-1}); alcohols and phenols (1365 cm^{-1}, 1234 cm^{-1}, 1204 cm^{-1}, 1158 cm^{-1}, and 1121 cm^{-1}); and epoxide groups (1051 cm^{-1}), as reported in the literature [20,25,27,28,32–38]. Also, Table S2 shows the assignments of the prominent bands in the infrared spectra for the GFT_H, rGO_H, GFT_E, and rGO_E materials.

2.2. Thermal and Textural Characterization of Carbon-Derived Materials

Thermogravimetric analyses (TGAs) were employed to assist in the study of the insertion of oxygenated groups into the structures of commercial graphites to form rGOs. Figure 3 presents the weight loss curves of the materials for the GFTH, rGOH, GFTE, and rGOE materials.

The GFT_H presented two weight loss events. The first, between 25 and 470 °C, was attributed to the loss of water molecules adsorbed on the surface and trapped between the layers of the material, resulting in a mass loss of 0.60%. The second event was assigned to decompose the more stable graphite layers (formed via sp^2 hybridized carbon), leading to a mass loss of 22.95%. GFT_H was the most stable among the materials, with a total mass loss of 23.55% [21,28,33,34,38,39]. Its reduced graphene-oxide-derived material also presented six weight loss events, with the first at 107 °C, which was attributed to the loss of surface-adsorbed water, resulting in a mass loss of 3.88%. At 228 °C, the second event was assigned to decomposing less stable functional groups in the rGO structure, with a weight loss of 12.07%. The third, at 367 °C, was attributed to the decomposition of more stable functional groups in the rGO structure, with a weight loss of 7.27%. The fourth, at 450 °C, was also assigned to decomposing more stable functional groups in the rGO structure, resulting in a weight loss of 6.65%. At 507 °C, the fifth event was attributed to the decomposition of less stable rGO layers (formed via sp^3 hybridized carbon); the most prominent weight loss event was 66%. Finally, at 663 °C, the sixth event was attributed to the decomposition of the more stable rGO layers (formed via sp^2 hybridized carbon), resulting in a weight loss of 2.71%. At the end of the analysis, only 0.69% of the original mass remained, indicating almost complete degradation of the rGO_H. The carpenter pencil presented four events, as seen in Figure 3c. The first, at 83 °C, was attributed to the loss of water molecules adsorbed on the surface of the graphite, with a weight loss of 0.28%. The second, at 150 °C, was assigned to the loss of water trapped between the layers of the graphite, resulting in a weight loss of 3.20%. The third, at 292 °C, was attributed to the pyrolysis of less stable and partially oxidized graphite layers, releasing CO, CO_2, and H_2O, with a mass loss of 6.80%. This result agrees with the infrared spectrum of GFT_E, seen in Figure 3c. The last thermal event, at 640 °C, was assigned to decomposing the more stable graphite layers (formed via sp^2 hybridized carbon), resulting in the most significant weight loss of 39.12%. However, the total mass loss of the GFT_E was 49.22%, making it the most stable compared to the other studied materials. In addition, the rGO_E material presented six events, seen in Figure 3d. The first, at 195 °C, was attributed to the loss of water molecules adsorbed on the surface of rGO_E, with a weight loss of 3.86%. At 258 °C, the second was assigned to decomposing less stable oxygenated functional groups in the rGO structure, with a weight loss of 1.67%. The third, at 308 °C, was attributed to the decomposition of more stable oxygenated functional groups in the rGO structure, resulting in a weight loss of 1.93%. The fourth, at 593 °C, was assigned to decomposing more stable functional groups in the rGO structure, leading to a weight loss of 12.65%. The fifth, at 615 °C, was attributed to the continuation of the decomposition of more stable functional groups in the rGO_E structure and sp^3 carbons, with a mass loss of 21.30%. The sixth and final event, at 703 °C, was assigned to the decomposition of rGO layers (sp^2 carbons), resulting in the most extensive mass loss of 34.68%. rGO_E showed a total weight loss of 76.09% of its initial mass [21,28,33,34,38,39].

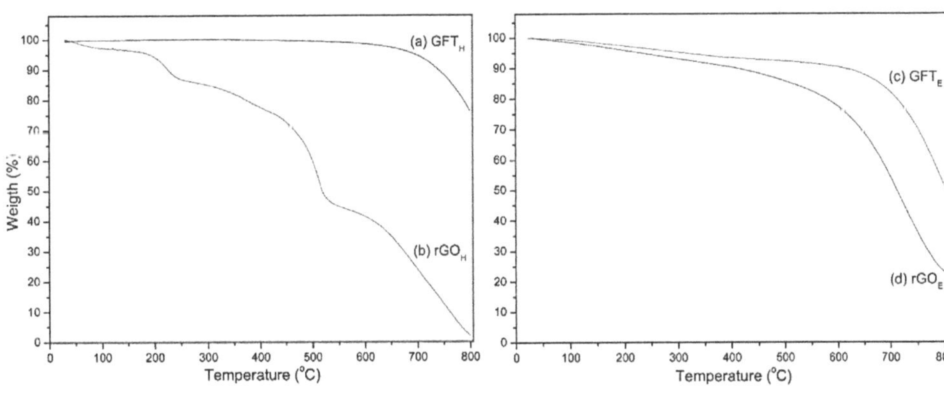

Figure 3. Thermogravimetric analysis of graphite flakes (**a**) GRF_H, (**b**) rGO_H, (**c**) GFT_E, and (**d**) rGO_E.

To test rGO_H and rGO_E as catalysts for the glycerol ketalization reaction [40], these materials and their precursors were also characterized via textural and porosity analysis using the B.E.T., t-plot, and B.J.H. methods, as well as via morphological analysis through scanning electron microscopy (SEM). Table 1 presents the results from the textural characterization for the GFTH, rGOH, GFTE, and rGOE materials via N_2 sorption analysis.

Table 1. Textural characterization of the GFT_H, rGO_H, GFT_E, and rGO_E materials.

	Materials			
	GRF_H	**rGO_H**	**GRF_E**	**rGO_E**
Specific superficial area (m^2 g^{-1})	5.1	31.0	4.3	21.3
Micropore volume (cm^3 g^{-1}) [1]	-	0.0008	-	0.0005
Mesopore volume (cm^3 g^{-1}) [2]	0.062	0.053	0.050	0.047
Average size pore (nm)	10.9	10.3	10.9	10.2

[1] Calculated using t-plot method; [2] calculated using B.J.H. method.

2.3. *Morphological Characterization of Carbon-Derived Materials*

The specific surface areas of GFT_H and GFT_E were 5.1 and 4.3 m^2 g^{-1}, respectively, which are very close to the values found in the literature, $\approx$4 m^2 g^{-1} [41,42]. The rGO_H showed a specific surface area of 31.0 m^2 g^{-1}; in rGO_E, this area was 21.3 m^2 g^{-1}. The electrochemical process typically produces rGO with a lower specific surface area than the exfoliation–oxidation carried out via the traditional Hummers method [43,44]. Both rGO_H and rGO_E exhibited larger specific surface areas than their precursor materials, suggesting that the exfoliation process was effective. This exfoliation is better verified with the aid of morphological analyses conducted using scanning electron microscopy, as shown in Figure 4, which displays the micrographs of the GFT_H, rGO_H, GFT_E, and rGO_E materials.

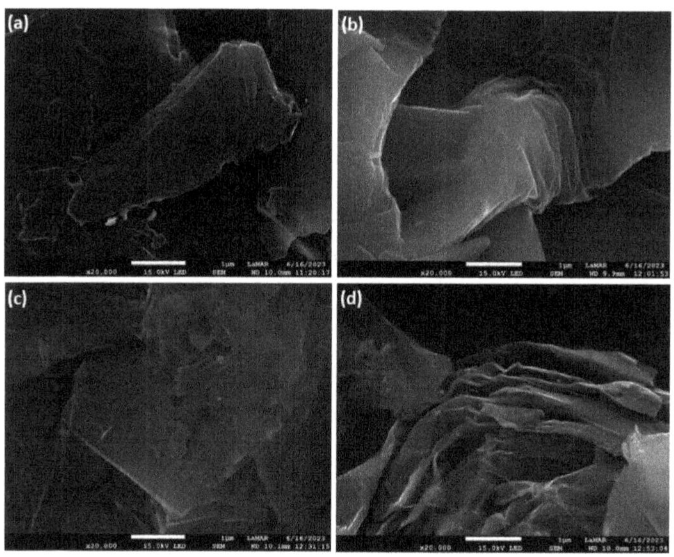

Figure 4. SEM images of graphite flakes (**a**) GRF_H, (**b**) rGO_H, (**c**) GFT_E, and (**d**) rGO_E.

Figure 4a,c display the SEM images of GFT_H and GFT_E, respectively, illustrating the morphology of graphite flake and carpenter's pencil graphite. These images reveal a morphology consisting of densely packed graphene sheets. Upon oxidation and exfoliation processes to form rGO_H and rGO_E, as shown in Figure 4b,d, there was an observable increase in spacing and disorganization between these sheets. In a morphological analysis conducted by Basso and coworkers [20], they observed similar morphology changes during the transition from graphite to rGO due to oxidation [20].

Energy Dispersive X-ray Spectroscopy (EDS) analyses were conducted during SEM analysis to investigate the materials' surface chemical compositions. The results of the surface chemical compositions of the materials are presented in Table 2.

Table 2. EDS results of surface chemical analysis of the GFT_H, rGO_H, GFT_E, and rGO_E materials.

	Elements (wt.%)				
Materials	**C**	**O**	**S**	**Si**	**Al**
GRF_H	98.0	2.0	-	-	-
rGO_H	79.6	19.6	0.8	-	-
GRF_E	73.5	21.0	-	4.8	0.7
rGO_E	68.3	24.9	0.5	4.8	1.5

rGO_H exhibited an oxygen content of 19.6%, whereas its precursor material showed only 2%, thus indicating a low oxidation level of commercial graphite. In contrast, the commercial GRF_E already had a high oxidation level, with 21.0% oxygen on its surface. After its exfoliation/oxidation, the oxygen content increased to approximately 5% in rGO_E, representing 24.9%. Both rGO_E and GFT_E exhibited impurities like Si and Al. The rGO_H material and rGO_E showed 0.8% and 0.5% sulfur content, respectively, attributed to the sulfuric acid used in both rGO production processes.

Considering these characteristics, acidity tests were also conducted on rGO_H and rGOE. However, due to their low thermal stability, acidity was determined through acid–base titration, as presented in Table S3. These values indicate that the exfoliation/oxidation process used to obtain rGO_H and rGO_E makes these materials significantly acidic for application as catalysts in the glycerol ketalization reaction with acetone for solketal production.

2.4. Catalytic Activity and Kinetic Studies of Glycerol Ketalization

Table 3 presents the results obtained from preliminary catalytic tests, including control (blank/no catalyst), the homogeneous *p*-Toluenesulfonic acid (PTSA) catalyst, and the heterogeneous rGO_H and rGO_E catalysts, in terms of glycerol conversion (X_A).

Table 3. Results of preliminary catalytic tests for glycerol conversion (X_A%), solketal selectivity (SS_5), dioxan selectivity (SD_6%), solketal yield (YS_5%), and dioxan yield (YD_6%).

Reaction	X_A (%)	SS_5 (%)	SD_6 (%)	YS_5 (%)	YD_6 (%)	TOF (h^{-1})
Control	0.20 ± 0.02	80.39 ± 0.02	19.61 ± 0.02	0.16 ± 0.02	0.04 ± 0.02	-
PTSA	35.45 ± 0.64	97.71 ± 0.64	2.49 ± 0.64	64.32 ± 0.64	1.64 ± 0.64	1.80 ± 0.37
rGO_H	66.18 ± 0.45	98.38 ± 0.45	1.62 ± 0.45	51.00 ± 0.45	1.55 ± 0.45	52.36 ± 0.32
rGO_E	63.97 ± 0.86	80.73 ± 0.86	19.27 ± 0.86	51.64 ± 0.86	12.33 ± 0.86	57.44 ± 0.43

(Molar ratio of glycerol/acetone = 1:4; 5% catalyst based on the mass of glycerol; 60 °C; 500 rpm; 120 min with autogenous pressure; conducted in a 100 mL Parr reactor).

The influence of an acidic catalyst on the glycerol ketalization reaction with acetone is evident from the negligible glycerol conversion (X_A = 0.20%) in the control reaction. The homogeneous PTSA catalyst exhibited low activity for solketal conversion (35.45%) but high selectivity towards solketal (97.71%). However, the heterogeneous catalysts showed increased activity, with 66.18% and 63.97% conversion rates for rGO_H and rGO_E, respectively. Nonetheless, the rGO_E catalyst synthesized using the electrochemical method showed lower selectivity towards solketal (80.73%) than the other catalysts, where the selectivity for rGO_H and PTSA was 98.38% and 97.71%, respectively. However, it is essential to note that PTSA is a homogeneous catalyst that causes equipment corrosion and waste generation. In addition, the conventional graphene-derived material rGO_H is prepared via a non-sustainable procedure and produces much waste. The electrochemically produced graphene-derived material rGOE, despite presenting lower solketal selectivity, must be highlighted for its greener production.

The TOF values for PTSA, rGO_H, and rGO_E are presented in Table 3. The number of acid sites for the heterogeneous catalysts was obtained through the acidity determination via the acid–base titration technique (Table S3). The highest TOF value was 57.44 h^{-1}, achieved with the rGO_E catalyst after 2 h of reaction. The lowest TOF value was 1.80 h^{-1}, obtained with the PTSA catalyst. No significant difference was observed when comparing the TOF value for the heterogeneous catalysts, highlighting the potential application of rGO_E as a more sustainable catalyst.

Figure 5 presents the results obtained for glycerol conversion tests (X_A%) over time using the homogeneous PTSA catalyst and the heterogeneous rGO_H and rGO_E catalysts.

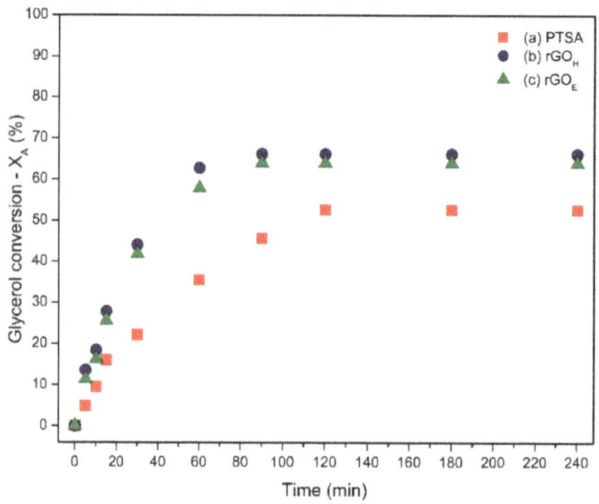

Figure 5. Results of glycerol conversion tests (X_A%) over time for the homogeneous catalyst PTSA and the heterogeneous catalysts rGO_H and rGO_E. (Glycerol/acetone molar ratio = 1:4; 5% catalyst relative to the mass of glycerol; 60 °C; 500 rpm with autogenous pressure; conducted in a 100 mL Parr reactor).

It can be observed that the rGO_H catalyst exhibited higher initial conversion values, converting approximately 19% of the glycerol in the first 10 min of the reaction. For comparison, rGO_E converted about 16% of the available glycerol simultaneously. Meanwhile, the homogeneous catalyst PTSA transformed 9% during the initial 10 min of the reaction. After 90 min of reaction, the glycerol conversion stabilized for the heterogeneous catalysts. This suggests that the reaction reached a steady state, with 66% and 64% conversion for rGO_H and rGO_E, respectively. The homogeneous catalyst PTSA was constant after 120 min of reaction.

Two approaches were considered to estimate kinetic parameters (k_1 and k_2) for the glycerol ketalization reaction in the presence of acidic catalysts: the pseudo-homogeneous and heterogeneous kinetic models.

Using the Python programming language and the data from Figure 5, through GOOGLE COLAB (https://colab.research.google.com/?utm_source=scs-index&hl=en, accessed on 5 November 2023), it was possible to estimate the values for the rate constants k_1 and k_2 for three different catalytic systems (PTSA, rGO_H, and rGO_E), and the estimated values are shown in Table 4.

Table 4. Estimated values for the rate constants k_1 and k_2 for three different catalytic systems (PTSA, rGO_H, and rGO_E). Conditions: 500 rpm; molar ratio 1:4 (G:A); 5% catalyst relative to the mass of glycerol; 60 °C; 240 min.

Catalysts	k_1	k_2	K	X_A	X_{Aeq}	Q	R^2
PTSA	0.0015	0.0077	0.1953	0.53	0.55	0.0024	0.9929
rGO_H	0.0037	0.0089	0.4184	0.66	0.67	0.0019	0.9966
rGO_E	0.0033	0.0092	0.3621	0.63	0.65	0.0010	0.9980

The pseudo-homogeneous model showed a good fit for all three reaction systems where PTSA, rGO_H, and rGO_E were used as catalysts, as shown in Figure 6. Based on the values of the rate constants, forward (k_1) and reverse (k_2), as presented in Table 4, it is possible to suggest that this reaction tends to shift the equilibrium towards forming the reactants. The constant k_1 is related to the formation of solketal, and k_2 is associated with the reversibility of the reaction, with $k_2 > k_1$, meaning that the rate of product formation is slower than the rate that signifies the reversibility of the reaction towards the formation of the reactants. However, the difference in the equilibrium constant (K) values indicates that the reaction had not yet reached equilibrium, as evident when examining the calculated X_{Aeq} values from Equation (S9). The squared residuals (Q) below 0.01 indicate a low deviation between the experimentally obtained X_A data and the values of X_A calculated by the model; $X_{A\,EXP}$ and $X_{A\,CAL}$ are very close, validating the applied kinetic model. The determination coefficient values (R^2) being close to 1 indicate that the pseudo-homogeneous model can describe the experimentally obtained results, as shown in Figure 6.

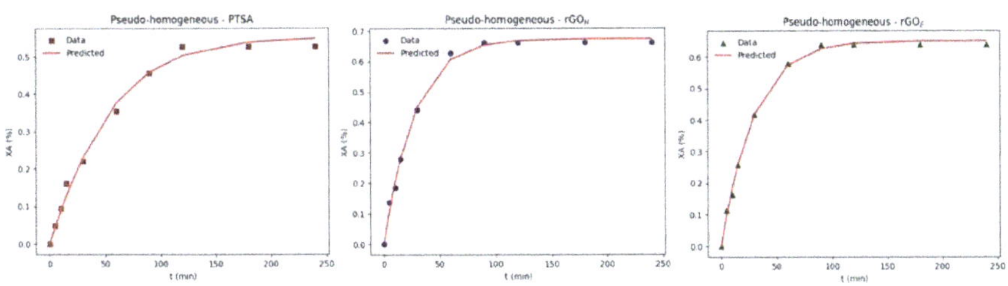

Figure 6. Pseudo-homogeneous models (data and predicted) applied to PTSA, rGO_H, and rGO_E were used as catalysts to produce the homogeneous catalyst (PTSA) and the heterogeneous catalysts (rGO_H and rGO_E). Conditions: glycerol/acetone molar ratio = 1:4; 5% catalyst relative to the mass of glycerol; 60 °C; 500 rpm with autogenous pressure; conducted in a 100 mL Parr reactor for 240 min.

The reversible model simplifies the complex kinetics of heterogeneous reactions by reducing the parameters so that k_1 and k_2 encompass all kinetic terms for the products and reactants, respectively. The pseudo-homogeneous model fits well with the heterogeneous rGO_H and rGO_E catalysts, but this does not imply that the catalysis is genuinely homogeneous. This happens because the surface of the heterogeneous catalyst is saturated with one of the reactants, in this case, acetone, as it is the excess reagent (glycerol/acetone molar ratio = 1:4). When there is an excess of reactant in the reactor, this excess can be much higher on the catalyst's surface. Thus, the model can be called pseudo-homogeneous. The literature has limited examples of modeling and kinetic studies using homogeneous or pseudo-homogeneous models for the glycerol ketalization reaction with acetone, and these are only for heterogeneous catalysts.

However, Esteban and coworkers [45] proposed a pseudo-homogeneous model and concluded that the reaction of $k_2 > k_1$ uses sulfonated resin as a catalyst. We observed the same conclusions regarding $k_2 > k_1$ in a study involving H-BEA zeolites in 2017, using a pseudo-homogeneous model programmed in Fortran, utilizing R2W [9].

The second approach used was heterogeneous kinetic modeling, aiming to gain insights into the rate-controlling step of the glycerol ketalization reaction with acetone (adsorption, chemical reaction, or desorption) when using heterogeneous catalysts of the rGO_H and rGO_E types. Once again, the reaction was considered to be of a bimolecular and reversible type. Dioxan was also disregarded in this study in an attempt to simplify the already complex heterogeneous model.

However, before proceeding with the modeling to estimate the kinetic parameters, some catalytic tests were conducted to determine whether the reaction was exclusively under kinetic control and free from external and internal diffusion limitations under the employed reaction conditions. The agitation (rpm) was varied, and glycerol conversion was monitored after 120 min of reaction, as shown in Figure 7.

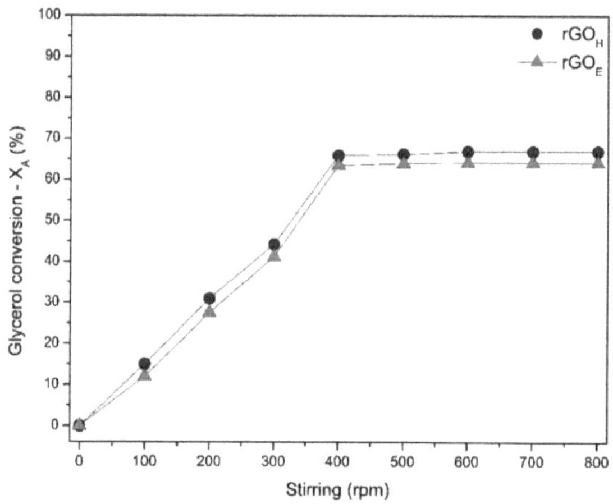

Figure 7. Results of glycerol conversion tests (XA%) as a function of agitation. (Glycerol/acetone molar ratio = 1:4; 5% catalyst based on glycerol mass; 60 °C; 100–800 rpm with autogenous pressure; conducted in a 100 mL Parr reactor for 120 min).

Figure 7 shows that between 400 and 800 rpm, glycerol conversion is free from external diffusion limitations. Additionally, internal diffusion limitations can also be disregarded within the agitation range of 500 rpm, as the kinetic diameters of the reactants and products of this reaction [9] are between 0.43 and 0.51 nm, and the average pore diameter of the catalysts used is around 10 nm, as indicated in Table 1.

The results of Figure 5 can be used to proceed with heterogeneous kinetic modeling using the rGO_H and rGO_E catalysts. In Nanda's studies (2014), for the same reaction, it was also observed that above 400 rpm, there was no influence of mass transfer [46]. The kinetic parameter results for each mechanism and the created models (Equations (S11)–(S23)) were estimated using Python programming language and the data from Figure 5.

Using Google Colab/Python language, it was possible to estimate the values for the rate constants, thermodynamic equilibrium constants, and the adsorption equilibrium constants of the reactants, as well as the desorption constants for the products, for two different catalytic systems (rGO_H and rGO_E). The estimated values are shown in Tables 5 and 6; LHHW and ER models, respectively.

Table 5. Results of the Estimated Kinetic Parameters for the glycerol ketalization reaction with acetone using the LHHW mechanism, considering the five proposed models (Equations (S13)–(S17)), using rGO_H and rGO_E catalysts. Conditions: 500 rpm; molar ratio 1:4 (G:A); 5% catalyst relative to the mass of glycerol; 240 min.

Catalyst	**Mechanism: LHHW**				
rGO_H	**Steep Control**				
Kinetic Parameters	**ADS, A**	**ADS, B**	**REACTION**	**DES, C**	**DES, D**
	Model 1	**Model 2**	**Model 3**	**Model 4**	**Model 5**
k (g_{cat} L mol^{-1} min^{-1})	0.0253	0.0027	0.5602	n	n
K	10.6544	0.1120	0.4063	n	n
K_A (L mol^{-1})	−3.3140	−0.5401	0.5473	n	n
K_B (L mol^{-1})	−0.0923	−0.2942	0.4391	n	n
K_C (L mol^{-1})	0.6902	0.5170	0.2468	n	n
K_D (L mol^{-1})	0.6912	0.5170	0.2468	n	n
Q	0.0007	0.0025	0.0017	-	-
R^2	0.9889	0.9940	0.9979	-	-

Table 5. *Cont.*

Catalyst	Mechanism: LHHW				
rGO_E	Steep Controlling				
Kinetic Parameters	ADS, A Model 1	ADS, B Model 2	REACTION Model 3	DES, C Model 4	DES, D Model 5
k (g_{cat} L mol^{-1} min^{-1})	0.0302	0.0028	0.5234	n	n
K	9.5613	0.1097	0.3512	n	n
K_A (L mol^{-1})	−1.7279	−0.4974	0.4851	n	n
K_B (L mol^{-1})	−0.0418	−0.2463	0.3522	n	n
K_C (L mol^{-1})	0.3599	0.3973	0.2150	n	n
K_D (L mol^{-1})	0.3602	0.3973	0.2150	n	n
Q	0.0007	0.0011	0.0008	-	-
R^2	0.9988	0.9975	0.9985	-	-

n = did not converge.

Table 6. Results of the Estimated Fitting Parameters for the glycerol ketalization reaction with acetone using the ER mechanism, considering the 6 proposed models (Equations (S18)–(S23)), using rGO_H and rGO_E catalysts. Conditions: 500 rpm; molar ratio 1:4 (G:A); 5% catalyst relative to the mass of glycerol; 240 min.

Catalyst	Mechanism: ER					
rGO_H	Steep Controlling					
Kinetic Parameters	ADS, A	ADS, B	SR, A	SR, B	DES, C	DES, D
	Model 6	Model 7	Model 8	Model 9	Model 10	Model 11
k (g_{cat} L mol^{-1} min^{-1})	0.0583	0.0089	0.0633	0.1693	n	n
K	10.6660	0.1100	0.4062	0.4063	n	n
K_A (L mol^{-1})	−7.2409	-	0.1535	-	n	n
K_B (L mol^{-1})	-	−0.1497	-	0.2658	n	n
K_C (L mol^{-1})	3.1890	-	0.0209	-	n	n
K_D (L mol^{-1})	-	0.1051	-	−0.0191	n	n
Q	0.0007	0.0031	0.0017	0.0017	-	-
R^2	0.9989	0.9944	0.9971	0.9971	-	-
Catalyst	Mechanism: ER					
rGO_E	Steep Controlling					
Kinetic Parameters	ADS, A Model 6	ADS, B Model 7	SR, A Model 8	SR, B Model 9	DES, C Model 10	DES, D Model 11
k (g_{cat} L mol^{-1} min^{-1})	0.0408	0.0093	0.0581	0.1531	n	n
K	9.5616	0.1092	0.3512	0.3512	n	n
K_A (L mol^{-1})	−2.3098	-	0.1479	-	n	n
K_B (L mol^{-1})	-	−0.3816	-	0.2745	n	n
K_C (L mol^{-1})	1.0152	-	0.0215	-	n	n
K_D (L mol^{-1})	-	1.5644	-	−0.0048	n	n
Q	0.0007	0.0012	0.0008	0.0008	-	-
R^2	0.9988	0.9975	0.9985	0.9985	-	-

n = did not converge.

The parameters' physical realism estimated through nonlinear regression was first considered to assess the proposed models' adequacy. This means that models that yielded negative values for any parameter and/or models that did not converge were discarded. To validate the model, the method chosen was the minimization of the sum of squared residuals (Q); see Equation (S10).

Upon observing Tables 5 and 6, it was found that the heterogeneous models created could only describe the kinetics of the heterogeneous reaction for the rGO_H and rGO_E

catalysts. Only one model was validated for each mechanism, as the other models resulted in one or more negative parameter values or did not converge and were thus discarded.

For the Langmuir-Hinshelwood-Hougen-Watson (LHHW) mechanism [47,48], the rate-controlling step of the reaction was the chemical reaction between the adsorbed reactants (A) glycerol and (B) acetone on the catalyst surface ($R^2 = 0.9979$; $Q = 0.0017$ for rGO_H and $R^2 = 0.9985$; $Q = 0.0008$ for rGO_E). This step can be considered a plausible rate-controlling step for the glycerol ketalization reaction with acetone.

As for the Eley-Rideal (ER) mechanism [45,49,50], the rate-controlling step of the reaction was the chemical reaction with only (B) acetone being adsorbed on the catalyst surface ($R^2 = 0.9971$; $Q = 0.0017$ for rGO_H and $R^2 = 0.9985$; $Q = 0.0008$ for rGO_E), which can also be considered a plausible rate-controlling step for the glycerol ketalization reaction with acetone. It is worth noting that the heterogeneous kinetic models were employed to predict which step serves as the rate-determining step for each of the mechanisms (LHHW and ER) used in the glycerol ketalization with an acetone reaction, not to estimate the value of each parameter involved physically.

It can be said that Model 3 (LHHW), Equation (S15), Model 8 (ER), and Equation (S20) all showed good fits for both the rGO_H and rGO_E catalysts. The squared residuals (Q) below 0.01 indicate a low deviation between the experimentally obtained X_A data and the X_A values calculated using the model. $X_{A\ EXP}$ and $X_{A\ CAL}$ are very close, validating the applied kinetic model. The values of the determination coefficients (R^2) very close to 1 indicate that the heterogeneous models can be employed to describe the experimentally obtained results, as shown in Figure 8.

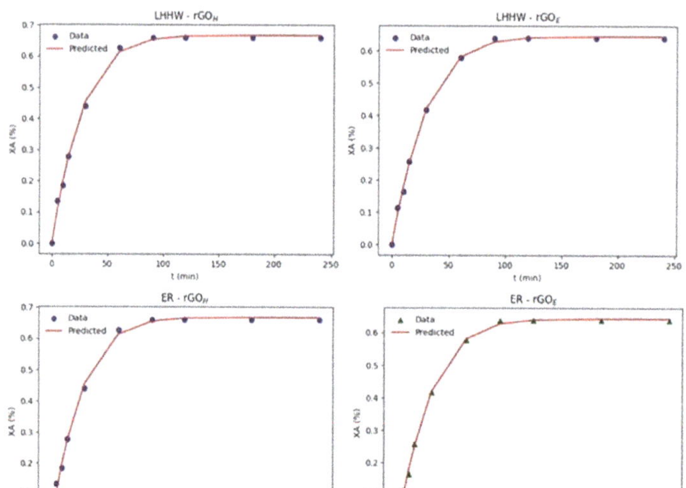

Figure 8. Kinetic heterogeneous models (data and predicted) applied to rGO_H and rGO_E were used as catalysts. Model 3 shows that the LHHW rate-controlling step of the reaction was the chemical reaction between the adsorbed reactants (A) glycerol and (B) acetone on the catalyst surface, and Model 8 shows that the ER rate-controlling step of the reaction was the chemical reaction with only (B) acetone being adsorbed on the catalyst surface. Conditions: glycerol/acetone molar ratio = 1:4; 5% catalyst relative to the mass of glycerol; 60 °C; 500 rpm with autogenous pressure; conducted in a 100 mL Parr reactor for 240 min.

The application of computational molecular modeling to study the mechanism of the glycerol ketalization reaction with acetone, using rGO_H and rGO_E as catalysts, would enable the confirmation of the mechanism and the rate-controlling step for this case. Studies on the kinetics of the glycerol ketalization reaction with acetone using heterogeneous catalysts were conducted by the research group led by Nanda (2014) and by the group led by Esteban (2015) [45]. Nanda (2014) used Amberlyst 35 resin as the catalyst and employed the LHHW model to describe the glycerol ketalization reaction with acetone, with the chemical reaction as the rate-limiting step, which, with the adaptation of the data presented in this work, represents a plausible model [46]. On the other hand, the Esteban group (2015) concluded that the Eley–Rideal mechanism describes the glycerol ketalization reaction with acetone, with the chemical reaction involving the adsorption of only one of the reactants as the rate-limiting step of the reaction [45].

2.5. Study of Reuse of Heterogeneous Catalysts

Table 5 presents the results of the reuse tests for the rGO_H and rGO_E catalysts, and Table S4 displays the acidity results of the catalysts before and after the reuse tests. It is worth noting that the catalysts were initially separated from the reaction medium via filtration, without washing and drying, and were reused four times.

Table 5 presents the deactivation (reuse) study for the rGO_H and rGO_E catalysts, which exhibited the highest catalytic activities among the heterogeneous catalysts. There was a significant reduction in glycerol conversion, dropping from 66% to 23% for the rGO_H catalyst and from 64% to 30% for the rGO_E catalyst. This fact indicates that deactivation of the acidic sites of the material may have occurred, hindering the catalysis of the reaction. Additionally, the selectivity for solketal remained constant, averaging 80% for the rGO_E catalyst. However, the selectivity toward solketal for the rGO_H catalyst decreased from 98% to 91%, indicating an increase in dioxan selectivity.

Table S4 shows a significant decrease in the concentration of acidic sites in the rGO_H and rGO_E materials, which was directly proportional to their deactivations, as seen in Table 7. The highest deactivation for glycerol conversion was observed for the reduced graphene oxide prepared via Hummers method, i.e., a decrease from 66.18 to 23.18% after the fourth reuse (64.97% of total deactivation). The electrochemically exfoliated material presented a conversion decrease from 63.97% to 30.37% (a total deactivation of 52.68%). When the yields of solketal were considered, the former catalyst presented an overall deactivation of 67.47%, and the latter showed 54.12%, thus demonstrating the superiority of the rGO_E material as a catalyst for glycerol conversion. As possible causes of the deactivation, we can cite the water formation during the glycerol conversion and, consequently, the hydration of the acid sites (as observed via the reduction in the overall acidity, as seen in Table S4). We can also cite the possible sulfur leaching that the reaction reuses. However, despite this deactivation, we must highlight that both materials presented good conversions and yields even after the third reuse reaction.

Table 7. Results of reuse tests for the rGO_H and rGO_E catalysts for the glycerol conversion (X_A%), solketal selectivity (SS_5), and dioxan selectivity (SD_6), as well as the solketal yield (YS_5) and dioxan yield (YD_6).

Reactions	X_A (%)	SS_5(%)	SD_6 (%)	YS_5 (%)	YD_6 (%)
rGO_H	66.18 ± 0.45	98.38 ± 0.45	1.62 ± 0.45	65.11 ± 0.45	1.07 ± 0.45
rGO_H R_1	57.37 ± 0.85	95.60 ± 0.85	4.40 ± 0.85	54.85 ± 0.85	2.53 ± 0.85
rGO_H R_2	55.18 ± 0.45	94.38 ± 0.45	5.62 ± 0.45	52.08 ± 0.45	3.10 ± 0.45
rGO_H R_3	45.37 ± 0.85	94.62± 0.85	5.38 ± 0.85	42.93 ± 0.85	2.44 ± 0.85
rGO_H R_4	23.18 ± 0.45	91.38 ± 0.45	8.62 ± 0.45	21.18 ± 0.45	2.00 ± 0.45
Reactions	**X_A(%)**	**SS_5(%)**	**SD_6 (%)**	**YS_5 (%)**	**YD_6 (%)**
rGO_E	63.97 ± 0.86	80.73 ± 0.86	19.27 ± 0.86	51.64 ± 0.86	12.33 ± 0.86
rGO_E R_1	60.54 ± 0.43	80.64 ± 043	19.36 ± 0.43	48.82 ± 0.43	11.72 ± 0.43
rGO_E R_2	58.25 ± 0.32	81.03 ± 0.32	18.97 ± 0.32	47.20 ± 0.32	11.05 ± 0.32
rGO_E R_3	53.12 ± 0.28	79.27 ± 0.28	20.73 ± 0.28	42.11 ± 0.28	11.01 ± 0.28
rGO_E R_4	30.27 ± 0.13	78.25 ± 0.13	21.75 ± 0.13	23.69 ± 0.13	6.58 ± 0.13

(Glycerol/acetone molar ratio = 1:4; 5% catalyst based on glycerol mass; 60 °C; 500 rpm with autogenous pressure; 120 min; conducted in a 100 mL Parr reactor).

3. Materials and Methods

3.1. Materials

Glycerol (99.5% P.A.) and acetone (99.5% P.A.) were purchased from Proquímios (Rio de Janeiro, Brazil)®; solketal (98%) was purchased from Sigma-Aldrich® (St. Louis, MO, USA) and was used without further purification. Flake graphite, to produce rGO_H (synthesized via the Hummers method) [24,38,51], was purchased from Sigma-Aldrich and a carpentry pencil, to synthesize rGO_E (using the electrochemical process) [44,52], was acquired from the BIC® (São Paulo, Brazil).

3.2. Synthesis of Graphene-Derived Materials

3.2.1. Synthesis of Partially Reduced Graphene Oxide Using a Conventional Approach—rGO_H

Reduced graphene oxide (rGO_H) was synthesized using the modified Hummers method [24,38,51]. First, 100 mL of H_2SO_4 was added to a beaker with a volume of 1 L under an ice bath, and then 4 g of graphite flakes and 2 g of $NaNO_3$ were added and continually stirred at 450 rpm for 4 h. After this time, 12 g of $KMnO_4$ was slowly added, and the reaction mixture was kept for another 2 h, still under an ice bath. Then, the ice bath was removed, and the reaction system was heated at 35 °C for 30 min. Next, 160 mL of distilled and deionized water was dropped into the reaction mixture, and the reaction system was stirred for another 30 min. At this moment, the system's temperature increased to 90 °C. To finish the reaction, 160 mL of a 30% (*v/v*) H_2O_2 solution was added, and the suspension was then stirred for 12 h and subsequently washed and purified with distilled and deionized water, HCl (30% *v/v*), and ethanol until the supernatant pH reached 7.0. The suspension was filtered and the reduced graphene oxide was dried in an oven at 50 °C for 12 h. This material produced via a conventional approach was named rGO_H.

3.2.2. Synthesis of Partially Reduced Graphene Oxide via Electrochemical Method—rGO_E

The synthesis of the reduced graphene oxide was based on a study reported by Ambrosi and Pumera (2016) [20] in an acid medium. First, 1.0 mol L^{-1} of sulfuric acid was prepared. Afterward, a reaction system consisting of a 25 mL beaker and two graphite electrodes was set up. Some physical parameters were standardized, such as the distance between the graphite electrodes (3.5 cm), the height of each electrode at the bottom of the beaker (1 cm), the temperature (25 °C), the volume of sulfuric acid solution (20 mL), the voltage (10 V), and the time (1 h). After this time, the oxidized residue at the bottom of the beaker was removed from the reaction system, filtered, and washed until it reached the pH of distilled water, equal to 6.0, due to dissolved CO_2 [44,52]. Finally, the partially reduced graphene oxide was dried in an oven at 80 °C and named rGO_E.

3.3. Material Characterization

The materials rGO_H and rGO_E, along with their respective precursor graphites, were characterized using X-ray diffraction and a Bruker D2 Phaser (Rheinstetten, Germany) apparatus utilizing CuKα radiation (λ = 0.154 nm) with a Ni filter, scanning at intervals of 0.05° within the 2θ range of 5–50°. A convergent slit of 0.6 mm, a current of 10 mA, and a voltage of 30 kV were used with a Lynxeye detector (Oxford, Germany). Raman spectra were obtained using a Witec Alpha 300 R Raman Microscope (Oxford, Germany), with a wavelength of 532 nm (green light). Infrared spectra were collected using a Fourier-transform infrared spectrophotometer, Nicolet model Magna-IR 760 (Spectralab, Canada). Samples were analyzed via Attenuated Total Reflection (ATR) within the range of 400–4000 cm^{-1}, with 32 scans and a resolution of 4000 cm^{-1}.

Thermal analysis was conducted using a TA Instruments SDT Q 600 (Waters, Milford, MA, USA) instrument in the 25–800 °C temperature range, with a heating rate of 10 °C min^{-1} and an airflow of 50 mL min^{-1}. Textural analyses were performed on a Micromeritics Tristar 3000 Surface Area and Porosimetry Analyzer. (Micromeritics, Norcross, GA, USA) Samples were subjected to thermal drying at 90 °C under a vacuum of 5×10^{-3} torr for

24 h, and then cooled to room temperature and re-weighed to initiate analysis at −196 °C. Specific surface area was determined using the B.E.T. (Brunauer, Emmet, and Teller) method, while specific volume and pore diameter were found using the B.J.H. method from the adsorption/desorption isotherm [53].

Brønsted acidity was quantified through acid–base titration following the procedure described by Ventura and coworkers [40]. The surface acidity of the potential catalysts, rGO_H and rGO_E, was evaluated through acid–base titration. For the measurements, 50 mg of the material was weighed, and then 20 mL of a 0.05 mol L^{-1} sodium hydroxide solution was added. The mixture was stirred continuously for 24 h at room temperature. After this period, the solid was separated via centrifugation and the supernatant solution was titrated with hydrochloric acid (0.05 mol L^{-1}), using phenolphthalein as an indicator. The measurements were performed in triplicate. The precursor graphites were not analyzed in this step. The number of acidic sites on the catalyst was determined using Equation (1) [40]:

$$\text{Acidity} = [\text{Initial amount of NaOH added (mol)} - \text{Amount of HCl consumed (mol)}]/\text{mass of the material (g)}. \quad (1)$$

3.4. Catalytic Evaluation

The reactions were carried out in a Parr model 4848B stainless steel autoclave reactor (Parr Instrument Company, Moline, IL, USA) with a reaction volume of 100 mL. The reactor has a thermocouple, transducer, temperature controller, external heating mantle, and an agitation system; see Figure S1. Initially, a control test was conducted without a catalyst (blank). Subsequently, tests were performed using the homogeneous industrial catalyst p-Toluenesulfonic acid (PTSA) and the heterogeneous catalysts synthesized in this study, rGO_H, and rGO_E.

The reactor was loaded with 10 g of glycerol (0.1186 mol) and 25.2 g of acetone (0.4343 mol). Thus, the glycerol/acetone molar ratio (G:A) was 1:4. Catalyst loading of 5% by mass relative to the glycerol mass was used, corresponding to 500 mg. The agitation was set at 500 rpm, and the reaction was conducted at 60 °C with autogenous pressure for 120 min. The catalytic testing utilized a batch reactor with a thermocouple within the reaction vessel. This reactor, functioning as an autoclave-type system with precise temperature control, ensures the even dispersion of the catalyst within the fluid. The temperature measurement reflects the entire reaction system, which requires heating due to the endothermic nature of the reaction [46]. At the end of the reaction, 2.5 g of sodium bicarbonate was added to the reaction mixture to neutralize the catalyst (in the case of the homogeneous catalyst), and then the reaction mixture was filtered. Subsequently, the samples were stored at 15 °C for gas chromatography analysis. All reactions were performed in triplicate [9,54,55].

3.5. Study of Glycerol Conversion as a Function of Time

Since the reaction conditions had already been optimized in a study conducted by Rossa and coworkers [9,54,55] using other catalysts, the conversion study as a function of time was carried out under the same conditions as in Section 2.4. Samples were taken at intervals of 0, 5, 10, 15, 30, 60, 90, 120, 180, and 240 min. Subsequently, the samples were stored at 15 °C for gas chromatography analysis. All reactions were triplicated using PTSA, rGO_H, or rGO_E as catalysts.

3.6. Determination of Kinetic Parameter Models

3.6.1. Pseudo-Homogeneous Kinect Model

Tests for determining kinetic parameters in pseudo-homogeneous catalysis [9,56] were conducted using the rGO_H and rGO_E catalysts based on the time and glycerol conversion results in Section 3.5. (See "Pseudo-homogeneous Kinetic Model Evaluation" in Supplementary Materials.)

3.6.2. Heterogeneous Kinect Models (LHHW and ER)

In the heterogeneous model, initially, eight experiments were conducted under the same conditions as described in Section 3.4 for 60 min, only varying the agitation (100–800 rpm), and glycerol conversion was monitored to investigate the influence of external mass transfer and then predict how the internal mass transfer study was to be conducted. All reactions were performed in triplicate, and the samples were stored at 15 °C until the moment of analysis using GC-FID (Shimadzu, Japan).

The kinetic study of heterogeneous catalysis involves various physical steps (external and internal mass transfer) and chemical steps (adsorption, surface-catalyzed reaction, and desorption of products) according to the proposed mechanism. The study was conducted under reaction conditions free from diffusion limitations (external and internal mass transfer) in this stage. Tests for determining kinetic parameters in heterogeneous catalysis were performed using the rGO_H and rGO_E catalysts, based on the results obtained in Section 3.5, to predict the rate-limiting step of the reaction, both in the LHHW mechanism and in the ER mechanism.

We developed eleven heterogeneous kinetic models derived from two distinct Mechanism: LHHW (Langmuir-Hinshelwood-Hougen-Watson) and ER (Eley-Rideal) [57,58]. An adapted methodology from Shekara (2011) was employed [48]. (See "Heterogeneous Kinetic Model Evaluation" in Supplementary Materials.)

3.7. *Study of Heterogeneous Catalyst Reuse*

Deactivation/reuse tests [9,55] of the rGO_H and rGO_E catalysts were conducted under the same reaction conditions described in Section 2.4. The reuse tests involved using the same catalyst employed in 4 consecutive cycles without washing or calcination steps. All reactions were performed in triplicate, and the samples were stored at 15 °C until the analysis using GC-FID (Shimadzu, Japan) [9].

3.8. *Product Analysis*

The products of the glycerol ketalization reaction were quantitatively analyzed using a Shimadzu GC-FID equipped with a column Carbowax (30 m × 0.25 mm × 0.25 μm polyethylene glycol) and employing the external standard method [9,55]. Analysis conditions: total injection volume: 1 μL; injector: split/splitless; injection: automatic; linear velocity: 45 cm s^{-1}; split ratio: 1:50; linear velocity: 45 cm s^{-1}; sampling rate: 40 ms; carrier gas: helium; flow control mode: linear velocity; detector temperature: 250 °C; injector temperature: 250 °C. The heating ramp was set at 50 °C for 5 min, followed by a temperature increase from 50 °C to 180 °C at a rate of 16 °C per min. It was held at 180 °C for an additional 2 min, then increased from 180 °C to 230 °C at a rate of 20 °C per min, and then held at 230 °C for another 2 min. Total analysis time: 20 min.

The glycerol conversion and product selectivity were obtained using Equation (2).

$$X_A(\%) = \left(\frac{C_{A_0} - C_A}{C_{A_0}}\right) \times 100 \quad (2)$$

where C_{A0} is the initial molar concentration (mol L^{-1}) of glycerol and C_A is the final molar concentration (mol L^{-1}) of glycerol at the end of the reaction.

Solketal Selectivity (S_5, dioxolan):

$$\text{Sel}(S_5)\% = \left(\frac{AS_5}{AS_5 + AD_6}\right) \times 100 \quad (3)$$

where AS_5 is the area of solketal and AD_6 is the area of dioxan, obtained from the chromatograms. The areas of all of the compounds were previously adjusted using their respective correction factors.

Dioxan Selectivity (D_6):

$$Sel(D_6)\% = \left(\frac{AD_6}{AS_5 + AD_6}\right) \times 100 \quad (4)$$

D_6 is the area of dioxan and AS_5 is the area of solketal, obtained from the chromatogram. The areas of all compounds were previously adjusted using their respective correction factors.

Yield—Y (%):

$$Y(\%) = (X_A \times S_i) \times 100 \quad (5)$$

S_i refers to the products S_5 or D_6, and X_A is the glycerol conversion (not in percentage) multiplied by the desired products' selectivity [9,55].

To calculate Turnover Frequency, Equation (6) was used.

$$TOF = \frac{\frac{n_{A_0} - n_A}{n_{cat}}}{t} \quad (6)$$

where

n_{A_0} is the initial amount of glycerol (mol);
n_A is the amount of glycerol (mol) at the end of the reaction, after 2 h;
$n_{cat}$is the number of acid sites on the catalyst (mol $g_{cat}{}^{-1}$);
t is the time in hours (h);
TOF is the Turnover Frequency (h^{-1}).

4. Conclusions

This study effectively demonstrated the feasibility of replacing the homogeneous catalyst (PTSA) with the proposed heterogeneous catalysts (rGO_H and rGO_E). Remarkably, both rGO_H and rGO_E catalysts exhibited superior performance, achieving significant glycerol conversion and TOF values, outperforming PTSA. Additionally, the rGO_E catalyst demonstrated reusability over five consecutive cycles, showing the highest activity without the need for post-reaction washing or treatment. This suggests that an effective catalyst for the glycerol ketalization reaction with acetone should possess specific characteristics, including small crystal size, porosity, and acidity. The synthesis of rGO_E used a more environmentally friendly methodology, enabling its future utilization in various catalytic processes and providing the potential for chemical modifications, such as impregnation with transition metals and other compounds, to enhance its acidic properties.

The pseudo-homogeneous kinetic model successfully described the ketalization reaction of glycerol with acetone for both the homogeneous (PTSA) and heterogeneous (rGO_H and rGO_E) scenarios, with the model assuming a pseudo-homogeneous form for the heterogeneous catalysts. Among the eleven heterogeneous kinetic models explored to elucidate the mechanism and controlling step of heterogeneous kinetics, both of the rGO_H and rGO_E catalysts demonstrated positive results. Notably, Model 3 (LHHW) and Model 8 (ER) provided good fits for both of the rGO_H and rGO_E catalysts, offering valuable insights into the mechanism and controlling step of the heterogeneous catalytic reaction.

Supplementary Materials: The following supporting information can be downloaded at https://www.mdpi.com/article/10.3390/catal13111427/s1: Figure S1. Batch reactor used in all ketalization reactions (Parr Instruments Inc.—Model 4848B), made of stainless steel, with a useful volume of 100 mL, in a simple batch system, with a maximum working pressure of 200 bar. This reactor has a temperature and pressure controller. Also controlled agitation and external blanket for heating; Figure S2. Infrared spectra of graphite flakes (a) GRFH and reduced graphene oxide, (b) rGOH, obtained by the Hummers method; carpenter's pencil graphite (c) GFTE and reduced graphene oxide, (d) rGOE, obtained by the electrochemical method from carpenter's pencil graphite; Table S1. Estimated values for the distance between layers, average crystallite size, and number of layers for

GFTH, rGOH, GFTE and rGOE materials; Table S2. Assignments of the main bands in the infrared spectra for the GFTH, rGOH, GFTE and rGOE materials; Table S3. Acidity results obtained through acid-base titration for the materials tested as catalysts; Table S4. Acidity results for the catalysts obtained by acid-base titration before and after five reactions; Pseudo-Homogeneous Kinetic Model Evaluation; Heterogeneous Kinetic Model Evaluation; Langmuir-Hinshelwood-Hougen-Watson Mechanism (LHHW); Eley-Rideal Mechanism.

Author Contributions: Conceptualization, V.R., F.F.M. and V.G.d.C.M.; methodology, V.R., S.d.C.V., G.C.D., J.d.A.R., J.P.R.M., V.G.d.C.M. and F.F.M.; formal analysis, V.R., S.d.C.V., G.C.D., J.d.A.R., V.G.d.C.M. and F.F.M.; investigation, V.R., T.d.M.L., Y.R.C., J.L.N. and Y.X.; resources, T.d.M.L. and Y.R.C.; writing—original draft preparation, V.R. and J.d.A.R.; writing—review and editing, T.d.M.L., E.A.P. and V.R.; supervision, T.d.M.L.; project administration, T.d.M.L.; funding acquisition, T.d.M.L. All authors have read and agreed to the published version of the manuscript.

Funding: This study was financed in part by the Coordenação de Aperfeiçoamento de Pessoal de Nível Superior-Brasil (CAPES)—Finance Code 001—and CNPq and FAPERJ (E-26/211.575/2019; E26/203.188/2019, E-26/201.358/2021; E-26/210.315/2022).

Data Availability Statement: Data are contained within the article and Supplementary Materials.

Acknowledgments: The authors also acknowledge CNPq, FAPERJ, CAPES, Donato Aranda (EQ/UFRJ) and Sibele Pergher (IQ/UFRN).

Conflicts of Interest: The authors declare no conflict of interest.

References

1. Li, M.; Zheng, Y.; Chen, Y.; Zhu, X. Biodiesel Production from Waste Cooking Oil Using a Heterogeneous Catalyst from Pyrolyzed Rice Husk. *Bioresour. Technol.* **2014**, *154*, 345–348. [CrossRef] [PubMed]
2. Datta, I.; Ghosh, A.; Acharjee, A.; Rakshit, A.; Saha, B. Overview on Biodiesel Market. *Vietnam J. Chem.* **2021**, *59*, 271–284.
3. Galadima, A.; Muraza, O. A Review on Glycerol Valorization to Acrolein over Solid Acid Catalysts. *J. Taiwan Inst. Chem. Eng.* **2016**, *67*, 29–44. [CrossRef]
4. Muraza, O. Peculiarities of Glycerol Conversion to Chemicals Over Zeolite-Based Catalysts. *Front. Chem.* **2019**, *7*, 233. [CrossRef] [PubMed]
5. Corrêa, I.; Faria, R.P.V.; Rodrigues, A.E. Continuous Valorization of Glycerol into Solketal: Recent Advances on Catalysts, Processes, and Industrial Perspectives. *Sustain. Chem.* **2021**, *2*, 286–324. [CrossRef]
6. Zhang, H.; Hu, Z.; Huang, L.; Zhang, H.; Song, K.; Wang, L.; Shi, Z.; Ma, J.; Zhuang, Y.; Shen, W.; et al. Dehydration of Glycerol to Acrolein over Hierarchical ZSM-5 Zeolites: Effects of Mesoporosity and Acidity. *ACS Catal.* **2015**, *5*, 2548–2558. [CrossRef]
7. Fatimah, I.; Sahroni, I.; Fadillah, G.; Musawwa, M.M.; Mahlia, T.M.I.; Muraza, O. Glycerol to Solketal for Fuel Additive: Recent Progress in Heterogeneous Catalysts. *Energies* **2019**, *12*, 2872. [CrossRef]
8. Vojvodic, A.; Nørskov, J.K. New Design Paradigm for Heterogeneous Catalysts. *Natl. Sci. Rev.* **2015**, *2*, 140–143. [CrossRef]
9. Rossa, V.; Pessanha, Y.d.S.P.; Díaz, G.C.; Câmara, L.D.T.; Pergher, S.B.C.; Aranda, D.A.G. Reaction Kinetic Study of Solketal Production from Glycerol Ketalization with Acetone. *Ind. Eng. Chem. Res.* **2017**, *56*, 479–488. [CrossRef]
10. Kowalska-Kuś, J.; Held, A.; Nowińska, K. Solketal Formation in a Continuous Flow Process over Hierarchical Zeolites. *ChemCatChem* **2020**, *12*, 510–519. [CrossRef]
11. Alsawalha, M. Catalytic Activity and Kinetic Modeling of Various Modules HZMS-5 and Treated MCM-41 Catalysts, for the Liquid-Phase Ketalization of Glycerol with Acetone. *Front. Chem.* **2019**, *7*, 799. [CrossRef] [PubMed]
12. Rodrigues, R.; Santos, M.S.; Nunes, R.S.; Carvalho, W.A.; Labuto, G. Solvent-Free Solketal Production from Glycerol Promoted by Yeast Activated Carbons. *Fuel* **2021**, *299*, 120923. [CrossRef]
13. Chen, T.; Dai, L. Carbon Nanomaterials for High-Performance Supercapacitors. *Mater. Today* **2013**, *16*, 272–280. [CrossRef]
14. Yang, F.; Mao, J.; Li, S.; Yin, J.; Zhou, J.; Liu, W. Cobalt-Graphene Nanomaterial as an Efficient Catalyst for Selective Hydrogenation of 5-Hydroxymethylfurfural into 2,5-Dimethylfuran. *Catal. Sci. Technol.* **2019**, *9*, 1329–1333. [CrossRef]
15. Pei, S.; Wei, Q.; Huang, K.; Cheng, H.M.; Ren, W. Green Synthesis of Graphene Oxide by Seconds Timescale Water Electrolytic Oxidation. *Nat. Commun.* **2018**, *9*, 145. [CrossRef] [PubMed]
16. Achee, T.C.; Sun, W.; Hope, J.T.; Quitzau, S.G.; Sweeney, C.B.; Shah, S.A.; Habib, T.; Green, M.J. High-Yield Scalable Graphene Nanosheet Production from Compressed Graphite Using Electrochemical Exfoliation. *Sci. Rep.* **2018**, *8*, 14525. [CrossRef]
17. Marcano, D.C.; Kosynkin, D.V.; Berlin, J.M.; Sinitskii, A.; Sun, Z.; Slesarev, A.; Alemany, L.B.; Lu, W.; Tour, J.M. Improved Synthesis of Graphene Oxide. *ACS Nano* **2010**, *4*, 4806–4814. [CrossRef]
18. Song, J.; Wang, X.; Chang, C.-T. Preparation and Characterization of Graphene Oxide. *J. Nanomater.* **2014**, *2014*, 1–6. [CrossRef]
19. Kumar, S.; Wani, M.Y.; Koh, J.; Gil, J.M.; Sobral, A.J.F.N. Carbon Dioxide Adsorption and Cycloaddition Reaction of Epoxides Using Chitosan–Graphene Oxide Nanocomposite as a Catalyst. *J. Environ. Sci.* **2018**, *69*, 77–84. [CrossRef]

20. Maraschin, T.G.; Correa, R.d.S.; Rodrigues, L.F.; Balzarettid, N.M.; Malmonge, J.A.; Galland, G.B.; de Souza Basso, N.R.; Basso, N.R.d.S. Chitosan Nanocomposites with Graphene-Based Filler. *Mater. Res.* **2019**, *22*, 1–10. [CrossRef]
21. Schmal, M. *Heterogeneous Catalysis and Its Industrial Applications*, 1st ed.; Springer International Publishing: Cham, Switzerland, 2016; ISBN 978-3-319-09249-2.
22. Andonovic, B.; Temkov, M.; Ademi, A.; Petrovski, A.; Grozdanov, A.; Paunović, P.; Dimitrov, A. Laue Functions Model vs Scherrer Equation in Determination of Graphene Layers Number on the Ground of XRD Data. *J. Chem. Technol. Met.* **2014**, *6*, 545–550.
23. Andonovic, B.; Grozdanov, A.; Paunović, P.; Dimitrov, A.T. X-ray Diffraction Analysis on Layers in Graphene Samples Obtained by Electrolysis in Molten Salts: A New Perspective. *Micro Nano Lett.* **2015**, *10*, 683–685. [CrossRef]
24. Hummers, W.S.; Offeman, R.E. Preparation of Graphitic Oxide. *J. Am. Chem. Soc.* **1958**, *80*, 1339. [CrossRef]
25. Cobos, M.; González, B.; Fernández, M.D.J.; Fernández, M.D.J. Chitosan-Graphene Oxide Nanocomposites: Effect of Graphene Oxide Nanosheets and Glycerol Plasticizer on Thermal and Mechanical Properties. *J. Appl. Polym. Sci.* **2017**, *134*, 45092. [CrossRef]
26. Pattnaik, S.; Swain, K.; Lin, Z. Graphene and Graphene-Based Nanocomposites: Biomedical Applications and Biosafety. *J. Mater. Chem. B* **2016**, *4*, 7813–7831. [CrossRef]
27. Serhan, M.; Sprowls, M.; Jackemeyer, D.; Long, M.; Perez, I.D.; Maret, W.; Tao, N.; Forzani, E. Total Iron Measurement in Human Serum with a Smartphone. In Proceedings of the AIChE Annual Meeting, Orlando, FL, USA, 10–15 November 2019.
28. Kumar, A.S.K.; Jiang, S.-J. Chitosan-Functionalized Graphene Oxide: A Novel Adsorbent an Efficient Adsorption of Arsenic from Aqueous Solution. *J. Environ. Chem. Eng.* **2016**, *4*, 1698–1713. [CrossRef]
29. Sundramoorthy, A.K.; Vignesh Kumar, T.H.; Gunasekaran, S. *Graphene-Based Nanosensors and Smart Food Packaging Systems for Food Safety and Quality Monitoring*; In Graphene Bioelectronics; Elsevier: Amsterdam, The Netherlands, 2018; pp. 267–306.
30. Kumar, N.; Srivastava, V.C. Simple Synthesis of Large Graphene Oxide Sheets via Electrochemical Method Coupled with Oxidation Process. *ACS Omega* **2018**, *3*, 10233–10242. [CrossRef]
31. Malard, L.M.; Pimenta, M.A.; Dresselhaus, G.; Dresselhaus, M.S. Raman Spectroscopy in Graphene. *Phys. Rep.* **2009**, *473*, 51–87. [CrossRef]
32. Yang, X.; Tu, Y.; Li, L.; Shang, S.; Tao, X. Well-Dispersed Chitosan/Graphene Oxide Nanocomposites. *ACS Appl. Mater. Interfaces* **2010**, *2*, 1707–1713. [CrossRef]
33. Grande, C.D.; Mangadlao, J.; Fan, J.; De Leon, A.; Delgado-Ospina, J.; Rojas, J.G.; Rodrigues, D.F.; Advincula, R. Chitosan Cross-Linked Graphene Oxide Nanocomposite Films with Antimicrobial Activity for Application in Food Industry. *Macromol. Symp.* **2017**, *374*, 1600114. [CrossRef]
34. Zuo, P.-P.; Feng, H.-F.; Xu, Z.-Z.; Zhang, L.-F.; Zhang, Y.-L.; Xia, W.; Zhang, W.-Q. Fabrication of Biocompatible and Mechanically Reinforced Graphene Oxide-Chitosan Nanocomposite Films. *Chem. Cent. J.* **2013**, *7*, 39. [CrossRef] [PubMed]
35. Muda, M.S.; Kamari, A.; Bakar, S.A.; Yusoff, S.N.M.; Fatimah, I.; Phillip, E.; Din, S.M. Chitosan-Graphene Oxide Nanocomposites as Water-Solubilising Agents for Rotenone Pesticide. *J. Mol. Liq.* **2020**, *318*, 114066. [CrossRef]
36. Han, D.; Yan, L.; Chen, W.; Li, W. Preparation of Chitosan/Graphene Oxide Composite Film with Enhanced Mechanical Strength in the Wet State. *Carbohydr. Polym.* **2011**, *83*, 653–658. [CrossRef]
37. He, L.; Wang, H.; Xia, G.; Sun, J.; Song, R. Chitosan/Graphene Oxide Nanocomposite Films with Enhanced Interfacial Interaction and Their Electrochemical Applications. *Appl. Surf. Sci.* **2014**, *314*, 510–515. [CrossRef]
38. Sabzevari, M.; Cree, D.E.; Wilson, L.D. Graphene Oxide–Chitosan Composite Material for Treatment of a Model Dye Effluent. *ACS Omega* **2018**, *3*, 13045–13054. [CrossRef]
39. Tene, T.; Usca, G.T.; Guevara, M.; Molina, R.; Veltri, F.; Arias, M.; Caputi, L.S.; Gomez, C.V. Toward Large-Scale Production of Oxidized Graphene. *Nanomaterials* **2020**, *10*, 279. [CrossRef]
40. Ventura, W.M. Preparação de Catalizadores Heterogêneos a Base de Nb_2O_5 e CeO_2 Para Oxidação da Anilina Em Fase Líquida. Master's Thesis, Universidade Federal de Ouro Preto, Ouro Preto, Brazil, 2017.
41. Grassi, P.; Drumm, F.C.; Georgin, J.; Tonato, D.; Foletto, E.L. Preparação, Caracterização e Avaliação Catalítica Do Compósito Fe_2O_3/Grafite Em Reação Foto-Fenton Preparation (Characterization and Catalytic Evaluation of Fe_2O_3/Graphite Composite in Photo-Fenton Reaction). *Matéria* **2019**, *24*, e12434.
42. Fernando, P.; Braga, A.; Junqueira, A.; Dutra, B. Grafita. In *Rochas & Minerais Industriais - Usos e Especificações*; Luz, A.B.d., Lins, F.A.F., Eds.; CETEM-MCT: Rio de Janeiro, Brazil, 2008; pp. 527–549. ISBN 978-85-61121-37-2.
43. Ramesh, A.; Jeyavelan, M.; Leo Hudson, M.S. Electrochemical Properties of Reduced Graphene Oxide Derived through Camphor Assisted Combustion of Graphite Oxide. *Dalt. Trans.* **2018**, *47*, 5406–5414. [CrossRef]
44. Ambrosi, A.; Chua, C.K.; Latiff, N.M.; Loo, A.H.; Wong, C.H.A.; Eng, A.Y.S.; Bonanni, A.; Pumera, M. Graphene and Its Electrochemistry-an Update. *Chem. Soc. Rev.* **2016**, *45*, 2458–2493. [CrossRef]
45. Esteban, J.; Ladero, M.; García-Ochoa, F. Kinetic Modelling of the Solventless Synthesis of Solketal with a Sulphonic Ion Exchange Resin. *Chem. Eng. J.* **2015**, *269*, 194–202. [CrossRef]
46. Nanda, M.R.; Yuan, Z.; Qin, W.; Ghaziaskar, H.S.; Poirier, M.-A.; Xu, C.C. Thermodynamic and Kinetic Studies of a Catalytic Process to Convert Glycerol into Solketal as an Oxygenated Fuel Additive. *Fuel* **2014**, *117*, 470–477. [CrossRef]
47. Hougen, O.A.; Watson, K.M. *Chemical Process Principles Part There Kinetics and Catalysis*; John Wiley: New York, NY, USA, 1959.
48. Chandra Shekara, B.M.; Ravindra Reddy, C.; Madhuranthakam, C.R.; Jai Prakash, B.S.; Bhat, Y.S. Kinetics of Esterification of Phenylacetic Acid with p -Cresol over H-β Zeolite Catalyst under Microwave Irradiation. *Ind. Eng. Chem. Res.* **2011**, *50*, 3829–3835. [CrossRef]

49. Eley, D.D.; Rideal, E.K. Parahydrogen Conversion on Tungsten. *Nature* **1940**, *146*, 401–402. [CrossRef]
50. Miranda, C.; Urresta, J.; Cruchade, H.; Tran, A.; Benghalem, M.; Astafan, A.; Gaudin, P.; Daou, T.J.; Ramírez, A.; Pouilloux, Y.; et al. Exploring the Impact of Zeolite Porous Voids in Liquid Phase Reactions: The Case of Glycerol Etherification by Tert-Butyl Alcohol. *J. Catal.* **2018**, *365*, 249–260. [CrossRef]
51. Kauling, A.P.; Seefeldt, A.T.; Pisoni, D.P.; Pradeep, R.C.; Bentini, R.; Oliveira, R.V.B.B.; Novoselov, K.S.; Castro Neto, A.H. The Worldwide Graphene Flake Production. *Adv. Mater.* **2018**, *30*, 1–6. [CrossRef]
52. Ambrosi, A.; Pumera, M. Electrochemically Exfoliated Graphene and Graphene Oxide for Energy Storage and Electrochemistry Applications. *Chem. - A Eur. J.* **2016**, *22*, 153–159. [CrossRef]
53. Thommes, M.; Kaneko, K.; Neimark, A.V.; Olivier, J.P.; Rodriguez-Reinoso, F.; Rouquerol, J.; Sing, K.S.W. Physisorption of Gases, with Special Reference to the Evaluation of Surface Area and Pore Size Distribution (IUPAC Technical Report). *Pure Appl. Chem.* **2015**, *87*, 1051–1069. [CrossRef]
54. Rossa, V.; Treichel, H.; Alexandre, D.; Aranda, G. Transformation of Glycerol Into Solketal By Beta and Ferrierite Acid. *Perspect. EREXIM* **2017**, *41*, 101–112.
55. Rossa, V.; Díaz, G.C.; Muchave, G.J.; Aranda, D.A.G.; Berenice Castellã Pergher, S. Production of Solketal Using Acid Zeolites as Catalysts. In *Glycerine Production and Transformation—An Innovative Platform for Sustainable Biorefinery and Energy*; IntechOpen: London, UK, 2019; pp. 1–18.
56. Câmara, L.D.T.; Aranda, D.A.G. Reaction Kinetic Study of Biodiesel Production from Fatty Acids Esterification with Ethanol. *Ind. Eng. Chem. Res.* **2011**, *50*, 2544–2547. [CrossRef]
57. Fogler, H.S. *Elementos de Engenharia Das Reações Químicas*, 3rd ed.; Editora LTC: Rio de Janeiro, Brazil, 2002.
58. Hill, C.G., Jr. *An Introduction to Chemical Enginnering Kinetics & Reactor Design*; John Wiley & Sons: New York, NY, USA, 1997.

molecules MDPI

Article

Hydraulic Retention Time as an Operational Tool for the Production of Short-Chain Carboxylates via Anaerobic Fermentation of Carbohydrate-Rich Waste

Kaoutar Aboudi [1], Silvia Greses [1] and Cristina González-Fernández [1,2,3,*]

1 Biotechnological Processes Unit, IMDEA Energy, Avda. Ramón de la Sagra 3, 28935 Madrid, Spain; silvia.greses@imdea.org (S.G.)
2 Department of Chemical Engineering and Environmental Technology, School of Industrial Engineering, University of Valladolid, Dr. Mergelina, s/n, 47002 Valladolid, Spain
3 Institute of Sustainable Processes, Dr. Mergelina, s/n, 47002 Valladolid, Spain
* Correspondence: cristina.gonzalez@imdea.org

Abstract: The carboxylate platform is a sustainable and cost-effective way to valorize wastes into biochemicals that replace those of fossil origin. Short-chain fatty acids (SCFAs) are intermediates generated during anaerobic fermentation (AF) and are considered high-value-added biochemicals among carboxylates. This investigation aimed to produce SCFAs through the AF of sugar beet molasses at 25 °C and semi-continuous feeding mode in completely stirred tank reactors. A particular focus was devoted to the role of hydraulic retention time (HRT) variation in SCFAs production and distribution profile. The highest SCFAs concentration (44.1 ± 2.3 gCOD/L) was reached at the HRT of 30 days. Caproic acid accounted for 32.5–35.5% (COD-concentration basis) at the long HRTs of 20 and 30 days due to the carbon chain elongation of shorter carboxylic acids. The findings of this study proved that HRT could be used to steer the anaerobic process toward the targeted SCFAs for specific uses. Furthermore, the successful operation at low-temperature conditions (i.e., 25 °C) makes the process economically promising.

Keywords: anaerobic fermentation; hydraulic retention time; short-chain fatty acids; chain elongation; caproic acid

1. Introduction

The carboxylate platform is a cost-effective way to valorize a wide range of residual biomass (e.g., agricultural wastes and by-products, animal manure, urban and food wastes, sewage sludge, etc.) into biofuels and biochemicals able to replace petrochemicals at some extent. Carboxylates are short-chain fatty acids (SCFAs) with industrial demand that might be produced via waste anaerobic fermentation (AF). AF can be regarded as an altered conventional anaerobic digestion (AD) by inhibiting methanogenic archaea to promote the acidogenesis step, leading to SCFAs accumulation [1,2]. Compared to biomethane, SCFAs have higher economic value and a wider range of applications through biological (i.e., oils, polyhydroxyalkanoates) or chemical conversion (i.e., esters, polymers) for their use in pharmaceutical, cosmetic, or food industries [3].

AF occurs through a series of biochemical reactions in which acidogenic bacteria ferment the hydrolyzed wastes into SCFAs (mainly acetate C2, propionate C3, and butyrate C4), alcohols, together with hydrogen and carbon dioxide [4]. Under suitable operational conditions, the microbial consortia could further produce longer fatty acids (i.e., caproate C6 and caprylate C8) through the carbon chain elongation (CCE) process. CCE of SCFAs occurs in an energy-rich environment and reducing agents [5,6].

Like conventional AD, various factors influence AF, including temperature, pH, hydraulic/solid retention time (HRT/SRT), organic loading rate (OLR), etc. Concerning pH, slightly acidic conditions (5.5–6.5) have been determined as the most suitable range for SCFAs production, especially at low-temperature conditions (i.e., 25 °C) [1,7].

HRT, equal to SRT in completely stirred tank reactors (CSTRs), refers to the average time microorganisms are in contact with the substrate in an anaerobic reactor. In this sense, an appropriate HRT must be set up to avoid the washout of microorganisms. Whereas low HRTs imply a low residence time of microbes inside reactors, high HRTs allow microbes to have enough time to grow and degrade organic matter more efficiently. Considering that the growth rate of methanogenic archaea is lower than that of acidogenic bacteria [8], HRT control can be used to force the washout of methanogens from the reactor, enabling the selection of the most suitable populations for SCFAs accumulation. It cannot be neglected that each AF process could have a different response to HRT variability according to the nature of the substrate used and the implemented operational conditions (i.e., temperature, pH, reactor configuration, etc.) [8,9]. Indeed, AF for SCFAs production has been intensively investigated using different waste streams (mainly food waste and sewage sludge) and process configurations to either increase production or obtain a specific distribution profile. Proof thereof is the wide number of reviews addressing the AF technology [10–13].

This investigation selected carbohydrate-rich waste (sugar beet molasses, SBM) as a substrate to be valorized via AF to produce SCFAs. SBM is considered a valuable by-product of the sugar production industry. The most common use of this waste is as a supplement for animal feed. However, previous renewable energy directives, such as the Renewable Energy Directive 2009/28/EC [14], classified molasses as an advanced biofuel and bio-based chemical feedstock. Since SBM could be alternatively valorized as SCFAs and contribute to greening the chemical industry, this waste was considered a suitable feedstock for AF. Furthermore, since the HRT implemented in the reactors ultimately affects the production yields and operational costs, the novelty of this investigation lies in the identification of the suitable HRT that can be applied without compromising the yields of this waste into SCFAs at low temperatures, as well as the targeted individual SCFAs. In addition, perspectives on using SBM for SCFAs versus the current use as an animal feed supplement were addressed.

2. Results and Discussion

2.1. Performance of the Hydrolytic and Acidogenic Activities in the AF of SBM

Figure 1 shows the solubilization and acidification degrees of SBM. SBM was expected to be a readily fermentable substrate (SCOD/TCOD ratio of 94.3%). Therefore, the solubilization degree, which assesses the hydrolytic activity in the reactors, was high at all the HRTs tested, reaching values ranging from 85 to 95%. The highest solubilization performance was observed in the reactors operating at HRTs of 30 and 20 days (95.8 ± 2.5 and 90.6 ± 1.4%, respectively). In CSTR, long HRTs (i.e., low flow rates of the feeding) enable longer reaction times for microorganisms to solubilize the organic material of substrates. When using complex substrates for AF, the hydrolysis step often takes a long time and is regarded as the rate-limiting stage to obtain soluble monomers, such as amino acid, long-chain fatty acid, monosaccharide, etc. [15]. In contrast, for highly soluble substrates, such as SBM, the acidogenesis step gains relevance because this process occurs faster than with complex substrates. This fact is the main reason of the high SCOD percentage reached regardless of the HRT evaluated.

It should be considered that acidogenesis is a fast step and, thereby, a high content of metabolites can be released altering the medium properties. pH is one of the parameters that most frequently gets affected by this release, decreasing its value towards acid values. Since methanogens are the most sensitive microorganisms in the context of low pHs, such a circumstance can be considered as an advantage in AF because SCFAs would be accumulated and not consumed for methane production [16].

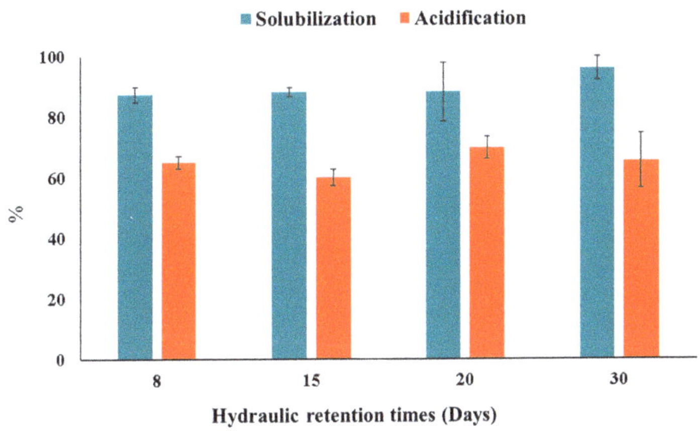

Figure 1. Solubilization and acidification degrees achieved in the SBM-AF at different HRTs.

Nevertheless, an excessive pH drop can also lead to an acidogenesis inhibition, resulting in a SCFAs concentration decrease. Previous studies reported that a decrease in pH values below 5.5 might result in the accumulation of primary fermentation products such as lactic acid and/or ethanol, which hinders the involved pathways for SCFAs generation [3,17]. Thus, AF process stability is strictly related to the pH values in the culture broth, affecting the SCFAs concentration achieved [17,18]. In the present research, pH fluctuations were higher at low HRTs than at the longest ones (Supplementary Material). At 20 and 30 days, pH dropped below the desired limit (5.5–6.5) only at the start-up period, while fluctuations were more pronounced for short HRTs. In this regard, the high consumption of the alkali reagent (i.e., NaOH) for the pH adjustment in reactors operating at short HRTs throughout the experiments may increase the operational cost, particularly at a large scale. In contrast, proper pH control without excessive reagent consumption allowed working at prolonged HRTs (i.e., 20–30 days), avoiding acidogenic system instability. The same pH-related issue was reported by Jankowska et al. [18] when working at different HRT. The rapid pH decrease in reactors fed with sugar-rich substrates is a common feature given the fast hydrolysis rates of carbohydrates compared to other macromolecules such as proteins and lipids [19]. In this sense, HRT selection (or any other operational parameter) should be selected not only in terms of SCFAs concentrations and yields but also in terms of process stability.

Concerning the acidification degree, all HRTs resulted in high values (60–70%, Figure 1) comparable to other studies dealing with carbohydrate-rich substrates [2,17]. Greses et al. [20] obtained a high acidification degree (83%) for AF of a carbohydrate-rich waste (79% of carbohydrates) by increasing the process temperature (55 °C). Nevertheless, the temperature increase involved a high-energy input that did not offset 13 percentual points of acidification enhancement. Overall, these results evidenced that AF of SBM can reach stable and high acidogenic performance regardless of the HRT when the values varied between 8 and 30 days.

2.2. Evaluation of the AF in Terms of SCFAs Production

As shown in Figure 2, increasing the HRT concomitantly increased the total concentration of SCFAs produced. Reactors operating at HRTs of 30 and 20 days showed the highest SCFAs accumulations, namely 44.1 ± 2.3 g COD/L (27.6 ± 1.6 g SCFAs/L) and 33.2 ± 1.1 g COD/L (20.4 ± 0.5 g SCFAs/L), respectively, whereas operating at short HRTs led to SCFAs concentrations of 21.2 ± 1.3 g COD/L (11.5 ± 0.6 g SCFAs/L) and 14.1 ± 1.1 g COD/L (8.5 ± 0.6 g SCFAs/L) for HRTs of 15 and 8 days, respectively (Figure 2 and Table 1).

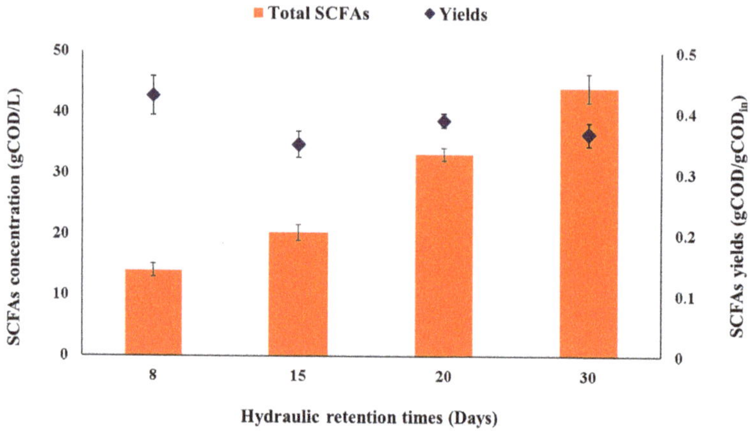

Figure 2. Total SCFAs productions and the obtained yields from AF of SBM at different HRTs.

These concentrations represented high values compared with previous investigations using carbohydrate-rich waste as feedstock. For instance, Onodera et al. [21] reported SCFAs concentrations of 15.6 and 25.9 g COD/L at HRTs of 13.1 and 18.2 days, respectively, when performing AF of diluted cane molasses (with high content of sucrose) at 35 °C. The high concentrations in the present research at similar HRTs revealed that 25 °C was a more suitable temperature than 35 °C when feedstocks with high readily organic matter content are subjected to AF. Moreover, this fact is relevant when energy costs are considered. Likewise, Bolaji and Dionisi [3] found that for the mesophilic AF of carbohydrates-rich food waste, the highest SCFAs concentration of 19.4 g COD/L was obtained at 30 days of HRT. That value was comparably lower than the one attained herein (44.1 ± 2.3 g COD/L for HRT at 30 days), highlighting the optimum conditions implemented other than retention time. The positive effect of low AF temperature can be thus confirmed by comparing with previous studies since only those conducting AF of carbohydrate-rich waste at 25 °C reported such high SCFAs concentrations [2,22].

In a conventional AD process, high HRTs are commonly applied for methane-rich biogas production, while short HRTs have been shown to benefit hydrogen-rich biogas generation. This is mainly due to the slow growth rate of methanogens compared to hydrolytic/fermentative acidogens [8]. Specifically, HRT has been shown to affect the relative abundance of each microbial community [16,23]. The underlying reason for this is based on the fact that HRT determines the daily flow rate of substrate fed to the reactor, which affects the available time for microorganisms to degrade the organic material. Llamas et al. [16] reported that short HRTs led to a 75% decrease in SCFAs concentration (i.e., from 5.2 to 1.3 g COD/L at HRTs of 10 and 4 days, respectively) due to the high flow rate, which triggers the wash out of the hydrolytic/acidogenic microorganisms from the system.

Along with SCFAs concentrations, high SCFAs yields ranging from 0.35 ± 0.02 g SCFA/g COD_{in} (0.49 ± 0.05 g SCFA/g VS_{in}) to 0.43 ± 0.03 g SCFA/g COD_{in} (0.66 ± 0.05 g SCFA/g VS_{in}) were reached at all HRTs, with the highest value observed at 8 days of HRT (Figure 2 and Table 1). This could be due to the need to dilute the organic material content of the daily feedstock (inlet flow rate) to keep the OLR constant at 4 g COD/L d. The obtained SCFAs yields were in line with the values reported for carbohydrate-rich substrates. For instance, Greses et al. [2] obtained SCFAs yields of 0.43–0.46 g SCFA/g COD_{in} in AF of highly carbohydrate-rich vegetable waste (i.e., 75.7–80.4% of carbohydrates) at pH values around 5.6–5.8 and 25 °C. In the case of using a substrate with low carbohydrate content (i.e., vegetable wastes with 5.8% of carbohydrates), AF at 37 °C and pH values of 5.6–5.8, the SCFAs yield dropped to 0.21 g COD/g VS_{in} compared to 0.45 g COD/g VS_{in} from AF of a vegetable waste with a carbohydrate content of 45% [24]. Thereby, it can be suggested that as carbohydrate content in the feedstock decreases, the carboxylates yields also do. Conversely, AF of protein-rich substrates such as microalgae led to low SCFAs yields, as reported by Magdalena et al. [6], highlighting that high SCFAs yields are aligned with carbohydrate-rich substrates.

Table 1. Characteristics of the effluents from AF operated at different HRT.

Parameters (Units)	HRT 8 d	HRT 15 d	HRT 20 d	HRT 30 d
pH	5.4 ± 0.3 **	5.4 ± 0.2 **	5.6 ± 0.3 *	5.8 ± 0.2 *
TS (g/L)	16.6 ± 3.4	25.2 ± 2.6	37.1 ± 4.6	54.4 ± 4.4
VS (g/L)	10.3 ± 1.7	17.2 ± 2.3	22.6 ± 1.9	34.4 ± 2.6
TCOD (g/L)	24.5 ± 3.1	41.5 ± 4.2	53.2 ± 0.9	70.5 ± 0.8
SCOD/TCOD	0.8 ± 0.1	0.8 ± 0.1	0.9 ± 0.1	0.9 ± 0.5
Total SCFA (g/L)	8.5 ± 0.6	11.5 ± 0.6	20.4 ± 0.5	27.6 ± 1.6
Caproic acid yield ($gCOD/gCOD_{in}$)	0.07 ± 0.01	0.09 ± 0.01	0.13 ± 0.01	0.013 ± 0.010
SCFAs yield $gCOD/gVS_{in}$	0.6 ± 0.1	0.5 ± 0.1	0.5 ± 0.1	0.6 ± 0.1
SCFAs yield $gCOD/gCOD_{in}$	0.43 ± 0.03	0.35 ± 0.02	0.39 ± 0.01	0.37 ± 0.02
COD removal (%)	23.8 ± 9.2	31.8 ± 6.8	37.9 ± 1.1	41.6 ± 0.7
H_2 yield ($ml/gCOD_{in}$)	12.6 ± 1.2	36.9 ± 5.9	32.6 ± 8.6	55.9 ± 8.9

*: pH control only at the startup; ** pH fluctuations were pronounced, and the control was throughout the experiment.

2.3. SCFAs Distribution Profiles upon Different HRT

The distribution profile of carboxylic acids is paramount in AF as it determines the possible end-uses of the harvested SCFAs. Figure 3 depicts SCFAs profile distribution according to HRT variation (expressed as SCFA/Total SCFAs × 100).

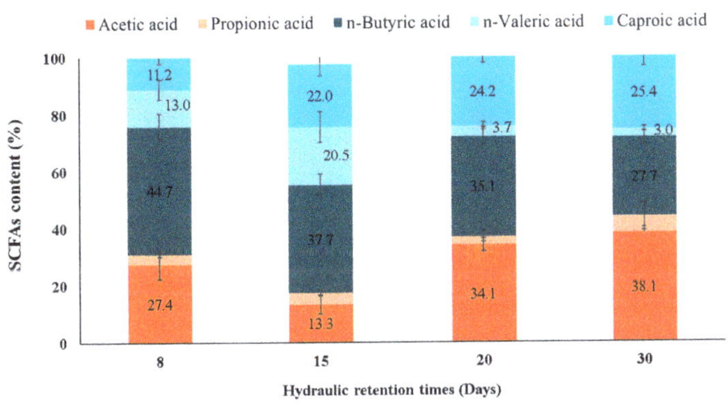

Figure 3. SCFAs distribution profile achieved in the SBM-AF at different HRTs.

As a general trend, it can be observed that longer carbon chain carboxylates were favored at longer HRTs. This pattern was in agreement with other studies [2,25].

A similar SCFAs distribution was observed at the longest HRTs (30 and 20 days). Acetic acid (38.2 ± 2.3% and 34.1 ± 3.2%), butyric acid (27.7 ± 2.1% and 35.1 ± 3.2%), and caproic acid (25.4 ± 2.9% and 24.2 ± 1.2%) were the primary carboxylic acids at HRTs of 30 days and 20 days, respectively. At 15 days of HRT, butyric acid presence prevailed with values of 37.6 ± 4.5%, while valeric and caproic acids accounted for almost the same (20.5 ± 1.1% and 21.9 ± 1.8%, respectively). For the short HRT of 8 days, butyric acid made up 44.7 ± 3.7%, followed by acetic acid at 27.4 ± 4.1%, whereas valeric and caproic acids exhibited almost equal proportions (11.2–13.1%). Propionic acid proportions were negligible in all reactors.

As stated in the previous section, HRT variation not only affected microbial population abundance but it may affect as well the predominant pathway of each individual SCFA [8,23] and, therefore, the final concentrations. Indeed, high HRTs have been reported to provide higher concentrations of more reduced compounds (i.e., long molecules such as butyrate, valerate, and caproate) rather than the more oxidized ones (i.e., short molecules such as acetate) [25,26].

The AF of carbohydrate-rich substrates usually produces even-chain carboxylic acids, whereas protein-rich substrates result in odd-chain carboxylates [17,22]. The molasses used as feedstock in the present investigation is considered a sugar-rich substrate, which justified the prevalence of even-chain carboxylates (acetic, butyric, and caproic acids). In addition to the substrate type (e.g., prevailing in carbohydrates, proteins, or lipids), operational parameters such as temperature, HRT, hydrogen partial pressure (pH_2), and inoculum type, among others, also affect the distribution of yielded SCFAs [8,11]. For instance, the effect of temperature on SCFAs production and profiles was studied by Greses et al. [27] in combination with the macromolecular composition of the organic material. Those researchers found that the temperature setup at 25 °C increased caproic acid content for carbohydrate-rich substrates, given that a low temperature hampers the conversion of long fatty acids into shorter ones.

Regarding pH_2, in a conventional AD process aiming to produce methane, the pH_2 must be low enough for a thermodynamically favored degradation of SCFAs into methane precursors [13]. For instance, Cazier et al. [28] reported that a high pH_2 inhibited the acetogenesis stage in solid-state mesophilic AD and lowered methane production. Likewise, the syntrophic oxidation of alcohols and SCFAs greatly depends on H_2. In this regard, the oxidation of SCFAs is only thermodynamically feasible at low pH_2 values [28]. The syntrophic oxidation can only spontaneously occur when H_2 consumption takes place, being the presence of H_2-consumers a requirement [29]. Because the Gibbs free energy of SCFAs oxidation reactions are positive under standard conditions, the process becomes thermodynamically favorable when methanogens maintain pH_2 lower than 10^{-4} atm [30]. In this investigation, the H_2 production attained (Table 1) could be responsible of the syntrophic oxidation limitation. This fact would justify the prevalence of longer SCFAs (butyric and caproic acids) over acetic acid.

Metabolite production in open, mixed cultures processes does not usually occur via one single pathway, given the co-occurrence of several bioreactions.. This fact gains relevance when non-standard conditions are applied since the metabolic role of microorganisms can also be altered. In this regard, one of the main singularities presented herein was the high accumulation of caproic acid at long HRTs (Figure 1). While some authors described this phenomenon as a consequence of an acetogenesis inhibition [6], recent investigations determined CCE as main pathway for caproic production when AF was performed at acid pH and long HRTs [2,31]. CCE enables the conversion of acetic and/or butyric acid into caproic acid in presence of an electron donor (ED), namely lactic acid or ethanol. Both EDs are primary intermediate metabolites of AF, mostly promoted at a low pH value. Thus, the prevalence of caproic acid at long HRTs compared to the short ones was likely due to the availability of ethanol and/or lactate produced via the primary fermentation of SBM. Moreover, it should be taken into account that CCE has been described as a hydrogenogenic pathway [2]. As can be seen in Table 1, the observed pattern of SCFAs and the H_2 production determined at long HRTs confirmed that caproic acid formation was likely attained via CCE.

It is essential to highlight that those EDs are normally added to the fermentation broth as an external supply to promote the CCE. For instance, Contreras-Dávila et al. [32] obtained 10.8 g COD/L of caproic acid from AF of food waste by adding lactate, and Roghair et al. [33] achieved a higher concentration of 51.6 g COD/L by adding ethanol to the anaerobic system. In contrast, in the present investigation, high caproic acid concentrations were attained without adding any external chemicals. During the steady state of all CSTRs, particularly those operating at long HRTs, lactic acid was not detected in the fermentation broth. However, when pH correction was necessary to avoid the drop below the threshold of 5.5, lactic acid was observed, although in small concentrations (0.10–0.35 g/L), corroborating the probable use of lactic acid as an ED when the CCE occurred at suitable pH values (i.e., 5.5–6.5). Previous studies found that process temperature also play an important role in the CCE since 25 °C seems to favor the process, while increasing fermentation temperatures to 55 °C may present a thermodynamic impediment [2,20]. These results showed that in situ CCE can be promoted by combining low temperatures (25 °C) with slightly acid pHs (5.5–6.0). Although ethanol and lactic acid exhibit similar Gibbs free energy as ED in CCE [34], it has been reported that lactic acid is preferred as ED [35]. This fact, along with the detection of lactic acid in the CSTR, might indicate that caproic acid accumulation was related to the use of lactic acid as ED for CCE.

CCE is a novel process gaining much attention recently due to the high economic value of the longest carboxylates in the material and chemical markets [4,33]. For instance, caproic acid is an increasingly growing biochemical platform due to its high economic value, which reached $3815/ton, compared to $600/ton for acetic acid [33]. Caproate can be used in various industrial applications. Some of those include the production of flavor compounds and antimicrobials [5] or biofuels such as diesel and jet fuels due to its similarities to gasoline in storage and transport [36].

Nevertheless, the present investigation produced a high concentration of caproic acid (e.g., 15.4 ± 1.5 g COD/L at HRT of 30 days) in situ without providing any external ED, reducing exogenous chemicals input in the fermenters and operating costs. Moreover, the high percentage of a specific fatty acid in the fermentation broth makes its separation easier for recovery. The high percentage and the fact that this acid exhibits high energy density and hydrophobicity enhance its insolubility and, therefore, its separation [36].

2.4. *Organic Matter Removal (COD) in the AF upon the Different HRT Tested*

COD removal in the CSTRs operated at different HRTs is depicted in Table 1. COD removal was higher (31–41%) at long HRTs than at the short HRT of 8 days (23%). This was aligned with previous AF studies. Bolaji and Dionisi [3] reported that for AF of carbohydrate-rich substrates, COD removal increased when increasing HRTs, with the highest COD removal of 35% at HRT of 30 days. In the present investigation, the observed COD removal values were partially attributed to the chain elongation process that was promoted at long HRT due to caproic acid formation at these retention times. Being caproic acid a hydrogenogenic process that releases hydrogen when lactic acid is involved in the elongation process [37], the gas release could be partially related to the high COD removal attained at long HRT. In this study, hydrogen was observed during AF of SBM as a co-product of SCFAs (Table 1). These hydrogen yields were lower than the reported in the literature from similar substrates where the experiments were directed toward hydrogen production rather than SCFAs, where favored microbial pathways were different [2,37].

In a complex biological process such as AF, the organic matter removal (e.g., COD) from the system could be attributable to several microbiological pathways than its conversion into SCFAs (e.g., CH_4, H_2, N, and P removal, etc.). In this study, the high COD removal attained without observing significant gas production suggested that some solubilized COD fractions were probably used by other microorganisms through different pathways. For instance, previous authors showed that adding a fermentation broth rich in SCFAs to a nutrient removal system significantly enhanced the process. This further supports the versatility of SCFAs as a high-value-added product in various applications [38].

Since the present research aimed to maximize SCFAs production by intervening only in applying different HRTs under specific temperature and pH conditions, determining which metabolic pathway was responsible for the COD consumption was out of the scope. However, it was highly relevant to know that long HRT diverted COD towards other pathways rather than the targeted COD conversion into SCFAs. In this way, when aiming at SCFAs accumulation, shorter HRTs were suggested to be more profitable since these HRTs resulted in the lowest COD consumption.

2.5. *Economic Perspectives of the Results and Future Scope*

Worldwide sugar beet production was around 270.16 million tons in 2021 [39]. Per each ton of sugar beet processed, around 40 kg of beet molasses are generated, producing about 10,8 million tons of global molasses [40]. Molasses are usually used as a supplement in animal nutrition [39] or in the production of fuels (i.e., bioethanol), as well as in the production of yeasts, vitamins, and other components for the food and beverages industries [39]. However, the low economic value of molasses as an animal feed supplement limits interest in this use. Furthermore, thick sugar juices generated during sugar production, starch, or glucose-based substrates are preferred feedstock in the fermentation industry rather than molasses [41]. In this context, producing other value-added products from molasses is promising and is currently attracting interest from the sugar industry stakeholders to find new outlets as a driver for the sector's competitiveness [40]. The chemical characteristics of molasses make them an attractive substrate in biotechnological processes, such as anaerobic digestion to produce biogas or carboxylic acids in the anaerobic fermentation process [42–44]. The global value of molasses is about seven billion USD [41]. In the EU, beet molasses cost 125–152 $/tons in 2019 [41]. Considering the average price of 2 $/kg of molasses as animal feed, the revenues from the molasses used for this purpose in 2021 could be around $21,600 billion [41].

On the other hand, considering the highest SCFAs concentration reached in the present research, an estimation of the economic revenue from the process performed herein was addressed (Figure 4). Given the daily input of molasses into this reactor, the total SCFAs concentration obtained at HRT 30 days was 13.4 $g/g_{Molasses}$. The prevailing SCFAs in this reactor were acetic, n-butyric, and caproic acids, with percentages of 38.2 $\pm$ 2.3%, 27.7 $\pm$ 2.1%, and 25.4 $\pm$ 2.9%, respectively. SCFAs with low proportions in the fermentation broth were excluded from this estimation. The current SCFAs prices considered were adapted from Agnihotri et al. [45], being $600, $2686, and $3815, per ton of acetic, butyric, and caproic acids, respectively. As can be seen in Figure 4, a total revenue of $281.61 billion can be generated from SCFAs production through AF of beet molasses at 25 °C, which is largely superior to the income from selling molasses as an animal feed supplement. Notwithstanding that, SCFAs extraction/separation from the fermentation broth is a subsequent process to AF to be considered when calculating the costs of the overall process [45,46]. Nevertheless, for this calculation, only the product costs are considered, namely animal feed vs. carboxylates as biochemicals. In this regard, a detailed techno-economic design of a full-scale SBM-AF process should be conducted to assess its real applicability.

The second approach to evaluate the effect of HRTs on the revenues from the AF of SBM in the studied conditions was adopted from Montecchio et al. [47], where the parameter "specific digester productivity" refers to SCFAs revenues and was calculated as GSCFAs = (Q $\times$ SCFAs)/Vol reactor. SCFAs are expressed as revenues ($) and GSCFAs is, therefore, defined as $\$/m^3\ day^{-1}$. The reactor working volume was 0.001 m^3, as described in the M&M section. The SCFAs concentrations considered for this calculation (expressed as g/L) were those of the predominant carboxylates in each condition and, their prices were according to Agnihotri et al. [45], as in the previous estimation. As can be seen in Figure 4B, the highest revenue could be reached concomitantly with the HRTs increase (from 8 to 30 days). Therefore, an almost linear relationship could be established between the two parameters.

In future research, other systems allowing shortening HRT while retaining biomass-like sequencing batch reactors (SBR) could also be considered to further optimize the productivity of SCFA from beet molasses at 25 °C. Another possible approach is using hybrid bioreactors, in which the CSTR configuration is modified by including support material inside the reactors to promote microorganisms' growth and retention (increasing the cellular retention time).

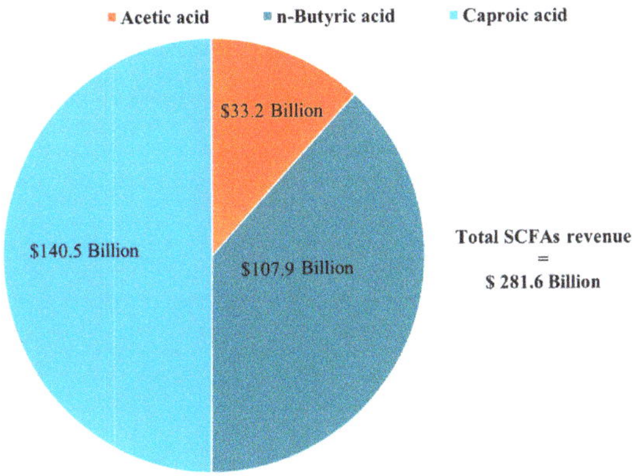

(A)

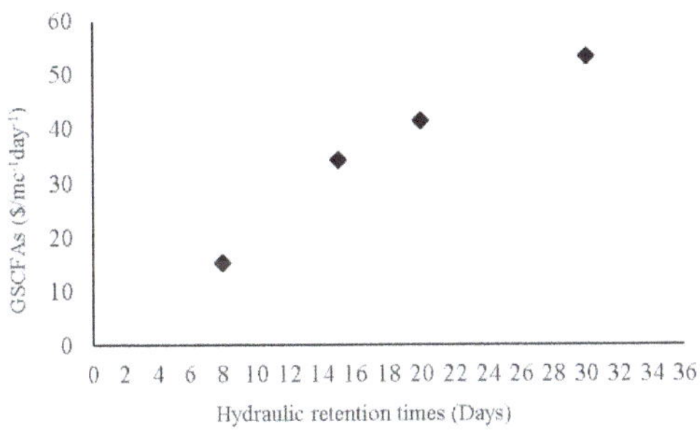

(B)

Figure 4. (**A**) Estimation of SCFAs revenue from AF of SBM at HRT of 30 days and 25 °C temperature condition. (**B**) Specific digester productivity at different HRTs.

3. Materials and Methods

3.1. Substrate and Inoculum

SBM (COMPAÑIA DE MELAZAS S.A., Madrid, Spain) was the primary substrate for SCFAs production. The characteristics of SBM are depicted in Table 2. SBM mainly contained sucrose (above 95% of the composition); thus, it was expected to be a suitable raw material for SCFAs production through AF. This substrate was selected based on its high carbohydrate content and since it does not exhibit low hydrolytic efficiencies (as happens with particulate wastes) to produce SCFAs via AF. In this way, by ensuring an optimum hydrolytic stage, this research can be entirely devoted to investigating the acidogenesis stage.

The inoculum was an anaerobic sludge collected from a local wastewater treatment plant (El Soto-Móstoles, Madrid, Spain). The mesophilic (35 °C) anaerobic sludge had the following physicochemical characterization: pH 7.3 ± 0.1; TS = 14.5 ± 1.3 g/L; VS = 8.2 ± 1.9 g/L; N-NH_4^+ = 0.8 ± 0.1 g N/L and alkalinity = 4.3 ± 0.1 g $CaCO_3$/L.

3.2. Experiments Setup

The AF of SBM was carried out in semi-continuous reactors (working volume of 1 L) at 25 °C. The temperature was maintained using a thermostatic water bath (F12-ED v2.0, Julabo, Seelbach, Germany). The mixing was performed by magnetic stirring using agitators (MR Hei-MixL, Heidolph Instruments, Schwabach, Germany). The applied low temperature (25 °C) was based on previous studies on AF for SCFAs production and is economically promising compared to higher temperatures where an energy supply for heating is needed [2,22]. A fixed OLR of 4 g COD/L·d was used, and different HRTs of 30, 20, 15, and 8 days were tested to identify the most suitable. Each reactor was operated for at least three times its HRT and until stable output concentrations were achieved.

Table 2. Physicochemical characterization of SBM (mean ± SD).

Parameters (Units)	SBM
pH	6.7 ± 0.1
TS (%)	76.4 ± 0.2
VS (%)	66.6 ± 0.9
TCOD (g/L)	1131.0 ± 4.5
SCOD/TCOD	0.9 ± 0.1
TCOD/VS	1.7 ± 0.1
Total sugars (g/L)	695.6 ± 1.8
Sucrose (g/L)	665.6 ± 0.4
Glucose (g/L)	30.0 ± 1.4
N-NH_4^+ (g N/L)	0.5 ± 0.1
Alkalinity (g $CaCO_3$/L)	1.7 ± 0.2
Carbohydrates (%) *	56.6 ± 0.8
Ash (%) *	9.8 ± 0.6
Lipids (%) *	20.7 ± 0.2
Proteins (%) *	12.9 ± 0.3

* Percentage calculated based on dry matter content. TS: Total solids; VS: Volatile solids; N-NH_4^+: Ammonium nitrogen; TCOD: Total chemical oxygen demand; SCOD: Soluble chemical oxygen demand.

Since SBM is a carbohydrate-rich material, the fast hydrolysis and acidification of sugars might lead to a fast pH decrease. To avoid a drastic pH drop that would result in the accumulation of primary metabolites (e.g., ethanol and lactic acid) instead of SCFAs [17], pH was monitored daily and adjusted to slightly acidic values (5.5–6.5) using 5 M of NaOH. This methodology was based on previous studies that identified this pH range as optimal for AF performance [2,17,31].

3.3. *Analytical Methods*

SBM used as feedstock and effluents from CSTRs were analyzed in terms of total solids (TS), volatile solids (VS), and chemical oxygen demand (COD), according to Standard Methods [48]. The samples for analyzing SCFAs, soluble chemical oxygen demand (SCOD), and ammonium (N-NH_4^+) were filtered by a 0.45 µm filter membrane. SCOD and N-NH_4^+ were analyzed using colorimetric commercial kits (ISO 7150-1, and ISO 15705, respectively. Merck, Darmstadt, Germany). Samples for SCFAs were further filtered by a 0.22 µm Teflon filter and analyzed via liquid chromatography (HPLC) (HPLC-RID 1260, Agilent, Santa Clara, CA, USA) with a refractive index detector. HPLC was equipped with an ion exclusion column (Aminex HPX-87H with 300 × 7.8 mm internal diameters) using the conditions described by Llamas et al. [16]. The analyzed SCFAs were lactic, acetic, propionic, isobutyric, butyric, isovaleric, valeric, and caproic acids. The content of ethanol was also analyzed using the same methodology described for SCFAs. Similarly, sucrose and fructose content in SBM was determined in the HPLC using a CARBOSep CHO-682 column (Transgenomic, Omana, NE, USA) under the conditions detailed by Cubas-Cano et al. [49], after filtering the samples by a 0.22 µm Teflon filter. Total Kjeldahl nitrogen (TKN) was determined according to the method 4500-N of the standard methods [50], and protein content was obtained from the value of the TKN content and applying a conversion factor (i.e., 6.25) [43,49]. Carbohydrates were measured using the phenol-sulfuric acid method [51]. Biogas composition in terms of H_2, CO_2, and CH_4 was determined using gas chromatography according to Greses et al. [2].

3.4. Calculations for the Evaluation of Process Performance

SCFAs yield was determined as the net SCFAs concentration (expressed as grams of COD_{out}) divided by the total COD of the feedstock ($TCOD_{in}$):

$$\text{SCFAs Yield (g/gCODin)} = \frac{\text{COD(SCFAs)out}}{\text{TCODin}}$$

The conversion factors of SCFAs concentrations into COD correspond stoichiometrically to the following values: 1.067 for acetic acid, 1.514 for propionic acid, 1.818 for n-butyric acid (the sum of iso-butyric and butyric), 2.039 for n-valeric (the sum of iso-valeric and valeric), and 2.207 for caproic acid. Thereby, concentrations in g/L were multiplied by each factor to attain g COD/L.

The solubilization and acidification degrees refer to the efficiencies of solubilizing (hydrolysis stage) and acidifying (acidification stage) the organic material fed (COD_{in}) to the reactor, respectively. Those percentages can be calculated as follows:

$$\%\ \text{Solubilization degree} = \frac{\text{SCOD}}{\text{TCOD}} \times 100$$

$$\%\ \text{Acidification degree} = \frac{\text{COD(SCFAs)}}{\text{SCOD}} \times 100$$

COD removal refers to the organic material, in terms of total COD, that has been removed from the system and transformed into one or various end-products other than SCFAs. This percentage can be calculated as follows:

$$\%\ \text{COD removal} = \frac{(\text{CODin} - \text{CODout})}{\text{CODin}} \times 100$$

4. Conclusions

The HRT effect on AF of SBM at 25 °C revealed a lack of significant variations on acidogenesis degree and SCFAs yields when this parameter ranged from 8 to 30 days. However, an apparent effect on microbial metabolisms was detected since the SCFAs profile was altered. Long HRTs (i.e., 30 and 20 days) were identified to be the most suitable to obtain long carbon chain carboxylates via CCE since high caproic acid concentrations (15.4 ± 1.5 g COD/L) were reached at the HRT of 30 days. Furthermore, the operation at long HRTs led to a more stable process, which is crucial for scaling up the technology. This investigation provided an efficient strategy to valorize SBM wastes for SCFAs production, which exhibits higher economic revenue than its uses for animal feed preparation. Aiming to decrease society's dependency on petrochemicals, SCFAs are essential building blocks in different industrial sectors.

In terms of SCFAs yield, the HRT of 8 days showed the highest values in terms on g SCFAs/g COD_{in}. However, the HRT of 30 days exhibited the highest stability and allowed the obtaining of a distribution profile rich in caproic acid.

Supplementary Materials: The following supporting information can be downloaded at: https://www.mdpi.com/article/10.3390/molecules28186635/s1, Figure S1: pH evolution throughout the experiment at the different HRTs. Red lines refer to the upper and lower suitable values for anaerobic fermentation aiming to produce SCFAs.

Author Contributions: K.A.: Investigation, Data curation, Writing—original draft, Writing—review and editing; S.G.: Conceptualization, Data curation, Writing—review and editing; C.G.-F.: Conceptualization, Data curation, Writing—review and editing, Funding acquisition. All authors have read and agreed to the published version of the manuscript.

Funding: This work received funding from the State Research Agency, MCIN/AEI /10.13039/501100011033, through the grant with reference number CEX2019-000931-M received in the 2019 call for "Severo Ochoa Centres of Excellence" and "María de Maeztu Units of Excellence" of the

State Programme for Knowledge Generation and Scientific and Technological Strengthening of the R&D&I System.

Institutional Review Board Statement: Not applicable.

Informed Consent Statement: Not applicable.

Data Availability Statement: Not applicable.

Conflicts of Interest: The authors declare no conflict of interest.

Sample Availability: Not applicable.

References

1. Agler, M.T.; Wrenn, B.A.; Zinder, S.H.; Angenent, L.T. Waste to Bioproduct Conversion with Undefined Mixed Cultures: The Carboxylate Platform. *Trends Biotechnol.* **2011**, *29*, 70–78. [CrossRef]
2. Greses, S.; Tomás-Pejó, E.; González-Fernández, C. Short-Chain Fatty Acids and Hydrogen Production in One Single Anaerobic Fermentation Stage Using Carbohydrate-Rich Food Waste. *J. Clean. Prod.* **2021**, *284*, 124727. [CrossRef]
3. Bolaji, I.O.; Dionisi, D. Acidogenic Fermentation of Vegetable and Salad Waste for Chemicals Production: Effect of PH Buffer and Retention Time. *J. Environ. Chem. Eng.* **2017**, *5*, 5933–5943. [CrossRef]
4. Spirito, C.M.; Richter, H.; Rabaey, K.; Stams, A.J.M.; Angenent, L.T. Chain Elongation in Anaerobic Reactor Microbiomes to Recover Resources from Waste. *Curr. Opin. Biotechnol.* **2014**, *27*, 115–122. [CrossRef] [PubMed]
5. Zagrodnik, R.; Duber, A.; Łężyk, M.; Oleskowicz-Popiel, P. Enrichment Versus Bioaugmentation—Microbiological Production of Caproate from Mixed Carbon Sources by Mixed Bacterial Culture and Clostridium Kluyveri. *Environ. Sci. Technol.* **2020**, *54*, 5864–5873. [CrossRef] [PubMed]
6. Magdalena, J.A.; Greses, S.; González-Fernández, C. Impact of Organic Loading Rate in Volatile Fatty Acids Production and Population Dynamics Using Microalgae Biomass as Substrate. *Sci. Rep.* **2019**, *9*, 18374. [CrossRef] [PubMed]
7. Aboudi, K.; Álvarez-Gallego, C.J.; Romero-García, L.I.; Fernández-Güelfo, L.A. Biogas, Biohydrogen, and Polyhydroxyalkanoates Production from Organic Waste in the Circular Economy Context. *Sustain. Biofuels Oppor. Chall.* **2021**, 305–343. [CrossRef]
8. Cavinato, C.; Da Ros, C.; Pavan, P.; Bolzonella, D. Influence of Temperature and Hydraulic Retention on the Production of Volatile Fatty Acids during Anaerobic Fermentation of Cow Manure and Maize Silage. *Bioresour. Technol.* **2017**, *223*, 59–64. [CrossRef]
9. Kuruti, K.; Nakkasunchi, S.; Begum, S.; Juntupally, S.; Arelli, V.; Anupoju, G.R. Rapid Generation of Volatile Fatty Acids (VFA) through Anaerobic Acidification of Livestock Organic Waste at Low Hydraulic Residence Time (HRT). *Bioresour. Technol.* **2017**, *238*, 188–193. [CrossRef]
10. Strazzera, G.; Battista, F.; Tonanzi, B.; Rossetti, S.; Bolzonella, D. Optimization of Short Chain Volatile Fatty Acids Production from Household Food Waste for Biorefinery Applications. *Environ. Technol. Innov.* **2021**, *23*, 101562. [CrossRef]
11. Vázquez-Fernández, A.; Suárez-Ojeda, M.E.; Carrera, J. Review about Bioproduction of Volatile Fatty Acids from Wastes and Wastewaters: Influence of Operating Conditions and Organic Composition of the Substrate. *J. Environ. Chem. Eng.* **2022**, *10*, 107917. [CrossRef]
12. Tomás-Pejó, E.; González-Fernández, C.; Greses, S.; Kennes, C.; Otero-Logilde, N.; Veiga, M.C.; Bolzonella, D.; Müller, B.; Passoth, V. Production of Short-Chain Fatty Acids (SCFAs) as Chemicals or Substrates for Microbes to Obtain Biochemicals. *Biotechnol. Biofuels Bioprod.* **2023**, *16*, 96. [CrossRef] [PubMed]
13. Pandey, A.K.; Pilli, S.; Bhunia, P.; Tyagi, R.D.; Surampalli, R.Y.; Zhang, T.C.; Kim, S.-H.; Pandey, A. Dark Fermentation: Production and Utilization of Volatile Fatty Acid from Different Wastes—A Review. *Chemosphere* **2022**, *288*, 132444. [CrossRef] [PubMed]
14. EU Legislation. Renewable Energy Directive 2009/28/EC. 2009. Available online: https://lexparency.org/eu/32009L0028/ (accessed on 10 September 2023).
15. Gottardo, M.; Micolucci, F.; Bolzonella, D.; Uellendahl, H.; Pavan, P. Pilot Scale Fermentation Coupled with Anaerobic Digestion of Food Waste—Effect of Dynamic Digestate Recirculation. *Renew. Energy* **2017**, *114*, 455–463. [CrossRef]
16. Llamas, M.; Greses, S.; Tomás-Pejó, E.; González-Fernández, C. Carboxylic Acids Production via Anaerobic Fermentation: Microbial Communities' Responses to Stepwise and Direct Hydraulic Retention Time Decrease. *Bioresour. Technol.* **2022**, *344*, 126282. [CrossRef]
17. Greses, S.; Tomás-Pejó, E.; González-Fernández, C. Assessing the Relevance of Acidic PH on Primary Intermediate Compounds When Targeting at Carboxylate Accumulation. *Biomass Convers. Biorefinery* **2022**, *12*, 4519–4529. [CrossRef]
18. Jankowska, E.; Chwialkowska, J.; Stodolny, M.; Oleskowicz-Popiel, P. Volatile Fatty Acids Production during Mixed Culture Fermentation—The Impact of Substrate Complexity and PH. *Chem. Eng. J.* **2017**, *326*, 901–910. [CrossRef]
19. Alibardi, L.; Cossu, R. Effects of Carbohydrate, Protein and Lipid Content of Organic Waste on Hydrogen Production and Fermentation Products. *Waste Manag.* **2016**, *47*, 69–77. [CrossRef]
20. Greses, S.; Tomás-Pejó, E.; González-Fernández, C. Food Waste Valorization into Bioenergy and Bioproducts through a Cascade Combination of Bioprocesses Using Anaerobic Open Mixed Cultures. *J. Clean. Prod.* **2022**, *372*, 133680. [CrossRef]

21. Onodera, T.; Sase, S.; Choeisai, P.; Yoochatchaval, W.; Sumino, H.; Yamaguchi, T.; Ebie, Y.; Xu, K.; Tomioka, N.; Mizuochi, M.; et al. Development of a Treatment System for Molasses Wastewater: The Effects of Cation Inhibition on the Anaerobic Degradation Process. *Bioresour. Technol.* **2013**, *131*, 295–302. [CrossRef]
22. Greses, S.; Tomás-Pejó, E.; Gónzalez-Fernández, C. Agroindustrial Waste as a Resource for Volatile Fatty Acids Production via Anaerobic Fermentation. *Bioresour. Technol.* **2020**, *297*, 122486. [CrossRef] [PubMed]
23. Lv, N.; Cai, G.; Pan, X.; Li, Y.; Wang, R.; Li, J.; Li, C.; Zhu, G. PH and Hydraulic Retention Time Regulation for Anaerobic Fermentation: Focus on Volatile Fatty Acids Production/Distribution, Microbial Community Succession and Interactive Correlation. *Bioresour. Technol.* **2022**, *347*, 126310. [CrossRef]
24. Zhang, Q.; Lu, Y.; Zhou, X.; Wang, X.; Zhu, J. Effect of Different Vegetable Wastes on the Performance of Volatile Fatty Acids Production by Anaerobic Fermentation. *Sci. Total Environ.* **2020**, *748*, 142390. [CrossRef] [PubMed]
25. De Groof, V.; Coma, M.; Arnot, T.; Leak, D.J.; Lanham, A.B. Selecting Fermentation Products for Food Waste Valorisation with HRT and OLR as the Key Operational Parameters. *Waste Manag.* **2021**, *127*, 80–89. [CrossRef]
26. Wang, J.; Yin, Y. Biological Production of Medium-Chain Carboxylates through Chain Elongation: An Overview. *Biotechnol. Adv.* **2022**, *55*, 107882. [CrossRef] [PubMed]
27. Greses, S.; Tomás-Pejó, E.; González-Fernández, C. Statistical Correlation between Waste Macromolecular Composition and Anaerobic Fermentation Temperature for Specific Short-Chain Fatty Acid Production. *Environ. Res.* **2022**, *206*, 112288. [CrossRef]
28. Cazier, E.A.; Trably, E.; Steyer, J.P.; Escudie, R. Reversibility of Hydrolysis Inhibition at High Hydrogen Partial Pressure in Dry Anaerobic Digestion Processes Fed with Wheat Straw and Inoculated with Anaerobic Granular Sludge. *Waste Manag.* **2019**, *85*, 498–505. [CrossRef]
29. Rocamora, I.; Wagland, S.T.; Villa, R.; Simpson, E.W.; Fernández, O.; Bajón-Fernández, Y. Dry Anaerobic Digestion of Organic Waste: A Review of Operational Parameters and Their Impact on Process Performance. *Bioresour. Technol.* **2020**, *299*, 122681. [CrossRef]
30. Müller, N.; Worm, P.; Schink, B.; Stams, A.J.; Plugge, C.M. Syntrophic Butyrate and Propionate Oxidation Processes: From Genomes to Reaction Mechanisms. *Environ. Microbiol. Rep.* **2010**, *2*, 489–499. [CrossRef]
31. Greses, S.; De Bernardini, N.; Treu, L.; Campanaro, S.; González-Fernández, C. Genome-Centric Metagenomics Revealed the Effect of PH on the Microbiome Involved in Short-Chain Fatty Acids and Ethanol Production. *Bioresour. Technol.* **2023**, *377*, 128920. [CrossRef]
32. Contreras-Dávila, C.A.; Carrión, V.J.; Vonk, V.R.; Buisman, C.N.J.; Strik, D.P.B.T.B. Consecutive Lactate Formation and Chain Elongation to Reduce Exogenous Chemicals Input in Repeated-Batch Food Waste Fermentation. *Water Res.* **2020**, *169*, 115215. [CrossRef] [PubMed]
33. Roghair, M.; Liu, Y.; Adiatma, J.C.; Weusthuis, R.A.; Bruins, M.E.; Buisman, C.J.N.; Strik, D.P.B.T.B. Effect of N-Caproate Concentration on Chain Elongation and Competing Processes. *ACS Sustain. Chem. Eng.* **2018**, *6*, 7499–7506. [CrossRef] [PubMed]
34. Cavalcante, W.d.A.; Leitão, R.C.; Gehring, T.A.; Angenent, L.T.; Santaella, S.T. Anaerobic Fermentation for N-Caproic Acid Production: A Review. *Process Biochem.* **2017**, *54*, 106–119. [CrossRef]
35. Zhu, X.; Tao, Y.; Liang, C.; Li, X.; Wei, N.; Zhang, W.; Zhou, Y.; Yang, Y.; Bo, T. The Synthesis of N-Caproate from Lactate: A New Efficient Process for Medium-Chain Carboxylates Production. *Sci. Rep.* **2015**, *5*, 14360. [CrossRef] [PubMed]
36. Harvey, B.G.; Meylemans, H.A. 1-Hexene: A Renewable C6 Platform for Full-Performance Jet and Diesel Fuels. *Green Chem.* **2014**, *16*, 770–776. [CrossRef]
37. Chen, C.C.; Lin, C.Y. Using Sucrose as a Substrate in an Anaerobic Hydrogen-Producing Reactor. *Adv. Environ. Res.* **2003**, *7*, 695–699. [CrossRef]
38. Feng, X.C.; Bao, X.; Che, L.; Wu, Q.L. Enhance Biological Nitrogen and Phosphorus Removal in Wastewater Treatment Process by Adding Food Waste Fermentation Liquid as External Carbon Source. *Biochem. Eng. J.* **2021**, *165*, 107811. [CrossRef]
39. Market Watch. Organic Molasses Market Research Report 2023—2030. 2023. Available online: https://www.marketwatch.com/press-release/molasses-market-research-2023-2030-2023-06-17 (accessed on 10 September 2023).
40. EU Beet Sugar Sustainability Partnership. Keep Molasses for Bioenergy and Bio-Based Products. 2017. Available online: http://www.sustainablesugar.eu/molasses (accessed on 10 September 2023).
41. S&P Global. Commodity Insights Molasses and Feed Ingredients Market Analysis. Available online: https://www.spglobal.com/commodityinsights/en/ci/products/food-commodities-food-manufacturing-softs-molasses-feed-ingredients.html (accessed on 10 September 2023).
42. Wang, B.; Li, Y.; Ren, N. Biohydrogen from Molasses with Ethanol-Type Fermentation: Effect of Hydraulic Retention Time. *Int. J. Hydrogen Energy* **2013**, *38*, 4361–4367. [CrossRef]
43. Khan, S.; Ahmad, R.; Phulpoto, I.A.; Kashif, M.; Shen, P. Effect of Rice Winery Wastewater as a Co-Substrate to Enhance Anaerobic Digestion of Molasses for Methane Production. *Bioresour. Technol. Rep.* **2022**, *18*, 101062. [CrossRef]
44. Albuquerque, M.G.E.; Torres, C.A.V.; Reis, M.A.M. Polyhydroxyalkanoate (PHA) Production by a Mixed Microbial Culture Using Sugar Molasses: Effect of the Influent Substrate Concentration on Culture Selection. *Water Res.* **2010**, *44*, 3419–3433. [CrossRef]
45. Agnihotri, S.; Yin, D.M.; Mahboubi, A.; Sapmaz, T.; Varjani, S.; Qiao, W.; Koseoglu-Imer, D.Y.; Taherzadeh, M.J. A Glimpse of the World of Volatile Fatty Acids Production and Application: A Review. *Bioengineered* **2022**, *13*, 1249–1275. [CrossRef] [PubMed]

46. Lee, H.; Im, S.J.; Kim, C.M.; Jang, A. Relative Selectivity of Short- and Medium-Chain Fatty Acids in Hydrophobic Membrane for Resource Recovery in Wastewater: Effects of Supported Liquid, Feed PH, and Membrane Pores. *Chem. Eng. J.* **2022**, *446*, 137258. [CrossRef]
47. Montecchio, D.; Braguglia, C.M.; Gallipoli, A.; Gianico, A. A Model-Based Tool for Reactor Configuration of Thermophilic Biogas Plants Fed with Waste Activated Sludge. *Renew. Energy* **2017**, *113*, 411–419. [CrossRef]
48. APHA; AWWA; WEF. *Standard Methods for the Examination of Water and Wastewater*, 23rd ed.; American Public Health Association: Washington, DC, USA, 2017; ISBN 9780875532356.
49. Cubas-Cano, E.; González-Fernández, C.; Tomás-Pejó, E. Evolutionary Engineering of Lactobacillus Pentosus Improves Lactic Acid Productivity from Xylose-Rich Media at Low PH. *Bioresour. Technol.* **2019**, *288*, 121540. [CrossRef]
50. Krul, E.S. Calculation of Nitrogen-to-Protein Conversion Factors: A Review with a Focus on Soy Protein. *J. Am. Oil Chem. Soc.* **2019**, *96*, 339–364. [CrossRef]
51. Dubois, M.; Gilles, K.A.; Hamilton, J.K.; Rebers, P.A.; Smith, F. Colorimetric Method for Determination of Sugars and Related Substances. *Anal. Chem.* **1956**, *28*, 350–356. [CrossRef]

catalysts

MDPI

Article

Sodium Silicates Modified Calcium Oxide as a High-Performance Solid Base Catalyst for Biodiesel Production

Shunpan Zhang [1,2,3,4], Junying Fu [1,2,3,*], Shiyou Xing [1,2,3], Ming Li [1,2,3], Xiaochun Liu [1,2,3,5], Lingmei Yang [1,2,3] and Pengmei Lv [1,2,3,*]

1 Guangzhou Institute of Energy Conversion, Chinese Academy of Sciences, Guangzhou 510640, China; zhangshunpan20@mails.ucas.ac.cn (S.Z.); xingsy@ms.giec.ac.cn (S.X.); liming@ms.giec.ac.cn (M.L.); liuxiaochun@mail.ustc.edu.cn (X.L.); yanglm@ms.giec.ac.cn (L.Y.)
2 CAS Key Laboratory of Renewable Energy, Guangzhou 510640, China
3 Guangdong Provincial Key Laboratory of New and Renewable Energy Research and Development, Guangzhou 510640, China
4 University of Chinese Academy of Sciences, Beijing 100049, China
5 School of Energy Science and Engineering, University of Science and Technology of China, Hefei 230026, China
* Correspondence: fujy@ms.giec.ac.cn (J.F.); lvpm@ms.giec.ac.cn (P.L.)

Abstract: Under the energy crisis and with greenhouse gases causing an ecological imbalance, biofuel has attracted worldwide attention due to its sustainability and low net-carbon emission. For years, the traditional biodiesel industry has been demanding a high-performance solid base catalyst. Its poor reusability is the bottleneck for a promising calcium-based solid-base catalyst. In this work, we successfully adopted a new silicate-strength strategy to improve the stability while preserving the activity of the catalyst. The newly synthesized catalyst, NCSO, had two main catalytic phases, Na_2CaSiO_4 and CaO, and showed a 98.2% FAMEs yield in 60 min at 80 °C with a methanol/oil molar ratio of 9:1 and 5 wt.% catalyst loading. After 12 consecutive reuses, a 57.3% FAMEs yield could still be achieved. The effect of the reaction temperature, methanol ratio, catalyst loading, and reaction time on the FAMEs yield was also investigated. With a combined characterization of XRD, XPS, and SEM, etc., we confirmed that Na_2CaSiO_4 and CaO showed a synergistic effect in catalyzing the transesterification reaction: the addition of the Na_2CaSiO_4 phase in NCSO could significantly improve the activity of CaO, while the CaO phase, in turn, helps to stabilize the Na_2CaSiO_4 phase. This silicate-strength strategy provides a new route to synthesize stable and highly active solid base catalysts.

Keywords: biodiesel; transesterification; solid base catalyst; silicate-strength strategy

1. Introduction

With the development of the world economy, energy demand is increasing daily. However, the limited reserve of traditional fossil fuels is causing energy insecurity, and the extra emission of CO_2 is challenging the global ecological environment [1–3]. The development of new, environmentally friendly, non-fossil fuels is quite urgent [4]. Biodiesel has attracted worldwide attention due to its lower sulfuric content/toxicity, better lubricity than petroleum diesel, and compatibility with traditional engines [5–7]. It can effectively replace or mix with fossil fuels, thereby reducing greenhouse gas emissions, benefitting the environment and alleviating the energy crisis. Generally, biodiesel can be synthesized by the transesterification of animal and vegetable oils, waste oils, microalgae, and microbial oils with short-chain alcohols (mainly methanol) [8]. The catalyst is the key component of this reaction, and traditionally, homogeneous base catalysts such as KOH and NaOH are often used in the transesterification reaction under mild reaction conditions. However, these catalysts cannot be recycled, thus reducing the biodiesel quality and causing unnecessary water waste (to remove residue catalyst in biodiesel), making by-product glycerol hard to refine [9]. Due to their recyclability advantages, heterogeneous solid base catalysts have received extensive attention from researchers. A lot of research has been conducted with CaO (calcium oxide), a potential solid base catalyst with relatively high activity, low cost, low solubility in methanol, low toxicity, and recyclability [10–13], to find a way to up-scale its application in the biodiesel industry. Some researchers have used base or other metal elements to dope CaO to improve its transesterification activity [14]. Teo et al. [15] synthesized CaO-MgO catalysts by the co-precipitation method and found that the mixed solid base oxides (Ca/Mg = 1, molar ratio) exhibited a relatively higher basicity and stability among all oof the researched samples. Rabie et al. [11] prepared a new type of basic heterogeneous catalyst with a CaO/MgO composite loaded on diatomite. They found that the diatomaceous earth support could significantly increase the total porosity and surface area of the catalyst, with a biodiesel yield as high as 96.47%. Kandis Sudsakorn et al. [16] synthesized strontium (Sr)-doped CaO/MgO catalysts and their results showed that the Sr^{2+} could provide many strong basic sites over CaO/MgO, and a FAME (fatty acid methyl ester) yield of 99.6% was obtained under the optimal conditions.

However, the bottleneck of CaO as a base catalyst is its deactivation by leaching during the reaction, thus shortening its working life [17]. To improve the reusability of solid base catalysts, supports such as nano-tubes, aluminate salts, and silicate salts have been investigated [18–24]. Mohd Lokman Ibrahim increased the basicity of the catalyst by loading sodium oxide onto carbon nanotubes. As a carrier, carbon nanotubes have super high compatibility with sodium oxide, and the best yield of FAME was 97.6% [22]. Behgam et al. [20] synthesized $MgAl_2O_4$ by the combustion method and then deposited MgO on the surface of the catalyst by the impregnation method. They found that $MgO/MgAl_2O_4$ showed a good reusability of six times and kept an 80% FAME yield. Wang et al. [25] found that the silicate phases, $Ca_2MgSi_2O_7$, in a gasified straw slag showed high reusability in catalyzing transesterification reactions. It could be reused for up to 50 cycles with over 50% of the maximum activity remaining. They also found that other bimetallic silicates also showed high reusability. However, these catalysts often required severe reaction conditions such as 160 °C or higher reaction temperatures, and 8 h or longer reaction times [26,27]. Liu et al. designed a Ca-based solid base catalyst modified by Na_2SiO_3 and the catalyst showed much better water and carbon dioxide resistance than CaO [8]. These results confirmed that the silicate modification could indeed improve the reusability of the Ca-based solid base catalyst, and the $[SiO_4]$ tetrahedron structures in the researched silicates seemed to be the main reason for improving the stability of the catalyst. However, research is still needed to further improve the activity of the silicate-modified Ca-based solid base. Romero et al. [21] synthesized Na_2ZnSiO_4 as a catalyst by doping sodium into ZnSi-O, and the catalyst showed a biodiesel yield higher than 99% within 45 min. Stefan M. Pavlović et al. synthesized calcium silicate loaded with CaO, which showed a 97.8% FAME yield in 30 min [28]. Although these studies claimed that the silicate modification could also show good activity, these catalysts could only last for five reused times. Thus, a better silicate modification method is needed to improve the reusability of the Ca-based solid base catalyst while preserving its activity.

In this study, a calcium sodium silicate doped CaO was synthesized as a high-performance solid base catalyst for biodiesel production. The effect of calcination temperature on the catalyst composition, texture, and performance was investigated. Reaction parameters such as the reaction temperature, reaction time, methanol-to-oil ratio, and catalyst loading were studied to understand its catalytic behavior. The combined effect of the calcium sodium silicate and calcium oxide phase in NCSO on its performance was investigated. Catalyst reusability was evaluated in a series of continuous transesterification reactions for biodiesel production.

2. Results and Discussion

2.1. Catalyst Characterization

To understand the crystalline composition of the catalyst, XRD and SEM analyses was carried out on samples calcinated at different temperatures and the results are shown in Figure 1a–d (with standard PDF card) and Figure 2, respectively. When the sample was calcinated at 600 °C, only the Na_2CO_3 and $NaNO_3$ phases were shown, and no calcium species were found, indicating the existence of amorphous calcium species. During catalyst preparation, the concentration of CO_3^{2-} and SiO_3^{2-} were both 0.483 mol/L before the 1.45 mol/L Ca^{2+}-containing solution was added. Because the K_{sp} (solubility product constant) of $CaCO_3$ is 4.8×10^{-9}, which is lower than that of calcium silicates (for example, $CaSiO_3$, 2.5×10^{-8}), $CaCO_3$ could be formed in favor of calcium silicates. Therefore, we deduced that the amorphous calcium species contains amorphous $CaCO_3$ and calcium silicate species. The SEM photo (Figure 2a) also prove that only irregular stacks of flake-like and granular solids were found. By increasing the calcination temperature to 800 °C, the Na_2CaSiO_4 phase emerged, a few Na_2CO_3 could still be found, and no other calcium silicates could be found. It could be deduced that the Na_2CaSiO_4 phase was formed by the solid-state reaction between calcium silicates or CaO (by decomposition of $CaCO_3$ at 600–700 °C) and sodium oxides, which was formed by the partial decomposition of Na_2CO_3. By comparing the SEM photo (Figure 2b), we also noted that some ordered layered structures started to form when the calcination temperature increased. When the calcination temperature reached 950 °C, only Na_2CaSiO_4 could be found, and its peak was higher than it was at 800 °C, indicating the full decomposition of Na_2CO_3. The SEM photo (Figure 2c,d) also revealed that the ordered layered crystal group found at 800 °C became a rectangular sheet stack with good crystallinity. Therefore, we concluded that the ordered layer crystal was the Na_2CaSiO_4 phase. This unique layered structure can increase the contact between the reactant and the active site of the catalyst during the transesterification process, which is beneficial to the transesterification reaction.

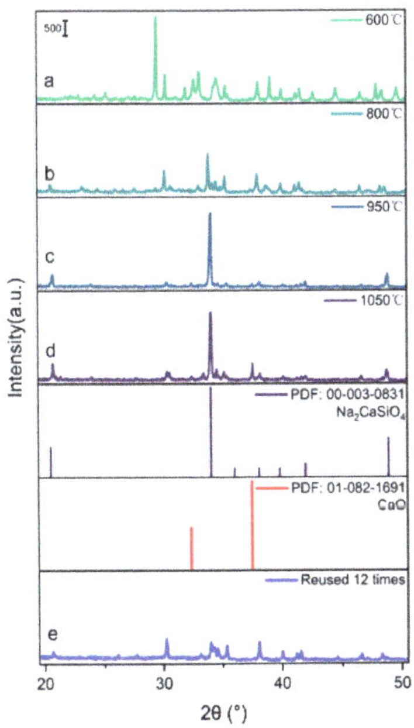

Figure 1. XRD spectrum of the NCSO catalyst calcinated at different temperatures (**a**–**d**) and the reused NCSO (**e**).

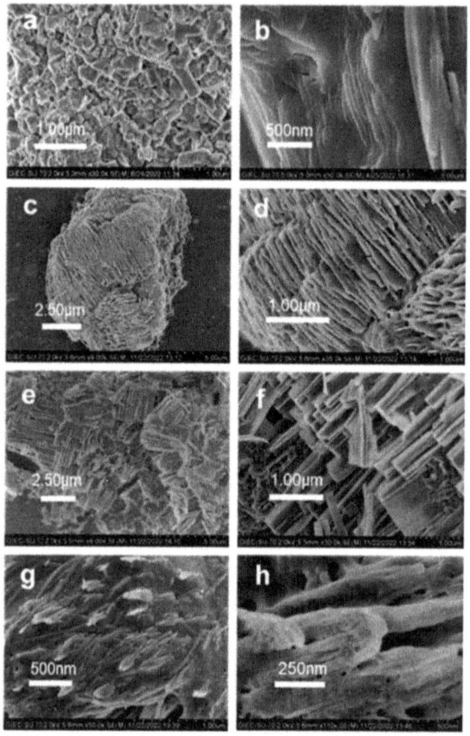

Figure 2. SEM images of the NCSO calcinated at different temperatures: (**a**) 600 °C; (**b**) 800 °C; (**c**,**d**) 950 °C; (**e**,**f**) 1050 °C; (**g**,**h**) NCSO reused for 12 cycles.

By further increasing the calcination temperature to 1050 °C, the CaO species emerged, and the peak Na_2CaSiO_4 was weaker than that at 950 °C. We also found that the sheets became thinner and sharper (Figure 2e,f) compared to that at 950 °C. Figure 1e presents the XRD result of the recycled catalyst, which was used for 12 batches of transesterification. The spectrum indicates the loss of the Na_2CaSiO_4 phase, consistent with its SEM photo: the edges of Na_2CaSiO_4 became blurred (Figure 2g,h), and the CaO species could also not be found.

Therefore, the main catalytic phase of this catalyst, which was named NCSO as follows, were Na_2CaSiO_4 and CaO.

2.2. *Influence of Reaction Conditions on Biodiesel Production*

To evaluate the effect of different reaction parameters on the NCSO-catalyzed conversion of soybean oil to biodiesel, different parameters including the reaction time, methanol/oil molar ratio, reaction temperature, and catalyst loading were studied. For experimental accuracy, all experiments were performed three times.

2.2.1. Influence of the Reaction Time

To investigate the effect of reaction time on the transesterification reaction of soybean oil and methanol, a series of reactions were carried out at a temperature of 80 °C, a molar ratio of methanol to oil of 9:1, and a catalyst content of 5 wt.%. The results are shown in Figure 3a. The transesterification reaction reached equilibrium at about 60 min with a biodiesel yield of 97.1%. Because the transesterification is reversible and trace water in the reaction media might cause the FAME saponification reaction in the base environment, further lengthening the reaction time could lead a fluctuation in the FAME yield [25]. Therefore, we chose 60 min as the reaction time in a typical transesterification reaction.

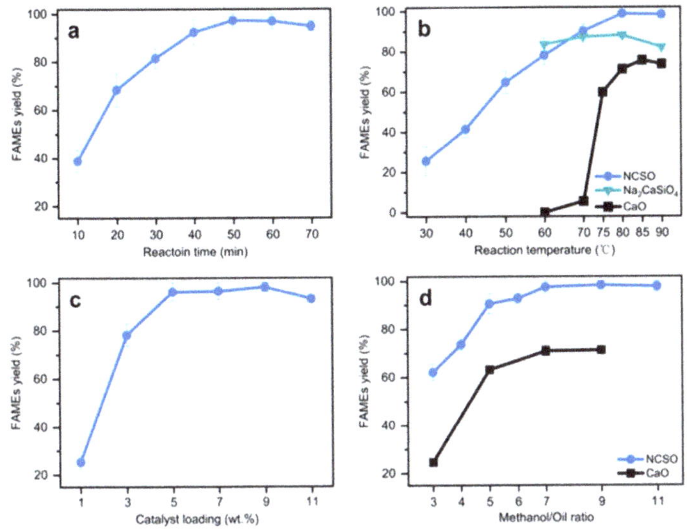

Figure 3. Effect of different reaction parameters on the transesterification reaction using NCSO: (**a**) reaction time; (**b**) reaction temperature; (**c**) catalyst loading; (**d**) methanol/oil molar ratio.

2.2.2. Influence of Reaction Temperature

The effect temperature on the transesterification reaction was studied at 30–90 °C, with 5 wt.% catalyst loading, and a 9:1 methanol/oil molar ratio, and the results are shown in Figure 3b. Due to the endothermic nature of this reaction, the FAME yield increased along with the temperature from 30 °C to 80 °C [29,30]. Additionally, increasing the temperature increased the molecular motion, leading to a greater chance to interact with the active site. On the other hand, an increase in temperature resulted in more dissolution of glycerol [31]. The biodiesel yield increased slowly when the temperature reached 60 °C or higher, and reached the maximum at 97.9% when the temperature was around 80 °C, indicating that the mass transfer resistance at the catalytic interface had largely been curbed.

To investigate the combined effect of Na_2CaSiO_4 and CaO in NCSO, Na_2CaSiO_4 was synthesized using the sol–gel method [32] (method and relative characterization mentioned in the Supplementary Materials, Figure S1. SEM, Figure S2. XRD) and used as a comparing catalyst for biodiesel production; the results are shown in Figure 3b. Under the same reaction conditions (catalyst loading 5 wt.%, methanol/oil ratio = 9:1), CaO (calcinated at 1050 °C) showed the lowest FAME yield. Especially at about 70–90 °C, both CaO and Na_2CaSiO_4 showed a lower FAME yield than NCSO. Therefore, we can conclude that there is a synergistic effect of CaO and FAMEs in catalyzing transesterification reactions.

2.2.3. Influence of Catalyst Loading

To explore the effect of the amount of catalyst on the transesterification reaction of soybean oil and methanol, a series of reactions was carried out by changing the amount of catalyst (1–11%) at a methanol/oil ratio of 9:1 and reaction temperature of 80 °C; the results are shown in Figure 3c. The biodiesel yield increased correspondingly from 1% to 5% catalyst loading, reaching 97.5% at a 5 wt.% catalyst loading, which may be attributed to the increased contact opportunities between the catalyst and reactant at higher catalyst concentrations [33–35]. However, if the catalyst biodiesel continued to increase beyond 5%, there was no significant increase, and if the catalyst increased beyond 9%, the biodiesel yield began to decline. It may be that the rise in the amount of catalyst may make the reactant mixture more viscous, and too much catalyst will lead to agglomeration [36], increasing the mass transfer resistance between the methanol–oil catalyst phase, resulting in difficulty in mixing the catalyst, reactants, and products, resulting in a lower oil conversion rate [37,38].

2.2.4. Influence of the Methanol/Oil Molar Ratio

According to a previous study on transesterification research, methanol activation on the solid base catalytic interface is the speed-limiting step in the transesterification reaction to produce biodiesel [39]. The effect of the methanol to oil molar ratio on the yield of FAMEs was also studied using a 5 wt.% catalyst at a reaction temperature of 80 °C in the range of 3:1 to 11:1 for 60 min; the results are shown in Figure 3d. The stoichiometric ratio of the reactant in the biodiesel transesterification reaction as methanol/oil = 3:1. Under this situation, the NCSO catalyst could even show a 62.1% FAME yield, but only a 24.6% FAME yield when using CaO. This could be explained by the layered sheet structure in NCSO being able to effectively activate methanol, even under low methanol concentrations. The FAME yield remained stable from 7:1 to 9:1, ranging from 94.8% to 98.2%. These results indicate that the diluting effect of extra methanol did not impact the catalytic efficiency of NCSO under the given reaction conditions.

2.3. *Reusability of NCSO*

As Figure 4 shows, after being reused six times, NCSO retained about 95% of its maximum activity. Compared to the pure Na_2CaSiO_4, only 12.9% FAME yield remained after five times of reuse. As shown in Figure 1d,e (XRD) and Figure 2g,h (SEM), the compositions of NCSO changed, especially the reduced crystalline structure of the Na_2CaSiO_4 and CaO, which was also confirmed by the blurring edge of the layered-sheet structure in the fresh catalyst (Figure 2e,f). To explain the stability of NSCO and its catalytic behavior in consecutive batches of reaction, several characterizations were conducted.

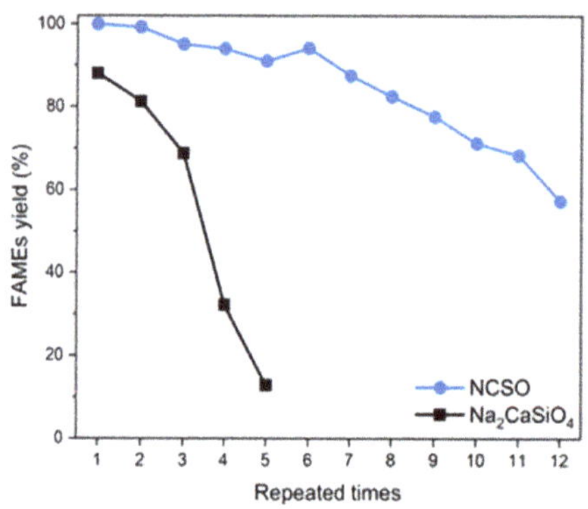

Figure 4. Reusability of the NCSO catalyst. Reaction conditions: molar ratio of methanol to oil = 9:1, agitation speed = 400 rpm, 9 wt.% catalyst loading, 80 °C, 60 min.

The XPS spectra of the fresh and recycled NCSO and pure Na_2CaSiO_4 were compared in Figure 5. According to previous reports, the binding energy (BE) of $Ca2p_{3/2}$ is 346.1 eV for pure CaO and 347.2 eV for $CaCO_3$, compared to 346.4 eV, 347.1 eV, and 346.4 eV for the fresh NCSO, recycled NCSO, and Na_2CaSiO_4, respectively (Figure 5a–c). It was proven that the active Ca species in fresh NCSO showed quite similar activity to Na_2CaSiO_4 and became less active after being reused several times (the recycled NCSO showed a similar $Ca2p_{3/2}$ BE to $CaCO_3$, which had been proven to be inactive for transesterification). In the crystal structure of Na_2CaSiO_4 (Figure S3), the Si species interact with the nearby O to form a tetrahedron [SiO_4], which, especially in Na_2CaSiO_4, does not bridge other [SiO_4] in the crystal. These oxygens in [SiO_4] turn to interact with its nearby Na and Ca cations. With CaO in NCSO, the Si species in NCSO (Figure 5d) showed a much higher binding energy ($Si2p_{3/2}$, 102.0 eV) than that in Na_2CaSiO_4 (Figure 5f, 100.8 eV), indicating a strong interaction between the CaO and the Si species from Na_2CaSiO_4 in NCSO [33]. The Si species in the recycled NCSO showed a much stronger binding energy (Figure 5e, 102.4 eV) than that in NCSO and Na_2CaSiO_4, indicating that some new silicates formed after being reused. O species in all researched samples showed O1s activity (Figure 5g–i), and O1s at 532.8 eV referred to the Si–O interaction as it remained stable after being reused. In contrast, O1s at 531.6–531.0 eV belonged to the metal–oxygen–silicon interaction. With CaO species in NCSO, the Ca–O–Si at the interface of CaO gave an O1s at 531.6 eV compared with 531.0 eV for Ca–O–Si or Na–O–Si, which dominated in the crystal phase of Na_2CaSiO_4. Thus, for the recycled group, O1s at 531.4 eV was in-between NCSO and Na_2CaSiO_4, indicating the loss of the CaO phase during batches of reaction. Figure 5j–l reflects the C species on all samples (except for the 284.8 eV C–C from contamination carbon). C1s at about 289.1 eV referred to the vertical linear absorption of CO_2 to cation M to form the M–O–C=O (M=Ca, Na) interaction, while the C1s at about 285.8 eV referred to the C–O species in the $CaCO_3$ [40,41].

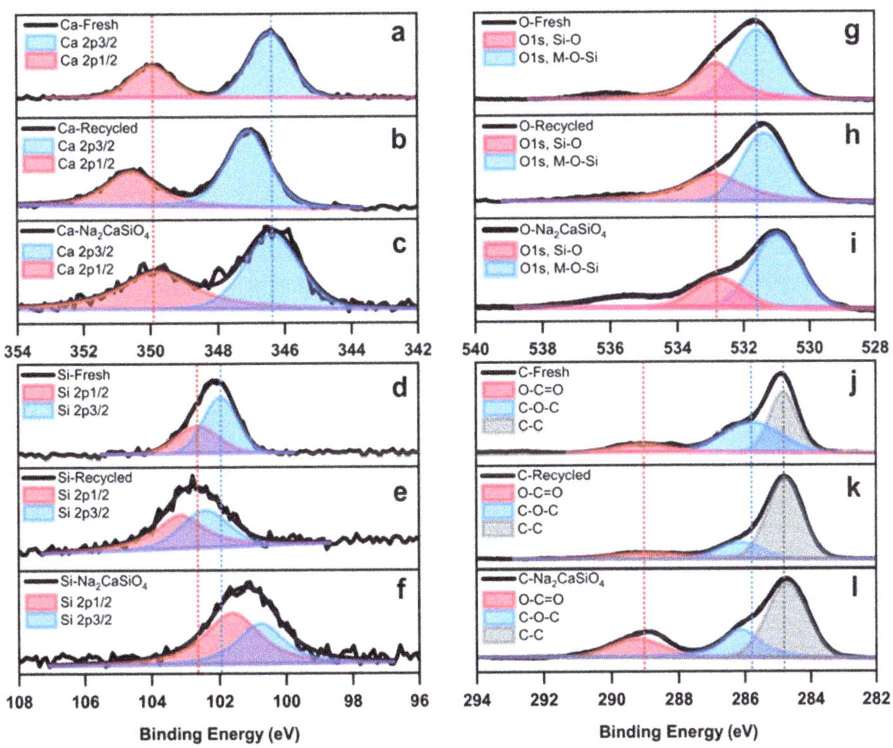

Figure 5. XPS analysis of the catalysts: (**a**,**d**,**g**,**j**) fresh NCSO; (**b**,**e**,**h**,**k**) recycled NCSO; (**c**,**f**,**i**,**l**) Na_2CaSiO_4.

The CO_2 absorbing behavior on the solid base mainly depends on the basicity of the catalyst sample. The results of the CO_2-TPD analysis of NCSO and Na_2CaSiO_4 are shown in Figure 6.

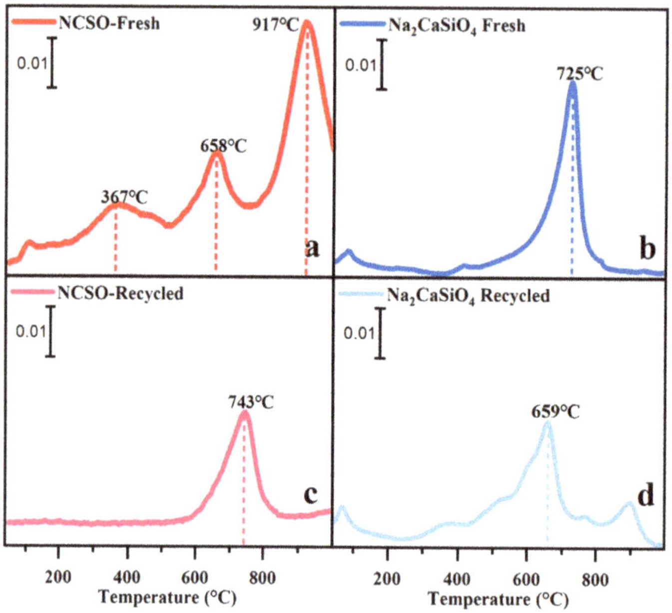

Figure 6. CO_2-TPD of NCSO and Na_2CaSiO_4: (**a**) fresh NCSO; (**b**) recycled NCSO (12 cycles); (**c**) fresh Na_2CaSiO_4; (**d**) recycled Na_2CaSiO_4 (5 cycles).

Fresh NCSO had two strong CO_2 desorption peaks at 658 °C and 917 °C, respectively, compared to only one at 725 °C for the Na_2CaSiO_4, and the total basicity was 2.22 mmol/g and 0.43 mmol/g, respectively. According to previous reports, CaO usually has a CO_2 desorption peak at 570–660 °C [42,43]. Therefore, we could confirm that the interaction between the CaO and Na_2CaSiO_4 phase provided stronger basic sites for NCSO than individually. The recycled samples showed an obvious loss of basic sites with the remaining total basic sites of 1.02 mmol/g and 0.23 mmol/g for NCSO (after 12 cycles, 57.3% FAME yield remaining) and Na_2CaSiO_4 (after five cycles, 12.9% FAME yield remaining), respectively, which also had weaker basicity than the fresh catalysts. This could indicate that the interaction between the [SiO_4] and CaO phase in NCSO, as confirmed in the XPS analysis, could help improve the stability of NCSO.

Although NCSO showed high activity and good reusability, the reason for its activity reduction should be discussed. First, regarding the leaching of CaO from NCSO, we noted that the recycled NCSO showed a similar basicity to the fresh Na_2CaSiO_4 (Figure 6b,c), proving that the CaO phase in NCSO could help to improve the basicity, and the loss of the CaO phase of NSCO was a reason for the reduced activity. XPS analysis on O1s revealed a Si–O–Ca BE shift, which could also prove the leaching of Ca species from NCSO. Second, mechanical stirring could make the catalyst leaching more serious. Therefore, more research is needed to strengthen the Ca–O species over Na_2CaSiO_4. Third, the destruction of the layered-sheet structure of NSCO during batches of reactions, as shown in Figure 2, led to a reduction in the activation efficiency of methanol, which was also a cause of the reduction in activity.

3. Materials and Methods

3.1. Materials

SiO_2 (600–800 mesh), CaO, and n-hexane were purchased from Shanghai Maclean Biochemical Technology Co. Na_2CO_3, NaOH, $Ca(NO_3)_2{\cdot}4H_2O$, and methanol wer purchased from Tianjin Fuchen Chemical Reagent Co. (Tianjin China), Tianjin Damao Chemical Reagent Factory (Tianjin China), and Tianjin Fuyu Fine Chemical Co. (Tianjin China). All of the chemicals used were of analytical grade and hence used without further purification. Refined soybean oil used for transesterification by Yihai Kerry Company (shanghai, China) was obtained from the local supermarket in Guangzhou, China.

3.2. Catalyst Preparation

As shown in Figure 7, a 0.145 mol SiO_2 powder was added to a 100 mL aqueous solution containing 0.725 mole of NaOH under vigorous stirring at 80 °C. After the SiO_2 powder was completely reacted with NaOH, 100 mL of an aqueous solution containing 0.145 mol of Na_2CO_3 was added. Then, a 100 mL aqueous solution containing 0.145 mol of $Ca(NO_3)_2{\cdot}4H_2O$ was slowly added to the above-mixed solution. The formed slurry-like product was dried at 80 °C overnight. Finally, the dried solid powder was thoroughly ground, placed in a 100 mL crucible, heated in a muffle furnace at 5 °C/min to 1050 °C, and kept for 2 h. After cooling to room temperature, the catalyst was ground to powder and kept in a vacuum drier before it was used for further research.

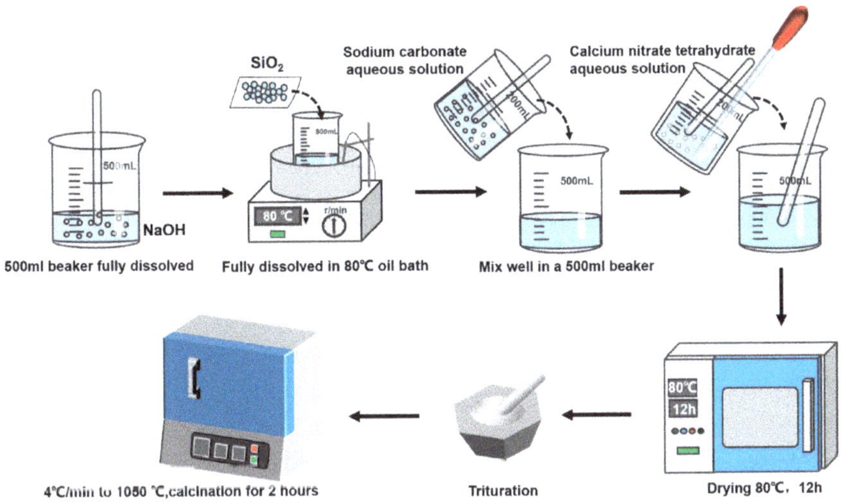

Figure 7. The NCSO catalyst preparation scheme.

3.3. Catalyst Characterization

The crystalline phases and the compositions of the catalysts were analyzed by X-ray diffraction (XRD, X'Pert Pro MPD, PANalytical, Almelo, The Netherlands) coupled with Cu Kα radiation (40 kV and 40 mA). The diffraction data of the catalyst were recorded with 0.0167° steps over the 10~90° angular range.

X-ray photoelectron spectroscopy (XPS) measurements were performed on a Thermo Scientific ESCALAB 250 Xi X-ray photoelectron spectrometer (Waltham, MA, USA). The working voltage of the XPS was 20 kV, and the current was 10 mA. The C1s photoelectron peak (284.8 eV) was employed for the binding energy calibration.

The surface morphology of the sample was observed by scanning electron microscopy (SEM, S-4800 FESEM, HITACHI, Tokyo, Japan).

The elemental composition of the catalyst was analyzed by wavelength dispersive X-ray fluorescence spectrometry (XRF, AXIOSmAX-PETRO, PANalytical) measurements.

Temperature-programmed desorption of CO_2 (CO_2-TPD, AutoChemII 2920, Micromeritics, Norcross, GA, USA) was conducted to determine the basicity and quantify the basic site of the samples. In a typical analysis, about 150 mg of solid catalyst is degassed in a helium flow. The sample is then saturated with CO_2 at ambient temperature. After being washed for 30 min with He flow, the catalyst is heated to 1000 °C at a heating rate of 10 °C/min.

3.4. Transesterification of Soybean Oil

The transesterification experiments were carried out in a 15 mL pressure-resistant flask, which was heated in an oil bath equipped with a magnetic stirrer. In a typical transesterification reaction, 0.6 g of methanol and 2.0 g of soybean oil were added into the reactor with 0.1 g of catalyst and a magnetic stirrer. Then, the reactor was put into the pre-heated oil bath at 80 °C with a stirring speed of 400 rpm; the reaction lasted 1 h. After the reaction, the mixture was cooled to room temperature in cold water and transferred into a 20 mL centrifuge tube. The biodiesel sample was separated using a centrifuge at 10,000 rpm for 5 min. The upper layer of liquid was washed with water to remove the methanol. The bottom layer of liquid, mainly glycerol, was also separated from the catalyst if the catalyst was meant to be recycled for further research. Finally, anhydrous calcium chloride was used to remove the residual water in the biodiesel sample before it was used in further analysis (biodiesel product and its GCMS spectral with molecular structure could be found in Figures S4 and S5 and Table S1 in the supplementary information, key properties of biodiesel product could be found in Table S2 in the supplementary information).

3.5. Product Analysis

The FAME content in the biodiesel products was quantified using GC (gas chromatography). A certain amount of the analyzed sample (ca. 20 mg) was added into a 5 mL volumetric flask and mixed with 1 mg of heptadecanoic acid methyl ester (dissolved in n-hexane) and the volume fixed with hexane to the scale line. The solution was filtered through a 0.25 μm needle filter and analyzed using a GC-2010 Plus (Shimadzu, Kyoto, Japan) with a flame ionization detector (FID) equipped with a capillary column (DB-WAX, 30 m × 0.25 mm × 0.25 μm). The chromatographic parameters were set as follows: the detector and injector temperatures were 300 °C and 250 °C, respectively; the column temperature started from 100 °C, rising at a rate of 20 °C/min until 185 °C and then 20 °C/min until 225 °C, before finally being increased to 240 °C at 20 °C/min and held for 3 min. Argon was used as a carrier gas with a flow rate of 40 cm/s. The yield of biodiesel is defined as:

$$Biodiesel\ yield = \frac{V \times \Sigma c_i}{m}$$

where m, c_i, and V are the mass of the product to be measured, the concentration of each fatty acid methyl ester, and the constant volume of the volumetric flask, respectively.

4. Conclusions

In this paper, a silicate-strength strategy was adopted to synthesize a solid base catalyst NSCO. The catalyst precursor was calcinated at 1050 °C to give NCSO two main catalytic phases, Na_2CaSiO_4 and CaO. Those two phases showed a synergistic effect, enabling NCSO to have high activity and good reusability. The layered-like-sheet structure in NCSO allowed for a higher methanol activation efficiency at a low methanol concentration when compared to CaO. A 98.2% FAME yield could be achieved in 60 min at 80 °C with a 9:1 methanol/oil molar ratio and 5 wt.% catalyst loading. A 94.3% FAME yield could still be achieved even after six consecutive reuses. This silicate-strength strategy provides a new way to improve the stability while retaining the activity of the calcium-based base catalyst.

Supplementary Materials: The following supporting information can be downloaded at: https://www.mdpi.com/article/10.3390/catal13040775/s1, Figure S1: SEM picture of the as-synthesized Na_2CaSiO_4; Figure S2: XRD result of the as-synthesized Na_2CaSiO_4; Figure S3: Crystal structure of Na_2CaSiO_4; Figure S4: GCMS spectra of biodiesel; Figure S5: Biodiesel product from soybean oil; Table S1: Molecular structures for biodiesel by GCMS; Table S2: Key properties of biodiesel product from soybean oil.

Author Contributions: Conceptualization, J.F.; Methodology, S.Z. and J.F.; Validation, J.F. and X.L.; Formal analysis, J.F. and S.Z.; Investigation, S.Z. and J.F.; Validation, S.Z. and J.F.; Resources, S.X. and J.F.; Data curation, S.Z. and J.F.; Writing—original draft, S.Z. and J.F.; Writing—review and editing, J.F. and M.L.; Supervision, L.Y. and P.L.; Funding acquisition, P.L. All authors have read and agreed to the published version of the manuscript.

Funding: This research was funded by the National Key Research and Development Project of China (2022YFE0207100); the Foundation of Key Laboratory of Renewable Energy, Chinese Academy of Sciences (2021000024); the National Natural Science Foundation of China (52206286, 32271822, 51903236); the Guangdong Basic and Applied Basic Research Foundation (2022A1515010440).

Data Availability Statement: Data are contained within the article and are available from the first corresponding author.

Conflicts of Interest: The authors declare no conflict of interest.

References

1. Zul, N.A.; Ganesan, S.; Hamidon, T.S.; Oh, W.-D.; Hussin, M.H. A review on the utilization of calcium oxide as a base catalyst in biodiesel production. *J. Environ. Chem. Eng.* **2021**, *9*, 105741. [CrossRef]
2. Munyentwali, A.; Li, H.; Yang, Q. Review of advances in bifunctional solid acid/base catalysts for sustainable biodiesel production. *Appl. Catal. A Gen.* **2022**, *633*, 118525. [CrossRef]
3. Orege, J.I.; Oderinde, O.; Kifle, G.A.; Ibikunle, A.A.; Raheem, S.A.; Ejeromedoghene, O.; Okeke, E.S.; Olukowi, O.M.; Orege, O.B.; Fagbohun, E.O.; et al. Recent advances in heterogeneous catalysis for green biodiesel production by transesterification. *Energy Convers. Manag.* **2022**, *258*, 115406. [CrossRef]
4. Jamil, U.; Husain Khoja, A.; Liaquat, R.; Raza Naqvi, S.; Nor Nadyaini Wan Omar, W.; Aishah Saidina Amin, N. Copper and calcium-based metal organic framework (MOF) catalyst for biodiesel production from waste cooking oil: A process optimization study. *Energy Convers. Manag.* **2020**, *215*, 112934. [CrossRef]
5. Demirbas, A. Biodiesel from waste cooking oil via base-catalytic and supercritical methanol transesterification. *Energy Convers. Manag.* **2009**, *50*, 923–927. [CrossRef]
6. Yildiz, I.; Acikkalp, E.; Caliskan, H.; Mori, K. Environmental pollution cost analyses of biodiesel and diesel fuels for a diesel engine. *J. Environ. Manag.* **2019**, *243*, 218–226. [CrossRef]
7. Qiu, F.; Li, Y.; Yang, D.; Li, X.; Sun, P. Biodiesel production from mixed soybean oil and rapeseed oil. *Appl. Energy* **2011**, *88*, 2050–2055. [CrossRef]
8. Liu, X.; Xing, S.; Yang, L.; Fu, J.; Lv, P.; Zhang, X.; Li, M.; Wang, Z. Highly active and durable Ca-based solid base catalyst for biodiesel production. *Fuel* **2021**, *302*, 121094. [CrossRef]
9. Lee, J.C.; Lee, B.; Ok, Y.S.; Lim, H. Preliminary techno-economic analysis of biodiesel production over solid-biochar. *Bioresour. Technol.* **2020**, *306*, 123086. [CrossRef]
10. Ma, Y.; Wang, Q.; Sun, X.; Wu, C.; Gao, Z. Kinetics studies of biodiesel production from waste cooking oil using $FeCl_3$-modified resin as heterogeneous catalyst. *Renew. Energy* **2017**, *107*, 522–530. [CrossRef]
11. Rabie, A.M.; Shaban, M.; Abukhadra, M.R.; Hosny, R.; Ahmed, S.A.; Negm, N.A. Diatomite supported by CaO/MgO nanocomposite as heterogeneous catalyst for biodiesel production from waste cooking oil. *J. Mol. Liq.* **2019**, *279*, 224–231. [CrossRef]
12. Chen, G.-Y.; Shan, R.; Yan, B.-B.; Shi, J.-F.; Li, S.-Y.; Liu, C.-Y. Remarkably enhancing the biodiesel yield from palm oil upon abalone shell-derived CaO catalysts treated by ethanol. *Fuel Process. Technol.* **2016**, *143*, 110–117. [CrossRef]
13. Kouzu, M.; Hidaka, J.-s. Transesterification of vegetable oil into biodiesel catalyzed by CaO: A review. *Fuel* **2012**, *93*, 1–12. [CrossRef]
14. Mazaheri, H.; Ong, H.C.; Amini, Z.; Masjuki, H.H.; Mofijur, M.; Su, C.H.; Anjum Badruddin, I.; Khan, T.M.Y. An Overview of Biodiesel Production via Calcium Oxide Based Catalysts: Current State and Perspective. *Energies* **2021**, *14*, 3950. [CrossRef]
15. Teo, S.H.; Rashid, U.; Choong, S.Y.T.; Taufiq-Yap, Y.H. Heterogeneous calcium-based bimetallic oxide catalyzed transesterification of Elaeis guineensis derived triglycerides for biodiesel production. *Energy Convers. Manag.* **2017**, *141*, 20–27. [CrossRef]
16. Sudsakorn, K.; Saiwuttikul, S.; Palitsakun, S.; Seubsai, A.; Limtrakul, J. Biodiesel production from Jatropha Curcas oil using strontium-doped CaO/MgO catalyst. *J. Environ. Chem. Eng.* **2017**, *5*, 2845–2852. [CrossRef]
17. Dias, A.P.S.; Ramos, M. On the storage stability of CaO biodiesel catalyst. Hydration and carbonation poisoning. *J. Environ. Chem. Eng.* **2021**, *9*, 104917. [CrossRef]
18. Malins, K. The potential of K_3PO_4, K_2CO_3, Na_3PO_4 and Na_2CO_3 as reusable alkaline catalysts for practical application in biodiesel production. *Fuel Process. Technol.* **2018**, *179*, 302–312. [CrossRef]
19. de Lima, A.L.; Ronconi, C.M.; Mota, C.J.A. Heterogeneous basic catalysts for biodiesel production. *Catal. Sci. Technol.* **2016**, *6*, 2877–2891. [CrossRef]
20. Rahmani Vahid, B.; Haghighi, M. Urea-nitrate combustion synthesis of $MgO/MgAl_2O_4$ nanocatalyst used in biodiesel production from sunflower oil: Influence of fuel ratio on catalytic properties and performance. *Energy Convers. Manag.* **2016**, *126*, 362–372. [CrossRef]

21. Rodríguez-Ramírez, R.; Romero-Ibarra, I.; Vazquez-Arenas, J. Synthesis of sodium zincsilicate (Na_2ZnSiO_4) and heterogeneous catalysis towards biodiesel production via Box-Behnken design. *Fuel* **2020**, *280*, 118668. [CrossRef]
22. Ibrahim, M.L.; Nik Abdul Khalil, N.N.A.; Islam, A.; Rashid, U.; Ibrahim, S.F.; Sinar Mashuri, S.I.; Taufiq-Yap, Y.H. Preparation of Na_2O supported CNTs nanocatalyst for efficient biodiesel production from waste-oil. *Energy Convers. Manag.* **2020**, *205*, 112445. [CrossRef]
23. Sánchez Faba, E.M.; Ferrero, G.O.; Dias, J.M.; Eimer, G.A. Na-Ce-modified-SBA-15 as an effective and reusable bimetallic mesoporous catalyst for the sustainable production of biodiesel. *Appl. Catal. A Gen.* **2020**, *604*, 117769. [CrossRef]
24. Rijo, B.; Fernando, E.; Ramos, M.; Dias, A.P.S. Biodiesel production over sodium carbonate and bicarbonate catalysts. *Fuel* **2022**, *323*, 124383. [CrossRef]
25. Wan, L.; Liu, H.; Skala, D. Biodiesel production from soybean oil in subcritical methanol using $MnCO_3/ZnO$ as catalyst. *Appl. Catal. B Environ.* **2014**, *152–153*, 352–359. [CrossRef]
26. Wang, J.; Yang, L.; Luo, W.; Yang, G.; Miao, C.; Fu, J.; Xing, S.; Fan, P.; Lv, P.; Wang, Z. Sustainable biodiesel production via transesterification by using recyclable $Ca_2MgSi_2O_7$ catalyst. *Fuel* **2017**, *196*, 306–313. [CrossRef]
27. Wang, J.; Wang, Z.; Yang, L.; Yang, G.; Miao, C.; Lv, P. Natural albite as a novel solid basic catalyst for the effective synthesis of biodiesel: Characteristics and performance. *Energy* **2017**, *141*, 1650–1660. [CrossRef]
28. Pavlović, S.M.; Marinković, D.M.; Kostić, M.D.; Janković-Častvan, I.M.; Mojović, L.V.; Stanković, M.V.; Veljković, V.B. A CaO/zeolite-based catalyst obtained from waste chicken eggshell and coal fly ash for biodiesel production. *Fuel* **2020**, *267*, 117171. [CrossRef]
29. Singh, V.; Hameed, B.H.; Sharma, Y.C. Economically viable production of biodiesel from a rural feedstock from eastern India, *P. pinnata* oil using a recyclable laboratory synthesized heterogeneous catalyst. *Energy Convers. Manag.* **2016**, *122*, 52–62. [CrossRef]
30. Samart, C.; Sreetongkittikul, P.; Sookman, C. Heterogeneous catalysis of transesterification of soybean oil using KI/mesoporous silica. *Fuel Process. Technol.* **2009**, *90*, 922–925. [CrossRef]
31. Bahador, F.; Foroutan, R.; Nourafkan, E.; Peighambardoust, S.J.; Esmaeili, H. Enhancement of Biodiesel Production from Chicken Fat Using MgO and MgO@Na_2O Nanocatalysts. *Chem. Eng. Technol.* **2020**, *44*, 77–84. [CrossRef]
32. Koppala, S.; Swamiappan, S. Glowing Combustion Synthesis, Characterization, and Toxicity Studies of Na_2CaSiO_4 Powders. *Mater. Manuf. Process.* **2015**, *30*, 1476–1481. [CrossRef]
33. Huang, J.; Zou, Y.; Yaseen, M.; Qu, H.; He, R.; Tong, Z. Fabrication of hollow cage-like CaO catalyst for the enhanced biodiesel production via transesterification of soybean oil and methanol. *Fuel* **2021**, *290*, 119799. [CrossRef]
34. Roschat, W.; Phewphong, S.; Thangthong, A.; Moonsin, P.; Yoosuk, B.; Kaewpuang, T.; Promarak, V. Catalytic performance enhancement of CaO by hydration-dehydration process for biodiesel production at room temperature. *Energy Convers. Manag.* **2018**, *165*, 1–7. [CrossRef]
35. Wang, J.-X.; Chen, K.-T.; Wu, J.-S.; Wang, P.-H.; Huang, S.-T.; Chen, C.-C. Production of biodiesel through transesterification of soybean oil using lithium orthosilicate solid catalyst. *Fuel Process. Technol.* **2012**, *104*, 167–173. [CrossRef]
36. Yan, B.; Zhang, Y.; Chen, G.; Shan, R.; Ma, W.; Liu, C. The utilization of hydroxyapatite-supported CaO-CeO_2 catalyst for biodiesel production. *Energy Convers. Manag.* **2016**, *130*, 156–164. [CrossRef]
37. Chen, G.-Y.; Shan, R.; Shi, J.-F.; Yan, B.-B. Transesterification of palm oil to biodiesel using rice husk ash-based catalysts. *Fuel Process. Technol.* **2015**, *133*, 8–13. [CrossRef]
38. Shan, R.; Zhao, C.; Yuan, H.; Wang, S.; Wang, Y. Transesterification of vegetable oil using stable natural diatomite-supported catalyst. *Energy Convers. Manag.* **2017**, *138*, 547–555. [CrossRef]
39. Tasić, M.B.; Stamenković, O.S.; Veljković, V.B. Cost analysis of simulated base-catalyzed biodiesel production processes. *Energy Convers. Manag.* **2014**, *84*, 405–413. [CrossRef]
40. Albuquerque, M.C.G.; Azevedo, D.C.S.; Cavalcante, C.L.; Santamaría-González, J.; Mérida-Robles, J.M.; Moreno-Tost, R.; Rodríguez-Castellón, E.; Jiménez-López, A.; Maireles-Torres, P. Transesterification of ethyl butyrate with methanol using MgO/CaO catalysts. *J. Mol. Catal. A Chem.* **2009**, *300*, 19–24. [CrossRef]
41. Black, L.; Garbev, K.; Stemmermann, P.; Hallam, K.R.; Allen, G.C. Characterisation of crystalline C-S-H phases by X-ray photoelectron spectroscopy. *Cem. Concr. Res.* **2003**, *33*, 899–911. [CrossRef]
42. Shan, R.; Chen, G.; Yan, B.; Shi, J.; Liu, C. Porous CaO-based catalyst derived from PSS-induced mineralization for biodiesel production enhancement. *Energy Convers. Manag.* **2015**, *106*, 405–413. [CrossRef]
43. Tang, Y.; Liu, H.; Ren, H.; Cheng, Q.; Cui, Y.; Zhang, J. Development KCl/CaO as a catalyst for biodiesel production by tri-component coupling transesterification. *Environ. Prog. Sustain. Energy* **2018**, *38*, 647–653. [CrossRef]

energies

MDPI

Article

Pyrolytic Conversion of Cellulosic Pulps from "Lignin-First" Biomass Fractionation

Charles A. Mullen *, Candice Ellison and Yaseen Elkasabi

USDA-Agricultural Research Service, Eastern Regional Research Center, 600 E. Mermaid Lane, Wyndmoor, PA 19038, USA
* Correspondence: charles.mullen@usda.gov

Abstract: Utilization of lignin is among the most pressing problems for biorefineries that convert lignocellulosic biomass to fuels and chemicals. Recently "lignin-first" biomass fractionation has received increasing attention. In most biorefining concepts, carbohydrate portions of the biomass are separated, and their monomeric sugar components released, while the relatively chemically stable lignin rich byproduct remains underutilized. Conversely, in lignin-first processes, a one-pot fractionation and depolymerization is performed, leading to an oil rich in phenolic compounds and a cellulosic pulp. Usually, the pulp is considered as a fermentation feedstock to produce ethanol. Herein, the results of a study where various cellulosic pulps are tested for their potential to produce valuable products via pyrolysis processes, assessed via analytical pyrolysis (py-GC), are presented. Samples of herbaceous (switchgrass) and woody biomass (oak) were subjected to both an acid-catalyzed and a supported-metal-catalyzed reductive lignin-first depolymerization, and the pulps were compared. Fast pyrolysis of the pulps produced levoglucosan in yields of up to about 35 wt %. When normalized for the amount of biomass entering the entire process, performing the lignin-first reductive depolymerization resulted in 4.0–4.6 times the yield of levoglucosan than pyrolysis of raw biomass. Pulps derived from switchgrass were better feedstocks for levoglucosan production compared with pulps from oak, and pulps produced from metal-on-carbon catalyzed depolymerization produced more levoglucosan than those from acid-catalyzed depolymerization. Catalytic pyrolysis over HZSM-5 produced aromatic hydrocarbons from the pulps. In this case, the yields were similar from both feedstocks and catalyst types, suggesting that there is no advantage to lignin fractionation prior to zeolite-catalyzed catalytic pyrolysis for hydrocarbons.

Keywords: levoglucosan; catalytic pyrolysis; lignin-first; depolymerization; pulp

1. Introduction

The largest share of both energy carriers and chemical industry products are derived from fossil fuels, leading to greenhouse gas emissions and other environmental problems. Efforts to decarbonize the production of chemicals and liquid fuels have focused on lignocellulosic biomass, the largest source of renewable carbon available [1]. While efforts over the last several decades have developed efficient methods for converting carbohydrates to various fuel and chemical products [2], lignin recalcitrance is a major hurdle preventing many biorefinery concepts from achieving economic viability. In biochemical, thermochemical, or hybrid conversion schemes, efficient conversion of lignin to renewable products has proven difficult. In standard biochemical based conversion models, lignin is a co-product released in the pre-fermentation process used to depolymerize complex carbohydrates to sugars [3,4]. Several types of pretreatments can be used, including dilute acid (usually H_2SO_4), alkaline (NaOH, KOH or NH_3), steam, liquid hot water, or organic solvents [5]. Each of these pretreatments modifies the biomass structure so that enzymatic hydrolysis can better release monomeric fermentable sugars. Often, the lignin remaining has little other use than to be burned for process energy. In thermochemical processes, a large fraction of the lignin carbon often winds up as part of solids (bio-char) or heavy ends of pyrolysis liquids and ultimately is poorly converted to deoxygenated hydrocarbons [6]. This fraction of bio-oil also largely contributes to coking of catalysts during downstream bio-oil upgrading processes.

Recent research has made several advances in the utilization of lignin [7]. Solvent-phase depolymerization of lignins has received a lot of attention as a means to produce phenolic monomers from isolated lignins [8–10]. These methods can be catalyzed by acids, bases, metal oxides, or supported active metals and may also be performed under oxidative or reductive conditions. Reductive depolymerizations have received the most attention recently due to the depolymerization efficiency of this approach. Like other lignin conversion methods, the yield of monomers produced is limited by two main factors: the relative stability of the non-β-O4 type linkages and the tendency for repolymerization reactions to occur. Because most isolation methods also cause structural changes that decrease the concentration of β-O4 type linkages and increase more stable C-C linkages compared with native structures, depolymerization of isolated lignins often has poor yields of monomers even if the average molecular weight is significantly decreased. Therefore, several researchers have recently proposed a "lignin-first" process where whole biomass is subject to solvent liquefaction conditions. Lignin is thereby separated and depolymerized to a monomer-rich phenolic oil via solvolysis and thermocatalytic depolymerization [11–18]. This combines the organic solvent-based pretreatment process mentioned above with lignin depolymerization in one pot. By acting on native lignin, the yield of monomers is generally increased because of the higher concentration of the labile β-O4 linkages. In this case, a cellulose-rich "pulp" is the byproduct.

In this study, we performed an in situ reductive polymerization via hydrogen transfer using Pd/C and Ru/C, without hydrogen gas, on switchgrass and oak biomass. Each of these catalysts has been proven effective for reductive lignin-first depolymerizations. Previously, Van den Bosch et al. reported monomer yields of 50 wt %, 44 wt %, 27 wt % and 21 wt % for Ru/C catalyzed lignin-first depolymerization from birch, poplar, miscanthus, and mixed softwoods, respectively [19]. Similarly, Galkin et al., 2016, reported monomer yields of greater than 30% with high selectivity to propylsyringol and propenylsyringol using Pd/C for reductive lignin-first depolymerization of birch wood [17]. We also tested using acid catalysis to aid in the depolymerization; both a solid (HZSM-5) and a soluble Lewis acid, aluminum triflate ($Al(OTf)_3$), were selected. Kramarenko et al. reported yields of phenolic monomers of up to 20 wt % using acidic zeolite (Hβ) catalysts for lignin-first depolymerization [14]. Two different types of lignocellulosic biomasses were chosen to compare the efficiency of the lignin fractionation and depolymerization process with the different types of lignin structures found in the woody and herbaceous feedstocks. Hardwoods have previously been found to perform more efficiently than softwoods and herbaceous feedstocks for lignin-first depolymerizations [19].

There is comparatively little work on the subsequent uses for the carbohydrate pulps, though most have considered this pulp as a feedstock to undergo hydrolysis to release fermentable sugars for subsequent production of ethanol or other fermentation products [11,15,16]. In contrast, there has been little attention towards this product as a potential feedstock for conversion via pyrolysis. Therefore, in this contribution, we report on the pyrolytic conversion of the cellulosic pulps produced from the different feedstock and catalyst combinations described above. The cellulosic pulps from these processes were tested for conversion to levoglucosan and related products via fast pyrolysis and aromatic hydrocarbons via catalytic pyrolysis over HZSM-5. These are two distinct options for thermochemical conversion of the cellulosic residues. Levoglucosan is a dehydrated form of glucose and is the major product of cellulose pyrolysis. Pure cellulose can give high yields (up to 60 wt %) of levoglucosan, but when cellulose remains in the biomass structure, a much less selective thermal depolymerization occurs. Levoglucosan is considered a potential important starting material for many fine chemical applications. In contrast, when catalytic pyrolysis is performed over zeolites such as HZSM-5, the primary oxygenates are converted via a series of reactions to a product mixture containing aromatic hydrocarbons, largely the important petrochemical BTEX mixture (benzene, toluene, ethyl benzene, and xylenes). The overall process considered in this study is summarized in Figure 1. The effect of the feedstocks and depolymerization catalysts were compared and contrasted for both the composition of the phenolic oils, the properties of the cellulosic pulps, and their pyrolysis products.

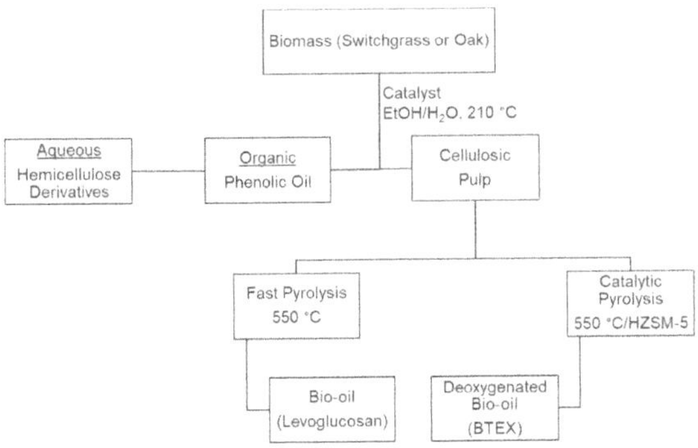

Figure 1. Schematic of biomass conversion process of this study.

2. Materials and Methods

2.1. Materials

Switchgrass was obtained from the McDonnell farm in East Greenville, PA, USA. Oak wood was recovered from a tree removed from the campus of the Eastern Regional Research Center, Wyndmoor, PA, USA. Each biomass was ground to 2 mm particles before use. Pd/C (5 wt %), Ru/C (5 wt %), and aluminum triflate Al(OTf)$_3$ were purchased from Sigma-Aldrich and used as received in powder form. HZSM-5 (SiO_2/Al_2O_3 = 23) was purchased from Zeolyst as NH_4ZSM-5 and calcined at 600 °C for 24 h before use to obtain the acid form. The same material was used for both lignin depolymerization and catalytic pyrolysis.

2.2. Lignin-First In Situ Depolymerization

A 100 mL stainless steel Parr reactor was loaded with 2 g biomass (switchgrass or oak) and 0.5 g catalyst. A quantity of 25 mL of ethanol and 25 mL of water was added. The reactor was purged of air by flushing with N_2 three times. The reactor was then pressurized to ~3.4 atm with N_2. The heating was started. When the temperature reached ±10 °C of the target temperature (210 °C), the reaction time was started. After 120 min, the heater was removed, and the reaction vessel cooled in an ice water bath. After cooling to room temperature, the reactor was opened, and the reaction mixture was filtered through a filter paper. The solid pulp and catalyst mixture was washed with ethanol/water, acetone, and then water, and air-dried. The filtrate was partially concentrated using a rotary evaporator and then underwent an extraction with dichloromethane and water. After separation, the aqueous layer was washed 3X with dichloromethane. The dichloromethane was removed by rotary evaporation to yield the phenolic oil. The dried pulp catalyst mixture (in case of heterogeneous catalysts) was subjected to sieving using a 150-mesh screen. The fine fraction was considered used catalyst and the coarse fraction recovered pulp.

GC/MS on the phenolic oil was performed using a Shimadzu GC-2010 (Columbia, MD, USA). A DB-1701(60 m × 0.25 mm and 0.25 μm) column was used. The GC oven was held at 45 °C for 4 min, then increased at 3 °C/min to 280 °C, and then held at 280 °C for 20 min. The injector temperature was 250 °C, and the split ratio was set to 30:1. The flow rate of the He carrier gas was 1 mL/min. The phenolic oil samples were prepared as 3 ± 1 wt % solutions in acetone.

Gel permeation chromatography (GPC) of the phenolic oils was performed on an Agilent Infinity 1260 Series HPLC equipped with two identical Agilent Oligopore GPC columns (polystyrene-divinyl-benzene copolymer, 300 × 7.5 mm) in series heated at 30 °C. Tetrahydrofuran (THF) was used as the mobile phase liquid and a flow rate of 1 mL/min was used. Samples were well dissolved (~1 mg/mL) in THF. Peak detection was done by a diode array UV detector. Several wavelengths were used but the GPC calculations were based on detection at 273 nm. Polystyrene standards in the MW range of 162–2900 were used for calibration.

A thermogravimetric analyzer (TA Instruments Q500 TGA) was used for proximate analysis of the pulp samples. For each material, an approximately 10 mg sample was heated to 800 °C at 20 °C/min under N_2 flow (60 mL/min), then the purge gas was switched to air to combust the remaining sample. The moisture, volatile matter, and fixed carbon concentrations were determined by the percent mass loss after heating to 150 °C, 800 °C under N_2, and 800 °C under air, respectively. The ash content was determined by the residual sample mass after combustion in air. Elemental analysis was performed using a Thermo Flash EA1112 CHNS/O analyzer by complete combustion of the material followed by GC quantification of combustion products. Oxygen was determined by difference after accounting for organic CHNS, ash, and water.

2.3. Pyrolysis-GC-Polyarc-FID

The pyrolysis volatiles were analyzed by means of a micropyrolyzer (Multi-Shot Pyrolyzer EGA/PY 3030D, Frontier Laboratories Ltd., Fukushima, Japan) coupled to a gas chromatograph (Shimadzu GC2010). Compounds from the GC were converted to methane via a catalytic microreactor (Polyarc, Activated Research Company, Eden Prairie, MN, USA) and detected by a flame ionization detector (FID). An identical analysis using Py-GC/MS (Frontier Laboratories, Shimadzu) was used for compound identification. Approximately 500 µg of biomass and 2000 µg of HZSM-5 (for catalytic pyrolysis) was used for each analysis. The micropyrolyzer temperature was 550 °C and the GC inlet temperature was 280 °C. Helium was the carrier gas (2.5 mL/min) and a split ratio of 30:1 was used. The column (Restek RTX-1701) was held at 45 °C for 4 min, heated to 280 °C at 3 °C/min, then held for 20 min. The Polyarc reactor was held at 300 °C and converted organic compounds exiting the GC to methane before detection by FID. The FID was held at 300 °C with H_2 flow of 1.5 mL/min and air flow of 350 mL/min. Py-GC-Polyarc-FID analysis was repeated in triplicate for each sample.

3. Results and Discussion

3.1. Lignin-First Fractionation and Depolymerization

The purpose of this study was to compare the fractionation depolymerization process from the two types of feedstocks (herbaceous grass and hardwood), and to test if the catalyst had effects on the composition and pyrolysis behavior of the cellulosic portion. The tests were performed at conditions similar to those previously known to be effective for production of bio-oil with relatively high concentrations of monomeric phenols [14,17,20,21]. The addition of water to the reaction mixture allowed for separation of hemicellulose-derived compounds from the lignin oil [18]. Oak, a hardwood, contained more lignin than switchgrass (Table 1). Additionally, the lignin structures are expected to be different between the two plant species; hardwoods generally have more of the dimethoxylated phenolic rings (syringyl (S)) than do grasses. Grasses contain appreciable levels of non-methoxylated (H) units, and some monomethoxylated (G) units, in addition to the S-units. Grasses are also expected to contain more C-O bonded linkages, and particularly, more β-O4 linkages, than hardwoods [22,23].

These differences in the composition of the biomass can be seen in the results of the lignin fractionation depolymerization results. Table 2 presents the mass balance for the solvent-phase lignin depolymerization step. Yields from oil range from 27 wt % to 42 wt % based on biomass input. The yields follow the trend of Pd/C < Ru/C < acids; however, the liquid products can contain impurities not derived from lignin, including solvent-derived products and any carbohydrate-derived products. This explains why some yields based on lignin are greater than 100%. Looking at a comparison of the GC/MS chromatograms of the liquids (Figure 2), one can see that peaks associated with various types of dimerization and oligomerization reactions of ethanol (generally appearing at RT < 32 min) are much more prevalent in the case of the acid catalysts and particularly so in the case of the soluble Lewis acid, $Al(OTf)_3$. In each case, a small peak for acetic acid can be observed at around a retention time of 10 min. Acetic acid is likely derived from the hydrolysis of acetylated hemicellulose, and much of it is retained in the aqueous phase of the reaction mixture. There appears to be a higher concentration of acetic acid in the oil produced from the combination of oak and Ru/C catalysis; why less of the acetic acid portioned into the aqueous phase in this case is unclear. For both the switchgrass and oak, the phenolic oil with the narrowest monomer distribution appears to be produced by the Pd/C catalyst. As can be seen in the chromatogram, the most prevalent phenolic monomers produced were 4-propylguaiacol (G3) and 4-propylsyringol (S3). 4-propenylguaiacol (isoeuguenol, G3′) and 4-propenylsyringol (S3′) were also present in the case of the Ru/C, and to a lesser extent over HZSM-5. The Pd/C were more selective for the phenolics with the saturated side chains, suggesting that Pd/C was a more effective transfer hydrogenation catalyst for C=C double bonds outside the aromatic ring. 4-ethylguaiacol (G2) was produced in smaller amounts than the C3 substituted phenolics for oak and switchgrass, while 4-ethylphenol (H2) was also observed for the oils produced from switchgrass with the metal catalysts reflective of the presence of H-type lignin. The Pd/C catalyst was more selective for the phenolic monomers with C2 saturated side chains; compounds with unsaturated side chains (e.g., vinyl phenol and vinyl guaiacol) were not observed in any case. These compounds are common products of thermal decomposition of lignin [24,25], and therefore the lack of their presence in the oil suggest that they are either hydrogenated to the saturated versions (Pd/C) or undergo oligomerization reactions, contributing to the oligomeric fraction of the phenolic oils which makes up the majority of the oil.

Table 1. Proximate, ultimate and biochemical analysis (wt %) of oak and switchgrass biomasses used in this study.

	Oak	Switchgrass
Moisture	6.6	6.4
VM, d.b.	81.0	87.7
FC, d.b.	17.4	9.3
Ash, d.b.	1.6	3.1
C	50.52	43.8
H [1]	6.24	4.7
N	0.12	0.5
O [1]	35.04	41.6
Cellulose	29.71	36.77
Hemicellulose	21.62	33.07
Lignin	37.08	27.42

[1] hydrogen and oxygen values are organic only, oxygen by difference.

Table 2. Mass balance (wt %) of lignin-first fractionation/depolymerization experiments in ethanol at 210 °C.

Biomass—Catalyst	Phenolic Oil Yield (from Biomass)	Phenolic Oil Yield (from Lignin)	Pulp Yield (from Biomass)	Pulp Yield (from Cellulose)
Oak—Pd/C	28.0	75.6	27.7	93.2
Oak—Ru/C	34.2	92.2	38.5	129.5
Oak—HZSM-5	39.5	106.6	36.3	122.3
SWG—Pd/C	27.1	98.9	31.1	84.6
SWG—Ru/C	41.6	151.9	33.4	90.8
SWG—Al(OTf)3	37.0	135.0	24.5	66.7

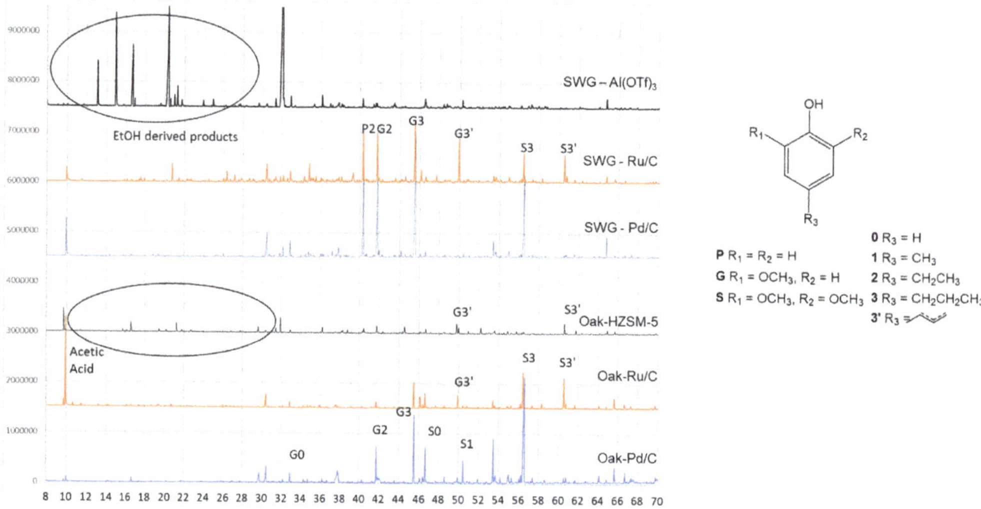

Figure 2. GC/MS chromatograms of phenolic oils produced from lignin-first fractionation and depolymerization in ethanol at 210 °C.

To understand the effect of the biomass and catalyst choices on the overall depolymerization rates of the lignin, the oils were analyzed by GPC analysis. The number average molecular weight (Mn) and weight average molecular weight (Mw) of the bio-oils are presented in Table 3. For lignin oils from oak, the Pd/C catalyst provided the lowest average MW, with those produced with Ru/C exhibiting a higher average MW. With HZSM-5, the average MW was in a similar range to the metal catalysts, despite a lower apparent production of monomers. For switchgrass, the Ru/C actually produced bio-oil with a lower MW than Pd/C did, but the MWs were more similar than for the oak lignin oils. The soluble $Al(OTf)_3$ acidic catalyst produced oils with a high average MW, indicating that it was either a poor depolymerization catalyst or was also a catalyst for repolymerization reactions. Given that it also produced a number of oligomerized products from ethanol, it is possible that some of those products such as diols acted as crosslinkers for phenolics, resulting in an increase in the observed average MW [10]. In general, for comparable catalyst/biomass combinations (Ru/C and Pd/C), the oak lignin oils had higher average molecular weights, which may be a reflection of fewer labile β-O4 linkages and more recalcitrant C-C linkages in their structures than in switchgrass lignin.

Table 3. The number average molecular weight (Mn) and weight average molecular weight (Mw) of phenolic oils produced via lignin-first fractionation/depolymerization experiments in ethanol/water at 210 °C.

	Mn (g/mol)	Mw (g/mol)
Oak—Pd/C	484	953
Oak—Ru/C	556	1128
Oak—HZSM-5	533	994
SWG—Pd/C	400	802
SWG—Ru/C	384	741
SWG—$Al(OTf)_3$	1141	1659

For oak, the yield of pulp ranged from 27.7 wt %, with the Pd/C catalyst to 38.5% for oak processed in the presence of Ru/C. Based on the estimated amount of cellulose contained in the oak, the obtained yields range from 93 to 122% of the theoretical cellulose recovery. For switchgrass, the yield of pulp ranged from 24.5% for that produced in the presence of $Al(OTf)_3$ to 31.1–33.4% for the metal catalysts, or 66.7% of theoretical cellulose from the acid and about 85–90% of possible cellulose with the metal catalysts. Yields of pulp above the theoretical maximum may be due to incomplete fractionation of lignin or hemicellulose or contamination with catalyst not separated during the sieving process. The comparably lower yield of pulp when $Al(OTf)_3$ is used is likely due to hydrolysis reactions providing breakdown of the cellulose into monomeric sugars and other water-soluble products that are lost to the aqueous phase.

The pulps were analyzed for proximate and ultimate analysis (Table 4). Interestingly, for oak, the fixed carbon amount of the pulps produced under the reductive conditions was similar to that of the parent biomass, but when the solid acid was used, the fixed carbon increased. However, for switchgrass, the fixed carbon rose dramatically in every case, with the largest increase coming from the use of $Al(OTf)_3$. This suggests that dehydration and polymerization reactions can occur to the point of producing a char-like structure, which can be initiated by the presence of the homogeneous acid. For the metal catalysts, since they are supported on carbon, this increase in fixed carbon could be related to catalyst contamination or a higher propensity for switchgrass to undergo dehydration and char-forming reactions compared with oak. Ash is concentrated in the pulp during the lignin separation process, with ash content of the pulps increasing from that of the parent biomass. Thereby, the ash content of the switchgrass pulps was higher than that of the oak pulps. The use of $Al(OTf)_3$ produced the pulp with the highest ash content. This is likely related to the lower yield and loss of additional organic material to the aqueous phase as mentioned above. The pulps generally contained more carbon, less oxygen, and less hydrogen than pure cellulose, which is consistent with the presence of some char-type materials or unfractionated lignin remaining with the pulp.

Table 4. Proximate and ultimate analysis (wt %) of cellulosic pulps isolated after lignin fractionation/depolymerization in ethanol at 210 °C using the biomass and catalyst combinations listed.

	Oak Pd/C	Oak Ru/C	Oak HZSM-5	SWG Pd/C	SWG Ru/C	SWG $Al(OTf)_3$
Moisture	3.2	3.2	5.6	2.6	2.4	5.3
VM, d.b.	80.9	76.9	70.98	66.8	76.5	42.9
FC, d.b.	16.6	19.4	28.0	28.0	17.9	47.4
Ash, d.b.	2.5	3.7	4.1	5.2	5.6	9.7
C	55.79	52.50	40.52	55.79	50.05	58.95
H [1]	5.72	3.82	4.03	3.88	3.58	3.84
N	0.63	0.55	0.25	0.62	0.20	0.20
O [1]	32.15	36.23	45.5	31.91	38.17	22.01
C/O (mol)	2.33	1.92	1.19	2.33	1.74	3.57
H/C (mol)	1.23	0.87	1.06	0.83	0.85	0.78

[1] hydrogen and oxygen values are organic only, oxygen by difference.

3.2. *Pyrolysis of the Cellulosic Pulps*

The main product of the pyrolysis of pure cellulose is levoglucosan, a dehydrated form of glucose. Thermal decomposition of isolated cellulose can provide levoglucosan in up to 60 wt % yield [26]. However, the yield of levoglucosan from cellulose during raw biomass pyrolysis is generally low, due to the interference of lignin, acid from decomposition of other biomass components, and especially the catalytic action of alkalis present in biomass. In biomass pyrolysis, more fragmentation reactions occur, producing compounds such as hydroxyacetaldehyde, along with various furans and cyclopentanone derivates, at the expense of levoglucosan [26,27]. Pyrolysis of the two parent biomasses used in this study resulted in 1.3 wt % and 2.8 wt % yields of levoglucosan from oak and switchgrass, respectively, at 550 °C. Using this pyrolysis methodology, pyrolysis of pure cellulose resulted in an average levoglucosan yield of about 50 wt %.

Contained in Figure 3 are the pyrolysis-gas chromatograms (pyrograms) from the pyrolysis of the pulps at 550 °C, allowing for the qualitative comparison of each pulp/catalyst combination and commercial cellulose. It is clear from the pyrograms that, much like the commercial cellulose, levoglucosan is the major product produced from most of the residual pulps, and many of the minor products are the same for pyrolysis of the different pulps. The obvious exception is the pulp produced from switchgrass over the soluble acid catalyst. Its levoglucosan peak is significantly smaller, and it also exhibits the production of phenolics, indicating that lignin is still present within this pulp.

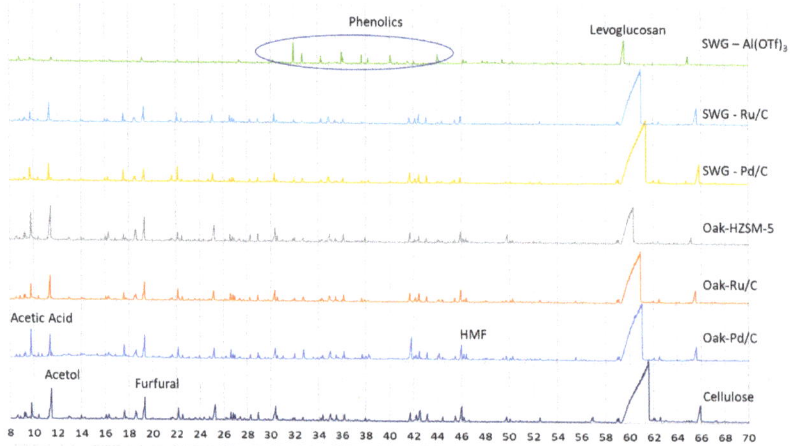

Figure 3. Pyrogram from Py-GC-FID of cellulosic pulps from lignin-first fractionation in ethanol at 210 °C. Pyrolysis at 550 °C.

Figure 4 presents the quantitative yields of levoglucosan based on the input mass of the pulps. Figure 5 presents this data normalized to the total amount of biomass, including the fractionation/lignin depolymerization step. As can be seen from both figures, the pulps generated in the presence of the metal catalyst (and, to a lesser extent, HZSM-5) produce a higher yield of levoglucosan upon pyrolysis than the raw biomass. While the yield is lower than that obtained from commercial cellulose, a significant amount of levoglucosan can be generated. The pulps produced from switchgrass using either metal catalyst produced greater than 30 wt % yield of levoglucosan or about 11–12 wt % yield after normalization to raw biomass. As discussed above, the pulp produced in the presence of the soluble acid catalyst has a significantly different composition, and therefore the yield of levoglucosan is predictably low. For oak, there was a bigger difference between the pulp produced in the presence of Pd or Ru, and the yields were lower compared with switchgrass. However, for both biomasses, the yield of levoglucosan (based on total biomass input) via pyrolysis of the pulps produced after lignin-first conversion was increased by about 4.0–4.7 times over the pyrolysis of raw biomass.

Table 5 presents the yields of some other selected pyrolysis products. Hydroxyacetaldehyde is a breakdown product that results from further fragmentation of cellulosic pyrolysis products and is often formed at the expense of levoglucosan [27]. Generally, the switchgrass pulps produced less hydroxyacetaldehyde than did commercial cellulose, while the oak pulps produced more, particularly that produced with HZSM-5. All of the pulps produced significantly less acidic acid than their parent biomass, which may make isolation or further conversion of the produced levoglucosan easier, as acid content and bio-oil stability tend to be negatively correlated. The pulp produced with $Al(OTf)_3$ produced the greatest concentration of phenolics, other than trace amounts of phenol. As mentioned above, it appears that that catalyst was less effective at net removal of lignin from the remaining solids.

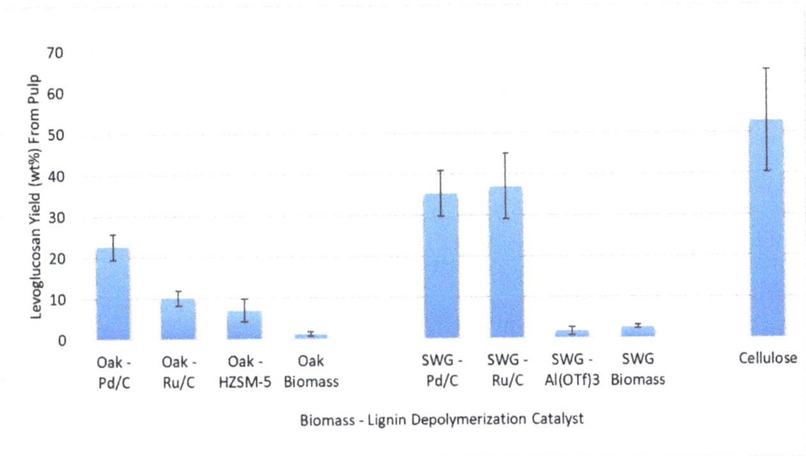

Figure 4. Levoglucosan yield (wt %) from pyrolysis of cellulosic pulps isolated after lignin-first fractionation in ethanol at 210 °C. Pyrolysis at 550 °C. Error bars are one standard deviation.

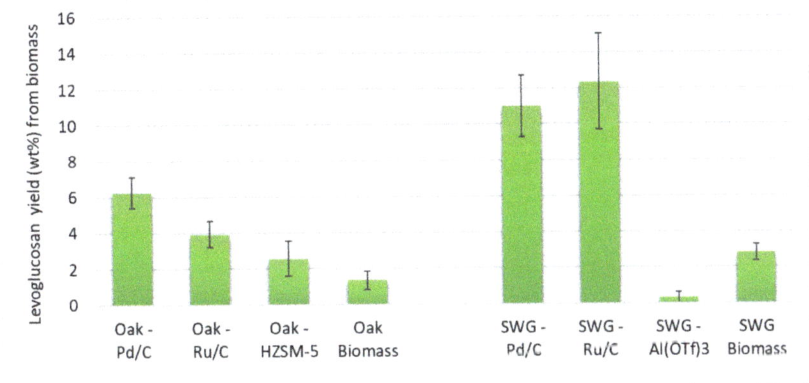

Figure 5. Levoglucosan yield (wt %) from pyrolysis of cellulosic pulps isolated after lignin-first fractionation in ethanol at 210 °C, based on total biomass input. Pyrolysis at 550 °C. Error bars are one standard deviation.

Table 5. Yields of selected compounds from pyrolysis of cellulosic pulps isolated after lignin-first fractionation in ethanol at 210 °C.

(wt % Yield)	Oak-Pd/C	Oak-Ru/C	Oak—HZSM-5	Oak—Biomass	SWG—Pd/C	SWG—Ru/C	SWG—Al(OTf)$_3$	SWG—Biomass	Cellulose
Hydroxy-acetaldehyde	2.95	1.29	5.28	2.41	0.55	1.51	0.08	1.63	1.98
Acetic Acid	0.47	0.37	1.10	6.76	0.69	0.54	0.17	3.60	0.32
Acetol	0.81	0.28	1.63	2.20	0.28	1.08	0.21	3.22	0.88
2-Cyclopenten-1-one	0.41	0.08	0.18	0.15	0.00	0.18	0.04	0.20	0.13
Furfural	0.82	0.38	0.86	0.76	0.35	0.86	0.13	0.75	0.52
Hydroxymethyl-furfural (HMF)	0.63	0.17	0.46	0.13	0.09	0.29	0.15	0.11	0.34
Phenol	0.03	0.08	0.08	0.06	0.00	0.07	0.20	0.14	0.01
Guaiacol	0.18	0.09	0.10	0.22	0.00	0.11	0.13	0.18	nd
Syringol	0.14	0.07	0.15	0.38	0.00	0.11	0.07	0.00	nd

3.3. Catalytic Pyrolysis over HZSM-5

Catalytic pyrolysis of biomass over HZSM-5 to produce aromatic hydrocarbons is another thermochemical conversion option for the cellulosic pulps [28]. At high temperatures, the microporous zeolite initiates a series of reactions, including decarbonylation, dehydration, oligomerization of alkenes, and aromatization via dehydrogenation. For pyrolysis vapors containing large molecules such as levoglucosan, cracking reactions occur on the surface of the zeolite, producing smaller molecules that can penetrate the pores. The specific shape selectivity of ZSM-5 makes it highly selective for aromatic hydrocarbons, specifically the BTEX compounds (benzene, toluene, ethyl benzene, xylenes). Although the yield of BTEX from biomass is generally low, the deoxygenated mixture is much easier and more efficient for refinement to finished fuels. Specifically, the aromatic compounds are a small but necessary fraction in gasoline and jet fuels, in addition to their many industrial uses. Aromatics from biomass via catalytic pyrolysis, combined with alkanes from vegetable oil processing, is a pathway to a 100% renewable sustainable aviation fuel.

Figures 6 and 7 show the yields of BTEX generated from catalytic pyrolysis of the pulps over HZSM-5 at a biomass-to-catalyst ratio of approximately 5:1, from the pulp and total input biomass, respectively. Under these conditions, BTEX yields ranged from 4.5–6 wt % for both the pulps and the raw biomasses. The exception was pulp generated with the $Al(OTf)_3$ catalyst. This was likely due to the high fixed carbon of this pulp, due to charring-like reactions with the soluble strong acid during the first step. Comparing both the pulps and the raw biomass, the normalized values show that the yield of BTEX is actually decreased by removing and depolymerizing the lignin first. This is because, in contrast to conversion to levoglucosan by fast pyrolysis, where lignin is not a viable feedstock, lignin can contribute to the production of BTEX via catalytic pyrolysis. Although not captured by this study, there may be an advantage to utilizing the pulps over raw biomass over time because lignin and hemicellulose are both strong sources of catalyst-deactivating coke [29,30].

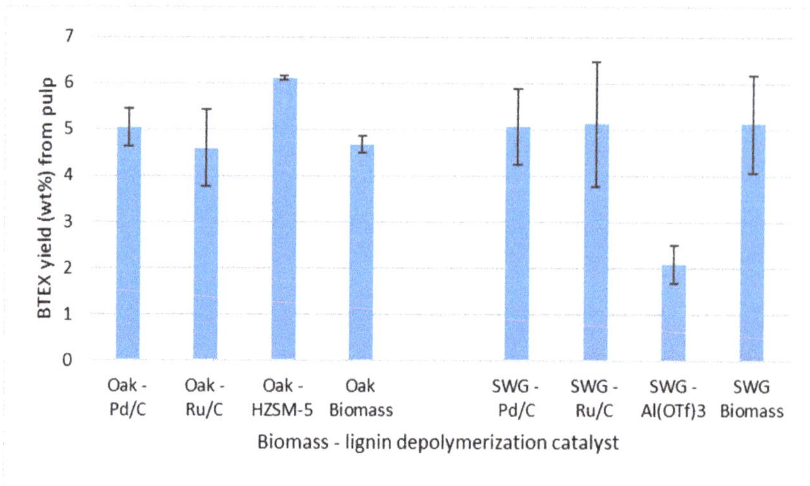

Figure 6. BTEX yield (wt %) from HZSM-5-catalyzed pyrolysis of cellulosic pulps isolated after lignin-first fractionation in ethanol at 210 °C. Pyrolysis at 550 °C, catalyst:biomass ratio of 5:1 (*w/w*). Error bars are one standard deviation.

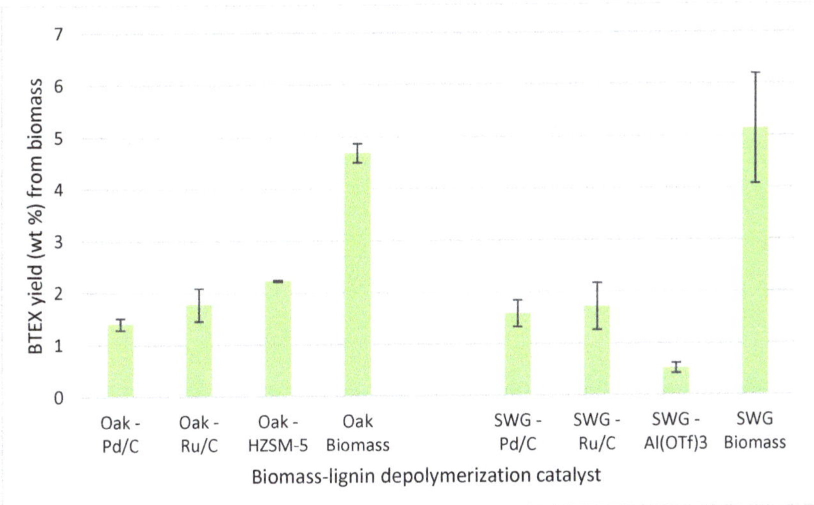

Figure 7. BTEX yield (wt %) from HZSM-5-catalyzed pyrolysis of cellulosic pulps isolated after lignin-first fractionation in ethanol at 210 °C based on total biomass input. Pyrolysis at 550 °C, catalyst:biomass ratio of 5:1 (*w/w*). Error bars are one standard deviation.

4. Conclusions

Oak and switchgrass were used in a lignin-first type biorefinery scheme where lignin was fractionated and depolymerized in a solvent phase reaction, and then the cellulosic pulps were tested for conversion to levoglucosan and BTEX via fast and catalytic pyrolysis over HZSM-5, respectively. For the initial step, reductive depolymerization catalyzed by both Pd/C or Ru/C with hydrogen transfer from ethanol and acid catalyzed depolymerizations (using HZSM 5 or $Al(OTF)_3$) were tested. Production of phenolic monomers and overall depolymerization rate, as measured by average molecular weight of the lignin oil, was better for the reductive conditions over the acidic conditions. Use of a soluble Lewis acid, $Al(OTf)_3$, led to particularly heavy lignin oil and also pulps containing both residual lignin and a high level of fixed carbon, making it a poor substrate for further conversion. The pulps from the reductive conditions proved to be a better feedstock for fast pyrolysis to produce levoglucosan. After normalizing for removal of lignin, the overall yield of levoglucosan was 4.0 to 4.6 times higher compared with pyrolysis of the raw biomasses. On the contrary, in catalytic pyrolysis over HZSM-5, the yield of BTEX was decreased compared with conversion of the raw biomass. The results show that initial solvent phase fractionation of lignin from biomass may be advantageous for thermal conversion of cellulosic portions of biomass to oxygenated chemicals such as levoglucosan, but, for selective catalytic conversion pathways for deoxygenated products, raw biomass may be a better feedstock.

Author Contributions: Conceptualization, C.A.M.; methodology, C.A.M., C.E.; investigation, C.A.M., C.E., Y.E.; data curation, C.A.M.; writing—original draft preparation, C.A.M., C.E., Y.E. All authors have read and agreed to the published version of the manuscript.

Funding: This research was supported by the United States Department of Agriculture—Agricultural Research Service.

Data Availability Statement: The data presented in this study are available in the article.

Conflicts of Interest: The authors declare no conflict of interest.

References

1. Langholtz, M.H.; Stokes, B.J.; Eaton, L.M. *2016 Billion-Ton Report: Advancing Domestic Resources for a Thriving Bioeconomy, Volume 1: Economic Availability of Feedstocks*; U.S. Department of Energy Office of Scientific and Technical Information: Oak Ridge, TN, USA, 2016.
2. Lynd, L.R.; Beckham, G.T.; Guss, A.M.; Jayakody, L.N.; Karp, E.M.; Maranas, C.; McCormick, R.L.; Amador-Noguez, D.; Bomble, Y.J.; Davison, B.H.; et al. Toward low-cost biological and hybrid biological/catalytic conversion of cellulosic biomass to fuels. *Energy Environ. Sci.* **2022**, *15*, 938–990. [CrossRef]
3. Devi, A.; Bajar, S.; Kour, H.; Kothari, R.; Pant, D.; Singh, A. Lignocellulosic Biomass Valorization for Bioethanol Production: A Circular Bioeconomy Approach. *Bioenergy Res.* **2022**, *15*, 1820–1841. [CrossRef]
4. Huang, C.; Jiang, X.; Shen, X.; Hu, J.; Tang, W.; Wu, X.; Ragauskas, A.; Jameel, H.; Meng, X.; Yong, Q. Lignin-enzyme interaction: A roadblock for efficient enzymatic hydrolysis of lignocellulosics. *Renew. Sustain. Energy Rev.* **2022**, *154*, 111822. [CrossRef]
5. Sun, S.; Sun, S.; Cao, X.; Sun, R. The role of pretreatment in improving the enzymatic hydrolysis of lignocellulosic materials. *Bioresour. Technol.* **2016**, *199*, 49–58. [CrossRef] [PubMed]
6. Patwardhan, P.R.; Brown, R.C.; Shanks, B.H. Understanding the fast pyrolysis of lignin. *ChemSusChem* **2011**, *4*, 1629–1636. [CrossRef]
7. Sharma, V.; Tsai, M.L.; Nargotra, P.; Chen, C.W.; Sun, P.P.; Singhania, R.R.; Patel, A.K.; Dong, C.D. Journey of lignin from a roadblock to bridge for lignocellulose biorefineries: A comprehensive review. *Sci. Total Environ.* **2023**, *861*, 160560. [CrossRef]
8. Mullen, C.A. Thermochemical and Catalytic Conversion of Lignin. In *Biomass Utilization: Conversion Strategies*; Springer International Publishing: Cham, Switzerland, 2022; pp. 133–200.
9. Patil, V.; Adhikari, S.; Cross, P.; Jahromi, H. Progress in the solvent depolymerization of lignin. *Renew. Sustain. Energy Rev.* **2020**, *133*, 110359. [CrossRef]
10. Mullen, C.A.; Strahan, G.D.; Elkasabi, Y. A comparison of the solvent liquefaction of lignin in ethanol and 1,4-butanediol. *J. Anal. Appl. Pyrolysis* **2022**, *164*, 105522. [CrossRef]
11. Abu-Omar, M.M.; Barta, K.; Beckham, G.T.; Luterbacher, J.S.; Ralph, J.; Rinaldi, R.; Román-Leshkov, Y.; Samec, J.S.M.; Sels, B.F.; Wang, F. Guidelines for performing lignin-first biorefining. *Energy Environ. Sci.* **2021**, *14*, 262–292. [CrossRef]
12. De Santi, A.; Galkin, M.V.; Lahive, C.W.; Deuss, P.J.; Barta, K. Lignin-First Fractionation of Softwood Lignocellulose Using a Mild Dimethyl Carbonate and Ethylene Glycol Organosolv Process. *ChemSusChem* **2020**, *13*, 4468–4477. [CrossRef]
13. Huang, X.; Zhu, J.; Koranyi, T.I.; Boot, M.D.; Hensen, E.J. Effective Release of Lignin Fragments from Lignocellulose by Lewis Acid Metal Triflates in the Lignin-First Approach. *ChemSusChem* **2016**, *9*, 3262–3267. [CrossRef]
14. Kramarenko, A.; Etit, D.; Laudadio, G.; D'Angelo, F.N. β-Zeolite-Assisted Lignin-First Fractionation in a Flow-Through Reactor. *ChemSusChem* **2021**, *14*, 3838–3849. [CrossRef]
15. Renders, T.; Van den Bosch, S.; Koelewijn, S.F.; Schutyser, W.; Sels, B.F. Lignin-first biomass fractionation: The advent of active stabilisation strategies. *Energy Environ. Sci.* **2017**, *10*, 1551–1557. [CrossRef]
16. Renders, T.; Van den Bossche, G.; Vangeel, T.; Van Aelst, K.; Sels, B. Reductive catalytic fractionation: State of the art of the lignin-first biorefinery. *Curr. Opin. Biotechnol.* **2019**, *56*, 193–201. [CrossRef] [PubMed]
17. Galkin, M.V.; Smit, A.T.; Subbotina, E.; Artemenko, K.A.; Bergquist, J.; Huijgen, W.J.; Samec, J.S. Hydrogen-free catalytic fractionation of woody biomass. *ChemSusChem* **2016**, *9*, 3280–3287. [CrossRef]
18. Liu, X.; Bouxin, F.P.; Fan, J.; Budarin, V.L.; Hu, C.; Clark, J.H. Recent Advances in the Catalytic Depolymerization of Lignin towards Phenolic Chemicals: A Review. *ChemSusChem* **2020**, *13*, 4296–4317. [CrossRef] [PubMed]
19. Van den Bosch, S.; Schutyser, W.; Vanholme, R.; Driessen, T.; Koelewijn, S.F.; Renders, T.; De Meester, B.; Huijgen, W.J.J.; Dehaen, W.; Courtin, C.M.; et al. Reductive lignocellulose fractionation into soluble lignin-derived phenolic monomers and dimers and processable carbohydrate pulps. *Energy Environ. Sci.* **2015**, *8*, 1748–1763. [CrossRef]
20. Huang, X.; Ouyang, X.; Hendriks, B.M.S.; Gonzalez, O.M.M.; Zhu, J.; Koranyi, T.I.; Boot, M.D.; Hensen, E.J.M. Selective production of mono-aromatics from lignocellulose over Pd/C catalyst: The influence of acid co-catalysts. *Faraday Discuss.* **2017**, *202*, 141–156. [CrossRef]
21. Deuss, P.J.; Lahive, C.W.; Lancefield, C.S.; Westwood, N.J.; Kamer, P.C.; Barta, K.; de Vries, J.G. Metal Triflates for the Production of Aromatics from Lignin. *ChemSusChem* **2016**, *9*, 2974–2981. [CrossRef]
22. Ahmed, R.; Liu, G.; Yousaf, B.; Abbas, Q.; Ullah, H.; Ali, M.U. Recent advances in carbon-based renewable adsorbent for selective carbon dioxide capture and separation-A review. *J. Clean. Prod.* **2020**, *242*, 118409. [CrossRef]
23. Lourenço, A.; Pereira, H. Compositional Variability of Lignin in Biomass. In *Lignin—Trends and Applications*; IntechOpen: London, UK, 2018.
24. Choi, H.S.; Meier, D. Fast pyrolysis of Kraft lignin—Vapor cracking over various fixed-bed catalysts. *J. Anal. Appl. Pyrolysis* **2013**, *100*, 207–212. [CrossRef]
25. Zhou, S.; Xue, Y.; Sharma, A.; Bai, X. Lignin Valorization through Thermochemical Conversion: Comparison of Hardwood, Softwood and Herbaceous Lignin. *ACS Sustain. Chem. Eng.* **2016**, *4*, 6608–6617. [CrossRef]
26. Wang, Z.; Zhou, S.; Pecha, B.; Westerhof, R.J.M.; Garcia-Perez, M. Effect of Pyrolysis Temperature and Sulfuric Acid During the Fast Pyrolysis of Cellulose and Douglas Fir in an Atmospheric Pressure Wire Mesh Reactor. *Energy Fuels* **2014**, *28*, 5167–5177. [CrossRef]

27. Hilbers, T.J.; Wang, Z.; Pecha, B.; Westerhof, R.J.M.; Kersten, S.R.A.; Pelaez-Samaniego, M.R.; Garcia-Perez, M. Cellulose-Lignin interactions during slow and fast pyrolysis. *J. Anal. Appl. Pyrolysis* **2015**, *114*, 197–207. [CrossRef]
28. Mullen, C.A. Upgrading of biomass via catalytic fast pyrolysis (CFP). In *Chemical Catalysts for Biomass Upgrading*; Wiley-VCH: Weinheim, Germany, 2020; pp. 1–33.
29. Du, S.; Valla, J.A.; Bollas, G.M. Characteristics and origin of char and coke from fast and slow, catalytic and thermal pyrolysis of biomass and relevant model compounds. *Green Chem.* **2013**, *15*, 3214–3229. [CrossRef]
30. Wang, K.; Kim, K.H.; Brown, R.C. Catalytic pyrolysis of individual components of lignocellulosic biomass. *Green Chem.* **2014**, *16*, 727–735. [CrossRef]

MDPI AG
Grosspeteranlage 5
4052 Basel
Switzerland
Tel.: +41 61 683 77 34

MDPI Books Editorial Office
E-mail: books@mdpi.com
www.mdpi.com/books

www.ingramcontent.com/pod-product-compliance
Lightning Source LLC
Chambersburg PA
CBHW061555220926
51850CB00005B/343